13th South African Conference on Computational and Applied Mechanics (SACAM 2024)

Stellenbosch, South Africa
22-23 January 2024

Editors:

J. Hoffmann
M.P. Venter

ISBN: 979-8-3313-0861-2

Printed from e-media with permission by:

Curran Associates, Inc.
57 Morehouse Lane
Red Hook, NY 12571

South African Association for Theoretical and Applied Mechanics (SAAM)
c/o Schalk Kok
9-21 Engineering I
Department of Mechanical Engineering
University of Pretoria
South Africa

Phone: 27 21 808 3554
Fax: 27 86 615 5206

Schalk.Kok@uj.ac.za

Additional copies of this publication are available from:

Curran Associates, Inc.
57 Morehouse Lane
Red Hook, NY 12571 USA
Phone: 845-758-0400
Fax: 845-758-2633
Email: curran@proceedings.com
Web: www.proceedings.com

Proceeding of the 13th South African Conference on Computational and Applied Mechanics

SACAM 2024

22-23 January 2024

Editors: J. Hoffmann and M.P. Venter

on behalf of the

South African Association for Theoretical and Applied Mechanics
(SAAM)

STIAS Wallenberg Centre, 10 Marais Road, Mostertsdrif Stellenbosch

Conference Title

13th South African Conference on Computational and Applied and Applied Mechanics.

Organising Istitute

University of Stellenbosch

Date and Location

Held on 22-23 January 2024, at STIAS Wallenberg Centre, 10 Marais Road, Mostertsdrif Stellenbosch.

Editors

J Hoffmann
MP Venter

SACAM 2024

Venue

The 13th South African Conference on Computational and Applied Mechanics (SACAM 2024) was held at the STIAS Wallenberg Centre, 10 Marais Road, Mostertsdrif Stellenbosch. SACAM is a premier international conference on computational and applied mechanics on the African continent, and attracts local and international participants. SACAM is held every two years under the auspices of the South African Association for Theoretical and Applied Mechanics (SAAM) as a forum to present current computational and applied mechanics developments. The conference was attended by 62 delegates of which 6 were from industry, and 7 countries were represented.

Welcome from the local organizing committee

Jaap Hoffmann

Martin Venter

The organizing committee wish you a warm welcome to the 13th South African Conference on Computational and Applied Mechanics in the beautiful town of Stellenbosch. We are honoured to host this gathering of students, scientists and engineers from across South Africa and abroad. With its stunning landscape, rich cultural heritage and warm hospitality, we believe your stay in Stellenbosch will be memorable.

Over the two days of the conference, we are looking forward to rigorous discussions, exchange of ideas, showcasing innovative research and lifelong connections between delegates. We hope that you will also make time to enjoy the many attractions Stellenbosch has to offer.

Finally, we also express our appreciation to our sponsors, Qfinsoft and Aerotherm who helped to make this conference a reality.

The ISBN number of conference proceedings is 978-0-7972-1914-4, and individual papers can be accessed via hyperlinks in the daily program. The proceeding is available from the SAAM website https://saam.africa/

Technical Review Committee

We were privileged to have the following keynote speakers at our conference

Prof. Corné Coetzee

Corne Coetzee is a professor in the Department of Mechanical and Mechatronic Engineering at Stellenbosch University with more than 20 years of experience in the modelling of granular flows. He grew up on a commercial fruit farm, and his family's earth-moving construction company ignited his interest in the behaviour of granular materials in general. His research focuses on the modelling of granular material or bulk solids using various numerical techniques. He uses this skill to predict the behaviour of granular materials for the design of improved bulk-handling equipment and processes. He has established a laboratory for observing and measuring material behaviour, and calibration and validation of these numerical models. He has also aligned his research with his interest in agriculture and is involved in the modelling and design of paper board packaging for the horticultural sector and he works closely with the faculty of Agri-Science.

Prof. Jat du Toit

Jat du Toit is an emeritus professor of Mechanical Engineering and the emeritus DSI/NRF SARChI Research Chair in Nuclear Engineering at the North-West University, Potchefstroom, South Africa. He obtained a B.Eng. (Hons.)(Civil) and a M.Eng. (Civil) from the University of Stellenbosch and a Ph.D. from the University of Cambridge. He has published more than 100 papers in international and national journals and conference proceedings and was the co-editor of the proceedings of two conferences. He is a registered professional engineer with practical experience in the fields of civil, mechanical, chemical, metallurgical, and nuclear engineering and focuses on the numerical modeling of thermal-fluid processes. He is a partner of the commercial engineering software development and consulting firm M-Tech Industrial which develops and markets the system CFD code Flownex. His current interests are the development of advanced thermal-fluid simulation models for high-temperature gas-cooled nuclear reactors and associated cooling systems using a systems CFD approach. He has been a senior technical consultant for the Pebble Bed Modular Reactor (PBMR) project. Along with his colleagues, he has made contributions to the influence of the packing structure of pebble bed on the thermal-hydraulic phenomena, as well as the interaction between the reactor unit and the reactor cavity cooling system for pebble bed reactor and prismatic block reactor-based systems.

Prof. Karin Wolf

A/Prof Karin Wolff has been in education for over two decades, and in engineering education since 2008. She has worked as a lecturer, curriculum specialist, student learning support manager, and in staff development at four South African universities. Her doctoral research (University of Cape Town) looked at the relationship between mathematics, physics and logic in complex engineering problem solving in the world of work. She draws on Social Realism, particularly Legitimation Code Theory, to analyse and design knowledge- and knower-building practices. She is the Teaching & Learning Advisor in the Faculty of Engineering, Stellenbosch University, and president of the South African Society for Engineering Education (www.sasee.org.za). Her current research focus is on facilitating community-of-practice development, and bridging the theory-practice divide in STEM fields in the African context. Her research website and publications are available at: https://sites.google.com/view/dr-karin-wolff-portfolio

Prof. Steeve Yuen

Steeve CHUNG KIM YUEN has worked in the field of structural impact for the past thirty years on various projects ranging from investigating the batting performance of cricket players to the response of "simple" structures to blast loads (5g to 26tons of explosives). His research involves a mix of material characterisation, experimental work, and finite element simulations and focuses on the survivability of the structures. Full-scale experiments are costly and limited. He has, therefore, focussed on small-scale testing investigating different blast loading scenarios which can result in the structure mitigating the high pressure or fragment damage after material failure. Recently he has worked on numerous projects studying the response of structures to blast load with focus on the effect of incorporating foreign objects in the explosive to simulate improvised explosive devices (IED) because of their increased use in conflicts. He has also considered the effects of buried and encased explosives which are associated with landmines, and confined blasts on different types of structures in an urban setting. The work carried out by Steeve is geared towards minimising life-changing injuries and saving lives.

Program (Day 1)

08:00		**Arrival, registration & coffee**	
09:00		**Welcome (Jaap Hoffmann)**	
09:15		**Opening address (Prof. Nico Wilke, UP) Session Chair: Prof. Martin Venter**	
10:00		**Keynote 1 (Prof. Jat du Toit, NWU) Session Chair. Prof. Jaap Hoffmann**	
10:45		**Break**	
		Track 1: CFD Session Chair Prof. Jan-Hendrik Kruger	**Track 2: Session Chair Prof. Wei Hua Ho**
11:00		Struwig, Kruger & Bosman; CFD Analysis of convection cooling of the NWU solar car PV wing	Ramatlo, Wilke & Loveday; Enhanced Data-Driven Hybrid Approach for Unavailable Damage Scenarios in Welded Rail Tracks with Multiple Reflections and Reflectors for Ultrasonic-Guided Wave Inspection
11:15		Kruger, Kruger & Bosman; Aerodynamic stability of the titling PV wing of the NWU solar car	Ellis & Bekker; Improved accelerometer placement for railway wheel flat detection using a single sensor (withdrawn)
11:30		Schutte, Bosman & Kruger; Aeroacoustic simulation and improvement of a propeller for an electric self launch sailplane	Mabaso; Enhancing Railway Condition Monitoring and Operational Efficiency through Real-Time Data Visualization (withdrawn)
		Track 3: CFD Session chair Prof. Jat du Toit	**Track 4: Session Chair: Dr. Willem Roos**
11:45		Gertenbach & Kruger; Development of a spillway and sluice gate demonstration system	Visser, Venter & Neaves; A Strain Gauge-Based Method for Correcting Out-of-Plane Motion Errors in 2D Digital Image Correlation
12:00			Blackwell et al; Temperature dependant mechanical property characterisation using digital image correlation and infrared thermography
12:15		Kruger & Le Grange; Experimental and numerical analysis of dilute gas-particle flow in a pipe section	Park, Venter & Du Plessis; Estimating Mechanical Properties of AM Lattices Using CT-Scans and FE Models
12:30		**Lunch**	
14:00		**Keynote 2 (Prof. Steeve Yuen, UCT) Session Chair. Prof. Gerhard Venter**	
		Track 5: CFD (environmental) Session Chair. Prof. Jaap Hoffmann	**Track 6: Session Chair Prof. Quinn Reynolds**
14:45		Nair, Boje & Verrinder: Three dimensional computational fluid dynamics modelling of pancake sea ice on waves	Koekemoer et al; Influence of the length of a photovoltaic single-axis tracker mounting rail (withdrawn)
15:00		Surajpal, Deidericks & Smit; Nearfield dispersion modelling of discharged effluents under various hydrodynamic conditions	Phillips & Conradie; Optical Laser Rail Wheel Wear Measuring and Detection Unit
15:15		Sponsor presentation: Qfinsoft	Sponsor presentation: Aerotherm
15:30		**Break**	
		Track 7: CFD (industrial) Session chair Prof. Craig Law	**Track 8: Session Chair. Prof. Nico Wilke**
15:45		Reynolds & Bogaers; Computational modelling of alternating current conduction effects in packed beds	Jones, Reynecke & Pambuka; Cost Effective Ballistic Protection for Vehicles
16:00		Surujhlal et al; Wake redirection techniques for wind farms	Meeser & Wilke; Optimizing driving behaviour for underground electric vehicles under varying loads
16:15		Strydom, Pretorius & Laubscher; Start-up transient response characteristics of natural draft direct dry cooling systems at various scales	Loveday & Fromme; Measurement of Lamb wave phase velocity using low-cost equipment
16:30		Matlakala; Optimization of the pumping capacity of the centrifugal pump taking in consideration the material properties (withdrawn)	Nickerson; Investigation of inverse model for the estimation of impulsive loads (withdrawn)
17:00		**Cocktail function**	

Day 2

08:30	**Arrival & coffee**		
09:15	**Keynote 3: Prof. Karen Wolff (SU) Session chair: Dr. Melody Neaves**		
		Track 9: Biomedics 1 Session Chair: Prof. Thanyani Pandelani	**Track 10 Session Chair Dr. Alfred Bogaers**
10:00		Goedhals, De Villiers & Smit; Numerical modelling of blood flow during syringing	Van Staaden; Making funding proposals more effective: plain language guidelines
10:15		Bougardt et al; In-vitro and in-silico modelling of arteriovenous malformations for surgical planning	Myburgh & Law; Angled expanding jet impingement in supersonic crossflow
10:30		Marais et al; Studying transvalvular aortic haemodynamics with simplified fluid-structure interaction and experiment	Du Toit, Boyes & Slabber; Thermal-hydraulic behavior of a micro nuclear reactor during a DLOFC transient
10:45	**Break**		
		Track 11: Biomedics 2 Session Chair. Dr. Andie de Villiers	**Track 12: Session Chair: Prof. Gerhard Venter**
11:00		De Wet & Kruger; The role of simulation in the development of a cost-effective biometric prosthetic hand	Moloi; Biomechanical analysis: Advancing snake robot locomotion in complex Environment (withdrawn)
11:15		Mabuda et al; Appropriate mode of storage of porcine heart myocardia without tempering with its passive mechanical properties	Lichthard & Venter; A Framework for the Design of Soft Pneumatic Actuators Using Computational Tools
11:30		Taylor, Bekker & Kruger; Augmentation of measurement procedures for motion sickness assessment on ships (withdrawn)	Mathebela et al; The effect of optimization model on the accuracy of constitutive model
		Track 13: Algorithms Session Chair Prof. Francois Smit	**Track 14 Session Chair: Prof. Martin Venter**
11:45		Malan & Nordström; A higher order stable continuous Galerkin scheme for smooth and discontinuous fields (withdrawn)	Botete, Blaine & Venter; Calibration of Laser Absorptivity in CFD Modelling of LPBF In-Situ Alloying with Ray Tracing
12:00		Oomar et al; A weakly-compressible two-phase formulation for hydrogen containment modelling (withdrawn)	Melim, Bekker & Kruger; Prediction of Ship Motion Using Forecast Weather (withdrawn)
12:15		Ngwangwa; Assessing regression capabilities of layer recurrent neural network with a scaled gradient conjugate training function (withdrawn)	Chandran & Kumar; Multilayered floating absorber for interfacial evaporation with excellent salt rejection for solar thermal applications (withdrawn)
12:30	**Lunch**		
14:00	**Keynote 4: Prof. Corne Coetzee (SU) Session Chair Prof Nico Wilke**		
		Track 15: Experimental fluids Prof. Attie Jonker	**Track 16 (DEM): Session Chair: Prof. Corne Coetzee**
14:45		Melman, Bosman & Jonker; Low-speed wind tunnel tests of two airfoils for standard class sailplane application	Moyo & Celik; Deep Reinforcement Learning in Frontier and Sampling-Based MAV Exploration of Unknown Environments
15:00		Aasa & Shote; Computational analysis of heat transfer augmentation through a rectangular dimple endwalls geometry (withdrawn)	Van Tonder, Bredell & Coetzee; Dynamic Simulation of a Nutating Grinding Mill (withdrawn)
15:15		Law & Nordström; Using experimental flow visualization techniques to interpret computational fluid dynamics data	Erasmus, Coetzee & Berry; Defining citrus pallet stability and failure modes: Developing tools for carton improvement
15:30	**Break**		
		Track 17: Modelling techniques Session Chair Prof. Quinn Reynolds	**Track 18 Session Chair: Prof. Phillip Loveday**
15:45		Bogaers et al; A novel fluid-structure interaction model for sea ice dynamics	Setshedi, Wilke & Loveday; Latent Space Perspicacity and Interpretation Enhancement (LS-PIE) Framework (withdrawn)
16:00		Bogaers et al; Including thermochemical equilibrium calculations into multiphyscis models - a finite volume based phase change model	Wilke; Multifidelity Surrogate Modelling: A New Data Fusion Perspective
16:15		Roos, Bogaers & Zietsman; Geometric acceleration of thermochemical equilibrium calculations for inclusion in multiphysics and process models	Wilke; Physics-Based, Hybrid, and Data-Driven Computational Mechanics Generative Modelling: Manifestation of Shared Concepts
16:40		**Closing Session (Prof. Marting Venter)**	
19:00	**Gala dinner**		

Full-length manuscript review

Full-length manuscripts presented at the conference were subjected to a double-blind review after the conference, and accepted papers are included in the Conference Proceedings. Two independent reviewers per manuscript were tasked to ensure that:

- The editorial standard of the paper is acceptable for an international conference
- The contents of the paper are of interest to the target audience
- The standard, methodology and quality of the work is appropriate

The two best papers, judged on both the manuscript and presentation, will be invited for publication in the R&D Journal of the South African Institution of Mechanical Engineering

(https://www.saimeche.org.za/page/rdjournal).

Some authors preferred to publish in technical journals and asked that their manuscripts be withdrawn from the proceedings. Hyperlinks in the daily program link papers that passed the review, and subsequent requests for improvement. For the sake of completeness, the book of abstracts includes all presentations at the conference.

Table of Contents

Full Papers

Page 1. Matthew Blackwell, Maxwell Vos, Sarah George, Melody Neaves, Thorsten H. Becker, "Temperature Dependant Mechanical Property Characterisation Using Digital Image Correlation and Infrared Thermography"

Page 9. Alfred E.J. Bogaers, Jaco J. Swanepoel, Willem A. Roos, Johan H. Zietsman, "Including Thermochemical Equilibrium Calculations into Multiphysics Models – A Finite Volume Based Phase Change Model"

Page 20. M.R. Botete, D.C. Blaine, G. Venter, "Calibration of Laser Absorptivity in CFD Modelling of LPBF In-Situ Alloying with Ray Tracing"

Page 31. Jordan Bougardt, Randall Paton, Malebogo Ngoepe, Cecil Pule, Karabo Mosiane, Jayandiran Pillai, Weihua Ho, "In-Vitro and In-Silico Modelling of Arteriovenous Malformations for Surgical Planning"

Page 39. Markus de Wet, Dr. Jan-Hendrik Kruger, "The Role of Simulation in the Development of a Cost-Effective Biomimetic Prosthetic Hand"

Page 50. CG du Toit, WA Boyes, JF Slabber, "Thermal-Hydraulic Behaviour of a Micro Nuclear Reactor During a DLOFC Transient"

Page 58. NJ Erasmus, CJ Coetzee, TM Berry, "Defining citrus pallet stability and failure modes: Developing tools for carton improvement"

Page 68. Mary A. Familusi, Sebastian Skatulla, Jagir R. Hussan, Olukayode O. Aremu, Daniel Mutithu, Ntobeko A. B. Ntusi, "Bi-ventricular Elastic Material Parameters Estimation Using 3D CMR Strains in RHD Patients"

Page 80. Niël Gertenbach, Dr. Jan-Hendrik Kruger, "Development of a Spillway and Sluice Gate Demonstration System"

Page 89. Jaimé Goedhals, Dr A. de Villiers, Prof. G.J.F. Smit, "Numerical Modelling of Blood Flow During Syringing"

Page 97. John D. Jones, J. David Reinecke, Lwazi Pambuka, "Cost Effective Ballistic Protection for Vehicles"

Page 112. Etienne Kruger, Dr. Jan-Hendrik Kruger, Dr. Johan Bosman, "Aerodynamic Stability of the Tilting PV Wing of the NWU Solar Car"

Page 120. Jan-Hendrik Kruger, Louis le Grange, Michael Roberts, "Experimental and Numerical Analysis of Dilute Gas-Particle Flow in a Pipe Section"

Page 131. Craig Law, Jan Nordström, "Using Experimental Flow Visualisation Techniques to Interpret Computational Fluid Dynamics Data"

Page 142. Philip Ligthart, Martin Venter, "On the Complexity of Internal Chambers in Soft Pneumatic Actuators"

Page 150. Philip W. Loveday, Paul Fromme, "Measurement of Lamb Wave Phase Velocity Using Low-Cost Equipment"

Page 158. I. Mabuda, T. Pandelani, H. Ngwangwa, F. Nemavhola, "Appropriate Mode of Storage of Porcine Heart Myocardia Without Tempering with its Passive Mechanical Properties"

Page 167. Eugene Marais, Martin P Venter, Ryno Laubscher, Philip Herbst, "Studying transvalvular aortic haemodynamics with simplified fluid-structure interaction and experiment: Lessons Learnt"

Page 178. Lebogang Mathebela, Rudzani Sigwadi, Tetelo Lekwana, Harry Ngwangwa, Thanyani Pandelani, Fulufhelo Nemavhola, "The effect of optimization model on the accuracy of constitutive model"

Page 186. R.F. Meeser, D.N. Wilke, N.J. Theron, "Optimizing Driving Behaviour for Underground Electric Vehicles Under Varying Loads"

Page 194. Tristan Melman, Dr. Johan Bosman, Prof. Attie Jonker, "Low-Speed Wind Tunnel Tests of Two Airfoils for Standard Class Sailplane Application"

Page 204. M.E. Miyambo, S. Kotole, M. Mokalane, T. Pandelani, A. Clarke, D. Reinecke, "Split Hopkinson Pressure Bar Compression Test Simulation and Experimental Validation of Commercially Pure Copper Specimens"

Page 214. Makhosazana E. Moyo, Turgay Celik, "Deep Reinforcement Learning in Frontier-Based Exploration of Unknown Environments"

Page 223. Sabrina G. Myburgh, Craig Law, "Angled Expanding Jet Impingement in Supersonic Crossflow"

Page 230. Anand Nair, Edward Boje, Robyn A. Verrinder, "Three Dimensional Computational Fluid Dynamics Modelling of Pancake Sea Ice on Waves"

Page 238. Minsol Park, Martin Phillip Venter, Anton Du Plessis, "Estimating Mechanical Properties of AM Lattices Using CT-Scans and FE Models"

Page 253. A.K. Phillips, P.D.F Conradie, "Optical Rail Wheel Wear Measurement Unit"

Page 262. Dineo A. Ramatlo, Daniel N. Wilke, Philip W. Loveday, "Enhanced Data-Driven Hybrid Approach for Unavailable Damage Scenarios in Welded Rail Tracks with Multiple Reflections and Reflectors for Ultrasonic-Guided Wave Inspection"

Page 270. Quinn G. Reynolds, Alfred E.J. Bogaers, "Computational modelling of alternating current conduction effects in packed beds"

Page 280. Karin Schutte, Dr. Johan Bosman, Dr. Jan-Hendrik Kruger, "Aeroacoustic Simulation and Improvement of a Propeller for an Electric Self-Launch Sailplane"

Page 294. Düter Struwig, Dr. Jan-Hendrik Kruger, Dr. Johan Bosman, "CFD Analysis of Convection Cooling of the NWU Solar Car PV Wing"

Page 304. Wian Strydom, Johannes Pretorius, Ryno Laubscher, "Start-Up and Transient Response Characteristics of Natural Draft Direct Dry Cooling Systems at Various Scales"

Page 312. T. L. Surajpal, G.P.J. Diedericks, G.J.F. Smit, "Nearfield Dispersion Modelling of Discharged Effluents Incorporating Far-Field Hydrodynamics"

Page 324. Kayla Surujhlal, Ayodimeji Biobaku, Nick Zang, Wei Hua Ho, "Wake Redirection Techniques for Wind Farms"

Page 332. Dr Antoinique van Staden, Dr Idette Noomé, "Making Funding Proposals More Effective: Plain Language Guidelines"

Page 340. Carl-Hein Visser, Gerhard Venter, Melody Neaves, "A Strain Gauge-Based Method for Correcting Out-of-Plane Motion Errors in 2D Digital Image Correlation"

Page 350. Daniel N. Wilke, "Multifidelity Surrogate Models: A New Data Fusion Perspective"

Abstracts

Page 359. SA Aasa, SA Shote, "Computational analysis of heat transfer augmentation through a rectangular dimple endwalls geometry"

Page 360. M Blackwell, M Vos, S George, M Neaves, TH Becker, "Temperature dependant mechanical property characterisation using digital image correlation and infrared thermography"

Page 361. AEJ Bogaers, JJ Swanepoel, WA Roos, JH Zietsman, "Including thermochemical equilibrium calculations into multiphysics models – a finite volume-based phase change model"

Page 362. MR Botete, DC Blaine, G Venter, "Calibration of Laser Absorptivity in CFD Modelling of LPBF In-Situ Alloying with Ray Tracing"

Page 363. J Bougardt, W Ho, RT Paton, J Pillai, "In-vitro and in-silico modelling of arteriovenous malformations for surgical planning"

Page 364. AK Chandran, SCS Kumar, "Multilayered floating absorber for interfacial evaporation with excellent salt rejection for solar thermal applications"

Page 365. MI Chibinyani, TC Dzogbewu, M Maringa, AM Muiruri, "Deformation Behaviour, Limitations in Design, Applications, and Additive Manufacturing of Hierarchical Honeycombs"

Page 366. M de Wet, JH Kruger, "The role of simulation in the development of a cost-effective biomimetic prosthetic hand"

Page 367. CG du Toit, W Boyes, JF Slabber, "Thermal-hydraulic behaviour of a micro nuclear reactor during a DLOFC transient"

Page 368. D Ellis, A Bekker, "Improved accelerometer placement for railway wheel flat detection using a single sensor"

Page 369. NJ Erasmus, CJ Coetzee, TM Berry, "Defining citrus pallet stability and failure modes: Developing tools for carton improvement"

Page 370. N Gertenbach, JH Kruger, "Development of a spillway and sluice gate demonstration system"

Page 371. J Goedhals, A de Villiers, GJF Smit, "Numerical Modelling of Blood Flow During Syringing"

Page 372. JD Jones, JD Reinecke, L Pambuka, "Cost Effective Ballistic Protection for Vehicles"

Page 373. JH Koekemoer, JR Bredell, G Venter, J Reinecke, "Influence of the length of a photovoltaic single-axis tracker mounting rail"

Page 374. E Kruger, JH Kruger, J Bosman, "Aerodynamic stability of the tilting PV wing of the NWU solar car"

Page 375. JH Kruger, L le Grange, "Experimental and numerical analysis of dilute gas-particle flow in a pipe section"

Page 376. Craig Law, J Nordström, "Using experimental flow visualisation techniques to interpret computational fluid dynamics data"

Page 377. PF Ligthart, MP Venter, "A Framework for the Design of Soft Pneumatic Actuators Using Computational Tools"

Page 378. PW Loveday, P Fromme, "Measurement of Lamb wave phase velocity using low-cost equipment"

Page 379. N Mabaso, "Enhancing Railway Condition Monitoring and Operational Efficiency through Real-Time Data Visualization"

Page 380. I Mabuda, T Pandelani, H Ngwangwa, F Nemavhola, "Appropriate Mode of storage of Porcine heart myocardia without tempering with its passive mechanical properties"

Page 381. NNJ Mahomed, A Bogaers, CH Rycroft, S Skatulla, "A novel fluid-structure interaction model for sea ice dynamics"

Page 382. AG Malan, J Nordstrom, "A High Order Stable Continuous Galerkin Scheme for Smooth and Discontinuous Fields"

Page 383. E Marais, MP Venter, R Laubscher, P Herbst, "Studying transvalvular aortic haemodynamics with simplified fluid-structure interaction and experiment"

Page 384. L Mathebela, R Sigwadi, T Lekwana, H Ngwangwa, T Pandelani, F Nemavhola, "The effect of optimization model on the accuracy of constitutive model"

Page 386. RF Meeser, DN Wilke, "Optimizing driving behaviour for underground electric vehicles under varying loads"

Page 387. M Melim, A Bekker, K Kruger, "Prediction of Ship Motion Using Forecast Weather"

Page 388. T Melman, J Bosman, A Jonker, "Low-speed wind tunnel tests of two airfoils for standard class sailplane application"

Page 389. BL Moloi, "Biomechanical analysis: Advancing snake robot locomotion in complex environment"

Page 390. ME Moyo, T Celik, "Deep Reinforcement Learning in Frontier and Sampling-Based MAV Exploration of Unknown Environments"

Page 392. S Myburgh, C Law, "Angled expanding jet impingement in supersonic crossflow"

Page 393. A Nair, E Boje, R Verrinder, "Three Dimensional Computational Fluid Dynamics Modelling of Pancake Sea Ice on Waves"

Page 394. HM Ngwangwa, "Assessing regression capabilities of layer recurrent neural network with a scaled gradient conjugate training function"

Page 395. BM Nickerson, "Investigation of inverse model for the estimation of impulsive loads"

Page 396. MY Oomar, AG Malan, MD Wright, LC Malan, S Zaleski, F Gambioli, "A Weakly-Compressible Two-Phase Formulation for Hydrogen Containment Modelling"

Page 397. M Park, MP Venter, A Du Plessis, "Estimating Mechanical Properties of AM Lattices Using CT-Scans and FE Models"

Page 398. AK Phillips, PDF Conradie, "Optical Laser Rail Wheel Wear Measuring and Detection Unit"

Page 399. DA Ramatlo, DN Wilke, PW Loveday, "Enhanced Data-Driven Hybrid Approach for Unavailable Damage Scenarios in Welded Rail Tracks with Multiple Reflections and Reflectors for Ultrasonic-Guided Wave Inspection"

Page 400. QG Reynolds, AEJ Bogaers, "Computational modelling of alternating current conduction effects in packed beds"

Page 401. WA Roos, AEJ Bogaers, JH Zietsman, "Geometric Acceleration of Thermochemical Equilibrium Calculations for Inclusion in Multiphysics and Process Models"

Page 402. K Schutte, J Bosman, JH Kruger, "Aeroacoustic simulation and improvement of a propeller for an electric self-launch sailplane"

Page 404. D Struwig, JH Kruger, J Bosman, "CFD analysis of convection cooling of the NWU solar car PV wing"

Page 405. W Strydom, JP Pretorius, R Laubscher, "Start-up and Transient Response Characteristics of Natural Draft Direct Dry Cooling Systems at Various Scales"

Page 406. TL Surajpal, GPJ Diedericks, GJF Smit, "Nearfield dispersion modelling of discharged effluents under various hydrodynamic conditions"

Page 407. K Surujhlal, WH Ho, A Biobaku, N Zang, "Wake Redirection Techniques for Wind Farms"

Page 408. NC Taylor, A Bekker, K Kruger, "Augmentation of measurement procedures for motion sickness assessment on ships"

Page 409. A van Staden, "Making funding proposals more effective: Plain Language guidelines"

Page 410. JJ van Tonder, JR Bredell, C Coetzee, "Dynamic Simulation of a Nutating Grinding Mill"

Page 411. CH Visser, G Venter, M Neaves, "A Strain Gauge-Based Method for Correcting Out-Of-Plane Motion Errors in 2D Digital Image Correlation"

Page 412. DN Wilke, "Physics-Based, Hybrid, and Data-Driven Computational Mechanics Generative Modelling: Manifestation of Shared Concepts"

Page 413. DN Wilke, "Multifidelity Surrogate Modelling: A New Data Fusion Perspective"

Additional

Page 403. II Setshedi, DN Wilke, PW Loveday, "Comparing Source-Driven and Variance-Driven Statistical Learning Models to Generate and Analyse Guided Wave Ultrasound Time Series Signals"

FULL LENGTH PAPERS

TEMPERATURE DEPENDANT MECHANICAL PROPERTY CHARACTERISATION USING DIGITAL IMAGE CORRELATION AND INFRARED THERMOGRAPHY

Matthew Blackwell[1,*], Maxwell Vos[2], Sarah George[2], Melody Neaves[1], Thorsten H. Becker[2]
[1]Department of Mechanical and Mechatronic Engineering, Stellenbosch University, Joubert street, Stellenbosch 7602, South Africa, *Email. matthewrsacape@gmail.com
[2]Centre for Materials Engineering, University of Cape Town, Library Rd, Cape Town, 7701, South Africa.

Abstract: Accurate mechanical property characterisation of high-temperature high-performance materials, such as additively manufactured (AM) Inconel 718 (IN718), is crucial to ensure the reliability of components. A sudden drop in strength accompanied by significant embrittlement occurs in wrought IN718 near 700 °C. It is not clear whether AM IN718 exhibits this behaviour.

The high cost of these materials and manufacturing methods necessitate the development of specimen-efficient testing techniques that allow material properties to be extracted simultaneously over a range of testing conditions. This paper presents a test setup, which is used to determine high-temperature mechanical properties over a range of operating temperatures by integrating digital image correlation (DIC) and full-field infrared thermography (IR).

A purpose designed specimen is resistively heated using a Gleeble 3800 thermomechanical simulator to create a thermal gradient across the length of the specimen. A high-temperature DIC setup and IR camera are used to capture displacement and full-field IR data, respectively, of the specimen surface. These datasets are temporally synchronised using a common triggering system and spatially correlated to allow for the extraction of temperature dependent properties for a range of up to 100 °C from a single tensile test. The setup is shown to successfully capture the strength drop-off and embrittlement of AM IN718.

Keywords: full-field, digital image correlation, infrared thermography, high-temperature, Inconel 718

1. Introduction

A recent push towards carbon reduction has necessitated an increase in the fuel efficiency of commercial jet engines [1]. Higher efficiencies can be achieved by increasing engine operating temperatures; however, this is often limited by material capabilities. Designers have started to incorporate cooling channels within parts to allow for higher engine gas temperatures. The design of these cooling channels is limited in part by the ability to manoeuvre the tooling within the turbine blade. This plays into the strengths of AM materials since AM technologies allow for the layer-wise addition of material rather than the subtraction of material. This layer-wise manufacturing technique allows built-in cooling channels, along with other design complexities, to be created during the building of a part rather than being machined after a part is formed [2].

IN718 is a nickel-based superalloy with substantial strength retention at temperatures up to 650 °C and excellent corrosion resistance, among other benefits. As such, it is widely employed in its wrought and cast forms in the aerospace industry and accounts for approximately 34 % of the total mass of jet turbine engines [3]. Heat treatments for wrought and cast IN718 are well established for high temperature use, and the microstructure has been extensively studied. On the other hand, the high temperature mechanical properties of IN718 produced by laser-powder bed fusion (LPBF), an AM technique, are not well documented. Due to the unique microstructure of LPBF materials, it is not possible to assume that the LPBF material will exhibit the same properties as the conventionally produced IN718.

Accurate characterisation of the high-temperature mechanical behaviour of LPBF IN718 is crucial to ensure the reliability and performance of the components. One notable material behaviour of wrought IN718 is a sudden drop in strength near 700° C [4], accompanied by embrittlement of the material between 650 °C and 800 °C [5]. It is unclear whether LPBF IN718 will exhibit a similar strength drop and embrittlement under the same conditions. Moreover, the high cost of these materials and manufacturing methods necessitates the development of specimen-efficient testing techniques that allow material properties to be extracted simultaneously in a range of testing conditions. Van Rooyen and Becker [6] experimented with the concept of measuring high temperature mechanical properties of steel from a power plant with a non-uniform temperature achieved through resistive heating. This heating method was used with a combination of high temperature DIC and IR to extract mechanical properties over a range of up to 120 °C from a single specimen. Their testing was limited to peak specimen temperatures of 600 °C. This paper presents an improved testing setup which incorporates automated temporal and spatial synchronisation of the data and applies the testing method to LPBF produced IN718.

2. Materials

The powder used for specimen manufacturing was gas atomised 15-53 μm IN718 powder manufactured by AP&C [7] with composition shown in Table 1. The particle size distribution (PSD) ranged between 21.25 μm and 51.64 μm ($D_{10} - D_{90}$).

Table 1: Elemental composition of IN718 powder* from AP&C [7].

Element	Ni	Cr	Nb + Ta	Mo	Ti	Al	Co	C	Fe
wt%	53.22	18.7	4.96	2.94	0.85	0.58	< 0.1	0.04	Balance

*Only main constituent elements are presented

Tensile specimens were built in cylindrical form in two orientations with the long axis either parallel or perpendicular to the build plate. A rectangular scan strategy was employed with the following machine settings: laser power of 190 W; laser spot size of approximately 70 μm; scan speed of 800 mm/s; hatch spacing of 120 μm; and powder layer thickness of 30 μm with a 45 ° scan offset and 90 ° scan path rotation between subsequent build layers. Printing was performed under an argon atmosphere with oxygen levels kept below 0.01 %. A stainless-steel build plate was used due to its ability to ensure strong metallurgical bonding and low cost of resurfacing between builds [8] and no build plate heating was applied. Two different heat treatment schemes, shown in Table 2, were applied after printing for comparative purposes. The stress relief was conducted while specimens were attached to the build plate.

Table 2: Heat treatment schemes applied to LPBF IN718 specimens in this research. Abbreviations: AC – air cooled, RT – room temperature, WQ – water quenched

Designation	Stress relief	Homogenisation	Ageing
LPBF – H1110	950 °C for 2 h, AC to RT	1110 °C for 2 h, WQ to RT	720 °C for 24 h, AC to RT
LPBF – H1020	950 °C for 2 h, AC to RT	1020 °C for 1 h, WQ to RT	720 °C for 24 h, AC to RT

3. Test setup

The test setup consists of three main interconnected elements. These elements, along with their relative physical positioning, are shown in Figure 1 and include the Gleeble 3800 thermomechanical simulator [9], stereo DIC system, and IR camera. The Gleeble performed two main functions during testing, namely displacement and temperature control, and thus took the role of both tensile machine and heating furnace. The electrical input used to heat the specimen was controlled via feedback from a K-type thermocouple spot welded to the top surface of the specimen centred along the length of the gauge region with an accuracy of ± 0.2 mm. Two additional thermocouples, each approximately 15 mm from the specimen centre, were welded to the specimen for IR image adjustment.

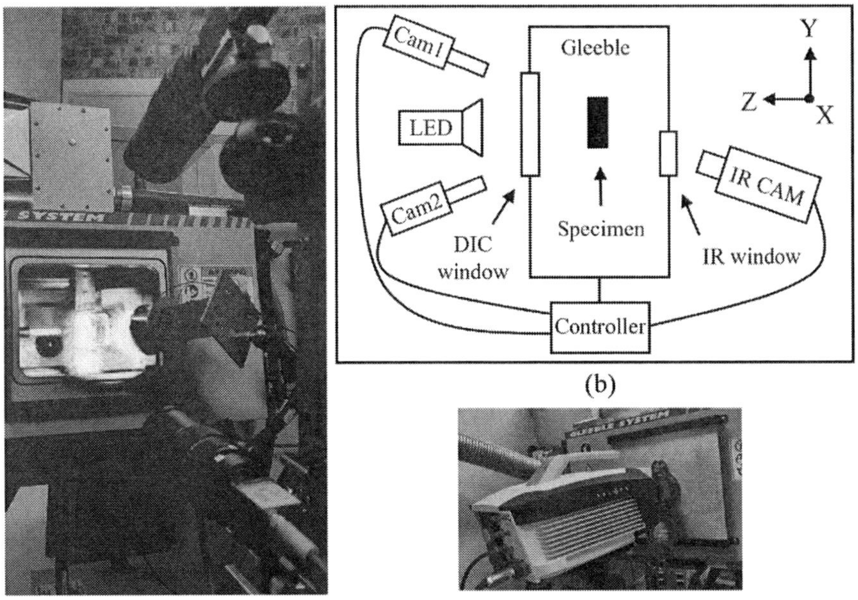

(a) (c)

Figure 1: (a) DIC system positioned at the front of the Gleeble. (b) The relative positioning of all major system elements. (c) Positioning of the IR camera at the back of the Gleeble.

The measurement accuracy of the Gleeble thermal control system is 1 ˚C and therefore the thermal control accuracy was limited to that of the K-type thermocouple which is 0.75 % or 3.4 ˚C and 6.4 ˚C at 450 ˚C and 850 ˚C respectively. The Gleeble can apply a vacuum of below 0.5 torr within the testing chamber during testing. This was used to limit specimen oxidation, discolouration of the paint used to create the speckle pattern, and heat haze that would negatively impact the DIC measurements. The hydraulic ram, which was used to apply the load at a constant grip displacement rate, has a displacement accuracy of 0.01 mm and the fitted 100 kN loadcell had a calibrated accuracy of 0.01 kN.

The geometry of the tensile specimen, shown in Figure 2(a), was based on a design by van Rooyen and Becker [6] that was designed for use with resistive heating to provide a parabolic thermal profile along the length of the specimen. The reduced cross-section gauge region was lengthened to 40 mm, instead of 25 mm, to increase the surface area available for DIC strain averaging at each temperature. Figure 2(b) shows the loading and electrical paths applied using H13 tool steel grip extenders which were necessary to mitigate grips blocking the views of the DIC and IR cameras.

A stereo DIC system was used for full-field displacement measurement through a borosilicate glass window on one side of the Gleeble. To achieve a suitable resolution, 75 mm double Gauss lenses were mounted to the cameras. To avoid image saturation due to radiation emitted at higher temperatures, narrow bandpass filters from Edmund Optics [10] were mounted on the lenses. These filters have a central wavelength of 470 nm and a full-width half-maximum width of 85 nm which blocks most visible radiation emitted by the specimen at 850 ˚C. These filters significantly reduce the overall incident light to the sensor and were therefore paired with a focused blue-light LED with a peak wavelength of 451 nm and rated power of approximately 20 W. DIC hardware parameters are reported in Table 3 according to the iDICs guidelines [11]. The speckle pattern consisted of both white and black speckles applied directly to the bare specimen surface rather than applying black speckles to a solid white paint background. This was found to improve paint adhesion during high strains.

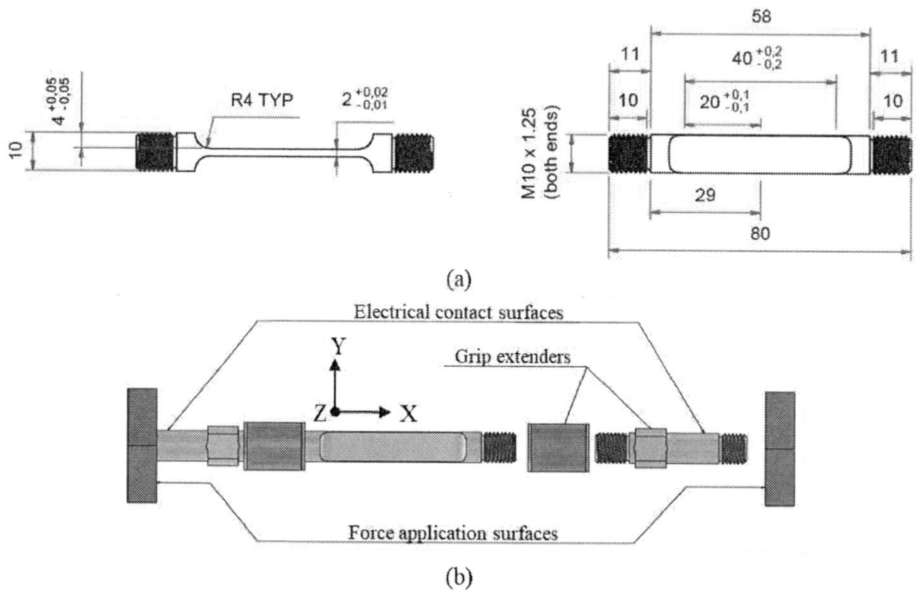

Figure 2: (a) Gleeble tensile specimen geometry (mm). (b) Partially exploded view of the tensile loading train including grip extenders showing locations of force and electrical contact and testing coordinate system.

Full-field IR measurement was achieved using a Telops FAST M350 [12] IR camera positioned at the back of the Gleeble. The IR camera captured images of the rear surface of the specimen which was painted with matte black very high temperature (VHT) flameproof paint [13] to provide constant emissivity across the entire surface and reduce reflections. The IR camera's measurable temperature range during testing was 133.7 - 1252.7 °C, with a thermal resolution of 20 mK, image resolution of 640 × 512 pixels and aperture of f/3. A Fluke ClirVu CV200 IR viewing window [14] was used for IR viewing access into the vacuum chamber.

Table 3: DIC hardware parameters reported according to iDICs guidelines [11]

Camera make and model	Baumer VCXU-50M
Image resolution (pixels)	2448 × 2048
Lens manufacturer and model	Edmund Optics 75 mm double Gauss
Focal length (mm)	75
Aperture (f/)	~ 18
Field-of-view (mm)	~ 69.49 × 58.13
Image scale (pixels/mm)	~ 35.23
Stereo angle (°)	~ 22
Stand-off distance (mm)	~ 650
Image acquisition rate (Hz)	0.125
Patterning technique	Black speckles on top of white speckles (no solid background)
Pattern feature size (pixels)	30

4. Temporal synchronisation, spatial calibration, and testing procedure

Temporal synchronisation was achieved during testing using a custom-built controller unit, which was designed around an Arduino Uno [15]. The controller unit was responsible for the simultaneous triggering of both DIC and IR systems at a preset acquisition rate. The controller unit was also responsible for saving and storing the synchronised data.

Both DIC and IR systems required spatial calibration to allow for overlay of the two datasets. To do this, the secondary DIC camera and IR camera needed to be spatially calibrated with respect to the first, master, DIC camera. To conduct a true calibration, all cameras need to face the same surface of the specimen. This was not possible due to space and IR window size limitations. For the purposes of this research, it was assumed that both specimen surfaces deformed equally during testing and therefore, an accurate calibration could still be achieved. Spatial calibration was achieved in the MatchID Calib DIC software (v2023.2.0) [16] using a custom calibration plate created by placing two similar calibration plates back-to-back. The front, DIC facing, surface of the calibration plate was an ordinary single plane MatchID calibration plate seen in Figure 3(a). The back, IR camera facing, surface of the calibration plate was machined from a plate of aluminium with holes corresponding to the MatchID calibration plate dots, as shown in Figure 3(b). The two faces of the calibration plate were then placed back-to-back and mounted on a stand to allow for two-axis rotation. A single calibration image was taken for the IR camera calibration with the calibration plate aligned with the specimen surface. Thereafter, images of the calibration plate were taken in various orientations for the stereo DIC calibration. In MATLAB (vR2022b) [17], the *imfindcircles*, *tform* and *imwarp* functions were used to identify corresponding dots on both calibration plates, calculate and apply the required transformation matrix to warp the IR image so that it would appear to be taken from the perspective of the second DIC camera, as shown in Figure 3(c). By doing this, the stereo calibration file that relates the DIC cameras can also be used to relate the IR camera.

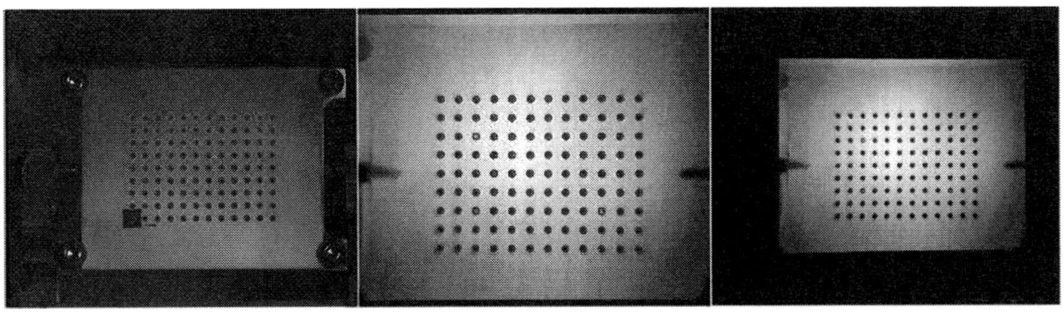

(a) (b) (c)

Figure 3: (a) DIC camera 2 image of the front calibration plate with marker locations identified in MATLAB. (b) IR image of the back calibration plate with marker locations identified in MATLAB. (c) Warped IR image.

After image calibration, the specimen was loaded into the Gleeble and a 400 N preload was applied to ensure strong electrical contact. A signal was sent from the Gleeble to trigger the controller to start image acquisition. Heating commenced at a rate of 10 °C/s. The thermal profile was allowed to stabilise for ten minutes upon reaching testing temperature. After the stabilisation period, the tensile test was automatically started by the Gleeble.

To meet quasi-static conditions, a constant displacement rate of 0.2 mm/min was chosen, which equals a strain rate of 0.005 ± 0.002 mm/mm/min. This meets the quasi-static strain rate requirements stipulated in ASTM-E21 [18] standard for high temperature testing of metallic materials. Due to the uneven thermal gradient of the specimen, the effective strain rate increased to a maximum of 0.0054 mm/mm/min in the hottest region during plastic deformation. This increased strain rate is allowed in ASTM-E21.

5. Post processing of test data

The RevealIR [12] IR camera software outputs data as a floating-point tagged image file format with pixel values corresponding to temperature values. The default temperature conversion formulas do not account for factors such as surface emissivity and transmissivity of mediums between the specimen and camera. These were corrected using measurements from the three thermocouples welded to the specimen using the correction factors (CF) in Equations 1 and 2.

$$IR_{image,\,CF1} = IR_{image} * CF1(imagenumber) \tag{1}$$

$$IR_{image,\,CF2}(x_{pixcolumn}) = IR_{image,\,CF1}(x_{pixcolumn}) * CF2(x_{pixcolumn}) \tag{2}$$

In these equations, CF1 corrected for the change in the emissivity caused by the degradation and darkening of the paint. As such, this correction factor was spatially constant but changed for each timestamp. This CF was calculated as the ratio of the IR and control thermocouple values for one image near the start of the test sequence and one image near the end of the test sequence. Intermediate CFs were determined by linear interpolation. CF2 corrected for the transmissivity of the IR viewing window. CF2 remained constant over time but varied spatially. To account for this, the three thermocouple temperatures were used to calculate the CF for three locations along the length of the specimen. Quadratic interpolation was used to determine the correction for each pixel along the specimen length (along $x_{pixcolumn}$ pixel vectors). The correction along the height of the sample was constant.

DIC analysis was performed on a single region of interest (ROI) which covered the entire specimen. Following this analysis, strain calculations were performed, and thermal images imported. The thermal image overlay process was automated by MatchID using the DIC calibration file. The data sets were subsequently merged into a single data set and exported in matrix format for further analysis.

The last steps of the processing workflow, which were completed in MATLAB, were data cleanup and extraction of temperature-dependent mechanical properties. Data cropping was performed, followed by smoothing using the *smoothdata* function with *movmean* (moving mean method) selected. Thereafter, working from the specimen centre outwards in both directions, the data was subdivided into smaller sub-ROIs, each approximating a span of 10 °C. This is shown in Figure 4(a), where the data points representing the sub-ROI bounds have been circled in red. In all subsequent operations, each sub-ROI was handled independently. The temperature was then averaged for each sub-ROI to give a single temperature representative of the sub-ROI. The strain for each sub-ROI at each time step was averaged to provide vectors, with each element corresponding to a time step.

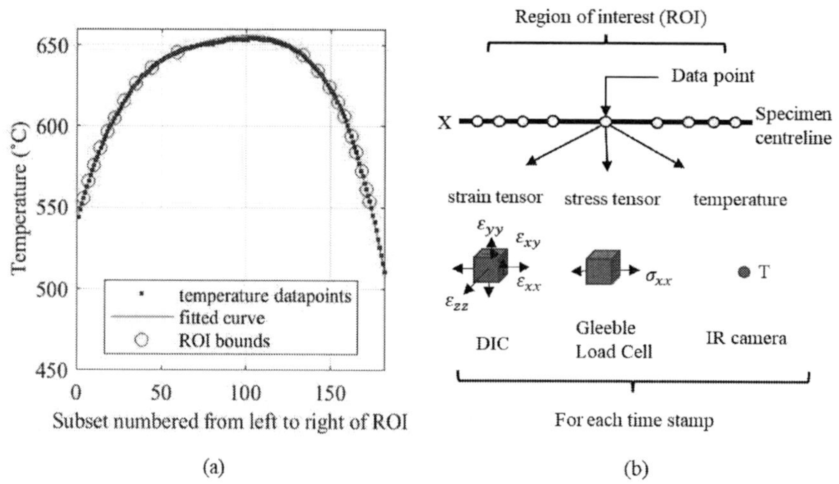

Figure 4: (a) Thermal profile of a specimen at 650 °C overlayed with data fit and sub-ROI bounds. (b) Overview of data components.

Figure 4(b) shows a summary of the resulting data. After ROI subdivision, a stress-strain curve was plotted for each sub-ROI and the yield strength and percentage elongation (%EL) were calculated. %EL was only calculated for the central sub-ROI at the highest temperature where the specimen failed.

6. Results

Figure 5 (a) shows the yield strength results from the two halves of a single specimen. The data shows close correlation between the two halves of each specimen as can be seen by the tight grouping of yield strength values at each temperature. Figure 5(b) shows the yield results of three specimens after averaging the results from both halves of each specimen. The data suggest excellent repeatability and consistency of both the manufacturing and testing methods with a variation of less than 15 MPa in yield strength between specimens.

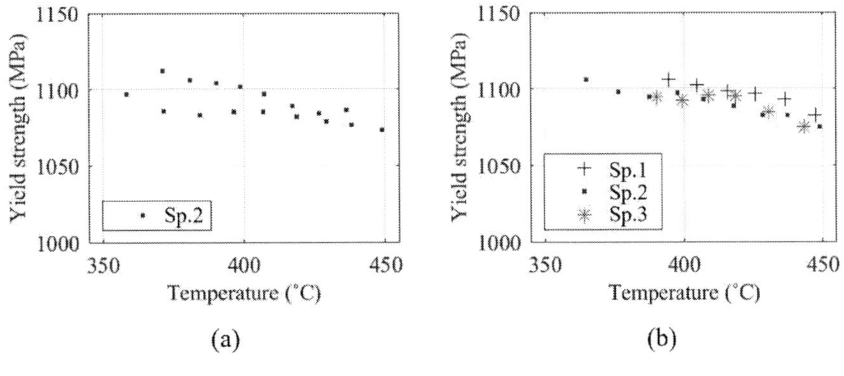

Figure 5: (a) YS data extracted from both halves of specimen 2 (Sp.2) subject to homogenisation at 1020 ˚C. (d) Comparative yield data after averaging the two specimen halves of three different tests.

Figure 6 shows the trends in yield strength and ductility as a function of temperature for the two different heat treatments and two print orientations. In Figure 6(a), it is observed that all LPBF IN718 specimens exhibit the same rapid strength decrease that is seen in wrought IN718 starting around 650 ˚C. The rate of strength decrease is roughly the same regardless of heat treatment and print orientation.

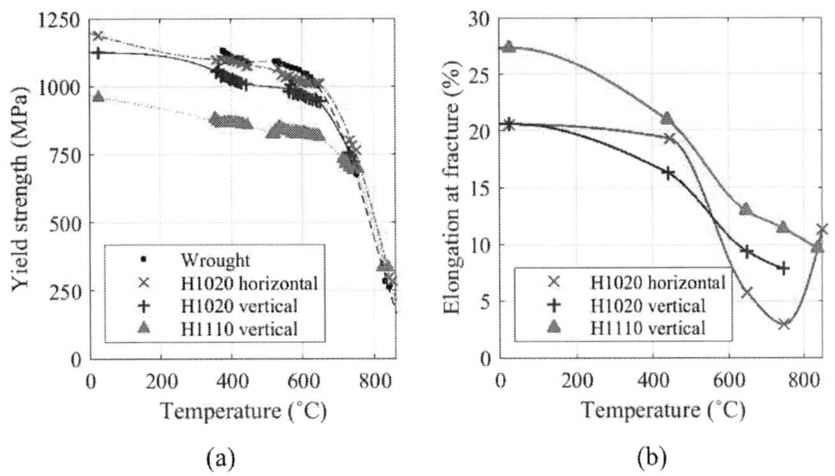

Figure 6: Comparative (a) yield strength and (b) elongation as a function of temperature.

Figure 6(b) shows that LPBF IN718 is not exempt from high temperature embrittlement as indicated by the drop in elongation values after 450 °C. All LPBF specimens perform significantly worse than wrought IN718 at high temperatures where the wrought material achieves an elongation of more than 15.3 % [4] at 650 °C compared to a mere 5.7 % in horizontally printed LPBF specimens. The test method, in its current state, also allows for the extraction of other temperature dependant properties including ultimate tensile strength, Young's modulus, Poison's ratio and hardening coefficients, and with slight adjustment can be used to determine thermal expansion coefficients.

7. Conclusion

This paper highlighted key shortfalls of LPBF IN718, specifically the high degree of embrittlement at high temperatures. Although this embrittlement also occurs in the wrought material, the degree of embrittlement is higher in the LPBF produced specimens which is detrimental to its potential use in the aerospace industry. This work also showed that the high temperature properties could be quantified in a specimen efficient manner using the synchronised Gleeble-DIC-IR technique. Future work is necessary to establish the mechanisms which cause the observed embrittlement.

Nomenclature

AC – Air cooled; AM – Additively manufactured; CF – Correction factor; DIC – digital image correlation; IN718 – Inconel 718; IR – Infrared; LPBF – Laser-powder bed fusion; PSD – particle size distribution; ROI – Region of interest; RT – Room temperature; UTS – Ultimate tensile strength; VHT – Very high temperature; WQ – Water quenched; %EL – Percentage elongation.

Acknowledgement

The financial assistance of the National Research Foundation (NRF) of South Africa (Grant Number 138432) towards this research is hereby acknowledged. Opinions expressed and conclusions arrived at, are those of the author and are not necessarily to be attributed to the NRF. The authors are also grateful to the Department of Science and Innovation through the Collaborative Program for Additive Manufacturing (CPAM) administered by the Council for Scientific and Industrial Research (CSIR) for the project funding, through contract number S004063.

References

[1] Perrut, M., Caron, P., et al. 2018. Comptes Rendus Physique. 19(8):657–671.
[2] General Electric. 2023. Additive Manufacturing: Aviation and aerospace industry. [Online]
[3] Gruber, K., Stopyra, W., et al. 2022. Journal of Manufacturing Processes. 73:642–659.
[4] Special Metals. 2007. Inconel alloy 718.
[5] Zheng, L., Schmitz, G., et al. 2012. Critical Rev. in Solid State and Mat. Sciences. 37(3) 181-214
[6] van Rooyen, M. & Becker, T.H. 2018. Journal of Strain Analysis for Eng. Design. 53(3):117–129.
[7] AP&C, Boisbriand, Canada
[8] Uhlmann, E., Düchting, J., et al. 2021. Progress in Additive Manufacturing. 6:385–394.
[9] Dynamic Systems Inc, Poestenkill, New York, United States
[10] Edmund Optics, Barrington, New Jersey, United States
[11] iDICs. 2018. A Good Practices Guide for Digital Image Correlation. E.M.C. Jones & M.A. Iadicola (eds.)
[12] Telops, Quebec City, Canada
[13] Dupli-Color Inc., Cleveland, Ohio, United States
[14] Fluke, Everett, Washington, United States
[15] Arduino, Lombardia, Italy
[16] MatchID, Gent, Belgium
[17] MathWorks Incorporated, Natick, Massachusetts, United States
[18] ASTM International. 2009. ASTM E21-09: Standard Test Methods.

INCLUDING THERMOCHEMICAL EQUILIBRIUM CALCULATIONS INTO MULTIPHYSICS MODELS – A FINITE VOLUME BASED PHASE CHANGE MODEL

Alfred E.J. Bogaers[1,2*], Jaco J. Swanepoel[1], Willem A. Roos[1], Johan H. Zietsman[1,3]
[1]Ex Mente Technologies, Pretoria, South Africa, * alfred.bogaers@ex-mente.co.za
[2]Department of Mechanical Engineering and Aeronautical Engineering, University of Pretoria, South Africa
[3]Department of Material Science and Metallurgical Engineering, University of Pretoria, South Africa

Abstract: In this paper we present two formulations of the energy equation which facilitates the direct coupling between computational fluid dynamics and computational thermochemistry, to simulate phase-change problems. The two formulations are designed to allow for treating the computational thermochemistry calculation step as a black-box operator, allowing for the methods described in this paper to be generically applicable to any computational thermochemistry tool. The energy equation formulations allow for stable and efficient, implicit in time, convergence behaviour. We present comparisons of the two energy equation formulations using two simple solid-liquid phase change test cases.

Keywords: solid-liquid phase change, computational thermochemistry, computational fluid dynamics

1. Introduction

Phase change is the process associated with matter changing from one state to another, such as a solid changing to liquid or gas, and it is a crucial part of most high temperature pyrometallurgical systems. Process material is introduced into a furnace as solids, allowed to melt, advect, diffuse and chemically react with one another to form the various products of interest. Modelling phase change in these systems require capturing the potentially large non-linearities in material thermochemical properties (e.g. heat capacity, enthalpy) and thermophysical properties (e.g. viscosity, thermal conductivity), as functions of temperature and chemical composition.

Numerous examples of computational fluid dynamics (CFD) models to simulate convection-diffusion problems, with phase change are readily available in open-literature [1-3]. Most of these models are formulated under the assumption that the temperature-enthalpy relationship of the material being modelled can in some way be approximated, or described, using linearised approximations of the change in enthalpy between known system states; for example approximating the change in enthalpy of a material between a known solidus temperature and a known liquidus temperature, assuming a fixed chemical composition. These simplifications are often overly restrictive when simulating the multitude of multi-component (multi-species), complex chemical compositions found in pyrometallurgical systems.

A far more generalised approach would be to approximate the system energy states by using computational thermochemistry. Tools such as ChemApp, FactSage and Thermo-Calc all allow for the approximations of system properties such as total enthalpy, heat capacity and phase compositions, by Gibbs energy minimisation [4]. Several attempts have been made to couple CFD with computational thermochemistry (or aspects of computational thermochemistry) [4-9]. These implementations are often limited in that they apply only one-way coupling, directly include a Gibbs free energy minimiser within the source code of the CFD tool, or use computational thermochemistry to extract system information and construct regression or interpolation sub-models.

The question of how to, in a truly generic fashion, couple CFD with computational thermochemistry, under the constraints of numerical stability and numerical efficiency remains an open-ended question. By 'truly generic fashion', we mean to imply, that the computational thermochemistry tool is treated as a true black-box operator. In other words, the thermochemistry tool is only used as an input-output operator, where, as input it receives system temperature and chemical compositions, and provides as output, estimates of system phase fractions, phase compositions, thermochemical properties, and thermophysical properties.

In this paper we present and compare two formulations of the energy equation which allows for the inclusion of computational thermochemistry to estimate solid and liquid phase fractions, heat capacity, and enthalpy. The first formulation is based on a slight modification of the popular source-based approach [3] where the second is based on the optimum enthalpy approach [1,10]. Despite the popularity and wide application of the source-based approach, we will demonstrate that this approach does lead to poor, unstable, numerical convergence behaviour when applied to problems similar to those found in pyrometallurgical systems. To improve this performance, we further propose augmenting the source-based approach by including dynamic under-relaxation.

The performance of the two energy equation formulations are compared using two simplified solid-liquid phase change test cases.

2. Governing Equations

The governing equations that describe the conservation of mass, momentum and energy, to solve for the flow velocity, $\boldsymbol{u}$, and system enthalpy, h, are given by

$$\frac{\partial \rho}{\partial t} + \nabla \cdot (\rho \boldsymbol{u}) = 0, \tag{1}$$

$$\frac{\partial \rho \boldsymbol{u}}{\partial t} + \nabla \cdot (\rho \boldsymbol{u}\boldsymbol{u}) = -\nabla p + \nabla \cdot \boldsymbol{\tau} + \rho \boldsymbol{g} + S_A, \tag{2}$$

$$\frac{\partial \rho h}{\partial t} + \nabla \cdot (\rho \boldsymbol{u} h) = \nabla \cdot (\kappa \nabla T), \tag{3}$$

where ρ is material density, $\boldsymbol{\tau}$ the fluid viscous stress tensor, $\boldsymbol{g}$ gravitational acceleration, and κ the thermal conductivity.

The material properties are all calculated as an average of the liquid and solid properties, linearly weighted with phase fraction, α, such that

$$\rho = \alpha \rho_s + (1 - \alpha)\rho_l \tag{4}$$

$$\kappa = \alpha \kappa_s + (1 - \alpha)\kappa_l \tag{5}$$

where subscripts l and s indicate fluid and solid regions respectively.

The source term, S_A, in Equation (2) is a porosity term, which forces the flow in solidifying regions to tend to 0. To ensure consistency with the simulation test case presented in Section 4.1, we define the porosity source term to be the same as the one presented in [3], given by:

$$S_A = D\frac{(1 - \alpha)^2}{\alpha^3 + \epsilon}, \tag{6}$$

where D is the Darcy coefficient, and ϵ is a small numerical constant to avoid division by 0 (set to $\epsilon = 1 \times 10^{-3}$). While the porosity-based approach has frequently been used in solid-liquid phase change problems, an alternative approach would be to use a solid-suspension viscosity model, similar to the

Krieger-Dougherty family of viscosity models [11]. These models approximate the influence of suspended solid particles on the effective flow viscosity, and typically exhibit an exponential increase in effective viscosity for solid fractions approaching 0.3.

3. Numerical Solution

The focus of the current paper is on proposing and testing different strategies for solving the energy equation, where the system enthalpy, or energy associated with phase change, is directly computed with computational thermochemistry. We will therefore limit our discussion to the solution of Equation (3) when directly coupling with numerical thermochemistry (and assume that appropriate solution schemes for Equations (1) and (2) are readily available).

The solid-liquid phase change problem is solved using an in-house developed finite-volume solver. The solver is based on a fixed-grid approach, where the solid and liquid phases are tracked using a phase fraction field, α, which ranges between 0 and 1, indicating solid and liquid regions respectively.

3.1. Numerical Thermochemistry

In this paper, the energy associated with phase change is calculated by directly incorporating ChemApp [4] into the finite-volume solver ChemApp is a thermochemical software library which has bindings for multiple programming languages, and facilitates multicomponent, multiphase complex equilibrium calculations based on thermochemical data from FactSage [4].

To perform an equilibrium calculation, ChemApp uses Gibbs energy minimisation to convert specified temperature and chemical composition, to phase fractions, phase compositions, and thermochemical properties of the system and individual phases.

With some loss of generality, we define the thermochemical equilibrium calculation step, as a black-box operator, $\boldsymbol{E}()$, defined as

$$h, C_p, \alpha = \boldsymbol{E}(T, x_1, x_2, \dots, x_n), \tag{7}$$

where x_i indicates the mole (or mass) fraction of the *i*-th chemical component. The $\boldsymbol{E}(T, \boldsymbol{x})$ operator must be performed for each computational cell, and if an implicit in time scheme is used to solve Equation (3), must be repeated several times per time step until convergence is obtained. Reducing the number of iterations per time step is therefore critical to the efficient solution of Equation (3).

The $\boldsymbol{E}(T, \boldsymbol{x})$ operator, as posed in (7), presents a challenge when solving Equation (3). Equation (3) requires solving for the rate of change of h, but h is computed by the thermochemical calculator. In the sections to follow we will present two possible approaches to solve the energy equation given this restriction. The first relates to the popular source-based approach, and the second to a gradient-based approach.

Finally, thermochemical equilibrium calculators do not provide chemical reaction kinetic rates, but rather provide the equilibrium state (the statistically most probable state) for a given chemical composition and energy state. In essence, the black-box operator, as posed in Equation (7), implies infinitely fast reaction kinetics. This is a reasonable assumption for most pyrometallurgical systems, which involve chemical reactions occurring at high temperatures. The assumption is not reasonable at cold, or near ambient conditions, which may require defining individual species reaction equations, along with associated Arrhenius-like reaction kinetic equations.

3.2. Source-Based Approach

The source-based approach, first popularised for modelling solidification phase change systems [1,3], is based on the idea of splitting the total enthalpy into sensible enthalpy and enthalpy of formation, Δh, such that

$$h = C_p T + \Delta h, \tag{8}$$

where C_p represents the specific heat capacity. Inserting Equation (8) into (3), then yields an energy equation in the form of

$$\frac{\partial(\rho C_p T)}{\partial t} + \nabla \cdot (\rho C_p \boldsymbol{u} T) - \nabla \cdot (\kappa \Delta T) = -\left(\frac{\partial(\rho \Delta h)}{\partial t} + \nabla \cdot (\rho \boldsymbol{u} \Delta h)\right). \tag{9}$$

Equation (9) can now be discretised and solved for T based on the, iteratively updated, explicit source terms related to Δh on the right-hand side of Equation (9). This approach has been used to great effect when assuming that the change in enthalpy can be approximated by a piece-wise continuous function, where enthalpy varies linearly between the material's solidus temperature, T_s, and liquidus temperature, T_l, where Δh can then be defined as

$$\Delta h(T) = \begin{cases} L, & T \geq T_l \\ L\alpha, & T_l > T > T_s, \\ 0, & T \leq T_s \end{cases} \tag{10}$$

where L represents the latent heat of fusion for a given material.

Real materials, with several chemical components (such as alloys and slags), or when the chemical compositions change, either due to transport phenomena or chemical reactions, do not obey this simplified relationship. For this reason, we propose an alternative form of Δh which accommodates the black-box operator posed in Equation (7), where

$$\Delta h^{k+1} = (h^k - h_{\text{ref}}) - C_{p,\text{ref}}(T^k - T_{\text{ref}}), \tag{11}$$

where superscript k indicates field values from the previous solution iteration, and h_{ref} and $C_{p,ref}$ indicates the reference total enthalpy and specific heat capacity, obtained from the thermochemical calculator, at a reference temperature, T_{ref}.

From numerical experiments, it appears that the choice in T_{ref} has little influence on convergence performance, provided the spatially varying chemical composition remains constant. However, to accommodate chemical composition changes (either due to transport phenomena, or chemical reactions), we set T_{ref} (and the corresponding h_{ref} and $C_{p,\text{ref}}$ fields) to be equal to the converged fields from the previous time step. The Δh source term is therefore always linearised locally around the current time step.

Since the terms associated with Δh feature as explicit source terms in Equation (9), convergence for a given time step is not guaranteed, especially for materials with highly non-linear temperature-enthalpy relationships. For this reason, under-relaxation is typically still required to assist with convergence, using a temperature update formula of the form

$$T^{k+1} = \omega T^{k+1,*} + (1 - \omega) T^k, \tag{12}$$

where ω represents the relaxation factor, $T^{k+1,*}$ represents the intermediate temperature solution obtained by solving Equation (9), and T^k represents the temperature field from the previous iteration. When including under-relaxation, the overall rate of convergence is highly dependent on an appropriate choice of ω, with no *a-priori* way to decide on an optimal value. To this end, we also propose using dynamic relaxation, which is further elaborated on in Section 3.4.

3.3. Gradient-Based Temperature-Enthalpy Approach

The second approach to solving the energy equation is based on the optimum enthalpy approach [1,10], originally developed for phase-change problems involving sensible enthalpy transport and constant thermophysical properties. The main idea is to linearise the phase change problem by applying a Taylor series expansion, relating temperature and enthalpy. To outline the method, we start with a semi-discretised form of the energy equation, given by

$$\rho\left(\frac{h_{n+1}^{k+1}-h_n}{\Delta t}\right)+\nabla\cdot\left(\rho\boldsymbol{u}h_{n+1}^{k+1}\right)=\nabla\cdot\left(\kappa_{n+1}^{k}\nabla T_{n+1}^{k+1,*}\right), \tag{13}$$

where superscripts k and $k+1$ refers to the previous and current solution iterations, when solving the energy equation implicitly in time. Subscript $n+1$ refers to the current time step being solved for, where the converged solutions from the previous time step are defined with subscript n. The superscript $*$ further refers to an intermediate solution of the temperature field, where we assume that ρ and $\boldsymbol{u}$ are frozen during the iterative solution of the energy equation.

The new enthalpy for iteration $k+1$ can then be approximated via a first order Taylor series expansion of the temperature-enthalpy relationship, given by

$$h_{n+1}^{k+1}=h_{n+1}^{\mathrm{k}}+\left.\frac{\partial h}{\partial T}\right|_{n+1}^{k}\left(T_{n+1}^{k+1,*}-\mathrm{T}_{n+1}^{k}\right). \tag{14}$$

Inserting (14) into (13), yields an energy equation of the form

$$\frac{\rho\left(h_{n+1}^{k}-h_n\right)}{\Delta t}+\rho\left.\frac{\partial h}{\partial T}\right|_{n+1}^{k}\left(\frac{T_{n+1}^{k+1,*}-\mathrm{T}_{n+1}^{k}}{\Delta t}\right)+\nabla\cdot\left(\rho\boldsymbol{u}h_{n+1}^{k}\right)$$
$$+\nabla\cdot\left(\rho\boldsymbol{u}\left.\frac{\partial h}{\partial T}\right|_{n+1}^{k}\left(T_{n+1}^{k+1,*}-\mathrm{T}_{n+1}^{k}\right)\right)=\nabla\cdot\left(\kappa_{n+1}^{k}\nabla T_{n+1}^{k+1,*}\right), \tag{15}$$

which can now be solved for the intermediate temperature field, $T_{n+1}^{k+1,*}$, by including explicit source terms based on the enthalpy from the previous solution iteration, h_{n+1}^{k}, and enthalpy from the previous time step, h_n. Equation (15) can further be simplified through the observation that $\left.\frac{\partial h}{\partial T}\right|_{\mathrm{n+1}}^{\mathrm{k}}=C_{p,\mathrm{n+1}}^{k}$, which is also an output obtained from the thermochemical calculator. At convergence, the second and fourth terms in Equation (15) tend to 0, since at convergence $T_{n+1}^{k+1,*}\approx\mathrm{T}_{\mathrm{n+1}}^{\mathrm{k}}$, satisfying the original equation, as posed in Equation (3).

3.4. Dynamic Relaxation

In the simulations to follow, we demonstrate that the source-based approach, even for a relatively simple temperature-enthalpy relationship, can fail to converge. Using fixed point iterations with under-relaxation improves this behaviour, but the optimal choice of the relaxation factor is problem dependent. For this reason, we propose making use of Aitken's dynamic relaxation, which has gained significant popularity to stabilise and improve the convergence rates in the field of partitioned fluid-structure interactions [12].

The idea behind Aitken's acceleration is to iteratively update the relaxation factor, ω, by using information from the preceding solution iterations. By defining a residual for the current iteration $k+1$ as

$$r^{k+1} = T^{k+1,*} - T^k, \tag{16}$$

Aitken's relaxation iteratively updates the relaxation factor by using the residuals from the current and preceding solution iterations:

$$\omega^{k+1} = -\omega^k \frac{(r_k)^T (r_{k+1} - r_k)}{(r_{k+1} - r_k)^T (r_{k+1} - r_k)}. \tag{17}$$

Despite its relative simplicity, Aitken's method has the potential to significantly reduce the number of required coupling iterations. In fact, it has been shown that Aitken's method can be interpreted as a quasi-Newton method, where the approximate system Jacobian is iteratively updated using information from the preceding three iterations [12].

4. Simulation Test Cases

4.1. 2D Buoyancy Driven Cavity Test Case

The first test case considers a square cavity with natural convection, buoyancy driven, phase change, first presented in [3]. The test case was the first reported numerical simulation of convection-diffusion phase-change, which considered an incongruently melting material.

The domain is a two-dimensional square, with a size of 1 unit x 1 unit (the original test case was presented without units). The top and bottom faces are insulating, where the left face is fixed to a cold temperature of $T_c = -0.5$, and a fixed hot face temperature of $T_h = 0.5$, applied to the right face. To ensure reproducibility with the results presented in [3], the temperature-enthalpy relationship is defined as a piece-wise continuous function of the form given by Equation (10) (in other words, this test case is not solved by directly coupling with ChemApp). The phase fraction field α is defined to vary linearly between the solidus temperature, $T_s = -0.1$, and liquidus temperature, $T_l = 0.1$.

The square domain is discretised into 40 x 40 equally spaced cells and solved using a time step size of $\Delta t = 10$, for a total of 100 time steps (final time of $t = 1000$). A convergence criterion of $T_{\text{tol}} = 1 \times 10^{-4}$ is used, where the remainder of the test case parameters are summarised in Table 1.

Table 1: 2D cavity test case parameters (all parameters are non-dimensional as per [3])

Property	Value
Initial temperature	$T_i = 0.5$
Hot wall temperature	$T_H = 0.5$
Cold wall temperature	$T_C = -0.5$
Liquidus temperature	$T_l = 0.1$
Solidus temperature	$T_s = -0.1$
Density	$\rho = 1$
Specific heat capacity	$C_p = 1$
Dynamic viscosity	$\mu = 1$
Thermal conductivity	$\kappa = 0.001$
Thermal expansion coefficient	$\beta = 0.01$
Gravitation acceleration vector	$\boldsymbol{g} = (0, -1000, 0)$
Latent heat of fusion	$L = 5$
Darcy constant	$D = 1.6 \times 10^3$

The simulation results at time 1000 are shown in Figure 1, showing the temperature and liquid fraction distributions. Three solid lines are added to show the solidus, liquidus and 50% solid fraction interfaces, along with a comparison to the results reproduced from [3] in Figure 1(c).

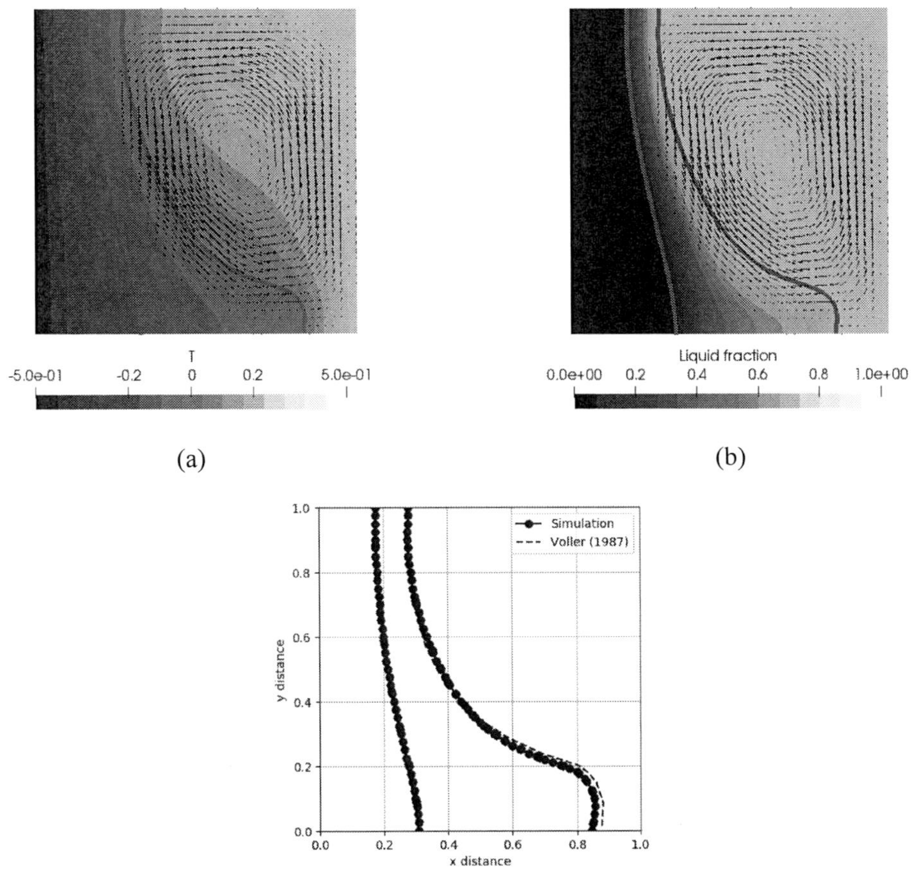

(c)

Figure 1: 2D cavity test case results at time step 1000, showing (a) temperature distribution, (b) liquid fraction distribution, and (c) a comparison of the solidus and liquidus interface lines, compared to results presented in [3].

Figure 2 shows the convergence performance of the source and gradient based approaches when applied to the 2D cavity test case. The plots compare the number of fixed-point iterations required for each of the time steps, for different under-relaxation factors, as well as when applying Aitken's dynamic relaxation. Despite the comparatively simple temperature-enthalpy relationship, the source-based approach fails to converge, within a limit of 1000 iterations, when no under-relaxation is used ($\omega = 1$) but benefits significantly from dynamic relaxation. Minor benefits are observed when combining dynamic relaxation with the gradient-based approach, which is approximately 3 times more efficient than the source-based approach (when also using Aitken's method).

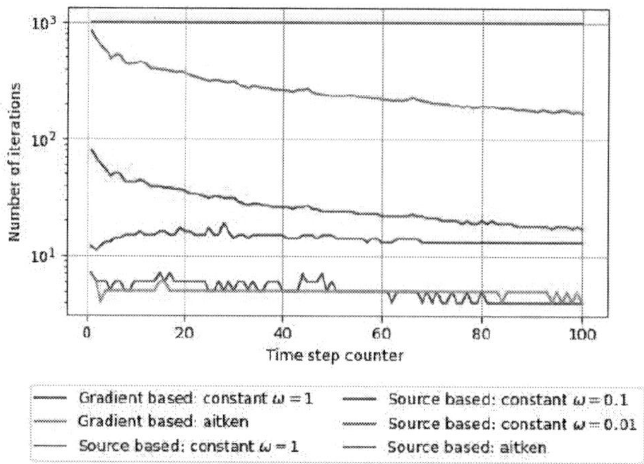

Figure 2: Convergence performance of the source and gradient based solution approaches applied to the 2D cavity problem, using different fixed point under relaxation factors.

4.2. Ilmenite particle submerged in slag

The second test case involves the melting and solidification behaviour of an ilmenite particle submerged within a slag bath (Figure *3*). The radius of the ilmenite particle is chosen to be representative of typical ilmenite feed, with $r_{\text{particle}} = 92\mu m$, surrounded by slag, modelled with a radius of $r_{\text{slag}} = 4r_{\text{particle}}$. The chemical composition of the two regions are presented in Table 2. The ilmenite particle is initially at a temperature of $T_{\text{initial particle}} = 27°\text{C}$ surrounded by slag with a temperature of $T_{\text{slag}} = 1627°\text{C}$.

The test case is solved by directly incorporating ChemApp, and for computational cost reasons is solved using a 1D spherical FVM formulation. Since the problem is solved in 1D, we omit mass transport, and therefore only solve the energy equation, assuming fixed chemical compositions. The simulation is performed with a total of 200 equally sized elements, with a time step size of $\Delta t = 1 \times 10^{-3} s$.

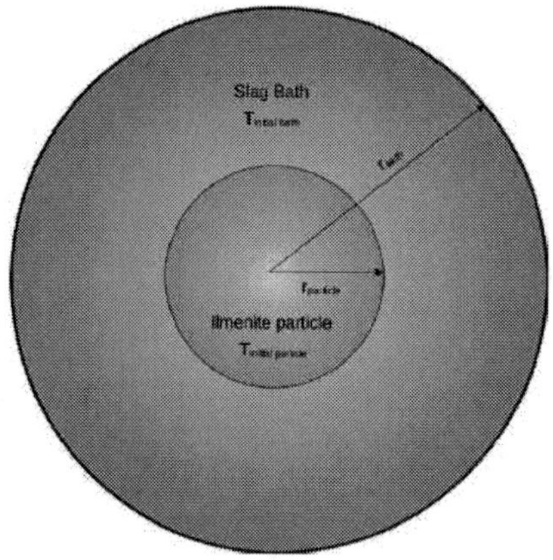

Figure 3: Model overview of an ilmenite particle submerged in slag.

Figure *4* shows a selection of the simulation results at four different time steps. The results show the temperature distribution over time (right), along with the liquid slag phase mass fraction (left). The particle starts off fully solid, with the slag bath fully liquid (Figure *4* (a)). The slag bath along the particle interface is rapidly cooled forming a solidified shell (Figure *4* (b)), consisting predominantly of a solid pseudobrookite phase along with a minor solid TiO_2 rutile phase (not shown). As the particle continues to heat up, it eventually starts melting, while the solidified shell remains intact ((Figure *4* (c). As the particle melts, the act of melting itself extracts additional energy from the surrounding slag, which in turn increases both the thickness and time to melt the solidified shell. The particle itself is fully molten before the solidified slag shell is completely removed (Figure *4*(d)) These results have interesting implications for ilmenite smelting processes. The total time to melt an ilmenite feed particle is governed by the time to melt the formed solidified shell, and not the melting rate of the particle itself. A comprehensive numerical investigation of ilmenite particle melting rates as a function of particle size, initial particle temperatures and material properties is reported on in [13].

Table 2: Chemical composition of the slag and ilmenite particle.

Phase	**Slag**	**Ilmenite particle**
$(FeO)(TiO_2)$	0	100
FeO	15	0
TiO_2	55	0
Ti_2O_3	30	0

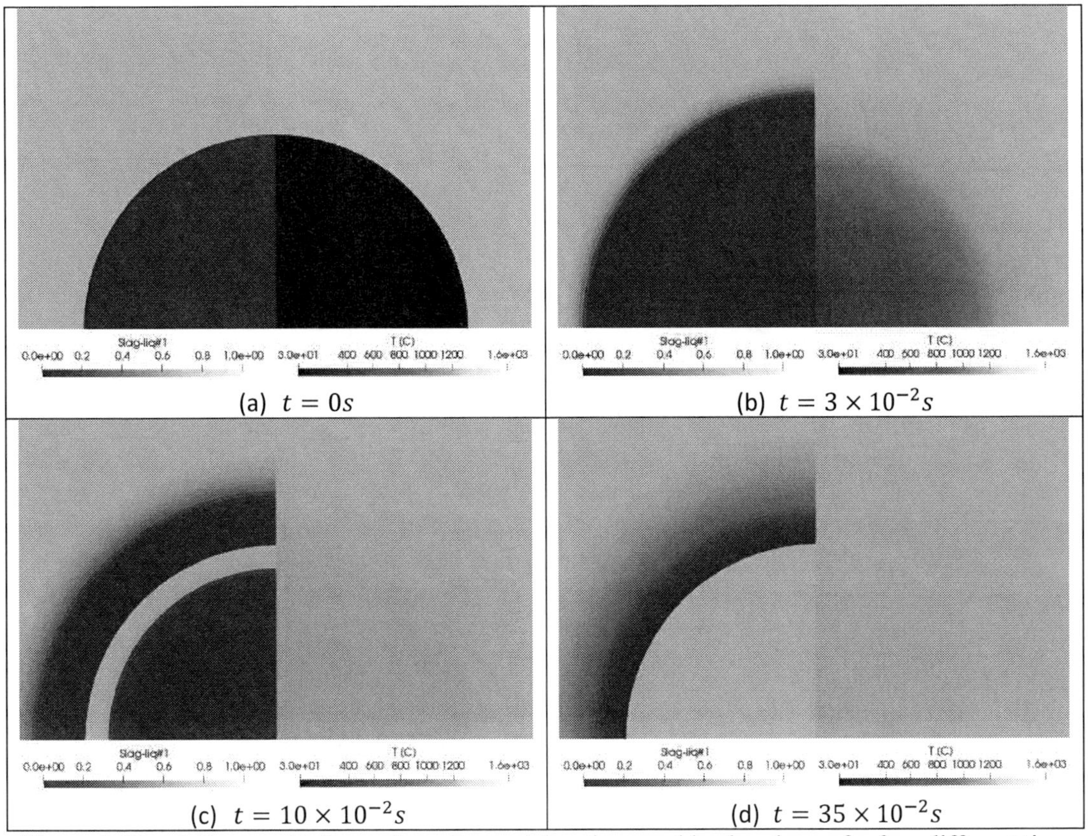

(a) $t = 0s$ (b) $t = 3 \times 10^{-2}s$

(c) $t = 10 \times 10^{-2}s$ (d) $t = 35 \times 10^{-2}s$

Figure 4: Simulation results for an ilmenite particle submerged in slag shown for four different time steps, illustrating the change in temperature over time and the formation and dissolution of a solidified shell.

Figure 5 presents the convergence behaviour of the two proposed solution schemes, when using Aitken's dynamic relaxation. The performance of the source-based approach does improve noticeably from time step 38 onwards, which coincides with the instance in time when the ilmenite particle is fully molten, significantly reducing the non-linearity of the system's temperature-enthalpy relationship. The gradient based approach is 3-10 times more efficient than the source-based approach over the first 38 time steps, before normalising towards the end of the simulation.

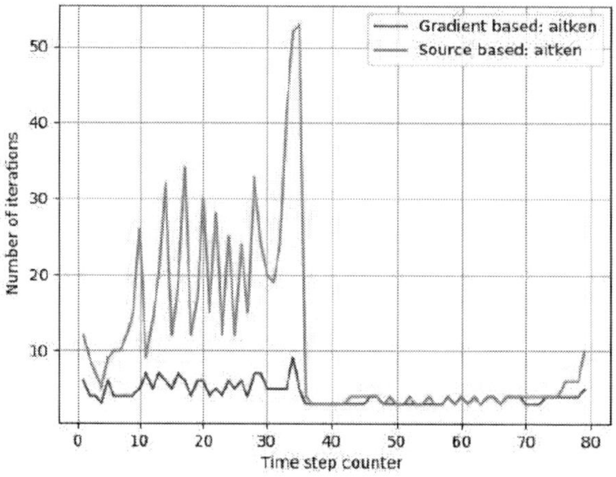

Figure 5: Number of energy equation iterations required for the ilmenite-slag particle simulation.

5. Computational Cost of Thermochemistry

The costs associated with computational thermochemistry can render the current solution approach computationally infeasible. To place this in perspective, for the three-component system (containing Fe, Ti, and O) solved in Section 4.2, ChemApp required on average 0.15s to complete a single thermochemical calculation. The one-dimensional problem, with only 200 cells, and solved for only 80 time steps, required close to 2 hours in computational time, compared to only a handful of seconds to solve the test case presented in Section 4.1, where no thermochemistry was used. The computational cost of direct thermochemical calculations is only further exacerbated when the number of chemical components increase, where a 10-component system can require as much as 15s for a single thermochemical calculation. Extending this to real systems, which often require spatial discretisations containing several million cells, is clearly infeasible. Recent developments in geometric thermochemical acceleration techniques [14], do however pose a promising avenue for future research. Early results suggest that acceleration factors of up to a thousand are currently possible, which if developed further has the potential of reducing computational times into the realm of feasibility.

6. Concluding Remarks

In this paper, we presented two potential approaches that lead to the stable numerical solution of the energy equation, when approximating the enthalpy change, associated with phase change, with the ChemApp computational thermochemistry library. While the solver presented in this paper is still in its infancy, the ability to directly incorporate thermochemistry in a computational fluid dynamics framework, opens several possibilities to improve the reliability and accuracy when modelling large pyrometallurgical systems.

References

[1] Faden, M., König-Haagen, A., Brüggerman, D., An optimum enthalpy approach for melting and solidification with volume change. *Energies.* **2019**, 12(5).

[2] Kewalramani, R.G., Pose, S., et al., Solid-liquid phase change in the presence of gas phase: Numerical modelling and validation. *Journal of Fluid Flow, Heat and Mass Transfer.* **2020**, 7, 1-11.

[3] Voller, V.R., Prahash, C., A fixed grid numerical modelling methodology for convection-diffusion mushy region phase-change problems. *Int. J. Heat Mass Transf.* **1987**, 30, 1709-1719.

[4] Petersen, S. and Hack, K, The thermochemistry library ChemApp and its applications. *International Journal of Materials Research*, **2007**, 98 (10), 935-945.

[5] Ersson M, Höglund L, Tilliander A, Jonsson L, Jönsson P. Dynamic coupling of computational fluid dynamics and thermodynamics software: applied on a top blown converter. *ISIJ international.* **2008**, 15;48(2):147-53.

[6] Jonsson L, Sichen D, Jönsson P. A new approach to model sulphur refining in a gas-stirred ladle–A coupled CFD and thermodynamic model. *ISIJ international.* **1998,** 15;38(3):260-7.

[7] Marin-Alvarado, Tanai, M.A. "Combining Multiphysics Modeling and Solution Thermodynamics Using M4Dlib, an External Library." COMSOL Conference, Boston. 2015.

[8] Roos W.A., Zietsman J.H,. Accelerating complex chemical equilibrium calculations—A review. *Calphad.* **2022** 1;77:102380.

[9] Scuro N.L., Poschmann M., Beneš O., Piro M.H. Progress in Coupling Computational Thermodynamics and Computational Fluid Dynamics to Support Molten Salt Reactor Applications. *In Proceedings of Fourth International Conference on Generation IV & Small Reactors*, G4SR4, **2022**, (pp. 1-15).

[10] Swaminathan, C.R. Voller, V.R., A general enthalpy method for modelling solidification processes. *Metall. Trans. B.* **1992**, 23, 651-664.

[11] Jonijn, B.J., Sanderink, O.B.J., Nicolaas, P, Experimental study of the viscosity of suspensions: Effect of solid fraction, particle size and suspending liquid. *Powder Technology*. **2014**, 266, 61-69.

[12] Bogaers, A.E.J., Kok, S., Reddy, B.D., Franz, T., Quasi-newton methods for implicit black-box FSI coupling, *Computer Methods in Applied Mechanics and Engineering*. **2014**, 279, 113-132.

[13 Bogaers, A.E.J., Swanepoel, J.J., Roos, W., Zietsman, J, Single particles entering a slag bath: How they influence process dynamics and furnace capacity, *Southern African Pyrometallurgy,* **2024**.

[14] Roos, W.A., Bogaers, A.E.J., Zietsman, J.H., Geometric acceleration of complex chemical equilibrium calculations – Performance in two-to five-component systems. *Calphad.* **2023**, 82.

Calibration of Laser Absorptivity in CFD Modelling of LPBF In-Situ Alloying with Ray Tracing

M.R. Botete[1, *], D.C. Blaine[1], G. Venter[1]
[1]Department of Mechanical and Mechatronic Engineering
Stellenbosch University, South Africa, *mrbotete@sun.ac.za

Abstract: This paper presents a cutting-edge approach for calibrating laser absorption in Laser Powder Bed Fusion (LPBF) simulations aimed at enhancing the accuracy of modelling the intricate interplay between lasers and materials during the additive manufacturing process. Leveraging a synergistic coupling of Computational Fluid Dynamics (CFD) with ray tracing, the simulations were designed to comprehensively capture the complexities of laser-material interactions. The calculation of laser absorption was accomplished by incorporating a complex refractive index model, which takes into account material composition. The study establishes a robust correlation between simulation results and experimental data, affirming the effectiveness of the proposed calibration method. However, recognizing the evolving nature of LPBF processes, the conclusion emphasizes the need for additional simulation cases to thoroughly calibrate the newly introduced parameter. The ongoing exploration across diverse scenarios is anticipated to further refine the accuracy and reliability of the calibrated model, providing a valuable tool for predicting and optimizing LPBF outcomes in various industrial applications.

Keywords: Absorptivity; LPBF; In-situ Alloying; Ray Tracing, Modelling

1. Introduction

Laser Powder Bed Fusion (LPBF) is a powerful additive manufacturing technique that allows the creation of complex geometries with superior mechanical properties. In-situ alloying during LPBF enables the fabrication of customised metal alloys with enhanced performance, by depositing different powders into the same build chamber [1, 2]. However, it is challenging to produce high-quality, defect-free microstructures using LPBF in-situ alloying technique [3]. The most common defects, in addition to defects inherent to the LPBF process [4–6], are micro-segregation, unmolten powders, and high levels of porosity [7]. These defects originate primarily from inhomogeneities within the feedstock mixture relative to the beam spot size. To increase the stability and adaptability of the in-situ alloying, as well as the broader LPBF process, it is essential to predict and control these defects. This requires a deep understanding of the physical processes involved and computational simulations offer great assistance in this regard.

Review studies have outlined the methodologies and applications of LPBF process simulations [8–12]. Most work has focused on three areas: the behaviour of the melt pool which forms during the laser-powder interactions, the evolution of microstructures during solidification and cooling, and the overall part distortion and stress distribution. Currently, particular emphasis has been placed on the simulation of the LPBF in-situ alloying process, particularly in the first focus area concerning melt pool behaviour [13–17]. Understanding the behaviour of the melt pool as driven by the complexities introduced by alloying is key for predicting material manufacturability and defects and optimising the LPBF in-situ alloying process.

In recent years, significant progress has been achieved regarding the mesoscale simulation of melt pool behaviour, effectively capturing the authentic physics associated with the LPBF process. This encompasses elements such as stochastic powder bed characteristics, melt pool dynamics, variations in surface tension, and vapour recoil pressure [11]. Nevertheless, achieving accurate simulation results remains a challenge due to the multitude of input parameters required. Among these parameters, laser absorptivity is the most critical, defining the fraction of incident laser energy absorbed by the material [18].

In laboratory settings, the laser absorptivity of metal powders can be measured using various techniques. However, these measurements may lack accuracy as they often disregard the dynamic conditions inherent to the LPBF process [19]. This is because LPBF surface conditions such as surface roughness and composition evolve during the LPBF process, altering absorptivity beyond standard laboratory conditions. Extensive studies have explored these effects [20–24].

This situation presents a significant challenge for the LPBF community, particularly as the majority of processed materials are alloys. These alloys consist of a mixture of elements, lacking a uniform, smooth surface, and often exhibiting varying compositions, especially in the case of LPBF in-situ alloying. Consequently, when modelling the LPBF process, absorptivity cannot be simplified as a single constant value; instead, it needs to be calibrated using experimental data.

This study introduces a composition-based modelling of laser absorption in Computational Fluid Dynamics (CFD) simulations of the LPBF process. The method uses the ray tracing technique to capture laser-powder interactions whereby the energy absorption is calibrated against experimental results. A specific case of Ti-6Al-4V alloy is studied and two calibration experimental techniques from literature are used to help align simulations with real-world LPBF conditions, where absorptivity varies due to changing surface properties and composition, ensuring more accurate and reliable results.

2. Basic Principle

During the LPBF process, a laser beam consisting of photons with a distinct wavelength and energy is emitted towards the metal surface. As the photons strike the metal, they interact with the electrons within the top 10 – 100 nm of the metal's surface. This interaction, known as electron-phonon collision, transfers energy into the metal, heating it up. The amount of energy transferred to the surface depends on the direction (incident angle θ_i) at which the photons strike, with some being absorbed while the remainder being reflected in the direction θ_r as illustrated in Figure 1. In the case of a non-uniform surface, such as powdered material, multiple reflections can occur as photons make contact with various surface points along their trajectory, releasing energy at each point.

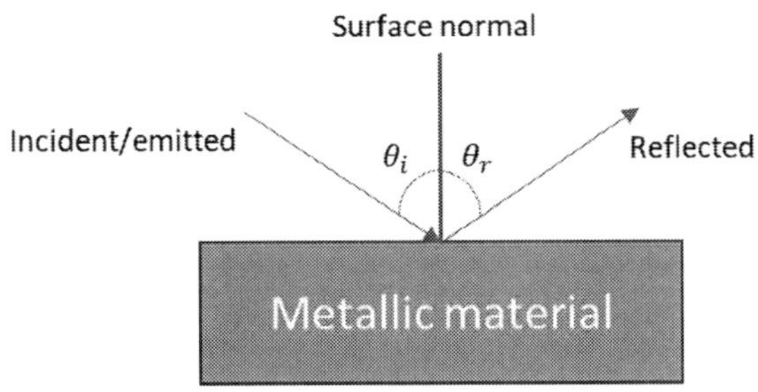

Figure 1: Incident light-metal interactions.

To fully capture this phenomenon within CFD simulations, the ray tracing technique is employed to trace the trajectory of each photon within the laser beam. This process involves determining not only the precise location and angle at which the photon impacts and subsequently reflects off the surface but also quantifying the fraction of energy that is absorbed.

The incident angle θ_i can be determined by computing the angle between the incident trajectory vector $(\vec{I})$, and the surface normal vector $(\vec{N})$ using the cosine rule as shown in equation (1). The inclusion of the negative sign is necessary to ensure that the surface normal points outward.

$$\theta_i = \cos^{-1}\left(-\frac{\vec{I}\cdot\vec{N}}{|\vec{I}||\vec{N}|}\right) \tag{1}$$

The vector representing the reflected trajectory can be derived by performing a Householder transformation of the incident trajectory with respect to the surface normal vector as shown in equation (2).

$$\vec{R} = \vec{I} + 2(-\vec{I}\cdot\vec{N})\vec{N} \tag{2}$$

The fraction of the energy absorbed is determined by considering two key wavelength-dependent optical properties of the metal: the refractive index (n) and the extinction coefficient (k). These optical constants characterise how electromagnetic waves, such as light, interact with materials. They account for the complex nature arising from the fact that materials can store and dissipate energy simultaneously due to their electrical conductivity [25]. Combining these optical constants into a complex refractive index $(\hat{n})$, as shown in equation (3), the absorption of light energy by metals can be described.

$$\hat{n} = n + ik \tag{3}$$

In principle, light waves are composed of electric and magnetic fields that oscillate perpendicular to the direction in which the light wave propagates. Taking this into account, absorptivity at different angles of incidence can be determined using Fresnel's equations (4) and (5) for S and P polarisation, respectively [25].

$$A_S = 1 - \left|\frac{\cos\theta_i - (\hat{n}^2 - \sin^2\theta_i)^{\frac{1}{2}}}{\cos\theta_i + (\hat{n}^2 - \sin^2\theta_i)^{\frac{1}{2}}}\right|^2 \tag{4}$$

$$A_P = 1 - \left|\frac{\hat{n}^2\cos\theta_i - (\hat{n}^2 - \sin^2\theta_i)^{\frac{1}{2}}}{\hat{n}^2\cos\theta_i + (\hat{n}^2 - \sin^2\theta_i)^{\frac{1}{2}}}\right|^2 \tag{5}$$

Normally, the angle of natural or unpolarised electromagnetic radiation constantly varies. For laser light, the overall absorptivity (A) of the materials is simplified as the average of S and P polarisation components, equation (6) [26].

$$A = 0.5A_S + 0.5A_P \tag{6}$$

Obtaining the optical constants n and k for metal alloys poses a significant challenge due to the limited availability of comprehensive data, which is typically only available and fully listed for pure metallic elements [27]. In this study, the effective complex refractive index of the alloy is estimated based on the composition of its constituents, using the Complex Refractive Index (CRI) model described by equation (7). Here, v_i represents the volume fraction of the constituent i, and p is a parameter influenced by the geometry and arrangement of the constituents within the alloy. In its general form, the CRI model is expressed with $p = 2$ for composite materials [28, 29]. The goal of the calibration process is to find the optimal value of p specifically tailored for metal alloys.

$$\hat{n}_{eff} = \left(\sum_{i=1}^{N} v_i \hat{n}_i^{\,p}\right)^{1/p} \tag{7}$$

For a specific case of Ti-6Al-4V alloy, the complex refractive index of the constituents at a laser wavelength of 1070 nm relevant to LPBF conditions are provided as follows: $\hat{n}_{Ti} = 3.35 + 3.29i$, $\hat{n}_{Al} = 1.22 + 10.47i$ and $\hat{n}_V = 3.10 + 3.67i$ [27]. Utilizing this information, Figure 2 illustrates the variation of absorptivity as a function of incident angle for random values of p. Notably, when $p = 1$, the absorptivity at an incident angle of 0° is greater for when $p = 2$ for composite materials with even distributions of the constituents. However, absorptivity increases again when p is increased to 3. This gives p-value in the range of $1 - 3$ for calibrations.

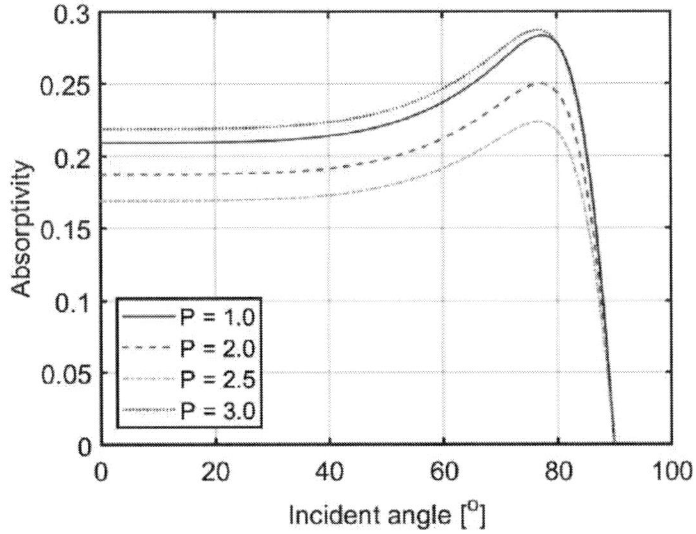

Figure 2: Absorptivity of Ti-6Al-4V alloy as a function of the incident angle for different values of p.

3. Calibration Experiments

This study uses experimental data obtained from the literature to calibrate laser absorptivity. The experiments follow two distinct methods as described below.

3.1. Experiment 1

This experiment uses a calorimetric technique to measure the thermal response of the powder under laser irradiation followed by explicit calculation of laser absorptivity for calibration. A precisely constructed setup of this experiment features a disc containing the powder with the desired layer thickness. This disc is strategically positioned on a purpose-built holder designed to minimize conductive cooling effects. The thermal response is measured using either a highly sensitive thermocouple or a precise pyrometer, as illustrated in Figure 3, (a) and (b) respectively.

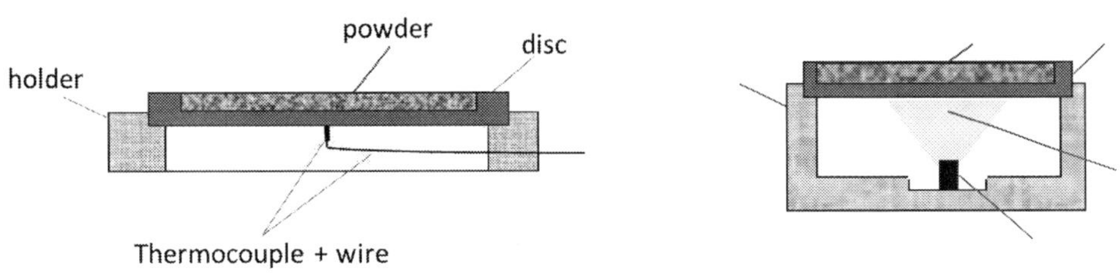

(a) (b)

Figure 3: Calorimetric setup for measuring thermal response using (a) thermocouple, and (b) pyrometer.

This setup allows for the in-situ measurement of thermal response during LPBF processing of single/multiple tracks on the disc or powder surface. The typical curve of temperature evolution throughout the deposition is shown in Figure 4. Absorptivity is calculated by first fitting the temperature decay function represented by equation (8), within the cooling region of the recorded temperature profile. Here, T_0 represent the temperature just before the laser is switched on, T_1 is the end temperature while the laser is still on, and t_1 is the time taken for the cooling process to reach thermal equilibrium.

$$T(t) = T_1 e^{-t/0.5t_1} + T_0 \tag{8}$$

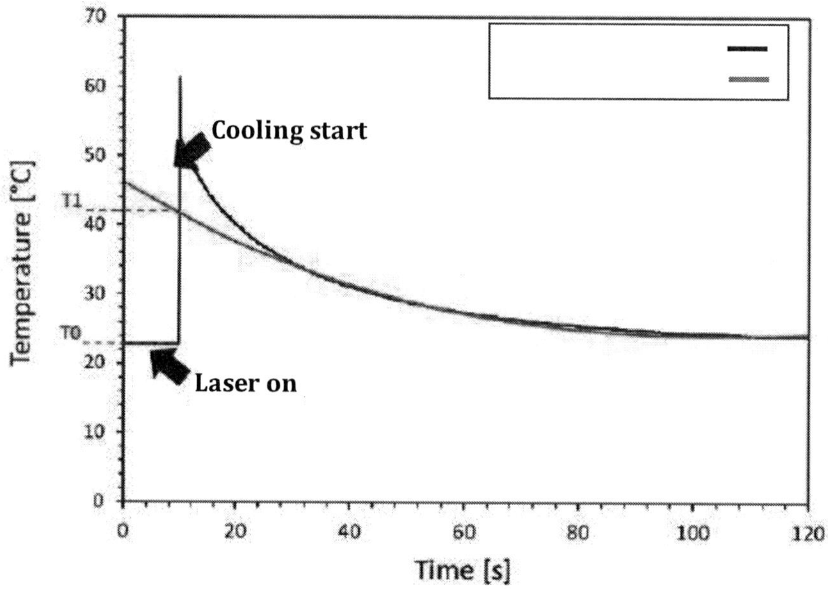

Figure 4: Typical temperature measurement curve used to evaluate net energy [30].

The effective absorptivity is then calculated by dividing the energy (E_H) necessary to uniformly heat the specimen from starting temperature T_0 to end temperature T_1 by the input energy (E_I) due to the laser irradiation as shown in equation (9). Here, C_p is the temperature-dependent specific heat capacity, m is the mass of the disc which includes the powder (if applicable), L is the total length of the laser track, v is the scanning speed and P is the nominal laser power.

$$A_{eff} = \frac{E_H}{E_I} = \frac{\int_{T_0}^{T_1} mC_p(T)}{L/vP} \tag{9}$$

Numerous studies in the literature have widely employed this method, utilising either thermocouples [24, 31–36] or pyrometers [30] to determine and calibrate the laser absorptivity for diverse materials during the LPBF process. Despite the extensive utilisation of thermocouples, certain studies have highlighted challenges in accurately capturing thermal losses over the disc before reaching thermal equilibrium [31]. To address this limitation, Trapp et al [31] proposed a technique that involves using two thermocouples placed a certain distance away from each other along the build direction. The end temperature T_1 is then extrapolated from fitting decay curves to the temperature profile of each thermocouple. Alternatively, Reynolds [30] introduced the use of a pyrometer ensuring that the track remains within a focused area as shown in Figure 3, enhancing precision in temperature measurements.

3.2. Experiment 2

The second experiment focuses on analysing the meltpool geometry of the actual as-built samples. The meltpool is systematically categorised into two distinct zones as depicted in Figure 5: the fusion zone, characterised by complete material melting, and the heat-affected zone (HAZ). This distinction allows for a detailed examination of both the fully molten region and the surrounding area influenced by the thermal effects of the laser. The comparison of these zones' dimensions with those simulated allows for further calibration of laser absorptivity, ensuring its alignment with actual LPBF conditions. Moreover, quantifying the dimensions as a function of laser intensity (I) defined by equation Figure 5 allows for capturing abrupt and significant changes attributed to the presence of a keyhole [35].

$$I = \frac{P}{\sqrt{v}} \tag{10}$$

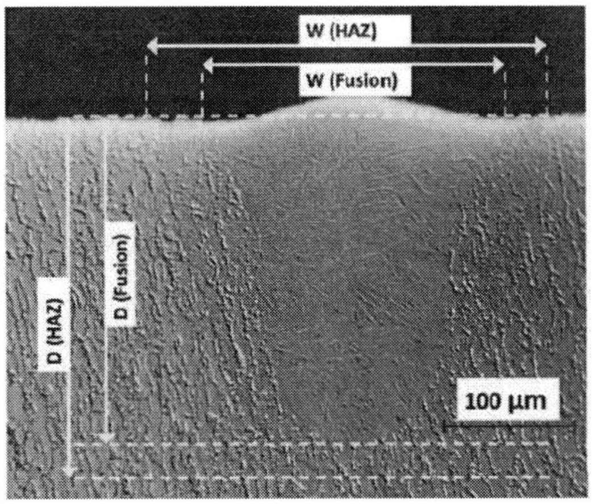

Figure 5: Typical meltpool geometry of the as-built LPBF samples [37].

A limited number of studies [37–39] have used this experimental method to determine and calibrate the laser absorptivity of materials during the LPBF process. Concerning a specific case of Ti-6Al-4V, the study by Cook and Ritchie [37], measured melt pool dimensions of LPBF tracks at a variety of laser powers and speeds as listed in Table 1. These tracks were built on a flat plate using a Concept Laser M2 Cusing machine under an inert argon atmosphere and will be the basis for the laser calibration in this study.

Table 1: Fusion and heat-affected zone (HAZ) measurements for single-track experiments as measured by [37].

Laser power [W]	Laser speed [m/s]	Fusion width [µm]	Fusion depth [µm]	HAZ width [µm]	HAZ depth [µm]
50	0.2	69.7	14.9	119.0	39.7
60	0.3	83.4	19.2	121.8	40.2
60	0.6	66.8	12.4	107.2	30.4
120	0.6	126.1	42.7	156.0	57.8
240	0.6	211.3	148.1	263.5	169.3
300	0.6	247.0	204.0	310.2	217.9
120	0.3	148.4	53.0	200.5	77.0
120	1.2	83.9	19.9	112.8	39.7
120	1.8	70.1	15.4	104.1	33.0
120	0.15	235.6	181.2	331.2	204.0
360	0.6	236.0	252.9	305.1	278.1

4. Simulations and Results

The CFD simulations are performed using the ANSYS Fluent 2023 R2 commercial software. The model is implemented to comprehensively integrate crucial elements, including melt pool formation, free surface movement, melt flow, evaporative heat loss, recoil pressure, and the Marangoni effect. The physical equations and material properties utilized in the model are thoroughly documented in the literature [37]. To accurately capture laser-material interactions outlined in Section 2, the incident beam is divided into a substantial number of rays. These rays are initialised to deliver energy with a Gaussian profile described by equation (11), where R represents the beam radius and r is the radial location of the laser ray. Each ray is subsequently tracked through successive reflections until less than 1% of the initial ray energy remains, or the ray exits the model domain. The effective absorptivity of the system was then computed by dividing the cumulative absorbed energy by the cumulative initial energy of the rays. This value is subsequently compared to that obtained in experiment 1 for calibrations, specifically to select the optimal p-value.

$$E_d = \frac{2P}{\pi R^2} \exp\left(-\frac{2r^2}{R^2}\right) \tag{11}$$

Conducting simulations on a flat Ti-6Al-4V plate with the laser power ($P = 120$ W) and laser scan speed of $v = 0.6$ m/s, Figure 6, illustrates the evolution of effective absorptivity for various p-values, with the steady-state average effective absorptivity indicated in brackets. Notably, when setting $p = 2$, an absorptivity of 0.26 is obtained aligning with the minimal value reported by Ye, et al [35] using the two-thermocouple approach of experiment 1 for flat and powdered samples. Opting for $p = 2.5$ yields an absorptivity of 0.27, mirroring the value reported by Cook and Ritchie [37] using experiment 2. However, their studies disregarded the stochastic nature of the absorptivity of each laser ray in their simulations. Nonetheless, this value remains acceptable, given that Simonds, et al [40], reported a maximum absorptivity of 0.33 using a light scattering experiment. This value is attributed to the presence of a keyhole causing an abrupt and significant change in absorptivity [35].

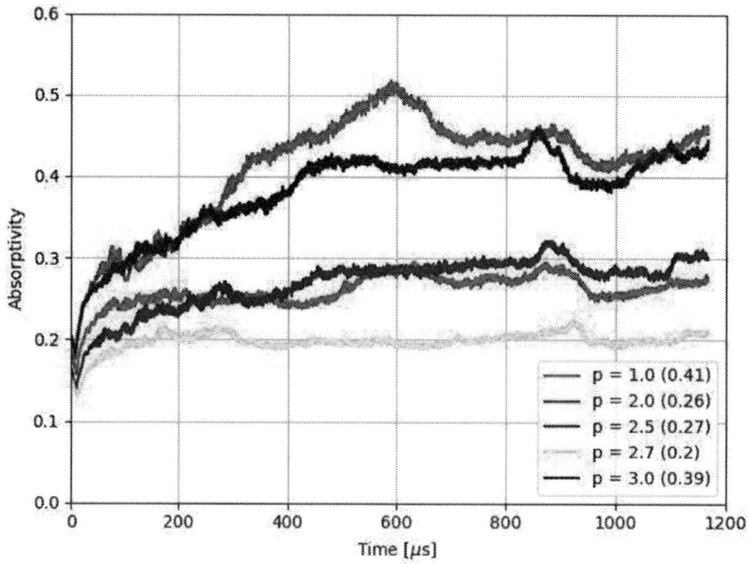

Figure 6: Absorptivity evolution on a Ti-6Al-4V flat plat at different values of p.

To calibrate further, the simulated meltpool dimensions are presented in Table 2. The dimensions obtained with a p-value of 2.5 closely match the experimental results at laser power ($P = 120$ W) and laser scan speed of $v = 0.6$ m/s. this further supports absorptivity of 0.27, reported by Cook and Ritchie [37].

Table 2: Fusion and heat-affected zone (HAZ) measurements for single-track simulations on a flat plate.

p-value	Fusion width [µm]	Fusion depth [µm]	HAZ width [µm]	HAZ depth [µm]
1.0	129.9	72.8	151.3	84.1
2.0	133.5	41.6	153.51	52.4
2.5	125.1	43.2	157.0	57.8
2.7	123.2	40.3	154.1	52.6
3	140.3	63.2	170.3	73.2

5. Conclusion

In conclusion, this study introduced a novel method for calibrating laser absorption in LPBF simulations. The approach involved the integration of CFD with ray tracing to comprehensively capture laser-material interactions. The computation of laser absorption was achieved by employing a complex refractive index model based on material composition. The obtained results demonstrated a robust agreement with experimental data; nonetheless, to achieve a thorough calibration of the newly introduced parameter, further exploration through a broader spectrum of simulation cases is imperative. The continued investigation across varied scenarios will enhance the reliability and applicability of the calibrated model in accurately representing the complex dynamics of LPBF processes.

Nomenclature

A Absorptivity
c_p Specific heat
E Energy
I Laser intensity
$\vec{I}$ Incident direction vector
k Excitation coefficient
m Mass
$\hat{n}$ Complex refractive index
n Refractive index
$\vec{N}$ Surface normal vector
p Material parameter
P Laser power
r Radial position
R Laser beam radius
$\vec{R}$ Reflecting direction vector
t Time
T Temperature
v Laser scan speed

ν Volume fraction
θ Incident angle

References

[1] Macpherson, A.; Li, X.; et al. (2017). Antibacterial Titanium Produced Using Selective Laser Melting, *JOM*, Vol. 69, No. 12, 2719–2724. doi:10.1007/s11837-017-2589-y

[2] Huber, F.; Papke, T.; et al. (2018). In Situ Formation of a Metastable β-Ti Alloy by Laser Powder Bed Fusion (L-PBF) of Vanadium and Iron Modified Ti-6Al-4V, *Metals*, Vol. 8, No. 12, 1067. doi:10.3390/met8121067

[3] Simonelli, M.; Aboulkhair, N. T.; et al. (2018). A comparison of Ti-6Al-4V in-situ alloying in Selective Laser Melting using simply-mixed and satellited powder blend feedstocks, *Materials Characterization*, Vol. 143, No. November 2017, 118–126. doi:10.1016/j.matchar.2018.05.039

[4] Matthews, M. J.; Guss, G.; et al. (2017). Denudation of metal powder layers in laser powder-bed fusion processes, *Additive Manufacturing Handbook: Product Development for the Defense Industry*, Vol. 114, 677–693. doi:10.1201/9781315119106

[5] King, W. E.; Anderson, A. T.; et al. (2017). Laser powder-bed fusion additive manufacturing of metals; physics, computational, and materials challenges, *Additive Manufacturing Handbook: Product Development for the Defense Industry* (Vol. 2), 461–506. doi:10.1201/9781315119106

[6] Sames, W. J.; List, F. A.; et al. (2016). The metallurgy and processing science of metal additive manufacturing, *International Materials Reviews*, Vol. 61, No. 5, 315–360. doi:10.1080/09506608.2015.1116649

[7] Simonelli, M.; Aboulkhair, N. T.; et al. (2018). A comparison of Ti-6Al-4V in-situ alloying in Selective Laser Melting using simply-mixed and satellited powder blend feedstocks, *Materials Characterization*, Vol. 143, No. November 2017, 118–126. doi:10.1016/j.matchar.2018.05.039

[8] Khairallah, S. A.; Anderson, A. T.; et al. (2016). Laser powder-bed fusion additive manufacturing: Physics of complex melt flow and formation mechanisms of pores, spatter, and denudation zones, *Acta Materialia*, Vol. 108, 36–45. doi:10.1016/j.actamat.2016.02.014

[9] Popova, E.; Rodgers, T. M.; et al. (2017). Process-Structure Linkages Using a Data Science Approach: Application to Simulated Additive Manufacturing Data, *Integrating Materials and Manufacturing Innovation*, Vol. 6, No. 1, 54–68. doi:10.1007/s40192-017-0088-1

[10] Parteli, E. J. R.; Pöschel, T. (2016). Particle-based simulation of powder application in additive manufacturing, *Powder Technology*, Vol. 288, 96–102. doi:10.1016/j.powtec.2015.10.035

[11] Markl, M.; Körner, C. (2016). Multiscale Modeling of Powder Bed–Based Additive Manufacturing, *Annual Review of Materials Research*, Vol. 46, No. 1, 93–123. doi:10.1146/annurev-matsci-070115-032158

[12] Hebert, R. J. (2016). Viewpoint: metallurgical aspects of powder bed metal additive manufacturing, *Journal of Materials Science*, Vol. 51, No. 3, 1165–1175. doi:10.1007/s10853-015-9479-x

[13] Gu, H.; Wei, C.; et al. (2020). Multi-physics modelling of molten pool development and track formation in multi-track, multi-layer and multi-material selective laser melting, *International Journal of Heat and Mass Transfer*, Vol. 151, 119458. doi:10.1016/j.ijheatmasstransfer.2020.119458

[14] Klassen, A.; Forster, V. E.; et al. (2017). Numerical simulation of multi-component evaporation during selective electron beam melting of TiAl, *Journal of Materials Processing Technology*, Vol. 247, 280–288. doi:10.1016/j.jmatprotec.2017.04.016

[15] Küng, V. E.; Scherr, R.; et al. (2021). Multi-material model for the simulation of powder bed fusion additive manufacturing, *Computational Materials Science* , Vol. 194, 110415. doi:10.1016/j.commatsci.2021.110415

[16] Chouhan, A.; Hesselmann, M.; et al. (2022). Numerical modelling of in-situ alloying of Al and Cu using the laser powder bed fusion process: A study on the effect of energy density and remelting on deposited track homogeneity, *Additive Manufacturing*, Vol. 59. doi:10.1016/J.ADDMA.2022.103179

[17] Samaei, A.; Sang, Z.; et al. (2023). Multiphysics modeling of mixing and material transport in additive manufacturing with multicomponent powder beds, *Additive Manufacturing*, Vol. 67. doi:10.1016/J.ADDMA.2023.103481

[18] Kim, J.; Lee, S.; et al. (2020). Calibration of Laser Penetration Depth and Absorptivity in Finite Element Method Based Modeling of Powder Bed Fusion Melt Pools, *Metals and Materials International*, Vol. 26, No. 6, 891–902. doi:10.1007/S12540-019-00599-3/FIGURES/15

[19] Cook, P. S.; Ritchie, D. J. (2023). Determining the laser absorptivity of Ti-6Al-4V during laser powder bed fusion by calibrated melt pool simulation, *Optics & Laser Technology*, Vol. 162, 109247. doi:10.1016/J.OPTLASTEC.2023.109247

[20] Kaplan, A. F. H. (2012). Absorptivity modulation on wavy molten steel surfaces: The influence of laser wavelength and angle of incidence, *Applied Physics Letters*, Vol. 101, No. 15, 21. doi:10.1063/1.4759126/150800

[21] Zhou, Y. H.; Zhang, Z. H.; et al. (2019). Selective laser melting of typical metallic materials: An effective process prediction model developed by energy absorption and consumption analysis, *Additive Manufacturing*, Vol. 25, 204–217. doi:10.1016/J.ADDMA.2018.10.046

[22] Jadhav, S. D.; Vleugels, J.; et al. (2020). Mechanical and electrical properties of selective laser-melted parts produced from surface-oxidized copper powder, *Material Design & Processing Communications*, Vol. 2, No. 2, e94. doi:10.1002/MDP2.94

[23] Lassègue, P.; Salvan, C.; et al. (2021). Laser powder bed fusion (L-PBF) of Cu and CuCrZr parts: Influence of an absorptive physical vapor deposition (PVD) coating on the printing process, *Additive Manufacturing*, Vol. 39, 101888. doi:10.1016/J.ADDMA.2021.101888

[24] Wadge, M. D.; Lowther, M.; et al. (2023). Tailoring absorptivity of highly reflective Ag powders by pulsed-direct current magnetron sputtering for additive manufacturing processes, *Journal of Materials Processing Technology*, Vol. 317, 117985. doi:10.1016/J.JMATPROTEC.2023.117985

[25] Krishnan, K. M. (2021). 6.9.1 p- and s-Polarized Light Waves, and Fresnel Equations of Reflection, *Principles of Materials Characterization and Metrology*, Oxford University Press

[26] Liu, B.; Fang, G.; et al. (2020). A new ray tracing heat source model for mesoscale CFD simulation of selective laser melting (SLM), *Applied Mathematical Modelling*, Vol. 79, 506–520. doi:10.1016/j.apm.2019.10.049

[27] Adachi, S. (2012). 2. Metal and Semimetal Elements, *Handbook on Optical Constants of Metals - In Tables and Figures*, World Scientific

[28] Nazarov, R.; Zhang, T.; et al. (2020). Effective Medium Theory for Multi-Component Materials Based on Iterative Method, *Photonics 2020, Vol. 7, Page 113*, Vol. 7, No. 4, 113. doi:10.3390/PHOTONICS7040113

[29] Razeghi Manijeh, B. A. N. (2021). Complex Refractive Index (CRI) Model, *Terahertz Emitters, Receivers, and Applications XII*, SPIE

[30] Reynolds, W. J. (2021, December). *Predicting Melt Pool Behaviour in LPBF through High Fidelity Modelling*University of Nottingham

[31] Trapp, J.; Rubenchik, A. M.; et al. (2017). In situ absorptivity measurements of metallic powders during laser powder-bed fusion additive manufacturing, *Applied Materials Today*, Vol. 9, 341–349. doi:10.1016/j.apmt.2017.08.006

[32] Gargalis, L.; Ye, J.; et al. (2021). Determining processing behaviour of pure Cu in laser powder bed fusion using direct micro-calorimetry, *Journal of Materials Processing Technology*, Vol. 294, 117130. doi:10.1016/J.JMATPROTEC.2021.117130

[33] Clare, A. T.; Reynolds, W. J.; et al. (2020). Laser calorimetry for assessment of melting behaviour in multi-walled carbon nanotube decorated aluminium by laser powder bed fusion, *CIRP Annals*, Vol. 69, No. 1, 197–200. doi:10.1016/J.CIRP.2020.04.053

[34] Ren, Z.; Fu, G.; et al. (2022). Thermal flow characteristics and the evolution of laser absorption in laser powder bed fusion of Cu-Cr-Zr alloy, *International Journal of Mechanical Sciences*, Vol. 216, 106957. doi:10.1016/J.IJMECSCI.2021.106957

[35] Ye, J.; Khairallah, S. A.; et al. (2019). Energy Coupling Mechanisms and Scaling Behavior Associated with Laser Powder Bed Fusion Additive Manufacturing, *Advanced Engineering Materials*, Vol. 21, No. 7. doi:10.1002/adem.201900185

[36] Ren, Z.; Zhang, D. Z.; et al. (2021). High-fidelity modelling of selective laser melting copper alloy: Laser reflection behavior and thermal-fluid dynamics, *Materials & Design*, Vol. 207, 109857. doi:10.1016/j.matdes.2021.109857

[37] Cook, P. S.; Ritchie, D. J. (2023). Determining the laser absorptivity of Ti-6Al-4V during laser powder bed fusion by calibrated melt pool simulation, *Optics & Laser Technology*, Vol. 162, 109247. doi:10.1016/j.optlastec.2023.109247

[38] Cook, P. S.; Murphy, A. B. (2020). Simulation of melt pool behaviour during additive manufacturing: Underlying physics and progress, *Additive Manufacturing*, Vol. 31, 100909. doi:10.1016/J.ADDMA.2019.100909

[39] Kim, J.; Lee, S.; et al. (2020). Calibration of Laser Penetration Depth and Absorptivity in Finite Element Method Based Modeling of Powder Bed Fusion Melt Pools, *Metals and Materials International*, Vol. 26, No. 6, 891–902. doi:10.1007/S12540-019-00599-3/FIGURES/15

[40] Simonds, B. J.; Tanner, J.; et al. (2021). The causal relationship between melt pool geometry and energy absorption measured in real time during laser-based manufacturing, *Applied Materials Today*, Vol. 23, 101049. doi:10.1016/J.APMT.2021.101049

IN-VITRO AND *IN-SILICO* MODELLING OF ARTERIOVENOUS MALFORMATIONS FOR SURGICAL PLANNING

Jordan Bougardt[1], Randall Paton[1], Malebogo Ngoepe[2], Cecil Pule[3], Karabo Mosiane[3], Jayandiran Pillai[3], Weihua Ho[1]

[1]School of Mechanical, Industrial and Aeronautical Engineering, University of the Witwatersrand, Johannesburg 2000, South Africa,* jordan.bougardt@gmail.com, *weihua.ho@uct.ac.za , * randall.paton@wits.ac.za

[2]Department of Mechanical Engineering, University of Cape Town, Rondebosch, Cape Town 7701, South Africa, * malebogo.ngoepe@uct.ac.za

[3]Department of Surgery, Donald Gordon Medical Centre, University of the Witwatersrand, Johannesburg, Gauteng, South Africa, * pule.cecil@yahoo.com, * 1473512@students.wits.ac.za, * jaypvascular@gmail.com

Abstract: Arteriovenous malformations (AVMs) are vascular abnormalities, presenting as direct connections between arterial and venous flow regimes, with symptoms such as severe ischaemia and cardiac overload. AVMs are most commonly treated through transcatheter coil embolisation, reported to have re-occurrence rates of as high as 25 %. This study aims to investigate these procedures to improve the likelihood of surgical success through the use of objective, quantitative metrics as opposed to the current subjective and qualitative measures of surgical efficacy. To achieve this aim, a computational fluid dynamics (CFD) framework of the pathology was developed to include dominant vascular system characteristics such as patient-specific geometry, hemorheology, pulsatile blood flow and clotting effects. Subsequently, the embolising coil is superimposed into this vascular replication. Experimental validation takes the form of flow tests performed on SLA printed flow phantoms. Current development focuses on coil modelling for embolisation and the development of a mechanical based clotting model.

Keywords: Arteriovenous malformation, embolisation, computational fluid dynamics, phantoms.

1 Introduction

AVMs generally manifest as congenital lesions (damaged tissue at birth) which allow the intermixing of high-pressure arterial and low-pressure venous blood flows. AVMs are structural deformities, classified according to the Yakes classification system [1], into four primary categories of increasing complexity. Examples of two such categories are exhibited in Figure 1, with red representing arteries, blue representing veins and purple representing aneurysmal (swollen) regions of the vessel.

Type I: A direct artery-to-vein attachment. Type IIIa: Multiple inflow arterioles shunting into a single aneurysmal draining vein.

Figure 1: Structural representation of type I AVM (left) and sub-type IIIa AVM (right) [1].

These AVMs exist in the absence of capillary beds, which are the sites of nutrient and gaseous exchange and are additionally the sites at which blood pressure is reduced during perfusion. The resultant symptoms are thus inevitably linked to an insufficient blood flow to various body tissues and include ischaemia (e.g., seizures, strokes), gangrene, haemorrhage, and heart failure [2].

The predominant treatment method for arteriovenous malformations is annihilation by way of transcatheter coil embolisation [3]. This takes the form of the catheter delivery of an embolising coil or wire into the feeder artery of the AVM in order to completely occlude (block) the flow of blood into the AVM. The introduction of the coil during the embolisation procedure can result in clotting in and around the coiled region as discussed in section 2.4. The current evaluation of the success of these procedures is subjective and is determined by visual inspection of contrast movement through the embolising coil and postoperative monitoring of the patient. Coil embolisation treatments are generally associated with a high rate of AVM revascularisation and re-occurrence which is reported to be as high as 25.5 % [4].

Pre-surgical planning of AVM embolisation may better inform clinicians of the necessary steps required to increase the likelihood of surgical success and thus improve patient outcomes. To achieve this a CFD model has been implemented to investigate post-embolisation AVM flow. A precursor to this is the replication of dominant vascular system characteristics which principally include patient-specific blood flow and vasculature data, hemorheology, clotting and vasodilation. For the purposes of this study vasodilation is neglected, however on the basis of this vascular replication, the application of an embolising coil and its flow effects are examined, the outcomes of which are presented herein.

2 Methodology

2.1 Patient-specific geometries and boundary conditions

Patient-specific morphologies have been derived from a variety of medical imaging modalities including Computed Tomography (CT) and Magnetic Resonance Imagery (MRI). The 3D reconstruction is performed through the use of segmentation techniques, which is a means of generating volumes from spatially encoded 2D images. Figure 2 depicts the simplified reconstruction of a type I AVM, comprised of a single feeder artery and draining vein.

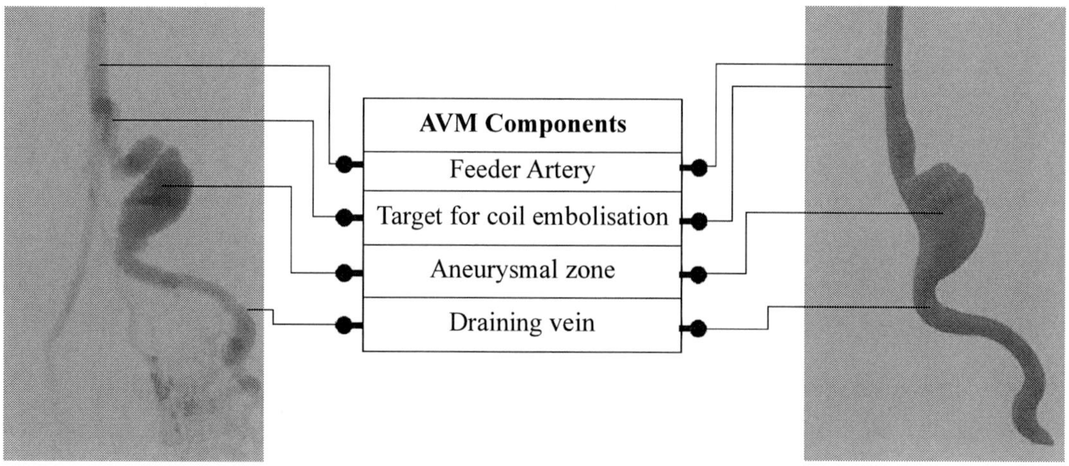

Figure 2: Patient-derived CT scan (left) and 3D reconstruction (right).

Boundary conditions may be informed by both literature derived waveforms and doppler ultrasound velocimetry. For the case introduced in Figure 2, provided by the Donald Gordon Medical Centre (DGMC), a pulsatile pressure waveform was taken from Gallo *et al* [5], for the particular feeder artery for this case i.e., the posterior tibial artery, and used as the time-varying inlet pressure boundary, while the outlet pressure was set as an average venous pressure of 8 mmHg (1066.576 Pa) [6]. The implementation of the waveform was accomplished through the use of piecewise Fourier fitting functions consolidated within a User-Defined Function (UDF). The output of this UDF is graphed in Figure 3 and shows good agreement with the measured pressure waveform with an R^2 value of 0.99997.

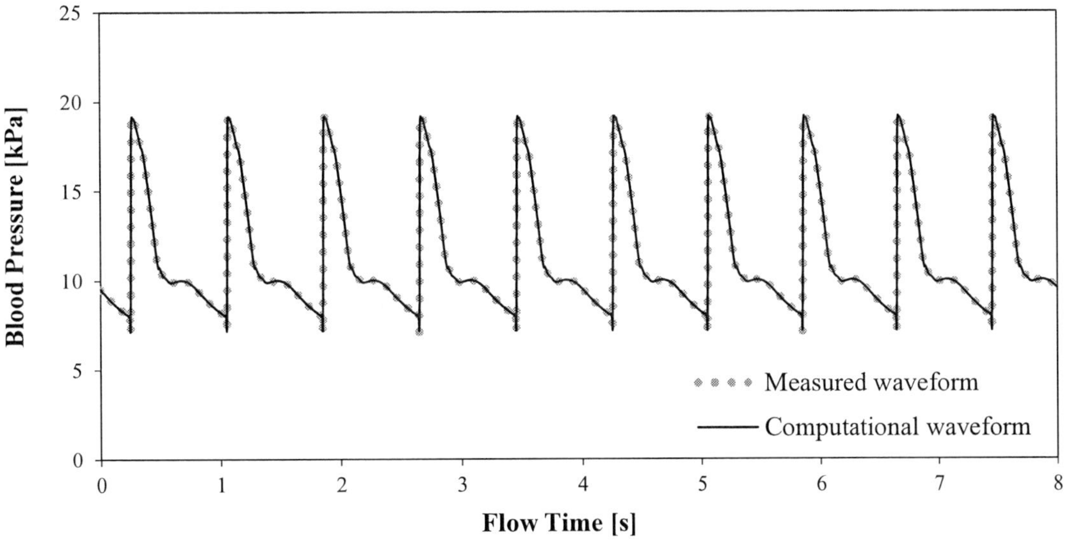

Figure 3: Pulsatile inlet pressure profile

2.2 Non-Newtonian haemodynamics

Whole blood may be described as a heterogenous fluid, which is composed of both cellular and liquid components. Whole blood, while having an approximate constant density of $1060\ \mathrm{kg/m^3}$, does not behave in the same way as a classically Newtonian fluid. Rather, whole blood exhibits non-Newtonian behaviour, more specifically it is shear-thinning and thus its viscosity reduces as shear rate increases. At low shear rates ($< 1\ \mathrm{s^{-1}}$) blood viscosity is as high as $0.0465\ \mathrm{Pa \cdot s}$, which thereafter reduces until a point at which the trend plateaus to a constant viscosity of $0.004\ \mathrm{Pa \cdot s}$ when the blood flow shear rate approaches $200\ \mathrm{s^{-1}}$.

This behaviour may be replicated computationally through the use of the power-law model in the general material description. The power-law model is of the form shown in equation 1, where μ is the viscosity of whole blood [$\mathrm{Pa \cdot s}$], $\dot{\gamma}$ is the shear rate of the flow [$\mathrm{s^{-1}}$], a ($= 0.0105$) is the scaling factor for the whole blood viscosity and b ($= -0.139$) is the negative exponent term which indicates the inverse relationship between shear rate and viscosity.

$$\mu = a\dot{\gamma}^b \qquad \dots\dots\dots\dots (1)$$

2.3 Coil implantation

The coils used as the basis for the implantation model are the Cook® Medical MReye® Embolisation Coils sourced from the Donald Gordon Medical Centre. The AVM case presented above utilises a coil scoped to be inserted into the region designated as the "Target for coil embolisation" in Figure 2, with an approximate diameter of 4 mm. The coil has a diameter of 2 mm and a coiled diameter of 4 mm.

Variable pressure head flow tests have enabled the derivation of the flow resistance (k – values) associated with the embolising coil, through the application of the simplified Darcy-Weisbach equation, given in equation 2.

$$\Delta P = kQ^2 \qquad \dots\dots\dots\dots (2)$$

Figure 4 depicts the setup of the flow rig used to ascertain the flow resistance values associated with the coil. The symbol z_1 refers to the height of the fluid within the reservoir, which may be adjusted to provide a variable pressure head, while z_2 refers to the height at which the outflow rate is determined.

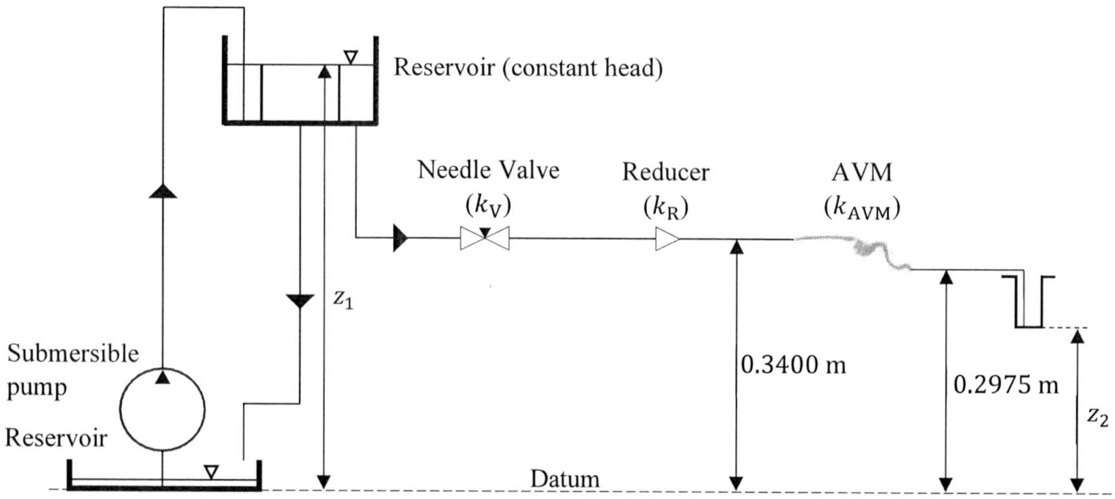

Figure 4: Variable pressure flow rig setup

Through the use of equation 2 and the additivity of the system's k values, experimentation at various inlet pressure heads was conducted to determine the flow resistance associated with the embolisation coil. Through the comparison of the flow rates in the open system (without an implanted coil) and the embolised system, the additive nature of flow resistance values within the system allowed for the isolation of the embolisation coil. The corresponding formulations are presented in equations 3 and 4.

$$\Delta P_{\text{open}} = (k_p + k_v + k_r + k_{AVM})Q_1^2 \quad \dots\dots\dots\dots (3)$$

$$\Delta P_{\text{embolised}} = (k_p + k_v + k_r + k_{AVM} + k_{\text{coil}})Q_2^2 \quad \dots\dots\dots\dots (4)$$

As demonstrated in equations 3 and 4, the flow resistance value associated with the 3D SLA printed AVM model (k_{AVM}) may be computed through subtraction of the total flow resistance for the embolised and open systems. Figure 5 illustrates the k value derivation from the gradient of the ΔP vs Q^2 plots.

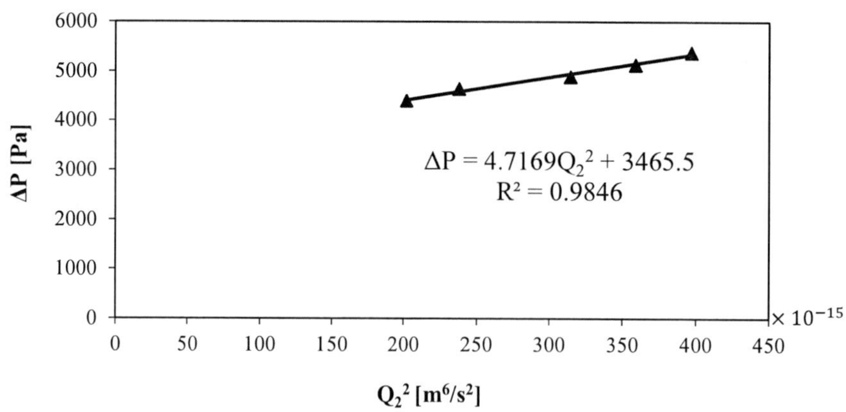

Figure 5: Flow resistance (k) derivation for the embolised system

The inertial resistance coefficient was then derived through application of the continuity equation and the manipulation of the momentum terms in the governing equations, wherein the v^2 term in equation 6, allows for direct comparison with the inertial resistance coefficient term with the neglect of the viscous resistance coefficient. This relationship is summarised in equation 5.

$$\Delta P = 735{,}071.4518v^2 = S_{\mathrm{i}} \qquad \dots\dots\dots\dots (5)$$

$$S_{\mathrm{i}} = \frac{1}{2}C_2\rho v^2\Delta n \qquad \dots\dots\dots\dots (6)$$

Regarding equations 5 and 6, S_{i} is the momentum sink term $[\mathrm{kg}\cdot\mathrm{ms}^{-1}]$, C_2 is the inertial resistance coefficient $[\mathrm{m}^{-1}]$, ρ is the fluid density, which in this case is water $[\mathrm{kg/m^3}]$, v is the physical velocity of the flow through the coil $[\mathrm{m/s}]$ and Δn is the length of the coil in the flow wise direction $[\mathrm{m}]$.

2.4 Clot modelling

Clotting is a complex function of both biochemical reactions and haemodynamics. The principal mechanisms for both high- and low-shear clotting, and the modelling thereof are summarised herein.

2.4.1 The low shear clotting regime

An injured blood vessel triggers a vascular spasm, which causes vasoconstriction to reduce blood flow. Blood platelets adhere to exposed collagen fibres at the site of injury, whereafter a series of biochemical reactants, most importantly, glycoprotein IIb–IIIa receptor (integrin $\alpha_{\mathrm{IIb}}\beta_3$) are exposed [7], to improve adherence and promote aggregation. A cascade of enzymatic reactions (coagulation) follows wherein, fibrin (insoluble protein forming a fibrous mesh) generation from the enzymatic action of thrombin on the soluble plasma protein fibrinogen solidifies into a stable clot structure.

2.4.2 The high shear clotting regime

The high shear clotting regime is dominated by platelet deposition as opposed to coagulation, which is primarily mediated by the multimer glycoprotein von Willebrand Factor (vWF) [8]. von Willebrand factor (vWF) binds to exposed collagen or artificial surfaces and changes from its globular to its elongated state, under high-shear conditions. Platelets are transported to the surface of the vWF which serves as a site of adhesion and undergo activation to release substantial amounts of vWF and activate integrin $\alpha_{\mathrm{IIb}}\beta_3$ on the platelets for firm adhesion, resulting in the formation of a stable clot structure.

2.4.3 Computational Model

A clotting potential model has been developed in a CFD software, namely Fluent (ANSYS, Pennsylvania, USA), through the implementation of biologically relevant conditions and shear rate limits. These conditions include the origination of clot formations from computational cells which make up the wall of the fluid domain, while interior cells may become clotted if an adjacent cell is clotted. Regarding the shear rate limits, Savage *et al* [7] and Casa *et al* [8], exhibit agreement on the shear rate limits for the clotting regimes, i.e., low shear clotting occurs at shear rates $\leq 50\ \mathrm{s}^{-1}$ and high shear clotting at $> 5000\ \mathrm{s}^{-1}$. Computational clot modelling takes the form of a cell-based porosity variance model, informed by experimentation from Diamond *et al* [9] which showed that stable, dense clot formations have an effective porosity of 0.75. The imposed shear rate limits and the associated porosity values are given in equation 7.

$$\varepsilon(\dot{\gamma}) = \begin{cases} 0.75, & 0 < \dot{\gamma} \leq 50 \\ 0.75, & \dot{\gamma} \geq 5000 \end{cases} \quad \ldots\ldots\ldots\ldots (7)$$

3 Results and Discussion

3.1 Coil implantation

The computational verification of resistance coefficients in modelling the embolisation coil was conducted through parametric mesh refinement. A key flow variable for comparison between experimental and computational domains was the outlet mass flow rate [kg/s]. In Figure 6, the error trend between experimentally and computationally derived values is illustrated. The error diminishes exponentially, reaching a constant percentage error of approximately 0.25%. The mesh metrics for the final utilised mesh are detailed in Table 1.

Table 1: Final mesh metrics

Mesh Details	Value
Number of mesh elements	1 181 188
Average aspect ratio	3.4381
Element quality	0.8337

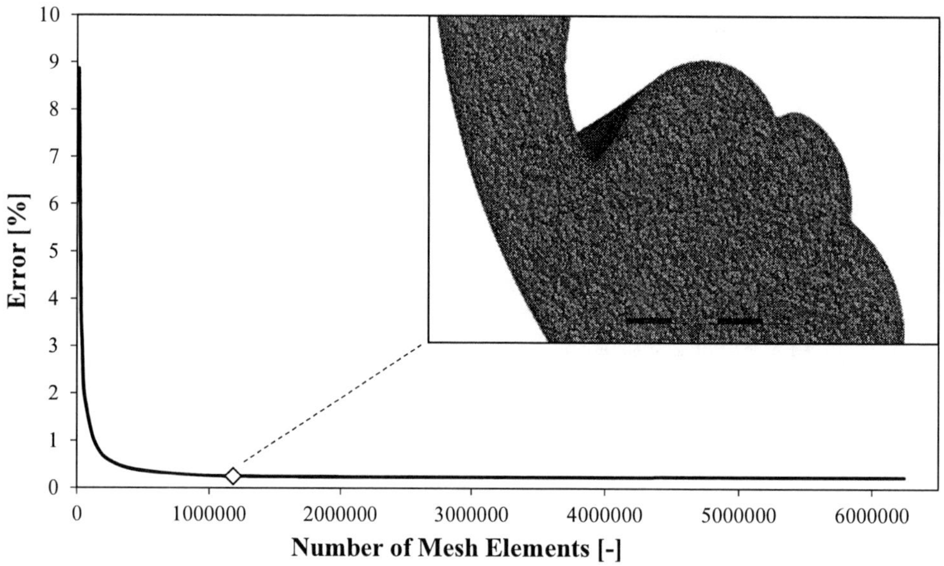

Figure 6: Outcome of parametric validation for embolisation coil simulation and cross-section of the final mesh.

Figure 7 elucidates the impact of embolisation coil implantation on the flow dynamics within the AVM structure. In both contours presented in Figure 7, the coil is clearly delineated. This delineation signifies the region where both quantities experience a marked and abrupt reduction, indicative of a significant decrease in the total energy within the working fluid.

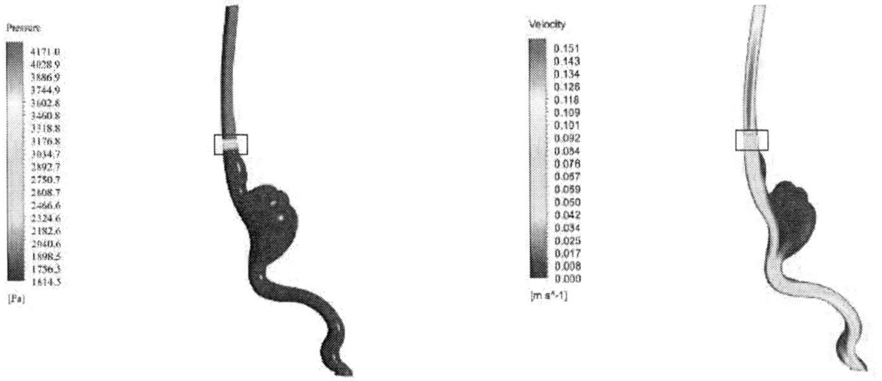

Figure 7: Total pressure contour (left) and velocity magnitude contour (right) for an embolised AVM structure.

The effects of an embolisation coil in a flow field may be accurately reproduced computationally via the derivation and application of resistance coefficients from experimental investigations. This method can replicate the overall flow caused by the implantation of an embolisation coil with an accuracy of as high as 0.25 %, as shown in Figure 6, which is typically considered to be a negligible discrepancy and can be associated with sources of error such as inconsistencies in the experimental working fluid or inadequacies of the applied turbulence model, which in this case was SST $k - \omega$. Furthermore, this method is able to allow for the capture of the changes in flow variables through the coil zone.

3.2 Clot Modelling

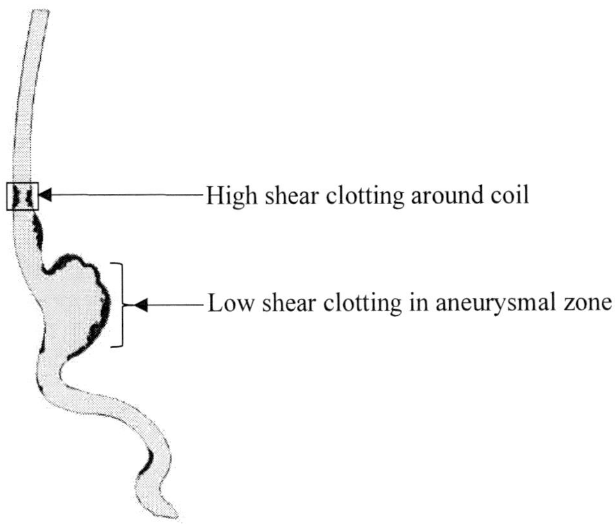

Figure 8: Clot modelling for an embolised AVM

Clotting at low shear rates is anticipated to occur predominantly in the enlarged, aneurysmal zone of the AVM. This propensity is attributed to the sudden expansion of the flow domain in this region, coupled with the presence of recirculation, which in turn results in a decrease in both the magnitude of the fluid velocity as well as the fluid shear rate. Figure 8 provides a visual representation of the clot formation observed within the aneurysmal zone of the embolised AVM which corresponds to this particular clotting regime.

High shear clotting is anticipated to occur in areas of vascular stenosis (narrowing) of the blood vessel or where there is a significant resistance to flow (as in the case of the embolising coil), which in turn results in the constriction of the fluid leading to greater fluid velocities and hence larger shear rates. Furthermore, the occurrence of the high shear clotting in the area constituting the coil zone, also shown in Figure 8, is further evidence that the effect of the coil can be captured without considering the intricacies of modelling its geometry. Clotting of this nature is expected to generate clot formations on or near artificial surfaces as in the case of the coil. This observation aligns with the intentional design of the coil, where the structure is tailored to promote clotting through vessel wall irritation.

The relationship between the embolisation coil implantation and the resultant clotting is significant in the context of any embolisation procedure. However, as the coil structure is not directly modelled computationally, there are fewer surfaces available for the simulation of platelet deposition and as a result clotting may be underestimated. This is supported by Umeda *et al* [10], wherein occlusion prediction, as characterised by Residual Flow Volume (RFV), was biased due to the application of uniform porous media modelling and a disregard for the coil diameter and distribution. As a result, the modelling of the embolisation coil in the computational environment is currently under investigation and the porous zone results as discussed herein are being used as the basis for their development.

4 Conclusion

The flow effects caused by an embolisation coil may be accurately replicated by applying a uniform resistance coefficient to a pre-determined zone of a blood vessel and allows for the modelling of a degree of high shear clotting expected around the coil. However, by neglecting the intricate shape of the coil in the vessel, blood flow shear rate as well as clot formation are likely to have a margin of error.

Ongoing research is directed towards employing models that can translate the physical characteristics of the coil into the computational domain. This approach aims to enhance the investigation of the intricate relationship between the coil structure and the clotting model.

References

1. Yakes, W., & Baumgartner, I. (2014). Interventional treatment of arterio-venous malformations. *Gefasschirurgie*, *19*(4), 325–330.
2. Uller, W., & Alomari, A. I., et al. (2014). Arteriovenous malformations. *Seminars in Pediatric Surgery*, *23*(4), 203–207.
3. Linh, D.-T. N., & Khanh, L., et al. (2022). Recurrence after treatment of arteriovenous malformations of the head and neck. *AIMS Medical Science*, *9*(1), 9–17.
4. Bokhari, M. R., & Bokhari, S. R. A. (2023). *Arteriovenous Malformation of the Brain.*
5. Gallo, C., & Ridolfi, L., et al. (2020). Cardiovascular deconditioning during long-term spaceflight through multiscale modeling. *Npj Microgravity*, *6*(1).
6. Shah, P., & Louis, M. A. (2023). Physiology, Central Venous Pressure. In *StatPearls*. StatPearls Publishing.
7. Savage, B., & Saldívar, E., et al. (1996). Initiation of Platelet Adhesion by Arrest onto Fibrinogen or Translocation on von Willebrand Factor. In *Cell* (Vol. 84).
8. Casa, L. D. C., & Ku, D. N., et al. (2017). *Thrombus Formation at High Shear Rates.*
9. Diamond SL. (1999(. Engineering design of optimal strategies for blood clot dissolution. Annu Rev Biomed Eng. 1999;1:427-62.
10. Umeda, Y., Ishida, F., Tsuji, M., Furukawa, K., Shiba, M., Yasuda, R., Toma, N., Sakaida, H., & Suzuki, H. (2017). Computational fluid dynamics (CFD) using porous media modeling predicts recurrence after coiling of cerebral aneurysms. *PLoS ONE*, *12*(12). https://doi.org/10.1371/journal.pone.0190222

13th South African Conference on Computational and Applied Mechanics (SACAM2024)
22 – 23 January, 2024, Stellenbosch, South Africa

THE ROLE OF SIMULATION IN THE DEVELOPMENT OF A COST-EFFECTIVE BIOMIMETIC PROSTHETIC HAND

Markus de Wet[1*], Jan-Hendrik Kruger[2†]

[*] Presenting author
[†]Corresponding author: jandhendrik.kruger@nwu.ac.za
[1,2]North-West University, School of Mechanical Engineering
Private Bag X6001, Potchefstroom, 2520

Abstract:
This project aimed to create a cost-effective upper limb prosthesis that restores key aspects of functionality. The resulting prosthetic hand features a fully articulated thumb and wrist along with abduction and adduction of the fingers, rare in commercial prosthetics. Powered by an Arduino microcontroller, the prosthetic uses 14 clutch servos for operational movement. Motion simulations using Siemens NX were performed on the phalangeal digits, ensuring that there were no collisions between the fingers and any other part of the prosthetic and to ensure adequate joint range of motion. FEM simulations were performed on the structural arm piece to verify the adequate load capacity. The GCODE produced by the additive slicing software was simulated before being run on the 3D printer to ensure that the correct parameters were used without errors. ISO-standard tests were performed on the prototype to evaluate functionality. During testing, the functional operation of the prototype and the finger loading tests showed acceptable results, validating the contribution of simulation as a design tool. However, it performed poorly on three tests: overall grip strength, wear characteristics, and wrist load bearing capabilities, due to premature tendon fatigue or servo failure in certain situations.

Key words: Prosthetic hand, Motion simulations, Finite element method

1 Introduction

Currently, there is a pressing need for affordable and accessible prosthetic limbs. Traditional prosthetic limbs available on the market are prohibitively expensive, require the speciality of a skilled technician to fit and maintain, and may not be suitable for all amputees [1]. There is a growing trend for inexpensive, accessible and easily manufacturable prosthetics, especially for the hand [2] [3].

The functionality of the human hand is especially difficult to replicate, consisting of 27 degrees of freedom and representing 13% of the bones in the human body per hand. Thus, this results in prosthetics that are usually complex and very expensive, or devices that are very simple approximations of the human hand and thus cannot adequately provide the functionality that was initially lost.

Using simulation software, the main author streamlined the processes that were undertaken to design and manufacture the prosthetic, ensuring timely and efficient manufacturing and assembly of the prototypes. A final version was successfully developed that satisfied the requirements for motion capabilities, strength, and weight.

2 Development process overview

2.1 Design objective

Develop a functional FDM-made prosthetic hand with an accurate biomimetic approximation of the functionality of the real human hand. The device needs to be easily manufacturable, maintainable, customisable, and affordable.

The prosthetic hand that was developed is shown in Fig. 1. It is capable of performing functions such as firmly grasping objects and has the ability to grab and carry bags and other containers. Ranges of movement include flexion, extension, abduction, and adduction of the fingers, full thumb articulation, palm articulation, and wrist rotation, flexion, extension, radial deviation, and ulnar deviation.

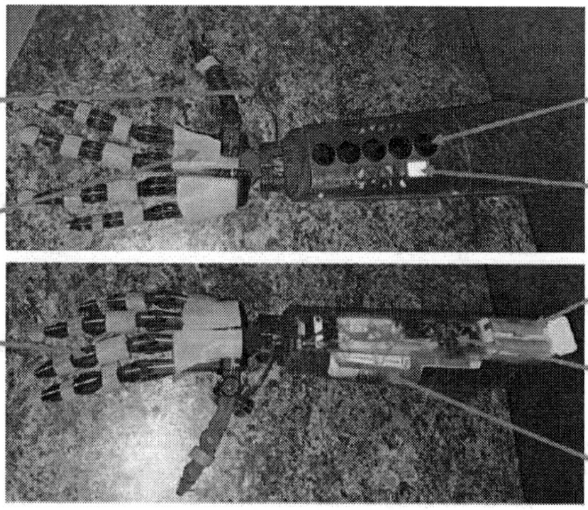

Figure 1: BioHand final prototype.

2.2 Simulation overview

During the conceptual design and final design development stages, as well as during the manufacturing of the physical prototypes, various simulations were performed. During the conceptual design stage, Siemens NX Motion simulations (NX v1980 [4]) were performed on the phalangeal digits of the conceptual design to validate that the digit joints exhibit a suitable range of motion (ROM), as well as to ensure that there are no collisions between parts.

With the final design version, an FEM simulation using Siemens NX was performed on the main arm to ensure adequate load bearing capacity and to reduce the weight of the part, as well as an FEM simulation on the phalangeal digits to ensure that the gripping of a load does not deform the parts.

3 FEM Simulation

3.1 Main Arm Structure FEM

3.1.1 Simulation Setup

The main structural piece of the prosthetic, the lower arm section, was optimised for weight using FEM simulation. The methodology used was to constrain the arm in the socket section, since that is the location where the prosthetic would be anchored to a person. A load of 100 N was then applied to the arm, which is double the amount that a prosthetic upper limb would normally be evaluated during a load bearing test. [1]

The model features of the FEM setup of the prosthetic housing that fits around the arm are shown in

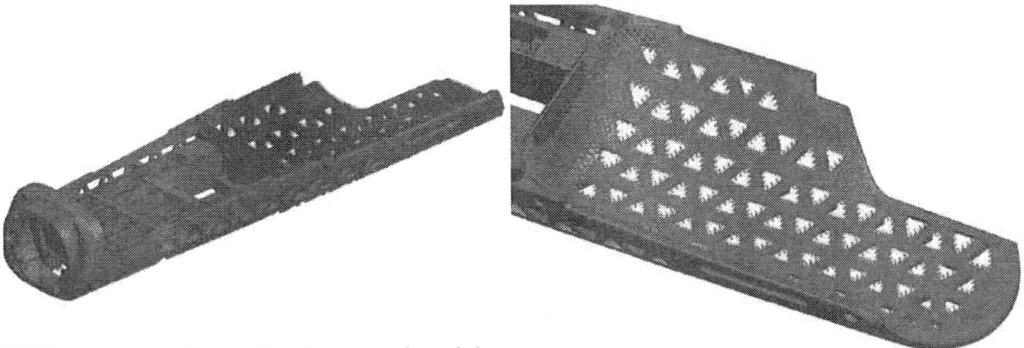

(a) FEM geometry of arm, showing general mesh layout.

(b) Detail of fixed constraint for housing.

Figure 2: Overall geometric features of FEM model for BioHand housing.

Fig. 2. The part shown in Fig. 2a was meshed using the CTETRA(10) element type, using an element size of 2.44 mm with the standard surface meshing method and maximum growth rate of 1.3.

In Fig. 2b the fixed constraint may be seen more clearly, highlighted in red, showing its distribution alongside the inner surface of the socket, approximating the support provided by the amputee residual stump.

In Fig. 3 the applied load may be seen. The load is applied along the surface of the inner main pocket of the wrist, which approximates the force that a load on the hand portion of the prosthetic would apply to the structure of the main arm. (Note the load direction.)

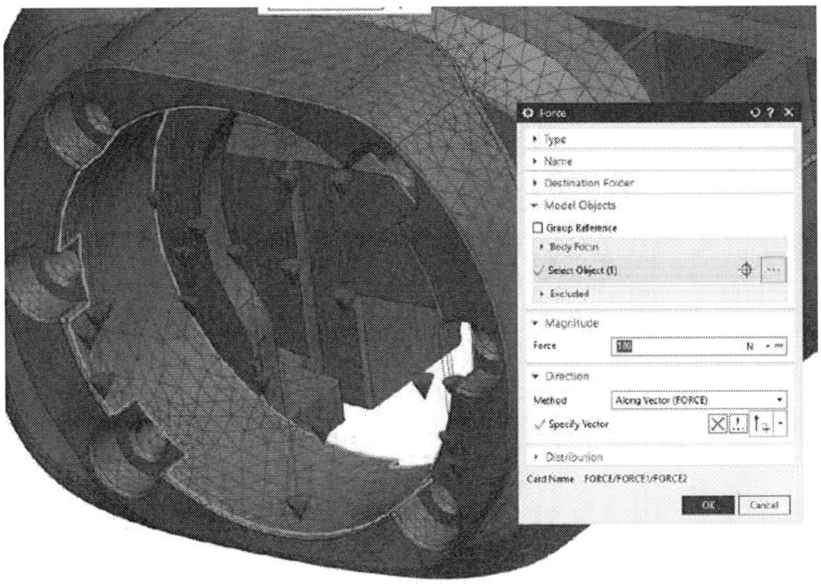

Figure 3: Load applied to the FEM model of the housing.

3.1.2 Results of housing simulation

From the FEM simulation performed, in Fig. 4 it can be seen that the arm developed a maximum stress of 65.81 MPa, which, with the eSUN CF-Nylon tensile strength of 140 MPa, would result in a safety factor of 2.13. This safety factor can be seen to be quite high; thus, further weight reduction may be possible in a future iteration of the design. Using the FEM simulation thus allowed for an optimised design to be created that is strong enough to allow the wearer to use the prosthetic without fear of damaging the device and also allowed the part to use less expensive carbon fibre-reinforced nylon, reducing the overall cost of the prosthetic.

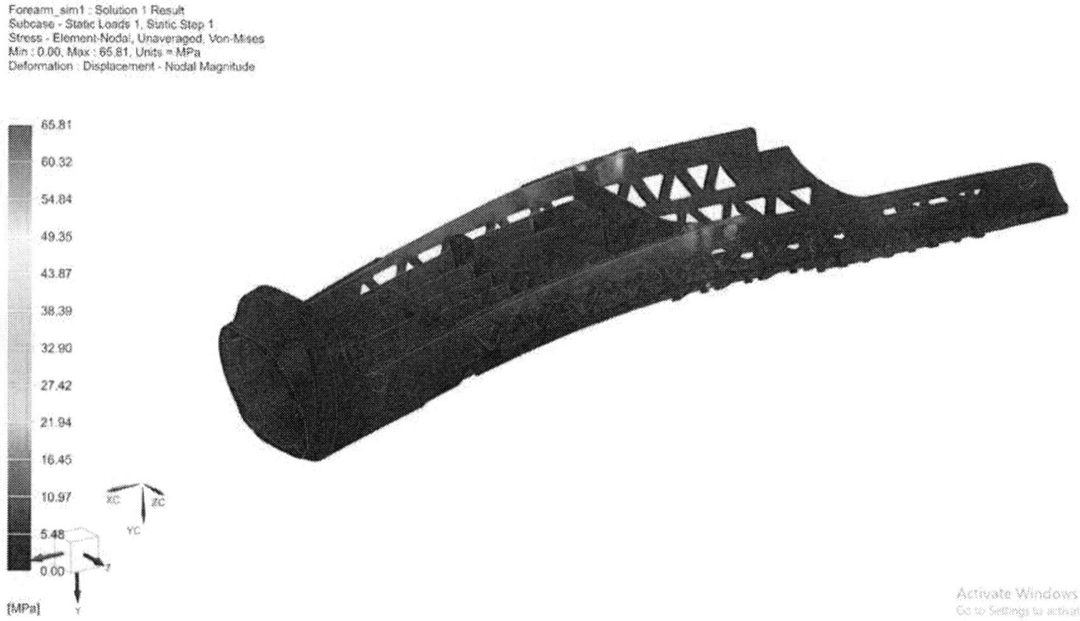

Figure 4: Strength calculation for housing showing stress distribution.

3.2 Phalangeal Digit FEM

3.2.1 Simulation Setup

A FEM simulation was performed on two types of load bearing phalangeal digits, namely the proximal phalanx and the metacarpal. The digit load diagram is shown in Fig. 5.

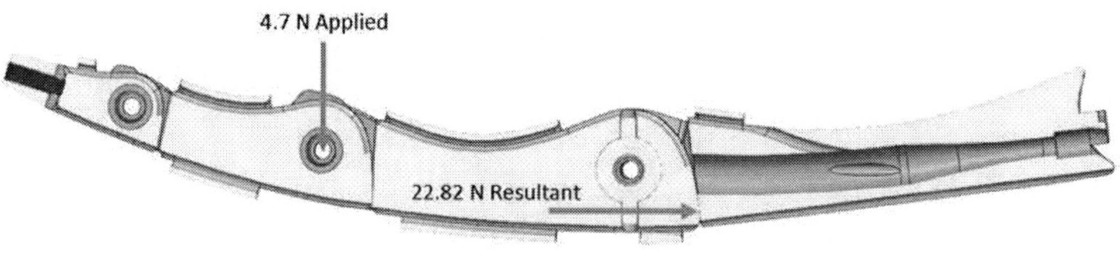

Figure 5: Phalangeal load diagram

A load of 4.7 N [5] was applied to the end of the proximal phalanx, producing a torque of 0.228 N.m transmitted to the metacarpal:

$$4.7\,N \times 0.04855\,m = 0.228\,N.m \tag{1}$$

This resulted in a reaction force of approximately 22.82 N:

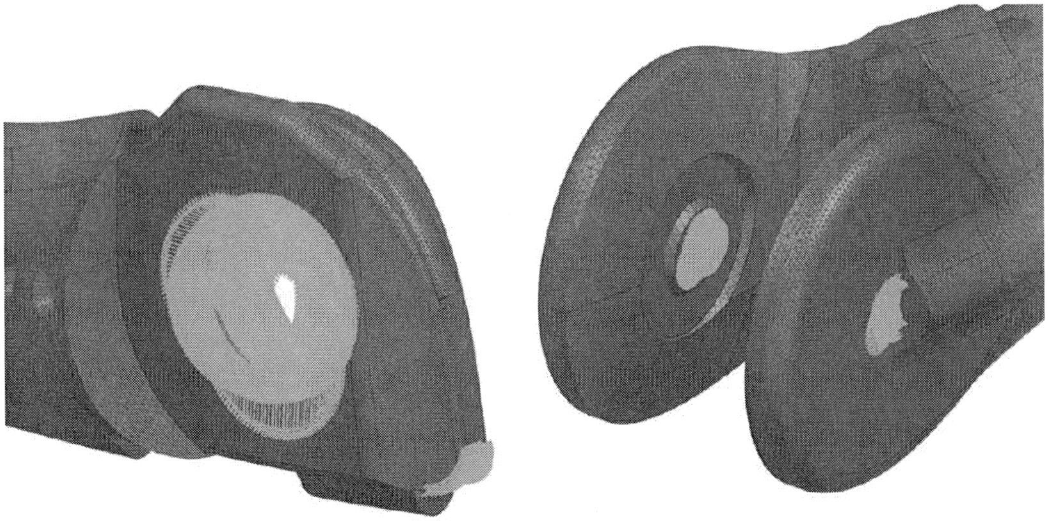

(a) Proximal phalanx (b) Metacarpal

Figure 6: FEM constraints for proximal phalanx and metacarpal geometries

$$\frac{0.228\ N.m}{0.01\ m} = 22.82\ N \tag{2}$$

Both segments used the same mesh detail, CTETRA(10) with an element size of 0.5 mm, with the standard surface meshing method and a maximum growth rate of 1.3.

For the proximal phalanx, a fixed constraint was added at the point where the resulting reaction force acts on the metacarpal, and a fixed translation constraint was applied that allows rotation of the proximal – metacarpal joint, as shown in Fig. 6a. For the metacarpal, a fixed constraint was added to the inside of the joint section, as shown in Fig. 6b.

For the proximal phalanx in Fig. 5, the 4.7 N load was applied directly to the joint through a hole that simulates the force applied to the part by a screw that holds the phalangeal digits together. For the metacarpal, the 22.82 N load was applied as seen in Fig. 7b, simulating the resulting contact force induced on the part by the proximal phalanx torque.

3.2.2 Results

For the proximal phalanx FEM, the maximum combined bending and compression stress developed in the part was 3.762 MPa and thus with the eSUN CF-Nylon tensile strength of 140 MPa, a safety factor of 37.2 was calculated; see Fig. 8a.

The maximum tension stress developed in the metacarpal is shown in Fig. 8b and was 19.48 MPa; therefore, with the tensile strength of eSUN CF-Nylon of 140 MPa, a safety factor of 7.2 was calculated.

Although these safety factors may be considered very high for both parts, it ensures that the phalangeal structures are durable enough to survive daily use with many grip-release cycles.

Thus, by using a FEM simulation, the phalangeal durability was verified, whilst also allowing for the identification of stress concentrations in the geometry that may lead to cracking of the additively manufactured component.

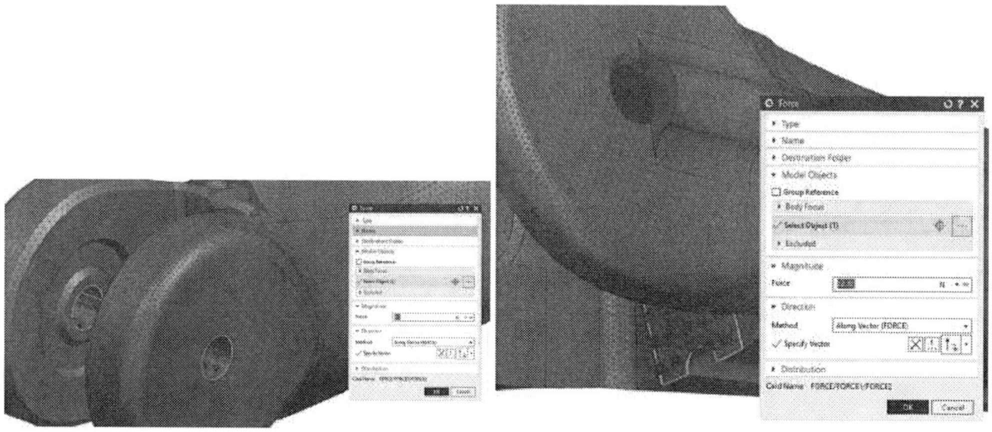

(a) Proximal phalanx (b) Metacarpal

Figure 7: Load application for proximal phalanx and metacarpal.

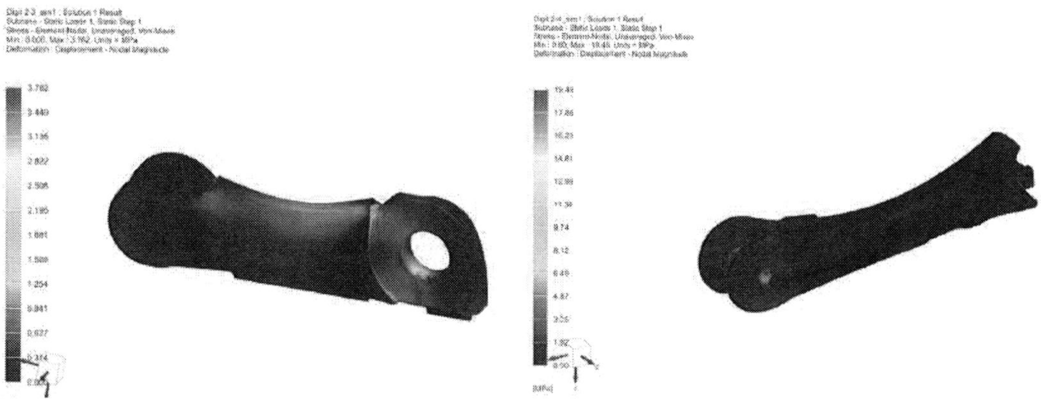

(a) Proximal phalanx (b) Metacarpal

Figure 8: Von Mises stress analysis of proximal phalanx and metacarpal.

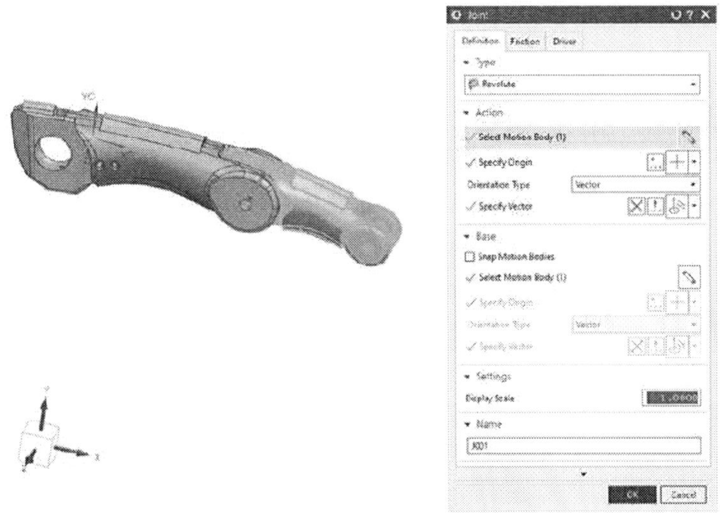

Figure 9: Motion Simulation revolute joint type setup, showing two motion bodies connected by one joint.

4 Motion Simulation

For the Motion Simulation the Siemens NX Motion Simcenter 3D application was used to simulate the movement of each joint. Only one joint will be shown as the simulation methodology is the same for all joints in the system.

For the analysis of the joint, a revolute type of joint was first set up to link the two motion bodies, as shown in Fig. 9. Next, a driver was configured to activate the joint, simulating the tendon that actuated the entire finger. This driver was defined to activate the joint for its entire range of motion in one second, which was the specified actuation velocity for the fingers of the prosthetic device. The actuation velocity is determined by the characteristic behaviour desired from the hand, as well as by the mechanical parameters of the servomotors.

To determine collisions, an interference function was set between the two motion bodies, used now in conjunction with an articulation run solution type. Fig. 10 shows the setup that allows the motion between the bodies to be calculated in a stepwise fashion until a collision is detected.

Figure 10: Articulation Run solution setup for two motion bodies connected by one joint.

Once a collision is detected, the simulation stops as defined in the setup and the final joint angular position may be discerned. When inspecting the interface shown in Fig. 11, the collision in this case was detected at 78°. Thus, for this specific joint, the ROM allowed by the geometry of the components is evidently 0° – 78°. (Keep in mind that additional factors such as the control algorithm and the reach of the servomotor also play a role in the final ROM of the mechanism.)

This motion analysis procedure was performed for all joints in the system. By utilising this method of simulation, any discrepancy in part geometry may be identified that may lead to a reduction in joint ROM or unwanted component collisions, allowing for the geometry to be modified in CAD before the part is manufactured and assembled. This avoids a situation where material and time may be wasted when a mechanism is assembled and found to have an insufficient range of motion during testing, requiring parts to be remanufactured and mechanisms reassembled.

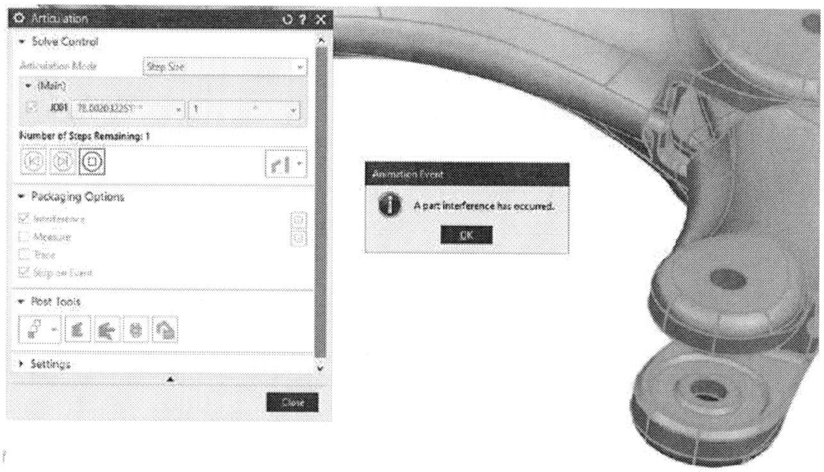

Figure 11: Motion Simulation Collision Detection

5 GCODE simulation for manufacturing

The slicer software that was used during the manufacturing process, Ultimaker Cura v5.6.0 [6], allows the GCODE that is produced of each part to be visualised and analysed using the GCODE simulation features of the software.

An overall visualisation of the predicted printed part can be seen that allows for inspection of correctly placed support structures (see Fig. 12a as an example), infill structures and layer heights, as well as identification of areas of the part that may require manual extra support placement or where support structures that were automatically generated are not necessary.

GCODE toolpaths can also be visualised (similarly to those shown in Fig. 12b), providing data such as the type of toolpath generated, as well as the speeds at which the toolpaths will be used for the 3D printing process. This ensures that the correct settings and parameters were applied for the component and the specific material in question before the actual printing is started.

Slicing software uses GCODE to also predict print parameters such as the time to finish the part, the approximate weight of the part, and the material costs associated with printing the part. All of this assists in the planning and management of the production of all of the parts, which is beneficial for a complex system consisting of many different and complex components that all need to be printed logically to allow for speedy assembly of the system.

6 Experimental validation of prototype

Five experimental tests were found in the existing literature to be the most applicable to this project. That is, to test the slippage characteristics of the prosthetic hand and glove, the test of wrist strength

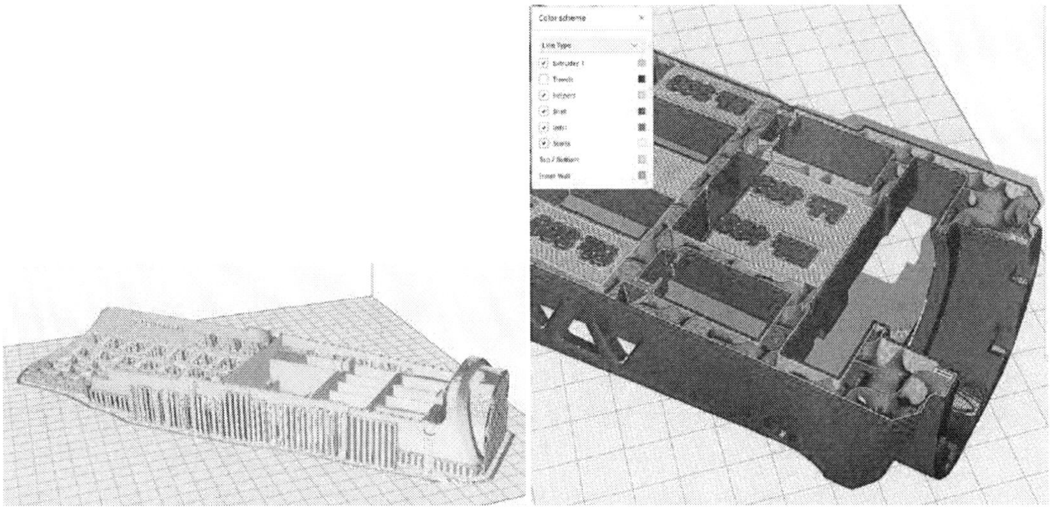

(a) GCODE Visualisation (b) GCODE Tool-paths

Figure 12: Visualisation of 3D printing manufacturing process using GCODE tool-paths.

against bending, the testing of worst-case failure conditions of an overloaded finger segment, the testing of component wear characteristics, and the testing of system battery life. Most of these experiments were derived from ISO standards used for the evaluation of upper limb prosthetics [7].

Important functional tests for prosthetic hands are:

- Grip Slippage Test, ISO 22523
- Wrist Strength Test, ISO 22523 – 2006
- Single Finger Failure Test, Non-ISO standard test
- Single Finger Cyclic Wear Test, ISO 14243 – 1
- System Battery Capacity Cyclic Test, Non-ISO standard test

During testing, the functional operation of the prototype was acceptable during the initial repetitions. Finger loading tests showed acceptable results, with the load carrying capacity considered sufficient, as well as the battery capacity of the prosthetic device being more than adequate.

However, three of the experiments showed that the prosthetic device in its current form does not perform satisfactorily when fatigue of the mechanism is encountered in high-repetitive use scenarios. This was mainly due to failures of two key components: the servo motors used as actuators and the nylon cord used as the tendon wire.

Firstly, servo motors do not have adequate torque to ensure satisfactory gripping performance due to plastic internal gearboxes that quickly wear out and deform, and secondly, nylon cord tendons do not have sufficient tensile strength when even slight amounts of frictional wear begin to set in.

More details of the test procedures and results can be found in the technical report of the project [8].

7 Conclusion

This project sought to create a prosthetic device that was easy and affordable to manufacture, with an easy method of operation, while providing form and function closer to that of the real human hand compared to its commercial competitors, while using simulation software to accelerate, streamline and optimise the development process. In the effort to create such a device, a successful design was created and also manufactured, although with drawbacks that somewhat offset the capabilities that were incorporated into the design.

During the prototype evaluation, the servomotors did not exhibit strong enough torque nor were they durable enough, with the plastic gear trains of the motors damaged. The nylon cord tendon wires lacked tensile strength and wear resistance, resulting in the failure of the wear resistance tests and the hand bending tests.

However, the prototype design had beneficial and successful aspects. The control system is easy to use, with the transducer dials providing an easy and quick way to pose the prosthetic thumb and wrist, and the range sensor alongside the touch button resulting in an intuitive method to grip and release objects. The prosthetic allowed for more degrees of freedom than commercial competitors, whilst providing an electronically actuated thumb and wrist.

The weight was considered acceptable, as was the durability of the fingers and the battery capacity. The device is also seen to be only a fraction of the cost of commercial competitors, costing roughly R6000.

The implementation of simulation software and methods was found to be a significant success, allowing the project to be completed in a manageable time-span, whilst also resulting in the prosthetic being optimised for weight, strength and motion capabilities.

Nomenclature

CF Carbon Fibre

FDM Fused Deposition Modelling

FEM Finite Element Modelling

GCODE Geometric code, programming language for computer numeric control machines

ROM Range of Motion

Acknowledgement

I would like to thank everyone who helped me during the duration of my project, especially my parents, for their emotional and technical support. The project produced many challenges for me to overcome and find solutions, which would not have been possible without the knowledge and financial support of my father. I also thank Dr. Jan-Hendrik Kruger from the North-West University for his guidance during the undertaking of this project.

References

[1] Ossur. *i-Limb Quantum*. 2023. URL: https://www.ossur.com/en-za/prosthetics/arms/i-limb-quantum#specificationContentAnchor.

[2] Open Bionics. *HeroArm*. 2023. URL: https://openbionics.com/en/hero-arm/.

[3] exiii Inc. *HACKberry*. 2023. URL: https://www.exiii-hackberry.com/.

[4] Siemens. *NX Simcenter Motion and multibody dynamics simulation*. 2021. URL: https://plm.sw.siemens.com/en-US/simcenter/simulation-test/motion-simulation/.

[5] Christian Pylatiuk Artem Kargov, Stefan Schulz Jan Martin, and Leonhard Döderlein. "A comparison of the grip force distribution in natural hands and in prosthetic hands". In: *Disability and Rehabilitation* 26.12 (2004), pp. 705–711. DOI: 10.1080/09638280410001704278.

[6] Ultimaker. *Cura*. 2023.

[7] Renato Mio, Midori Sánchez, and Quino Valverde. "[Regular Paper] Mechanical Testing Methods for Body-Powered Upper-Limb Prostheses: A Review". In: *2018 IEEE 18th International Conference on Bioinformatics and Bioengineering (BIBE)*. 2018, pp. 170–176. DOI: 10.1109/BIBE.2018.00040.

[8] M de Wet. *Design and Experimental Validation of a Biomedical Upper Limb Prosthesis*. Tech. rep. Private Bag X6001, Potchefstroom, 2520: North-West University, School of Mechanical Engineering, Oct. 2023.

THERMAL-HYDRAULIC BEHAVIOUR OF A MICRO NUCLEAR REACTOR DURING A DLOFC TRANSIENT

CG du Toit[1,*], WA Boyes[2], JF Slabber[2]
[1]North-West University, Potchefstroom, South Africa,[*] jat.dutoit@nwu.ac.za
[2]University of Pretoria, Pretoria, South Africa

Abstract: Using micro modular nuclear reactors integrated with renewable energy sources can supply energy on a decentralized basis. The High Temperature Advanced Micro Reactor (AMR) will be designed to produce 10MW of thermal power to address this need. Prismatic graphite blocks will be employed for structural and moderator material and helium coolant will enter the core at 320 °C and exit at 750 °C. Flownex SE has been used to set-up a systems network model of the reactor to study the behaviour of thermal-hydraulic phenomena under steady-state conditions and during a Depressurized Loss of Forced Cooling (DLOFC) transient event. This paper considers the fuel and solid temperatures and the associated heat fluxes, and the heat balance of the reactor as functions of time. It is shown how the solids initially release accumulated heat, only to start absorbing heat again and then later again release heat. It can also be seen how the temperatures and associated heat fluxes change from the steady-state conditions to set-up the required temperature profiles in the axial and radial directions to be able to reject the heat for cool down through the reactor cavity cooling system (RCCS).

Keywords: Prismatic block reactor, Gas-cooled reactor, Depressurized loss of forced coolant.

1. Introduction

Safe affordable base load can be achieved by integrating nuclear power with renewable energy sources on a decentralized basis. This can be achieved through the development of micro modular nuclear reactors, to supply amongst others energy to towns, small communities, mines, and processing plants [1],[2].

As part of a power cell concept [3], a High Temperature gas-cooled Advanced Micro Reactor (AMR) is under investigation with the philosophy to design for inherent safety, using technology that has been developed and validated in previous HTR programs [4]. It is intended that the AMR reactors be factory built and rolled out to various sites. The reactor would be returned to a licensed organisation for refurbishment and refuelling when the fuel has reached its end of life. The AMR will employ prismatic graphite blocks for structural and moderator material and will be designed to produce 10 MW of thermal power. The fuel assemblies will consist of silicon carbide tubes housing coated particle fuel, immersed in a lead-bismuth eutectic alloy (LBE). Each fuel assembly is to be contained in a boring within the graphite moderator that allows an annulus for coolant flow. The helium coolant will enter the core at 320 °C and exit at 750 °C.

In the design detailed computational fluid dynamics (CFD) models can be set-up to simulate the fluid flow and heat transfer [5]. However, the computational resources required precludes the simulation of the entire reactor and associated thermal-flow systems. System network codes [6],[7] are suited to simulate thermal-fluid analyses of reactor systems and can provide valuable insights into their performance.

This paper discusses a system network model of the AMR that had been set-up using Flownex SE [8]. The model is based on the principles developed by Sambureni [9], Khoza [10] and Nel [11]. The behaviour of the fuel and solid temperatures and the associated heat fluxes, and the heat balance of the reactor as functions of time during a DLOFC are considered in particular.

2. Flownex model of the AMR

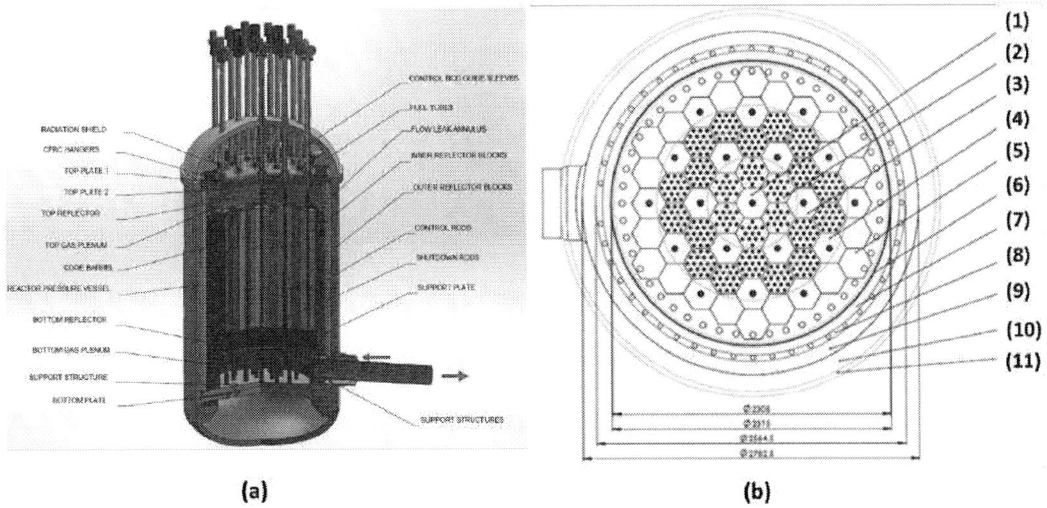

Fig. 1: (a) Cut view of reactor, (b) Cross-section through core [4].

A cut view of the AMR reactor is shown in Fig. 1(a) and a cross-section through the core is shown in Fig. 1(b). An axi-symmetric network model was constructed using the system code Flownex SE [8] to perform the thermal-hydraulic analysis for steady-state and transient conditions. The model does not include the upper and lower domes shown in Fig. 1(a).

In the axial direction the reactor has been discretised into 12 layers: (1) bottom support (BS); (2) lower plenum (LP); (3) bottom reflector (BR); (4) to (10) seven core layers (I – VII), the first and last being a quarter fuel block in height and the rest being a half a block block high comprising three layers of fueled blocks; (11) upper plenum (UP); and (12) top reflector (TR). In the radial direction the reactor has been discretised into 11 rings: (1) Inner reflector (IR) + control rod (CR); (2) First fuel ring (G2 + F2); (3) Second fuel ring (G3 + F3 + CR); (4) Third fuel ring (G4 + F4 + CR); (5) Outer reflector (OR + CR + risers); (6) Gap between OR and core barrel (CB); (7) CB; (8) Gap between CB and reactor pressure vessel (RPV); (9) RPV; (10) Cavity between RPV and Reactor Cavity Cooling System (RCCS); and (11) RCCS face plate. There are a total of 19 control rods and 420 fuel assemblies (FA).

The Flownex network consists of collections of one-dimensional (1D) components. Pipe components are used to model the flow in the risers, plenums, coolant channels and inlet and outlet pipes. The convection heat transfer between the solids and the gas is modelled using convection components. Conduction components are used to model the heat conduction heat transfer through the solids. Finally, radiation heat transfer components are used to model the radiation heat transfer across the annuli between the CR and FA tubes and the graphite, between the OR and the CB, between the CB and the RPV, and between the RPV and RCCS face plate. The helium in the gaps between the OR and CB, and CB and RPV are assumed to be stagnant and conduction through the helium is also considered. For each ring of blocks only a representative control rod, a representative fuel rod, a representative coolant channel, and a representative riser channel, as applicable, are modelled [11].

The network model consists of three interconnected networks. The first network models the graphite blocks, core barrel (CB), reactor pressure vessel (RPV), reactor cavity and reactor cavity cooling system (RCCS) face plate located around the RPV next to the concrete cavity wall. The network also accounts for the radiation and conduction across the gaps between the OR and CB, and the CB and RPV, as well as the radiation and convection across the cavity between the RPV and RCCS. The second network models the flow paths, control rods and the SiC tubes of the fuel. The third network models the conduction through the coated particle fuel and LBE mix inside the SiC tubes. The power that is generated by the fuel is distributed between the control volumes in the network. The details of the layouts of the networks can be found in Boyes et al. [4].

Under steady-state conditions helium enters the cold inlet, blue arrow in Fig. 1(a), at a flowrate of 4.4783 kg/s and a temperature of 320 ˚C. The helium moves up through the risers into the UP where it is distributed between the annuli around the CRs and FAs. The helium flows down through the core where it collects heat, through the holes in BR into the LP and then exits the hot outlet, red arrow, at a pressure of 4 MPa. The temperature at the outer surface of the RCCS face plate was prescribed as 65 ˚C [12]. The power generated in the fuel under steady-state and transient conditions was distributed in the axial direction according to a sine profile. The emissivities of the CB, RPV and RCCS were assumed to be 0.8 and the associated convection heat transfer coefficients to be 4.0 W/m^{2}K.

Flownex employs a finite volume-based implicit pressure correction method [13] to obtain the steady-state solution of the conservation equations and a non-iterative algorithm to obtain the transient solution.

The decay power Q_{DH} was modelled employing the Wigner-Way correlation [14] given by:

$$Q_{DH}(t) = Q_0 \times 0.06 \times [(t - t_1)^{-0.2} - (t - t_1 + t_s)^{-0.2}] \tag{1}$$

Here Q_0 is the nominal power, $t_1 = 5\,s$ time at which the reactor was shutdown and $t_s = 4.82112 \times 10^8\,s$ time in operation before start of the simulation.

The coolant mass flowrate $\dot{m}$ at the inlet after the scram was modelled to vary from 4.4783 kg/s to 0.001 kg/s over the period $t = 5\,s$ to $t = 10\,s$ according to a cosine shaped profile. Over the same period the outlet total pressure p_0 after the scram varied from 4000 kPa to 103 kPa also according to a cosine shaped profile.

The heat balance of the reactor during the DLOFC transient can expressed as:

$$Q_T(t) = [\dot{m}h_0(t)]_{out} - [\dot{m}h_0(t)]_{in} = Q_{DH}(t) + Q_{ST}(t) - Q_{RCCS}(t) \tag{2}$$

Q_T is the heat transferred to the coolant, Q_{ST} is the stored or accumulated heat in the solids released to the coolant and Q_{RCCS} is the heat rejected by the RCCS.

3. Results

Table 1: Map of steady-state temperature distribution and RCCS heat fluxes.

Level	Elevation	GRAPHITE TEMPERATURES					FUEL CENTRE TEMPERATURES				CB in	RPV in	RCCS HEAT		
		IR-rng1	FA1-rng2	FA2-rng3	FA3-rng4	OR-rng5	FA1-rng2	FA2-rng3	FA3-rng4		rng7	rng9	Radiation	Convec	Total
	[m]	[C]	[C]	[C]	[C]	[C]	[C]	[C]	[C]		[C]	[C]	[kW]	[kW]	[kW]
TR	2.4000	329.49	329.46	329.38	329.24	329.00					300.01	218.85	5.31	1.14	6.45
UP	2.2500					341.35					311.29	222.86	1.58	0.33	1.91
I	2.2000	367.66	360.07	360.19	357.92	345.15	405.24	405.11	404.15		313.40	223.81	2.92	0.62	3.53
II	1.8333	420.28	422.77	420.75	414.80	357.20	602.96	601.48	597.39		324.50	231.28	6.27	1.29	7.56
III	1.4667	518.04	535.57	529.46	515.65	372.79	833.41	829.47	820.35		337.27	239.31	6.76	1.35	8.11
IV	1.1000	625.92	661.81	650.25	626.81	387.20	984.56	977.30	962.07		348.89	246.43	7.21	1.40	8.61
V	0.7333	711.16	760.60	743.84	711.17	395.81	1027.66	1016.53	994.34		355.88	250.87	7.50	1.44	8.93
VI	0.3667	755.10	804.46	784.24	743.59	396.64	957.77	942.87	913.06		356.68	251.68	7.55	1.44	8.99
VII	0.0000	767.79	795.85	770.78	708.73	390.27	850.98	833.15	794.04		351.80	249.24	7.98	1.54	9.52
BR	-0.4250	697.96	753.61	757.93	702.40	385.15					346.70	244.42	4.10	0.80	4.91
LP	-0.7125					364.45					333.64	239.63	10.65	2.12	12.77
BS	-1.1	388.17	387.46	384.13	378.38	368.42					334.27	238.78	3.67	0.73	4.41

The distribution of the temperatures through-out the TR, IR, fuel blocks and fuel centres, OR, BR and BS, as well as on the inside surfaces of the CB and RPV are summarised in Table 1. Also included in Table 1 are the radiation and convection heat fluxes from the RPV to the RCCS. The outlet temperature of the coolant is 746.5 °C, the total amount of heat rejected by the RCCS is 85.7 kW, and 94.5% of the coolant flows down the annuli around the fuel assemblies. The maximum temperature for each radial ring is indicated in red in Table 1. The maximum fuel temperature is 1027.7 °C, the maximum graphite moderator temperature is 804.5 °C, and the maximum temperatures on the inside surfaces of the CB and RPV are 356.7 °C and 251.7 °C respectively. All the temperatures are below the maximum allowable temperatures for all the materials.

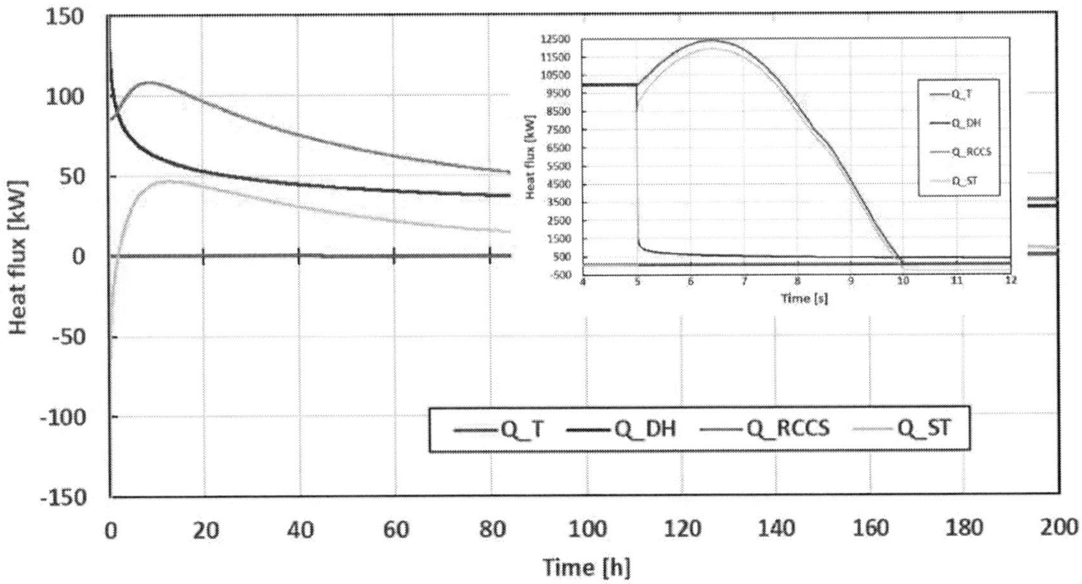

Fig. 2: Heat balance during a DLOFC as function of time.

Fig. 2 shows the variation with time of the heat balance during a DLOFC. The insert shows the variation for $4\,s < t < 12\,s$, while the main graph shows the variation for $t < 200\,h$. The values for Q_T, Q_{DH} and Q_{RCCS} were obtained from the simulation, whilst Q_{ST} was obtained employing eq. (2). The graphs provide in essence a view of the thermal behaviour of the reactor by looking from the outside.

The DLOFC and scram starts at $t = 5\,s$. The power Q_{DH} generated decreases sharply and at 6 s it has already dropped to 600 kW. Due to the thermal inertia of the coolant the heat released by the solids sharply increases from 0 kW to 8500 kW to fill the gap left by Q_{DH}. As a result of the pressure drop in the LP the mass flow rate through the core initially increases leading in an increase in the heat transferred by the solids to the coolant. Q_{ST} reaches a maximum value of 11922 kW and then decreases along with the decrease in mass flow rate. At 9.94 s Q_{ST} becomes negative as Q_{DH} exceeds the heat removed by the coolant and the RCCS. Q_{St} reaches a minimum value of -314.8 kW at 10.7 s. The magnitude of the rate at which the solids absorb heat then gradually decreases as Q_{DH} decreases and Q_{RCCS} increases. At 2 h Q_{DH} becomes less than Q_{RCCS} and Q_{ST} becomes positive and the solids again release heat. The rate at which the solids release heat reaches a maximum of 46.8 kW at 12.85 h after which the rate decreases. Following the scram Q_{RCCS} gradually increases until it reaches a maximum of 108.3 kW at 8.69 h after which it decreases.

The distributions of the temperatures and heat fluxes (indicated by the arrows) in the core region, i.e., excluding the TR, UP, LP and BR, at $t = 0\,h$ are shown in Table 2. Included in Table 2 are the heat fluxes from the OR to CB, the convection heat fluxes from the OR to the coolant in the risers and heat fluxes rejected by the RCCS. Not shown are the convection heat fluxes to the coolant flowing through the CR and FA annuli. The maximum temperature for each radial ring is now indicated by a red circle.

Table 2: Map of temperature (°C) and heat flux (kW) distributions for core region at t = *0 h*.

	IR		G2	F2		G3	F3		G4	F4		OR	OR-CB	RiserCnv	CB in	RPV in	RCCS
In	↓ 39.69		↓ 2521.44			↓ 3936.70			↓ 7713.20			↓ 269.15					→ 8.36
I	367.66	→ 0.49	360.07	405.24	← -0.02	360.19	405.11	→ 0.52	357.92	404.15	→ 3.31	345.15	→ 3.59	← -1.22	313.40	223.81	→ 3.53
	↑ -0.60		↑ -4.03			↑ -7.85			↑ -10.91			↑ -6.11					
II	420.28	← -0.26	422.77	602.96	→ 0.54	420.75	601.48	→ 2.53	414.80	597.39	→ 28.66	357.20	→ 7.47	→ 22.10	324.50	231.28	→ 7.56
	↑ -1.02		↑ -6.58			↑ -12.83			↑ -17.70			↑ -7.78					
III	518.04	← -1.63	535.57	833.41	→ 1.45	529.46	829.47	→ 5.25	515.65	820.35	→ 66.51	372.79	→ 8.22	→ 55.89	337.27	239.31	→ 8.11
	↑ -1.02		↑ -6.54			↑ -12.72			↑ -17.54			↑ -7.07					
IV	625.92	← -3.00	661.81	984.56	→ 2.45	650.25	977.30	→ 8.00	626.81	962.07	→ 104.36	387.20	→ 8.98	→ 90.14	348.89	246.43	→ 8.61
	↑ -0.74		↑ -4.66			↑ -9.00			↑ -12.22			↑ -4.17					
V	711.16	← -3.85	760.60	● 1027.66	→ 3.30	743.84	● 1016.53	→ 10.39	711.17	● 994.34	→ 131.15	395.81	→ 9.43	→ 115.47	355.88	250.87	→ 8.93
	↑ -0.36		↑ -1.97			↑ -3.70			↑ -4.49			↑ -0.40					
VI	755.10	← -3.72	● 804.46	957.77	→ 3.86	● 784.24	942.87	→ 12.59	● 743.59	913.06	→ 142.10	● 396.64	→ 9.44	→ 127.11	● 356.68	● 251.68	→ 8.99
	↑ -0.10		↓ 0.38			↓ 1.22			↓ 4.83			↓ 3.08					
VII	● 767.79	← -1.06	795.85	850.98	→ 2.41	770.78	833.15	→ 9.78	708.73	794.04	→ 66.47	390.27	→ 4.66	→ 135.06	351.80	249.24	→ 9.52
		← -1.23			→ 2.83			→ 11.51			→ 77.70		→ 5.02				
	↓ 0.51		↓ 1.64			↓ 1.03			↓ 0.78			↓ 2.16					
BR	697.96	← -2.54	753.61		← -0.50	757.93		→ 10.38	702.40		→ 84.68	385.15	→ 5.37	→ 68.50	346.70	244.42	→ 4.91
Out	↓ 56.50		↓ 4345.44			↓ 6596.37			↓ 12507.78			↓ 298.87					→ 17.18
	IR		G2	F2		G3	F3		G4	F4		OR	OR-CB	RiserConv	CB in	RPV in	RCCS

The first to be noticed is that due to the coolant heating up as it flows downwards the temperatures of the IR, fuel blocks and OR also increase in the downward axial direction resulting in upwards heat fluxes. The maximum fuel temperatures occur at layer V, the maximum temperatures for G2, G3, G4, OR, CB and RPV at layer VI and for the IR at layer VII. The OR-CB heat fluxes corresponds to the RCCS heat fluxes. There are large convection heat fluxes from the OR to the risers supplied primarily by the fuel ring G4/F4 because no heat is generated in the OR. The temperatures of the IR, except the temperature for layer I, are lower than the corresponding temperatures of fuel ring G2/F2 with the associated heat fluxes from G4 to the IR. This is due to the fact that no heat is generated in the IR and heat is transferred to the coolant flowing down the CR annulus in the IR.

Table 3: Map of temperature and heat flux distributions for core region at t = *20 min*.

	IR		G2	F2		G3	F3		G4	F4		OR	OR-CB	RiserCnv	CB in	RPV in	RCCS
In	↓ 0.02		↓ 0.41			↓ 0.75			↓ 1.60			↑ -8.37					→ 8.36
I	389.06	← -0.02	389.32	391.78	→ 0.61	384.90	387.45	→ 2.36	374.29	377.01	→ 6.80	347.75	→ 3.84	← -0.10	313.76	223.81	→ 3.53
	↑ -0.55		↑ -3.28			↑ -6.17			↑ -7.56			↑ -9.89					
II	438.77	← -0.31	441.82	450.72	→ 2.18	433.50	442.63	→ 8.20	414.09	423.63	→ 23.11	367.39	→ 9.40	← -0.14	326.43	231.31	→ 7.56
	↑ -1.04		↑ -6.10			↑ -11.51			↑ -13.65			↑ -13.33					
III	541.05	← -0.65	548.18	561.15	→ 3.85	531.80	545.29	→ 15.76	490.82	505.40	→ 44.88	394.59	→ 12.38	← -0.13	341.58	239.38	→ 8.11
	↑ -1.04		↑ -6.04			↑ -11.51			↑ -13.60			↑ -12.44					
IV	654.70	← -0.89	665.51	678.00	→ 5.27	640.68	653.88	→ 23.44	573.85	588.75	→ 67.32	420.79	→ 15.48	← -0.08	355.65	246.56	→ 8.61
	↑ -0.74		↑ -4.18			↑ -7.97			↑ -9.23			↑ -7.99					
V	742.41	← -0.89	753.91	763.45	→ 6.23	722.57	732.81	→ 29.21	634.17	646.17	→ 82.99	438.04	→ 17.68	← -0.02	364.45	251.04	→ 8.94
	↑ -0.31		↑ -1.54			↑ -2.78			↑ -2.61			↑ -1.99					
VI	● 781.37	← -0.50	● 788.01	● 793.39	→ 6.91	● 752.42	● 758.34	→ 32.61	● 651.79	● 658.96	→ 87.63	● 442.38	→ 18.39	→ 0.04	● 365.99	● 251.86	→ 9.00
	↓ 0.05		↓ 0.68			↓ 1.77			↓ 3.55			↓ 3.33					
VII	775.28	→ 0.10	772.77	775.18	→ 3.87	733.40	736.34	→ 17.40	627.85	631.74	→ 40.97	435.10	→ 9.15	→ 0.13	360.82	249.41	→ 9.52
		→ 0.11			→ 4.54			→ 20.50			→ 47.89		→ 9.87				
	↓ 0.23		↓ 1.04			↓ 1.41			↓ 2.40			↓ 5.66					
BR	743.20	← -0.14	746.36		→ 3.52	716.34		→ 21.01	609.65		→ 51.79	421.06	→ 9.48	→ 0.46	354.37	244.56	→ 4.91
Out	↓ 0.04		↓ 0.73			↓ 1.31			↓ 2.41			↓ 24.52					→ 17.18
	IR		G2	F2		G3	F3		G4	F4		OR	OR-CB	RiserConv	CB in	RPV in	RCCS

The temperature and heat flux distributions at t = *20 min* corresponding to those shown in Table 2 are summarized in Table 3. A pink coloured cell indicate that the control volume (CV) has heated up since the previous time instance considered and a green coloured cell indicate that the CV has cooled down since the previous time instance considered. When the DLOFC occurs, the removal of heat by the coolant becomes negligible. The reactor then has to rely on conduction, radiation, and convection (in the reactor cavity) only to move the heat to the RCCS to be rejected for the reactor to cool down. To reach a fully developed cool down state, in the axial direction the top part of the core must be heated up and the lower part of the core must be cooled down, centred around layer IV. This is due to the sine profile of the decay power. In the radial direction a negative temperature gradient must set-up to transport the heat to the RCCS. All the maximum temperatures now occur at layer VI. The decay power has decreased to 134.43 kW and the heat rejected by the RCCS has increased slightly to 85.72 kW. Thus heat is being absorbed in the solids as also indicated in Fig. 2.

In Table 3 it can be seen how the IR is heated because it lost its ability to remove heat, the top part of the core is heated up and the lower part of the core is cooled down. The large amount of convection heat transferred under steady-state conditions to the risers previously is now negligible and thus the OR must be heated by G4, aided by G3 and G2, up to reduce the temperature gradient between G4 and OR. The ability of the RCCS to reject more heat must be increased leading to the heating up of the CB and the RPV. It can also be seen that the temperatures of the fuel remain higher than the corresponding temperatures of the surrounding graphite.

Table 4: Map of temperature and heat flux distributions for core region at *t = 1 h*.

	IR		G2	F2		G3	F3		G4	F4		OR	OR-CB	RiserCnv	CB in	RPV in	RCCS
	↓ 0.01		↓ 0.40			↓ 0.75			↓ 1.61			↑ -11.42					→ 8.37
I	422.38	→ 0.15	419.93	421.88	→ 1.48	408.92	410.92	→ 4.09	390.06	392.15	→ 8.26	357.34	→ 4.55	← -0.12	317.34	224.05	→ 3.54
	↑ -0.41		↑ -2.36			↑ -4.35			↑ -5.69			↑ -12.22					
II	460.28	→ 0.16	458.68	465.59	→ 3.82	443.85	450.94	→ 9.82	420.39	427.73	→ 18.79	381.93	→ 11.08	← -0.17	334.07	232.03	→ 7.59
	↑ -0.87		↑ -4.86			↑ -8.75			↑ -10.40			↑ -16.22					
III	546.27	→ 0.20	544.06	554.45	→ 6.07	518.50	529.34	→ 15.50	478.70	490.19	→ 29.21	415.74	→ 14.38	← -0.17	355.12	240.87	→ 8.18
	↑ -0.91		↑ -5.00			↑ -8.93			↑ -10.23			↑ -15.47					
IV	645.02	→ 0.42	639.97	650.34	→ 8.53	600.94	611.97	→ 22.22	539.73	551.78	→ 39.93	449.23	→ 17.92	← -0.11	375.46	248.91	→ 8.73
	↑ -0.64		↑ -3.40			↑ -5.99			↑ -6.56			↑ -10.16					
V	719.46	→ 0.76	709.89	718.07	→ 10.35	659.90	668.74	→ 27.41	580.87	590.73	→ 46.55	471.93	→ 20.48	← -0.03	388.95	254.07	→ 9.09
	↑ -0.22		↑ -1.00			↑ -1.59			↑ -1.30			↑ -2.76					
VI	● 746.14	→ 1.17	● 731.12	● 735.90	→ 11.24	● 676.01	● 681.21	→ 29.78	● 589.24	● 595.00	→ 47.10	● 478.19	→ 21.28	→ 0.06	● 392.33	● 255.16	→ 9.17
	↓ 0.11		↓ 0.87			↓ 1.85			↓ 2.80			↓ 4.00					
VII	732.50	→ 0.79	712.44	714.84	→ 5.72	657.23	659.88	→ 14.97	571.34	574.16	→ 21.97	469.13	→ 10.51	→ 0.21	386.10	252.54	→ 9.70
		→ 0.92			→ 6.71			→ 17.64			→ 25.68		→ 11.33				
	↓ 0.12		↓ 0.74			↓ 1.64			↓ 3.24			↓ 11.43					
BR	715.67	→ 0.98	694.43		→ 6.89	638.55		→ 18.92	548.11		→ 30.37	439.89	→ 9.90	→ 0.55	371.54	246.74	→ 4.97
	↓ 0.04		↓ 0.68			↓ 1.23			↓ 2.27			↓ 27.01					→ 17.22
	IR		G2	F2		G3	F3		G4	F4		OR	OR-CB	RiserConv	CB in	RPV in	RCCS

Table 4 shows the temperature and heat flux distribution two time instances further at *t = 1 h*. The decay power has decreased to 105.67 kW and the heat rejected by the RCCS has increased to 86.56 kW resulting in heat still being absorbed in the solids. The top part of the core, the OR CB and RPV are still being heated. Following the two top layers of the core, the two top layers of the fuel have also been heated. It can also be observed that G4 has cool down more or heated up less than IR. The maximum temperatures occur all still at layer VI.

Table 5: Map of temperature and heat flux distributions for core region at *t = 7.5 h*.

	IR		G2	F2		G3	F3		G4	F4		OR	OR-CB	RiserCnv	CB in	RPV in	RCCS
	↓ 0.02		↓ 0.45			↓ 0.84			↓ 1.61			↑ -13.28					→ 9.55
I	455.44	→ 0.16	452.73	453.93	→ 1.32	442.56	443.79	→ 3.58	425.41	426.70	→ 7.47	394.52	→ 5.14	← -0.12	350.84	243.12	→ 4.18
	↑ -0.16		↑ -1.01			↑ -2.09			↑ -3.49			↑ -11.54					
II	471.06	→ 0.14	469.69	474.01	→ 2.51	459.83	464.22	→ 6.20	444.67	449.16	→ 12.21	418.80	→ 10.96	← -0.11	373.25	258.64	→ 9.49
	↑ -0.27		↑ -1.63			↑ -3.23			↑ -4.68			↑ -10.38					
III	497.73	← -0.01	497.83	504.89	→ 2.64	487.13	494.31	→ 6.37	471.12	478.49	→ 13.74	441.22	→ 11.83	← -0.08	393.08	273.57	→ 10.67
	↑ -0.20		↑ -1.16			↑ -2.28			↑ -3.26			↑ -7.35					
IV	517.66	← -0.06	518.28	526.16	→ 2.74	506.95	514.97	→ 6.59	490.02	498.26	→ 14.69	457.45	→ 12.34	← -0.02	407.98	285.65	→ 11.68
	↑ -0.03		↑ -0.13			↑ -0.32			↑ -0.60			↑ -2.46					
V	● 520.34	← -0.02	● 520.52	● 527.35	→ 2.61	● 509.72	● 516.66	→ 6.29	● 493.51	● 500.64	→ 13.72	● 462.95	→ 12.23	→ 0.04	● 414.22	● 292.13	→ 12.25
	↓ 0.15		↓ 0.90			↓ 1.66			↓ 2.12			↓ 2.85					
VI	505.71	→ 0.11	504.59	508.70	→ 2.28	495.27	499.44	→ 5.56	481.15	485.41	→ 11.14	456.58	→ 11.53	→ 0.09	410.45	291.39	→ 12.20
	↓ 0.22		↓ 1.32			↓ 2.57			↓ 3.62			↓ 7.03					
VII	484.36	→ 0.14	481.53	483.65	→ 1.04	473.27	475.41	→ 2.59	460.41	462.59	→ 4.47	441.08	→ 5.47	→ 0.18	397.99	283.65	→ 12.46
		→ 0.16			→ 1.22			→ 3.05			→ 5.22		→ 5.89				
	↓ 0.11		↓ 0.67			↓ 1.53			↓ 2.82			↓ 10.21					
BR	472.00	→ 0.22	468.24		→ 1.43	458.69		→ 3.95	442.31		→ 8.01	415.71	→ 5.66	→ 0.44	376.51	267.94	→ 5.94
	↓ 0.04		↓ 0.56			↓ 1.05			↓ 1.90			↓ 14.43					→ 19.54
	IR		G2	F2		G3	F3		G4	F4		OR	OR-CB	RiserConv	CB in	RPV in	RCCS

The temperature and heat flux distributions thirteen time instances further at *t = 7.5 h* are depicted in Table 5. The decay power has now decreased to 66.96 kW and the heat rejected by the RCCS has increased to 107.96 kW. The solids are, therefore, now releasing heat. The maximum temperatures now all occur at layer V. It can be seen in Table 5 that the top parts of IR and G2 have since the time instance considered in Table 4 reached their maximum temperatures and have started to cool down. The top layers of G3 and G4, the top two layers of the OR and CG, and the top five layer of the RPV are still heating up. It can also be observed that layers II to IV in G2 are transferring heat to the corresponding layers in IR. This is due to the fact that the G2 layers have cooled down at a slower rate as a result of the decay power being generated in F2 and the reduction in the heat transfer between G2 and G3.

Table 6: Map of temperature and heat flux distributions for core region at t = *17 h*.

	IR		G2	F2		G3	F3		G4	F4		OR	OR-CB	RiserCnv	CB in	RPV in	RCCS
	0.03		0.47			0.88			1.65			-8.49					10.63
I	437.43	0.12	435.45	436.48	1.00	427.83	428.88	2.72	415.01	416.09	5.61	391.98	4.51	-0.07	353.69	252.45	4.53
	-0.10		-0.65			-1.34			-2.22			-7.08					
II	447.10	0.09	446.17	449.88	1.95	438.67	442.42	4.86	427.06	430.88	9.75	406.75	9.41	-0.06	367.36	262.25	9.81
	-0.15		-0.92			-1.81			-2.60			-5.34					
III	461.34	-0.03	461.61	467.79	2.06	453.56	459.83	5.00	441.40	447.79	11.05	418.05	9.82	-0.02	377.36	269.97	10.42
	-0.07		-0.41			-0.81			-1.15			-2.30					
IV	467.81	-0.07	468.55	475.60	2.11	460.27	467.42	5.10	447.78	455.08	11.67	422.97	9.93	0.02	382.00	274.05	10.76
	0.07		0.43			0.79			1.01			1.61					
V	460.85	-0.05	461.31	467.51	1.95	453.70	459.98	4.74	442.16	448.57	10.69	419.53	9.61	0.06	379.79	273.55	10.72
	0.19		1.14			2.16			2.88			5.09					
VI	442.75	0.05	442.27	446.03	1.64	435.99	439.78	4.05	426.32	430.17	8.44	408.74	8.94	0.08	371.41	268.75	10.34
	0.22		1.31			2.53			3.54			7.05					
VII	422.70	0.09	420.98	422.97	0.71	415.65	417.65	1.81	407.20	409.22	3.23	394.01	4.24	0.12	359.22	260.98	10.51
		0.10			0.84			2.13			3.77		4.57				
	0.10		0.59			1.31			2.29			7.79					
BR	412.31	0.14	409.97		0.96	403.93		2.71	393.33		5.62	375.59	4.40	0.25	343.95	250.27	5.17
	0.03		0.52			0.97			1.76			10.27					17.65
	IR		G2	F2		G3	F3		G4	F4		OR	OR-CB	RiserConv	CB in	RPV in	RCCS

At approximately 17 h all the CVs in the reactor model are in a state of cooling down. The temperature and heat flux distributions at t = *17 h* are shown in Table 6. The decay power generated at this instant is 55.19 kW and the heat rejected by the RCCS is 100.55 kW. All the maximum temperatures now occur at layer IV. It can be deduced from Fig. 2 that although the reactor has reached a full cool down state, it still has to reject excess heat accumulated in the solids as evidenced by the Q_{ST} curve in Fig. 2 which also represents the gap between the Q_{RCCS} and Q_{DH} curves. It can also be deduced from Fig. 2 that it can be stated that at 200 h the reactor has reached an equilibrium cool down state. In the equilibrium state as the rate at which decay power is generated decreases the fuel temperatures also decrease. The solid temperatures have to decrease accordingly, and the solids release an associated amount of heat.

The results shown in Tables 2 to 6 provide in essence a view of the thermal behaviour of the reactor by looking from the inside. These results should be considered along with the results shown Fig. 2 to develop a full understanding of the thermal-hydraulic behaviour of the AMR reactor under DLOFC conditions.

4. Conclusions

In this paper a brief overview was given of the AMR prismatic block micro reactor that is currently being designed. A description was given of a Flownex SE network model that was constructed to analyse the thermal-hydraulic behaviour of the reactor under steady-state conditions and during a DLOCF transient.

In the steady-state analysis it was found that maximum fuel temperature is 1027.7 °C and the outlet temperature of the coolant is 746.5 °C. All the material temperatures are lower than the maximum allowable temperatures. The RCCS rejects heat at a rate of 85.7 kW.

A study of the heat balance of the reactor during the DLOFC, as depicted in Fig. 2, reveals that the heat rejected by the RCCS first increases reaching a maximum rate of 108.3 kW at 8.69 h and then decreases gradually. The solids initially release accumulated heat, then absorb heat and finally release the accumulated heat to reach a maximum rate of 46.8 kW at 12.85 h, before decreasing gradually.

Finally, an analysis was presented of the temperatures and heat fluxes distributions throughout the core at selected time instances during the DLOFC as depicted in Tables 2 to 6. It was shown how the top layers of the core, the IR, OR, CB and RPV heat up, whilst the lower layers of the core cool down. At 17 h the reactor has reached a fully developed cool down state although it still has to reject excess heat accumulated in the solids. At approximately 200 h the reactor has achieved an equilibrium cool down state when all the excess heat accumulated in the solids has been released.

Nomenclature

	Symbols	BR	Bottom reflector
h_0	Total enthalpy	BS	Bottom support
$\dot{m}$	Mass flow rate	CB	Core barrel
Q	Power / heat	CFD	Computational fluid dynamics
t	Time	CR	Control rod
	Subscripts	CV	Control volume
0	Nominal value	DLOFC	Depressurized loss of forced coolant
1	Start of DLOFC	FA	Fuel assembly
DH	Decay heat	IR	Inner reflector
in	Inlet	LBE	Lead-bismuth eutectic
out	Outlet	LP	Lower plenum
s	Time in operation	OR	Outer reflector
ST	Accumulated heat in solids	RCCS	Reactor cavity cooling system
T	Heat transferred to coolant	RPV	Reactor pressure vessel
	Acronyms	TR	Top reflector
AMR	Advanced micro reactor	UP	Upper plenum

References

[1] Westinghouse (2023), eVinci Microreactor, https://www.westinghousenuclear.com/energy-systems/evinci-microreactor (Accessed 6 November 2023).

[2] C. Nunez (2021), Micro-nuclear reactors: up to 20 MW, portable, safer, https://energypost.eu/micro-nuclear-reactors-up-to-20mw-portable-safer (Accessed 6 November 2023).

[3] J. Slabber, J. du Toit et al. (2023), The South African Power Cell Concept, Proceedings of the 11th CIGRE Southern Africa Regional Conference held in Pretoria, South Africa, 24-27 October 2023, Paper 82.

[4] W.A. Boyes, J. Slabber et al. (2023), Evaluation of the Basic Neutronic and Thermal-Hydraulics for the Safety Evaluation of the Advanced Micro Reactor (AMR), Nuclear Science, Vol. 8, pp. 8-29.

[5] N.I. Tak, S.N. Lee et al. (2014), Development of a core thermo-fluid analysis code for prismatic gas cooled reactors, Nuclear Engineering and Technology, Vol. 46, pp. 641-654.

[6] P.G. Rousseau, G.P. Greyvenstein (2002), Modelling of the HTTR in Flownet, Proceedings of the 1st International Topical Meeting on High Temperature Reactor Technology held in Petten, Netherlands, 22-24 April 2002.

[7] M.H. Kim, N.I tak et al. (2010), Thermal-fluid assessment of the design options for reactor vessel cooling in a prismatic core VHTR, Annals of Nuclear Energy, Vol. 37, pp. 1774-1782.

[8] M-Tech Industrial (2023), Flownex Simulation Environment, http://www.flownex.com, Version 8.15.0.5221.

[9] P. Sambureni (2015), Thermal fluid network model for a prismatic block in a gas-cooled reactor using Flownex, M.Eng. dissertation, North-West University.

[10] S.N. Khoza (2019), An integrated thermal hydraulic system CFD model of a prismatic block core using Flownex, Ph.D. thesis, North-West University.

[11] G.J. Nel (2022), Simulation of the heat transfer and coolant flow for a PMR200 reactor using a network approach in Flownex, Ph.D. thesis, North-West University.

[12] L.J. Lommers, B.E. Mays et al. (2014), Passive heat removal impact on AREVA HTR design, Nuclear Engineering and Design, Vol. 271, pp. 569-577.

[13] G.P. Greyvenstein (2002), An implicit method for the analysis of transient flows in pipe networks, International Journal of Numerical Methods in Engineering, Vol. 53, pp. 1127-1143.

[14] W.J. Garland, F. Saunders (1998), Decay heat estimates for MNR, Technical Report 1998-05, McMaster University, Canada.

DEFINING CITRUS PALLET STABILITY AND FAILURE MODES: DEVELOPING TOOLS FOR CARTON IMPROVEMENT

N.J. Erasmus[1*], C.J. Coetzee[1], T.M. Berry[2]
[1]Department of Mechanical and Mechatronic Engineering,
Stellenbosch University, *nico.erasmus77@gmail.com
[2]Citrus Research International, Department of Horticultural Sciences,
Stellenbosch University

Abstract:

This work aimed at developing experimental tools to measure the load distribution acting on individual cartons inside a stacked citrus pallet, as well as the whole pallet stability. The tools include the A15C Load Measurement Device (A15C-LMD) and the Pallet Lateral Acceleration Tester (PLAT). The A15C-LMD allowed the measuring of dynamic loads applied to cartons within the stacked pallets. The PLAT allowed repeatable dynamic test conditions due to being developed according to the EUMOS 40509 standard. These tools were used in unison to quantify the carton load conditions as well as the stacked citrus pallet stability for two stack configurations.

The dynamic tests indicated that the peak carton load was 3.6 kN and peak carton corner load was 1.7 kN for the standard configuration at a peak displacement of 171 mm. For the brick configuration the peak carton load was 3.8 kN and the peak carton corner load was 1.5 kN at a peak displacement of 54 mm. This correlated to a carton fatigue damage of 40 % for the standard configuration and 19 % for the brick configuration. Therefore, it was noted that the stack configuration has a direct impact on the carton load distribution as well as the stability of a stacked citrus pallet.

Keywords: Citrus pallet stability, Carton load conditions, Box Compression Test, Carton fatigue

1. Introduction

The export of horticultural produce, especially citrus, is an ever-evolving and dynamic industry with constant regulation changes in terms of carton design and post-harvest management. This dynamic, however, offers opportunities for carton design innovations.

Paper and paperboard packaging is the most commonly used material within the food packaging market, especially for fresh horticultural produce [1]. Reports indicate that corrugated paperboard is the most common export handling packaging for apples and pears [2]. It is also noted that over 90 % of the fruit industry's packaging in the USA, is in the form of corrugated cartons [3]. Thus, with the high amount of paper and paperboard packaging used by the horticultural industry, it is important to focus on the design of the cartons in terms of three aspects: (1) cooling ventilation allowed by the carton, (2) carton structural strength, and (3) stackability of the carton including stack stability. In terms of ventilation, [4] investigated the impact that vent hole placement and sizing have on the cooling efficiency and structural strength of a carton. This research has a direct impact on the stackability of the cartons, as [5] identified that 30 % of produce losses in the distribution chain, were due to collapsed lower levels in stacked pallets.

Previous research with a focus on corrugated paperboard packaging has been conducted by modelling the material as well as quantifying its structural properties [6-9]. This research aided in understanding the loading conditions and failure mechanisms of fruit packaging (cartons or boxes) for various ambient and humidity conditions and resulted in design improvements. Virtual prototyping of fruit packaging stacked on a pallet has been conducted to test and verify various packaging designs for adequate ventilation and heat transfer in the stacked system [10-12]. However, these studies were limited to static loading of a single carton in the case of structural analyses, and for a stacked pallet, only ventilation and heat transfer were analysed and not the structural integrity of the packaging. There is no published literature available on the dynamic loading of cartons when stacked on a pallet, nor on the stability of the whole stack under dynamic loads.

Therefore, this study aimed to identify and quantify the dynamic loads experienced by a single carton within a stacked pallet, as well as the stability of the whole stack under dynamic load conditions. Thus, the study required the development of a load measurement device as well as a dynamic test facility.

2. Materials and Methods

2.1 Load Measurement Device

The load quantification of a carton within a stacked citrus pallet was required for two reasons. The first was to determine the realistic load conditions that are applied to cartons placed within a stacked citrus pallet in terms of the total applied load as well as the load distribution. The second was to correlate the carton load conditions to the stability of the stacked pallet under lateral acceleration. Therefore, the A15C Load Measurement Device (A15C-LMD) was developed according to the objectives as described and is shown in Figure 1.

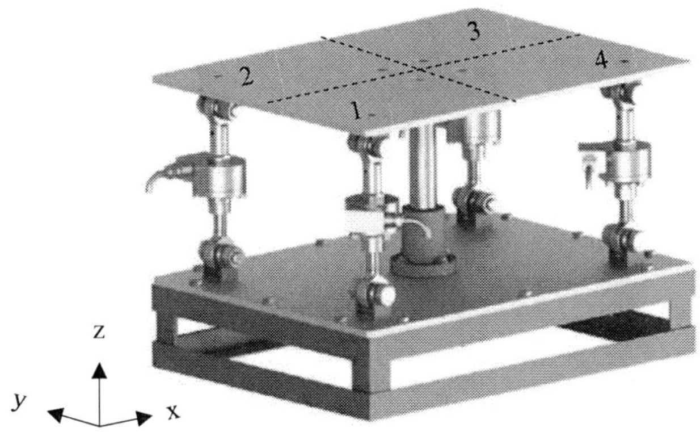

Figure 1: A15C-LMD

The A15C-LMD was developed with two operation requirements. The first was the ability to measure total carton loads within the stacked pallet. Therefore, the device was designed to the same geometric dimensions as a standard A15C carton filled with fruit. This enabled the device to substitute a carton within the stacked pallet without compromising the stack. Thus, accurate carton load measurements could be taken at various carton locations within the stacked citrus pallet. The second operation requirement was to measure the individual corner loads in order to accurately determine the load distribution on cartons. Thus, the device was designed with four load cells located at each corner, as shown in Figure 1. In order to achieve the correct load distribution a spherical bearing with a floating shaft was used to enable rotation of the load plate about the x- and y-axes.

2.2 Pallet Lateral Acceleration Tester

Repeatable dynamic test conditions were required to analyse and quantify the stability of a stacked citrus pallet under standard heavy vehicle acceleration and deceleration as described in the EUMOS 40509. Therefore, the Pallet Lateral Acceleration Tester (PLAT) was developed as shown in Figure 2. This facility enabled the repeatable dynamic movement of a stacked citrus pallet under various acceleration profiles and thus simulating heavy vehicle movement. This ensured that representative data could be captured of the pallets dynamic conditions.

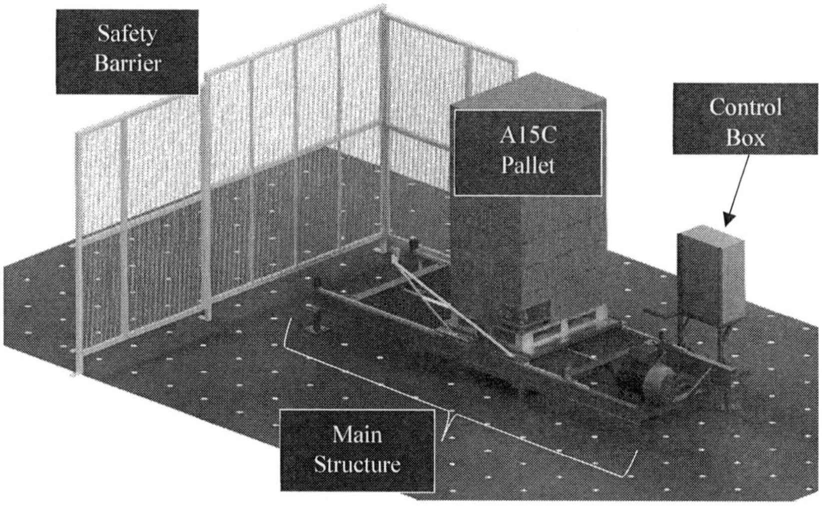

Figure 2: Pallet Lateral Acceleration Tester (PLAT)

2.3 Test Setup and Procedure

2.3.1 Stability and Load Quantification

The stability analysis of the stacked citrus pallet was completed with two stack configurations as shown in Figure 3. The first was the standard stack configuration as outlined by Citrus Research International [13], Figure 3-a. This configuration consisted of the first three layers being placed in phase in a column stack method and thereafter each alternating layer rotated by 180° in a brick stack manner. The second was the brick stack configuration, Figure 3-b. This configuration consisted of each alternating layer rotated by 180° in a brick stack manner. This variation in pallet stack configuration provided insight into the effect that a stack configuration has on the stability of a pallet as well as the load distribution.

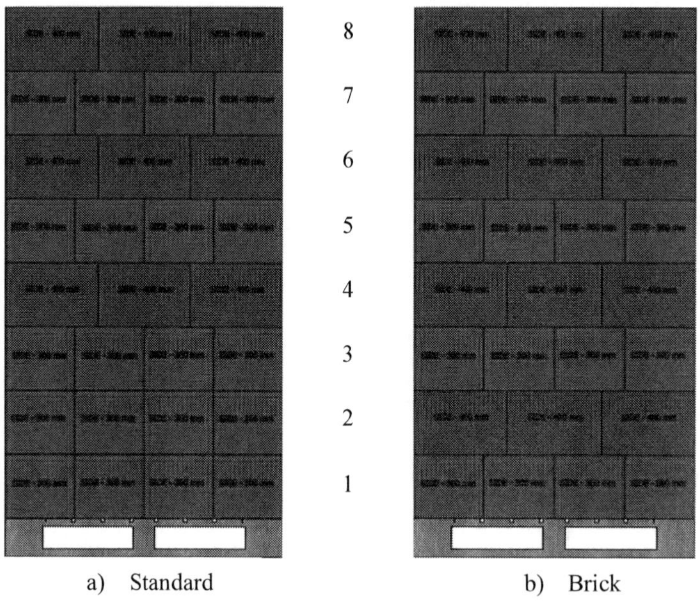

Figure 3: Pallet stack configurations

The carton load was conducted by placing the A15C-LMD at three positions per layer as indicated in Figure 4. The three positions were selected due to the carton support modes present at these positions. The positions included the middle (four wall supported), the side (three wall supported) and the corner (two wall supported). Thus, this selection of measurement positions could provide insightful data on carton loading as well as load distribution within the stack.

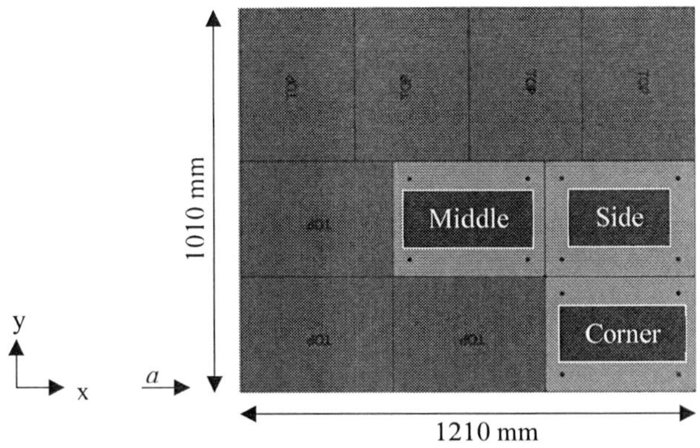

Figure 4: The A15C-LMD placement on a level to measure the carton positional and carton corner loads

The dynamic test setup is shown in Figure 5. The placement of the A15C-LMD at the corner position is indicated with the pallet placed on the PLAT. The GoPro used for video analysis of the stack movement during acceleration is also indicated. Tests were conducted for three acceleration profiles according to the EUMOS 40509 standard [14]. The three profiles were 0.3g, 0.4g and 0.5g (where 1g represents gravitational acceleration of the earth at 9.81 m/s^2). These profiles were selected on the basis of normal heavy vehicle transport conditions [15].

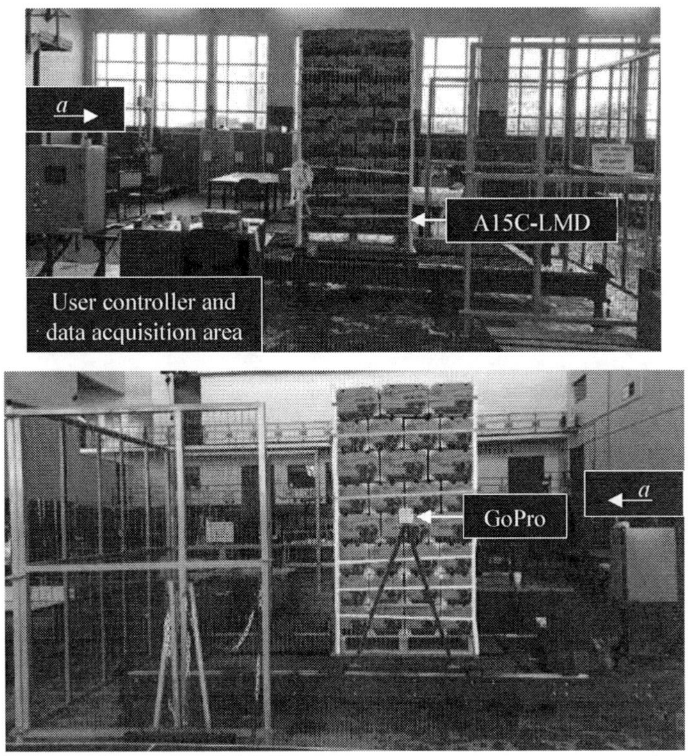

Figure 5: The dynamic test setup

2.3.2 Box Compression Tests

Box compression tests (BCTs) were required in order to quantify the dynamic fatigue that was exerted on the cartons used during dynamic testing. Two BCT variations were conducted as shown in Figure 6.

The standard full BCT (Figure 6-a) was conducted as a baseline method as commonly used in industry as described in ASTM D642 and ISTA 2A. The quarter BCT (Figure 6-b) was conducted according to observations made during the dynamic testing. It was observed that a single quarter of the carton can be exposed to high loads and thus possible local failure of a corner is possible. The BCTs were conducted with new-cartons as baseline and with used-cartons that were exposed to dynamic tests. This comparison enabled the quantification of the dynamic fatigue induced to cartons within both stack configurations under dynamic conditions. Take note that the cartons have not undergone any climatic conditioning as the focus was laid on the impact of the stack configurations and dynamic conditions.

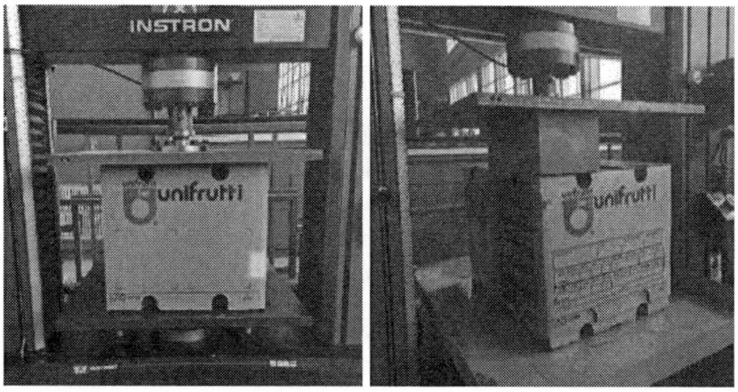

a) Standard full b) Quarter

Figure 6: BCT variations

3. Results and Discussion

3.1 Pallet Stability and Carton Loading

The stack movement of the citrus pallet under a 0.5g acceleration profile for both the (a) standard and (b) brick configuration is shown in Figure 7. The origin of the stack is indicated by the red dot with the peak displacement at acceleration and deceleration indicated by the yellow dots. It can be noted that the standard configuration has a visually high elastic displacement compared to the brick configuration.

a) Standard b) Brick

Figure 7: Stack movement of the (a) standard configuration and the (b) brick configuration under a 0.5g acceleration profile

The stack displacement data obtained from the dynamic tests as described, is summarized and presented in Table 1. Table 1 compares the permanent and elastic displacements of the standard and brick stack configurations, for the 0.5g acceleration profile, to the criteria set out in the EUMOS 40509 standard. Both stack configurations were within the limits of stability for permanent as well as elastic displacement. The permanent displacement for both the standard and brick configurations was less than 10 % of the EUMOS limit. Thus, this mode of displacement is not a concern. However, the elastic displacement of the standard configuration was 73 % of the EUMOS limit and that of the brick configuration was 23 % of the EUMOS limit. This indicates that the elasticity of the stacked pallet, and specifically the standard stack configuration, is a concern due to the highly dynamic movement that can induce carton fatigue failure.

Table 1: Displacement comparison of level 8 for the standard and brick configurations to the EUMOS standard under a 0.5g acceleration profile

Displacement [mm]	Standard	Brick	EUMOS
Permanent	7.1	4.2	116.3
Elastic	170.8	54	232.6

During the dynamic tests it was observed that the highest load acted on the carton in the side position. For this position, the carton loads, and corner loads are summarised in Table 2 for the standard and brick stack configurations respectively. Under static conditions, the brick configuration resulted in a carton position load of 2.56 kN. This was ± 1 kN larger than that of the standard stack configuration. The maximum carton corner load for the brick stack configuration was 0.69 kN and that of the standard stack configuration was 0.78 kN. The peak dynamic carton position load measured for the brick configuration was 3.8 kN and for the standard configuration it was 3.6 kN. The corresponding peak carton corner loads were 1.5 kN for the brick configuration and 1.7 kN for the standard configuration.

Table 2: Comparative load data at side carton position

Position or corner load	Load [kN]	
	Standard	Brick
Static		
Position	1.56	2.56
Peak Corner	0.78	0.69
Peak		
Position	3.6	3.8
Peak Corner	1.7	1.5

3.2 Carton Fatigue

The load-displacement graphs of the standard full carton and the quarter carton BCT, comparing the new empty, used standard stack configuration and the used brick stack configuration cartons are shown in Figure 8. It can be seen that the used cartons reach a lower peak load compared to the new cartons for both BCT setups, indicating the structural losses obtained by the cartons due to prior dynamic loads. Comparing Figure 8-a and Figure 8-b it is seen that the quarter carton BCT results do not necessarily translate to a quarter of the full carton BCT results. This indicates that uneven load distribution can have an influence on carton structural integrity and possible failure.

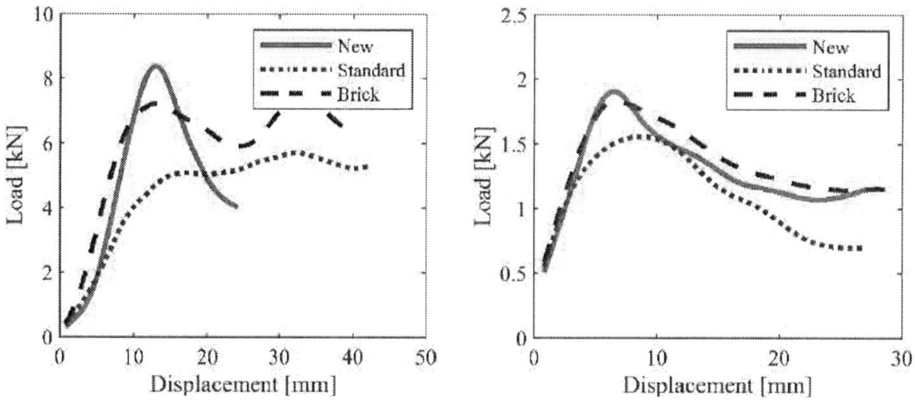

a) Full carton BCT b) Quarter carton BCT

Figure 8: BCT load-displacement graphs comparing new cartons to used cartons from a standard and a brick stack configurations in (a) Full carton and (b) Quarter carton test configurations

Table 3 outlines the full and quarter BCT strength for new cartons and cartons after being subjected to dynamic loading on the PLAT using the standard and brick stack configurations. From this information, it can be seen that the full carton in the standard stack configuration experienced a cyclic load of 3 kN over a period of 0.6 seconds. This resulted in a decrease in the BCT strength of 39.9 %. For the brick stack configuration, the cyclic load on the full carton was 2.7 kN over a period of 0.45 seconds, which resulted in a decrease in the BCT strength of 14.0 %.

When considering the load acting on a quarter carton in the standard stack configuration, it experienced a cyclic load of 1.7 kN over a period of 0.6 seconds, which resulted in a decrease in the quarter BCT strength of 18.4 %. In the brick stack configuration, the quarter carton experienced a cyclic load of 1.4 kN over a period of 0.45 seconds, resulting in a decrease in BCT strength of 4.2 %.

Thus, from this information, two correlations could be identified. Firstly, the cyclic (dynamic) load experienced by a full and quarter carton is related to the stack configuration. The cartons experienced a 10 % to 20 % higher cyclic load with the standard stack configuration compared to the brick configuration. Secondly, relating the reduction in BCT strength to carton structural damage, the standard stack configuration resulted in significant damage with a decrease in BCT strength ranging from 18 % to almost 40 %. On the other hand, the BCT strength was reduced by 4 % to 14 % in the brick stack configuration.

Table 3: Carton structural strength fatigue damage

Full/Quarter Load	New-BCT [kN]	Standard Configuration			Brick Configuration		
		Cyclic load [kN]	BCT [kN]	Reduction	Cyclic load [kN]	BCT [kN]	Reduction
Full	8.4	3.0	5.05	39.9 %	2.7	7.22	14.0 %
Quarter	1.9	1.7	1.55	18.4 %	1.4	1.82	4.2 %

Table 4 compares the BCT data of the used cartons to the peak test load measured for each stack configuration from the dynamic tests. The full carton BCT strength of the standard configuration was measured at 5.05 kN, while the peak test load measured for the full carton was 3.6 kN. This gives a fatigue safety factor of 1.4 which is a decrease of 40 % to the new safety factor for the full carton in a standard configuration. The full carton BCT strength of the brick configuration was measured at 7.22 kN, while the peak test load measured for the full carton was 3.8 kN. This gives a fatigue safety factor of 1.9 which is a decrease of 13 % to the new safety factor for a full carton in a brick configuration.

When inspecting the quarter carton results, the standard configuration measured a strength of 1.55 kN with a peak test load of 1.7 kN. This gives a fatigue safety factor (SF) of 0.91 which is a decrease of

16.5 % to the new safety factor for a quarter carton in a standard configuration. Whereas the brick configuration measured a quarter BCT strength of 1.82 kN and a peak test load of 1.5 kN. This gives a fatigue safety factor of 1.21 which is a decrease of 6.9 % to the new safety factor for a quarter carton in a brick configuration.

Thus, it can be concluded that the stack configuration had a notable impact on the fatigue safety factor of both the full and quarter cartons. This is due to the stability of each configuration and thus the dynamic fatigue exerted on the cartons. The brick configuration only presented a slight decrease in both safety factors from new to used. However, the standard configuration presented significant decreases in the safety factors from new to fatigue. The critical fatigue safety factor of 0.91 for the quarter carton indicates that (local) failure is likely to occur for a fatigued carton on a corner. Thus, due to the nature of corrugated paperboard, as previously noted, the whole carton will fail and buckle, and possibly lead to the collapse of the stacked pallet.

Table 4: Fatigue carton load comparison

Full/Quarter Load	**Standard Configuration**			**Brick Configuration**		
	Fatigue-BCT [kN]	**Peak Test Load [kN]**	**Fatigue-SF**	**Fatigue-BCT [kN]**	**Peak Test Load [kN]**	**Fatigue-SF**
Full	5.05	3.6	1.4	7.22	3.8	1.9
Quarter	1.55	1.7	0.91	1.82	1.5	1.21

4. Conclusion

The stability modes of the stacked pallet were identified for two stack configurations i.e. standard and brick. Both configurations were stable according to the EUMOS 40509 standard, however, the standard configuration showed an elastic displacement of more than double of that observed for the brick configuration. This behaviour was also observed in the measured carton loads as the standard configuration displayed much higher load shifts than the brick configuration.

The fatigue induced on cartons during the dynamic testing was more significant for the standard configuration than for the brick configuration. This was seen with the full BCT of the cartons where the standard configuration experienced a 40 % strength reduction compared to the brick configuration with a 14 % strength reduction. The fatigue safety factor identified that the quarter carton of the standard configuration had a critical safety factor of 0.91 which indicated that local failure of the carton corner is likely to happen. On the other hand, the brick configuration had a fatigue safety factor of 1.21 for the quarter carton, which was not deemed as problematic.

Thus, this study identified and quantified the dynamic loads exerted on a single carton in a stack, as well as the stability of the whole stacked pallet. This was conducted for two stack configurations i.e., standard and brick, and compared with one another. The tools developed and the observations made through this research, can be used to improve the design of corrugated cartons and thus ultimately decrease food losses throughout the distribution chain, with possible savings on the cost of the packaging.

Nomenclature

A15C-LMD	A15C Load Measurement Device
ASTM	American Society for Testing and Materials
BCT	Box Compression Test
EUMOS	European Safe Logistics Association
PLAT	Pallet Lateral Acceleration Tester
SF	Safety Factor

References

[1] T. Fadiji, T. M. Berry, et al., "Mechanical design and performance testing of corrugated paperboard packaging for the postharvest handling of horticultural produce," Biosyst Eng, vol. 17, pp. 128–140, Jul. 2018, doi: 10.1016/J.BIOSYSTEMSENG.2018.05.004.

[2] T. Berry, M. A. Delele, et al., "Geometric design characterisation of ventilated multi-scale packaging used in the South African pome fruit industry," Agricultural Mechanization in Asia, Africa, and Latin America, vol. 46, no. 3, pp. 34–42, 2015.

[3] C. R. Little and R. J. Holmes, Storage technology for apples and pears: a guide to production, postharvest treatment and storage of pome fruit in Australia. Victoria: Department of Natural Resources and Environment, 2000.

[4] T. M. Berry, T. S. Fadiji, et al., "The role of horticultural carton vent hole design on cooling efficiency and compression strength: A multi-parameter approach," Postharvest Biol Technol, vol. 124, pp. 62–74, Feb. 2017, doi: 10.1016/J.POSTHARVBIO.2016.10.005.

[5] D. Hellström, A. Olsson, et al., Managing packaging design for sustainable development: A compass for strategic directions. Chichester: John Wiley and Sons, 2016. doi: 10.1002/9781119151036.

[6] M. M. Starke, "Material and structural modelling of corrugated paperboard packaging," 2020. [Online]. Available: https://scholar.sun.ac.za

[7] M. E. Biancolini, "Evaluation of equivalent stiffness properties of corrugated board," Compos Struct, vol. 69, no. 3, pp. 322–328, Jul. 2005, doi: 10.1016/j.compstruct.2004.07.014.

[8] T. M. Berry, "Optimisation of multi-scale ventilated package design for next-generation cold chain strategies of horticultural produce," Doctor of Philosophy, Stellenbosch University, Stellenbosch, 2017. [Online]. Available: https://scholar.sun.ac.za

[9] T. Fadiji, A. Ambaw, et al., "Application of finite element analysis to predict the mechanical strength of ventilated corrugated paperboard packaging for handling fresh produce," Biosyst Eng, vol. 174, pp. 260–281, Oct. 2018, doi: 10.1016/J.BIOSYSTEMSENG.2018.07.014.

[10] M. Mukama, A. Ambaw, et al., "A virtual prototyping approach for redesigning the vent-holes of packaging for handling pomegranate fruit – A short communication," J Food Eng, vol. 270, Apr. 2020, doi: 10.1016/j.jfoodeng.2019.109762.

[11] A. Ambaw, M. Mukama, et al., "Fresh fruit packaging design verification through virtual prototyping technique," Food Packag Shelf Life, vol. 32, p. 100858, Jun. 2022, doi: 10.1016/j.fpsl.2022.100858.

[12] T. M. Berry, T. Defraeye, et al., "Designing ventilated packaging for the fresh produce cold chain," Food and Bioproducts Processing, vol. 134, pp. 121–149, Jul. 2022, doi: 10.1016/J.FBP.2022.04.005.

[13] D. Groenewald, "Packaging material specifications and palletisation protocols for," Jan. 2022.

[14] EUMOS, "EUMOS 40509:2020 - Test method for load unit rigidity," 2020.

[15] J. R. Billing, "Performance limits of heavy trucks," Ontario, Oct. 1998.

Bi-ventricular Elastic Material Parameters Estimation Using 3D CMR Strains in RHD Patients.

Mary A. Familusi [1,2*], Sebastian Skatulla [1], Jagir R. Hussan [3], Olukayode O. Aremu [4,5], Daniel Mutithu[4,5], and Ntobeko A. B. Ntusi [4,5,6]

[1] Computational Continuum Mechanics Research Group, Department of Civil Engineering, University of Cape Town, South Africa

[2] South African DST-NRF Centre of Excellence in Epidemiological Modelling and Analysis, Stellenbosch University, mary@aims.ac.za

[3] Auckland Bioengineering Institute, University of Auckland, New Zealand

[4] Division of Cardiology, Department of Medicine, University of Cape Town, Cape Town, South Africa

[5] Cape Universities Body Imaging Centre, Faculty of Health Sciences, University of Cape Town

[6] South African Medical Research Council Extramural Unit on Intersection of Noncommunicable Diseases and Infectious Diseases

Abstract

Insight into the biomechanical properties of myocardial material behavior in healthy and diseased hearts using finite element analysis could provide important information on cardiac function and potential treatment of the diastolic abnormalities responsible for ventricular hypertrophy and inflammation. To ensure a realistic myocardial constitutive behavior, inverse modeling based on the finite element approach combined with the Levenberg-Marquardt method is used for the material parameter optimization. The end-diastolic pressure-volume relationship (EDPVR) and strains were used to estimate the elastic material characteristics. Global circumferential-, longitudinal-, and radial strains were considered, and corresponding different multi-objective optimization problems were designed. Here we report on the sensitivity of elastic material parameters and muscle fiber orientation angles regarding the specific strain components included as targets in the objective function. Patient-specific computer simulations of EDPVR and strains for all objective functions showed good agreement with clinical data. The difference between the predicted and clinical parameters is less than 0.1%. We present qualitative and quantitative differences in stress and strain distributions for each of these optimization experiments. Model calibration utilizing EDPVR and strain could result in more accurate elastic material properties.

Keywords: Bi-ventricle; Elastic material parameters; Objective functions; Global strains; Rheumatic heart disease; Levenberg-Marquardt method.

1 Introduction

Rheumatic heart disease (RHD) remains the leading non-communicable cardiovascular disease in low-and middle-income countries (LMICs) ([13, 54]). RHD is a long-term sequela of acute rheumatic fever (ARF) which refers to the delayed cardiac damage caused by single severe or multiple recurrent episodes of ARF leading to mutilation of valves and chronic RHD [44]. Globally, RHD contributes to an overall burden of 15.6-19.6 million affected and over 233,000 deaths annually, accounting for nearly 2% of all deaths from cardiovascular disease [3, 31]. Hence, effective strategies for RHD prevention and treatment are necessary. Valvular dysfunction is the critical feature of RHD whereby the valves of the left ventricle are most commonly left in a state of permanent damage or complete failure [16, 30]. The valves most affected by rheumatic fever, in order, are the mitral, aortic, tricuspid, and pulmonary valves [44]. In most cases, the mitral valve is involved with one or more of the other three valves [30]. The failure mechanisms of RHD are not well studied in comparison to myocardial infarction which prevents effective treatment of this disease. An effective treatment of Acute rheumatic fever (ARF) and RHD depends on the understanding of the involved failure mechanisms. However, due to the improved socioeconomic conditions, the occurrence of ARF has been almost eradicated in the developed world and the research capacity invested in this field is lacking. Progress has been made, e.g. in heart valve replacement therapy allowing the heart to regenerate, but the main focus is on infectious and inflammatory aspects of ARF and less on the subsequent proliferative responses of the heart. In this sense, research efforts aimed specifically at the tertiary prevention and treatment of RHD are still rather under-represented [47].

Structural remodeling of the ventricles is an important process that takes place during the development of diastolic and systolic dysfunction and subsequent failure [2, 17]. Changes in physiological, geometrical, and hemodynamic loading conditions can lead to regional ventricular wall thickening and enhancement or degradation of regional muscle function. This remodeling process results in changes in muscle stiffness during filling, in the force-tension relationship, and in material properties, which are important determinants of the global and regional performance of the heart [28]. For patients with RHD, ventricular hypertrophy and eccentric remodeling of the myocardium are often found. A hypertrophic heart has undergone compensatory remodeling but eventually becomes incapable of working efficiently. This type of remodeling is characterized by an increased end-diastolic volume, relatively unchanged wall thickness and smaller wall-thickness-to-radius ratio [9]. Characterization of myocardial material properties has significant clinical applications [51, 6]. Knowledge of mechanical stress is necessary to comprehend the myocardium's mechanics. [53]. It is possible to evaluate myocardium stress if the mechanical properties of the tissue are quantified. It has been discovered that the consumption of oxygen within the myocardial tissue [42] and blood flow in the coronary artery is impacted by the myocardial stress distribution [22]. Also, left ventricle hypertrophy is caused by abnormal myocardial stress [14]. Unfortunately, experimentally invasive myocardial stress determination is not yet a reliable choice due to the high measurement errors inherent in the experimental approaches. [19, 36].

Cardiac magnetic resonance (CMR) has emerged as the gold standard for assessing myocardial structure and function, due to its superior spatial and temporal soft tissue resolution. In comparison to other imaging modalities, CMR is a valuable tool in the evaluation of patients with suspected or confirmed hypertrophic cardiomyopathy [32]. Amidst different CMR robust techniques, radiologists and cardiologists have become more interested in myocardial strain obtained from CMR feature tracking (CMR-FT), detail can be found in [43, 52, 33, 29]. Myocardium deformation is measured using myocardial strain, which is a more accurate imaging biomarker of myocardial disease when compared to the often utilized ventricular ejection fraction [1]. The most common type of strain is *Lagrangian strain*, which is calculated as $\epsilon = L/L_0$, where ϵ = strain, L = change in length, and L_0 = original length. Three main strain or deformation types-longitudinal, radial, and circumferential-have been used to characterize the mechanics of the myocardium at the regional level. Positive strain is defined as elongation, whereas negative strain is shortening.

In the study of Nikou et al.[34], optimisation was carried out by reducing the discrepancy between the strain and cavity volumes predicted by finite element method and those measured by MRI in the porcine hearts. To calibrate the passive and active mechanical parameters of the left ventricle (LV), Dabiri et al. [5] used LV (EDV and EDP) and speckle tracking 3D echocardiography segmental strains from beating hearts. Recently, we calibrated the mechanical properties of human ventricles using CMR data to construct a precise anatomical geometry [8]. Only the left and right ventricular volumes, which are global biventricular (BiV) measures, were used to determine the elastic material parameters in the healthy and RHD subjects. In determining a suitable constitutive equation, several studies have shown that the passive stress-strain relationship in the myocardium is exponential [48, 7]. The first constitutive equation in the form of an exponential strain energy formulation for passive myocardial behavior has been credited to Yuan-Cheng Fung [18, 20, 41]. Humphrey and Yin [21] presented the first constitutive model based on invariants, adding a new fiber-specific condition to account for the anisotropic material nature, but only for transverse isotropy. Costa et al. [4] created an extended orthotropic formulation that includes a fiber-specific coordinate system as well as principal material stiffness along the fiber, sheet, and normal directions. According to Usyk et al. [49], the myocardium can be modeled using three types of constitutive equations: isotropic, transversely isotropic, and orthotropic.

In this work, we used the constitutive equation described in [8, 49] and the in-house finite element method-based software SESKA, for model calibration. Inverse modeling which is based on the finite element approach combined with the Levenberg-Marquardt method is used for the material parameter optimization (LVM) [46, 15, 10]. We calibrated our previously published BV models [8] by targeting global strains and end-diastolic volumes. Due to the scarcity of the end-diastolic pressure recorded clinically, we assumed a population-based value. We scaled our calibration method to accurately match Klotz's relation for the passive filling [24]. Using these computational models, we formulate a multi-objective optimization problem, and investigate the sensitivity of elastic material parameters and fiber angles to change in the objective function. We qualitatively and quantitatively show the differences in stress and strain distributions. The inclusion of the transverse angles allows our models to be more flexible in the calibration processes.

2 Computational modeling

2.1 Finite element model of passive mechanics

We calibrate and analyze our previously described biventricular models of the fifteen RHD patients [8]. The passive tissue behavior was based on the mechanical model proposed by Usyk et al. [49] and reformulated in terms of the invariants of the Green strain tensor, **E**, by Legner et al. [27]. The model describes the myocardium as a nonlinear, orthotropic, and nearly incompressible hyperelastic material using the strain energy function given by:

$$\Psi = A/B\,(\exp^{BQ} - 1) + A_{\text{comp}}\,(J\ln J - J + 1), \tag{1}$$

where parameters A and B are stiffness factors and $J = \det \mathbf{F}$ is the Jacobian, with $\mathbf{F}$ being the deformation gradient tensor, quantifying volumetric deformation of cardiac tissue as linked to the numerical penalty parameter A_{comp} enforcing the near incompressibility of the myocardium. Additionally, Q is defined in terms of the Green strain tensor and the material directions defining structural tensors $\mathbf{M}_f$, $\mathbf{M}_s$, and $\mathbf{M}_n$ as follows:

$$\begin{aligned} Q := \; & a_1(tr(\mathbf{M}_f\mathbf{E}))^2 + a_2(tr(\mathbf{M}_s\mathbf{E}))^2 + a_3(tr(\mathbf{M}_n\mathbf{E}))^2 \\ & + a_4 tr(\mathbf{M}_f\mathbf{E}^2) + a_5 tr(\mathbf{M}_s\mathbf{E}^2) + a_6 tr(\mathbf{M}_n\mathbf{E}^2), \end{aligned} \tag{2}$$

where $a_i (i = 1, \ldots, 6)$ are the anisotropy coefficients associated with the three preferred material directions, namely fiber axis, $\mathbf{V}_f$, sheet axis, $\mathbf{V}_s$, and sheet normal axis, $\mathbf{V}_n$.

The boundary conditions that are used adhere to the methodology described in [8]. Basically, Dirichlet and Neumann's boundary conditions are required to model the diastolic filling phase of the heart. To prevent rigid body motion and enable connections between the base and the major blood vessels, Dirichlet conditions involve fixing ventricles at the base. In order to keep deformation at a physiologically acceptable level, Neumann's boundary conditions involve applying an elastic line force around the epicardial base. Under a weak Dirichlet boundary condition, an elastic line force is provided to the epicardial base, inhibiting rigid body motion but not obstructing ventricular wall thickening.

The inflow of blood into the ventricular cavities during the diastolic filling stage of the cardiac cycle imparts hydrostatic pressure on the endocardial walls. To model this, an incremental surface pressure load is prescribed on the endocardium of both ventricles as Neumann boundary conditions. The magnitude of the pressure increments is controlled by prescribed ventricular cavity volume increments [45]. Due to the lack of subject-specific ventricular pressure which requires invasive measurements, we assumed LVEDP as 3.0 kPa (22.50 mmHg) [11, 26, 50]. One-third of LV blood pressure was applied to the RV endocardium [12, 35].

2.2 Parameter optimization procedure

Combining the finite element (FE) method and the Levenberg-Marquardt algorithm (LVM), the eight material parameters of the orthotropic constitutive law described in Sec. 2.1 for the two ventricles of fifteen RHD patient's models, as well as the epi- and endocardial fiber orientation angles are patient-specifically obtained. In line with the methodology proposed by [25], the parameter optimization with respect to myocardial stiffness, anisotropy parameters, and fiber orientations was achieved by reducing difference between the Klotz curve [23] and the simulated EDPVR curve of each subject, beginning with the least-loaded LV volume (ESV) and progressing until the end-diastolic volume (EDV) was reached at the stipulated end-diastolic pressure (EDP). The approach by [46] was utilized for the direct inverse computation of the unknown unloaded configuration of the heart ventricles (V_0) used as the reference volume so that the residual stress and strain associated with the corresponding end-systolic pressure (ESP) was incorporated in the mechanical response of the BiV models.

We design various multi-objective optimization problems involving global end-diastolic volumes and strains to obtain simulated EDPVR by minimizing the difference between the clinical and computational end-diastolic volumes and strains in order to observe the sensitivity of elastic material parameters and muscle fiber orientation angles with respect to the particular strain components (see Tab. 1).

Table 1: Different optimization target.

	Optimization target
1	LV end-diastolic volume, RV end-diastolic volume, LV global circumferential strain
2	LV end-diastolic volume, RV end-diastolic volume, LV global longitudinal strain
3	LV end-diastolic volume, RV end-diastolic volume, LV global circumferential strain, LV global radial strain
4	LV end-diastolic volume, RV end-diastolic volume, LV global longitudinal strain, LV global radial strain

The following initial values of the myocardial material properties were chosen: $a_1 = -6.0$, $a_2 = -5.0$, $a_3 = 9.0$, $a_4 = 12.0$, $a_5 = 12.0$, $a_6 = -6.0$, $A_{comp} = 100$ kPa, $\theta_{epi} = -57^0$, $\theta_{endo} = 72^0$, $\theta_{epi} = 45^0$, and $\theta_{endo} = -10^0$. The initial anisotropy parameters a_i, $i = 1 \ldots 6$ used in the strain-invariant representation of the orthotropic material law (Eq. (2)) are taken from [38] and can be converted to corresponding strictly positive parameters as originally proposed by [49] using Eqs.5a - 5f in [8]. The value of the numerical penalty parameter $A_{comp} = 100$ kPa has been found most suitable to obtain near incompressible material behavior and avoid non-physiological deformation behavior and numerical instability for the chosen material model and finite element discretization [37, 40]. Also, the initial fiber angles were chosen according to the work of [37, 39]. A preliminary parameter discovery search was undertaken within a parameter space based on [49] for $A = [0.015; 1.5]$

3 Sensitivity to the material parameters in the constitutive model using different optimization target

In this section, we estimate the eight material parameters A, B, a_i ($i = 1 \ldots 6$) using data from RHD patients, following the objective functions described in Table 1. Twelve of fifteen patients were successfully calibrated using the first optimization target (OpT). Ten out of fifteen were successful in the second OpT, seven in the third OpT, and five in the fourth OpT.

EDVs and LV circumferential strains optimization target

In the calibration of passive material properties based on the end-diastolic pressure-volume relationship targeting experimental EDVs, and circumferential strain, the mean difference (the sum of the differences between the experimental and computational EDVs divided by the number of patients) between the analytical and experimental LV volumes was approximately 0.008 ml, and 0.019 ml for the RV volumes. The computational and experimental LV EDVs were 195.26 and 195.25 ml, respectively (experimental EDV is 0.003% lower). Similarly, the computational and experimental RV EDVs were 150.52 and 150.50 ml, respectively (experimental EDV is 0.013% lower). The mean differences between the experimental and computational circumferential strain is 0.0003, longitudinal strain is 10.03 and radial strain is 2.79 (see Tab. 3.

The calibrated fiber angle varies globally from -49.9^0(±17.30) (plus or minus its standard deviation) on the epicardium to +60.9^0(±22.50) on the endocardium. The minimum fiber angle is -80.0^0 on the epicardium and ±24.4^0 on the endocardium while the maximum fiber angle is -20.3 on the epicardium and +93.9^0 on the endocardium. Similarly, the transverse angle varies from +42.30(±5.40) on the epicardium to -10.6^0(±0.70) on the endocardium. The minimum transverse angle is +34.0^0 on the epicardium and -11.8^0 on the endocardium while the maximum fiber angle is +52.6^0 on the epicardium and -9.5^0 on the endocardium.

EDVs and LV longitudinal strains optimization target
In the estimated material parameters obtained from the second target, the computational LV EDV was 0.00047% smaller than the experimental volume, whereas the computational RV EDV was 0.00005% lower. The mean difference between experimental and computational circumferential strain is 5.36, 0.00002 for longitudinal strain, and 4.43 for radial strain.

The calibrated fiber angle varies globally from −28.0^0 (±6.40) on the epicardium to +39.9^0(±12.60) on the endocardium. The minimum fiber angle is −41.5^0 on the epicardium and +20.0^0 on the endocardium while the maximum fiber angle is −20.0^0 on the epicardium and +58.3^0 on the endocardium. Similarly, the transverse angle varies from +37.4^0(±4.10) on the epicardium to −9.0^0(±0.90) on the endocardium. The minimum transverse angle is +34.00 on the epicardium and −9.8^0 on the endocardium while the maximum fiber angle is +46.8^0 on the epicardium and −6.9^0 on the endocardium.

Table 2: The material parameters obtained from the four optimization targets.

Material parameters	OpT 1 (n=12)	OpT 2 (n=10)	OpT 3 (n=7)	OpT 4 (n=5)
	Median (IQR)	Median (IQR)	Median (IQR)	Median (IQR)
Right ventricle				
A (kPa)	0.56 (0.38 – 0.92)	0.61 (0.43 – 0.95)	0.55 (0.41 – 0.82)	0.58 (0.52 – 0.87)
B	0.80 (0.52 – 1.22)	0.76 (0.45 – 1.12)	0.91 (0.49 – 1.23)	1.07 (0.92 – 1.35)
Left ventricle				
A (kPa)	0.42 (0.30 – 0.58)	0.36 (0.27 – 0.43)	0.41 (0.24 – 0.43)	0.22 (0.21 – 0.32)
a1 (-)	-3.04 (-4.17 - (-2.85))	-3.00 (-3.99 - (-2.79))	-6.05 (-8.26 - (-4.00))	-7.25 (-8.13 - (-6.62))
a2 (-)	-2.97 (-3.75 - (-2.21))	-2.87 (-3.45 - (-2.55))	-4.03 (-5.44 - (-3.31))	-6.28 (-6.87 - (-5.34))
a3 (-)	4.23 (2.47 – 6.83)	4.92 (4.43 – 5.65)	10.28 (6.37 – 12.09)	8.15 (7.92 – 8.28)
a4 (-)	12.60 (9.59 – 16.13)	12.33 (9.15 – 20.55)	10.48 (8.22 – 11.34)	20.87 (18.08 – 22.27)
a5 (-)	9.00 (6.36 – 12.63)	9.63 (6.37 – 12.19)	9.66 (6.01 – 11.44)	10.94 (8.69 – 12.67)
a6 (-)	-2.82 (-3.76 - (-2.22))	-3.46 (-4.11 - (-3.00))	-4.67 (-5.31 - (-0.71))	-8.67 (-9.34 – (-7.46))
$\Theta endo$ (^{0})	62.08 (41.48 – 79.03)	23.47 (20.00 – 29.76)	73.01 (70.10 – 81.75)	22.76 (21.03 – 37.18)
θ_{epi} (^{0})	-46.3 (-63.7 - (-40.0))	-28.3 (-31.6 - (-24.1))	-61.2 (-76.4 - (-55.7))	-63.9 (-66.1 - (-59.9))
$Bendo$ (^{0})	-10.6 (-11.3 - (-10.2))	-9.2 (-9.5 - (-8.9))	-11.6 (-13.7 - (-11.4))	-11.9 (-12.6 - (-10.9))
β_{epi} (^{0})	41.52 (39.44 – 44.51)	35.96 (34.00 – 40.01)	34.83 (34.03 – 40.98)	43.90 (39.29 – 45.16)

OpT: Optimization target; IQR: Inter quartile range; A & B: stiffness parameters; a1 - a6: stiffness orientation-dependent parameters; θ_{endo}: fiber orientation at the endocardium; θ_{epi}: fiber orientation at the epicardium; β_{endo}: transverse fiber orientation at the endocardium; β_{epi}: transverse fiber orientation at the epicardium.

4 Comparison of the estimated material parameters across different optimization targets

The difference between computational and experimental longitudinal strain was 12.51 when both circumferential and radial strain were included in the objective function. Meanwhile, the difference between computational and experimental circumferential strain was 2.11 when both longitudinal and radial strains were included in the objective function. The disparity between experimental and computational strain is relatively modest when longitudinal and radial strains are targeted. The calculated material parameters *A, B,,* a_i and fiber orientations for the four targets are reported in Table 2. When the longitudinal strain is taken into account in the objective function, the left ventricular fiber angles are smaller, particularly at the endocardium (targets 2 and 4). This indicates how the fiber angles have a substantial impact on the longitudinal strain calibration. Across the models, the transverse angles are very much constant.

5 Model performance

In this section, we observe the performance of the constitutive model using our four optimization targets. Fig. 1 shows the BiV stress and strain distributions. As quantitatively presented in Tab. 2, results show that the overall distribution achieved from targets 1 and 3, and targets 2 and 4 are similar. There is an increased stress when longitudinal strain is present in the objective function. In the left and right ventricles, there is no difference in the average predicted stress and strain result. In the LV and RV longitudinal directions, the stress is approximately 4.0 kPa and 3.0 kPa respectively. Also, in the circumferential direction, the LV stress is approximately 7.0 kPa, and the RV stress is approximately 4.0 kPa. In the radial direction, the LV stress is approximately 0.8kPa and 0.7kPa in the RV. Tab. 3 shows the mean and maximum difference between the computational and experimental strains for different targets. It is observed that the difference in the circumferential experimental and computational strain is lower when longitudinal strain is targeted (OpT2 and OpT4) as compared to when not. Furthermore, using the Klotz curve as the benchmark, we see that the fourth optimization target which includes longitudinal and radial strains has a close mark with the Klotz curve (see Fig. 2).

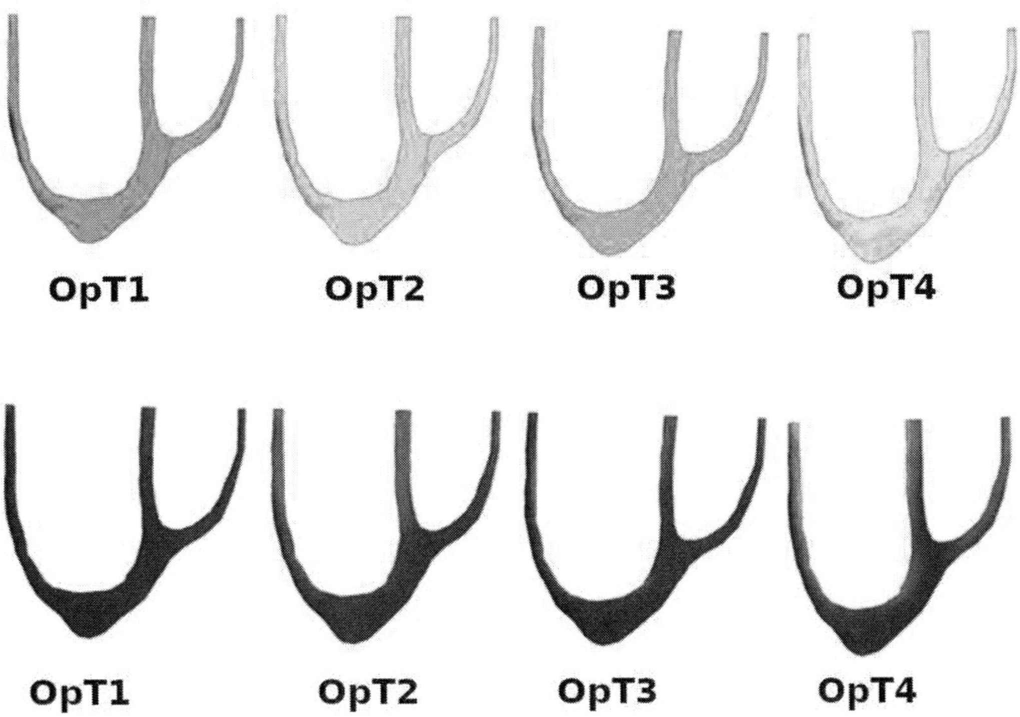

Figure 1: Stress (Top) and Strain (Both) distribution from the four optimization targets.

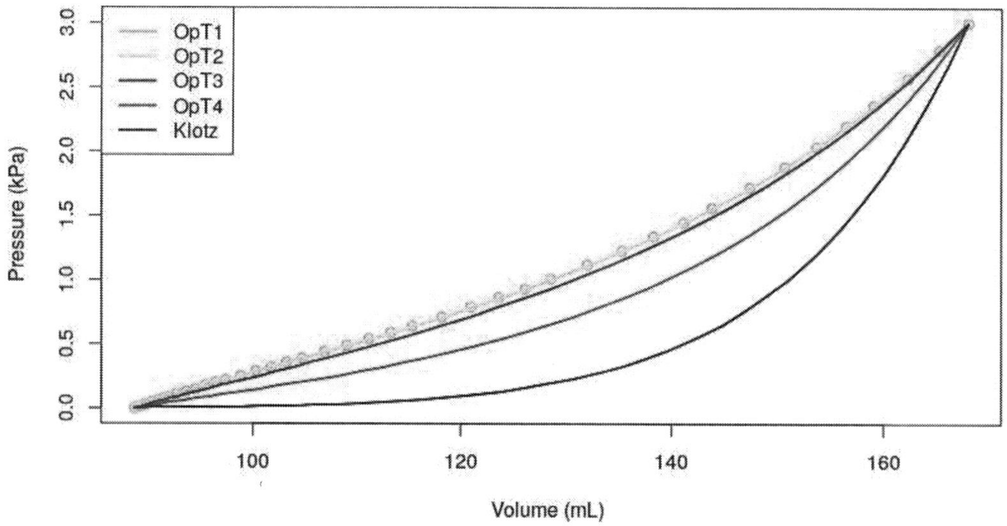

Figure 2: An exemplary graphical representation of the closeness between the empirical Klotz and the four optimization targets.

Table 3: The mean and maximum difference between the computational and experimental strains for different optimization targets described in Table 1

Optimization targets	Mean difference circumferential	Max difference circumferential	Mean difference longitudinal	Max difference longitudinal	Mean difference radial	Max difference radial
1	0.0003	0.0010	10.03	23.84	2.79	6.72
2	5.36	20.01	0.0002	0.0029	4.48	7.40
3	0.0005	0.0015	12.51	23.79	0.0003	0.0009
4	3.11	11.53	0.0005	0.0014	0.0001	0.0008
Familusi et al. (2023) [8]	3.36	12.53	10.48	21.48	4.02	9.44

6 Discussion

In this study, we used EDV, EDP, and speckle-tracking 3D CMR global strains from rheumatic heart disease patients beating hearts to calibrate the passive mechanical properties of biventricular models. To non-invasively obtain parameter estimation of the constitutive model from the routinely used clinical data, a combination of the FE simulations and LVM optimization schemes [15, 10, 39] was used. Using our earlier created ventricular geometries for the RHD patients [8], in this study we examined four categories of optimization targets. In each category, different sets of material parameters were obtained. First, we incorporate circumferential strain in the objective function in addition to the LV and RV EDVs as targets. Secondly, we included longitudinal strain and the EDVs. The third optimization targets are the circumferential strain, radial strain, and EDVs. The fourth category contains longitudinal strain, radial strain, and EDVs. We examined these categories to determine which of the strains helps to minimize the difference between the predicted and experimental measurements. We used the end-systolic phase of the extracted tagged CMR motion as the reference geometry.

In our previous study, we presented the elastic material parameters obtained for RHD patients and healthy controls while targeting LVEDV and RVEDV [8]. Our model validation method using clinically measured strain and simulated strain shows a closer agreement between the circumferential and radial strain but a substantial difference was observed in the measured and simulated longitudinal strain. In this current study, we decided to incorporate strains in the objective function to minimize the observed difference between the measured and simulated strain. Presently, our result shows a significant reduction in the difference between the experimental and computational strains. In the first category, where we calibrated passive material parameters targeting LVEDV, RVEDV, and circumferential strain, the difference between these targeted variables as compared to the clinically obtained values is approximately zero. In addition, the difference in the radial strain is also well reduced when compared to the previous study where only LVEDV and RVEDV were included in the objective function. Similarly, the mean difference between the targeted variables in the second category is well minimized. There is also a reduction in the radial strain as compared to when only LVEDV and RVEDV were included in the objective function. Lastly, in the third and fourth categories where two strains were included in the cost function with the LV and RV end-diastolic volume, the average difference in the longitudinal strain is higher in the third category as compared to the difference of the circumferential strain in the fourth optimization target (12.51 vs 3.11).

In the study of Dabiri et al. (2019)[5] that used segmental strain metrics from 16 segments, it was found that when both circumferential and longitudinal strains were targeted, the average difference between the predicted and experimental circumferential and longitudinal strains was higher than when only circumferential or longitudinal strains were used. In our study, it was impossible to calibrate targeting circumferential and longitudinal strain simultaneously, this could be due to substantial differences that might result from it because, our simulation software produces results of minimum difference of the target parameters. Alternatively, we included radial strain as the target, and we can also confirm that the difference was higher in the category having two strains. However, the difference in the experimental and computational strains of the targeted strain is approximately zero.

When circumferential strain is targeted, the calibrated fiber angles differ significantly from when longitudinal strain is targeted. The fiber angle distribution is observed to be small when longitudinal strain is included in the objective function as compared to when not. The variation in the calibrated transverse fiber angle compared with the start value is less significant when compared with the difference in the helix fiber angles especially at the endocardium. Although the inclusion of transverse fiber angles allows our models to be more flexible, it does not have a significant impact on the calibration result. This shows that strain calibration significantly depends on the fiber angles but less on the transverse angles. According to [50], their findings demonstrated that the fiber angle distribution can significantly affect how much stress is distributed within the left ventricular wall during diastole. They added that, in contrast, the sheet or transverse orientation, at least during diastole, has relatively minimal influence on stress and strain distributions.

6.1 Model Limitations

The differences between the model-predicted strains and the experimental strains not included in the objective function were noticeably large. The calibration of elastic material parameters targeting strain largely depends on fiber orientation. A more realistic model would include fiber orientation that could be approximated from diffusion tensor magnetic resonance imaging data. Another parameter that has a profound role in the calibration process is the end-diastolic pressure which was not available due to invasive measurement. A more realistic model would have a different transmural pressure for the LV septum that is less than that for the LV free wall (to take into account RV pressure).

7 Conclusion

A noninvasive solution to determine mechanical features of myocardial contractility in healthy and diseased hearts using finite element analysis could provide crucial information on heart function as well as therapeutic treatment of diastolic disorders that cause ventricular hypertrophy and inflammation. Furthermore, stress, which may be the catalyst for remodeling in pathological situations like heart failure can be ascertained if mechanical qualities are well understood. This study presented a methodology to determine the mechanical properties of myocardium in living hearts using the finite element method coupled with the Levenberg-Marquardt method. This methodology may lead to more accurate mechanical properties as it is based on experimental volume and strain measurement.

Author contributions

The study's idea and design were overseen by MF, SS, and NN. The computational modeling and statistical analyses were done by MF. The initial draft of the work was written by MF. All authors helped to revise the work and read and approved the final version.

Funding

This work was supported by the South African DST-NRF Centre of Excellence in Epidemiological Modelling and Analysis (SACEMA) and National Research Foundation (NRF) of South Africa (UID 142877). Opinions expressed and conclusions arrived at, are those of the authors and are not necessarily to be attributed to the SACEMA or NRF. Further financial support was provided by the Department of Medicine, Postgraduate Scholarship, University of Cape Town (UCT).

Conflicts of Interest

The authors declare no conflict of interest.

References

[1] Bernadette Brady, Gerard King, et al. Myocardial strain: A clinical review. *Irish Journal of Medical Science (1971-)*, 192(4):1649–1656, 2023.

[2] CG Brilla and B Maisch, et al. Regulation of the structural remodelling of the myocardium: from hypertrophy to heart failure. *European heart journal*, 15(suppl D):45–52, 1994.

[3] Jonathan R Carapetis, Andrew C Steer, et al. The global burden of group a streptococcal diseases. *The Lancet infectious diseases*, 5(11):685–694, 2005.

[4] Kevin D Costa, Jeffrey W Holmes, et al. Modelling cardiac mechanical properties in three dimensions. *Philosophical transactions of the Royal Society of London. Series A: Mathematical, physical and engineering sciences*, 359(1783):1233–1250, 2001.

[5] Yaghoub Dabiri, Kevin L Sack, et al. Method for calibration of left ventricle material properties using three-dimensional echocardiography endocardial strains. *Journal of biomechanical engineering*, 141(9), 2019.

[6] Yaghoub Dabiri, Kevin L Sack, et al. Relationship of transmural variations in myofiber contractility to left ventricular ejection fraction: Implications for modeling heart failure phenotype with preserved ejection fraction. *Frontiers in physiology*, 9:1003, 2018.

[7] Socrates Dokos, Bruce H Smaill, et al. Shear properties of passive ventricular myocardium. *American Journal of Physiology-Heart and Circulatory Physiology*, 283(6):H2650–H2659, 2002.

[8] Mary A Familusi, Sebastian Skatulla, et al. Model-based assessment of elastic material parameters in rheumatic heart disease patients and healthy subjects. *Mathematical and Computational Applications*, 28(6):106, 2023.

[9] Antonello Ganau, Richard B Devereux, et al. Patterns of left ventricular hypertrophy and geometric remodeling in essential hypertension. *Journal of the American College of Cardiology*, 19(7):1550–1558, 1992.

[10] Michael W Gee, Ch Förster, et al. A computational strategy for prestressing patient-specific biomechanical problems under finite deformation. *International Journal for Numerical Methods in Biomedical Engineering*, 26(1):52–72, 2010.

[11] Martin Genet, Lik Chuan Lee, et al. Distribution of normal human left ventricular myofiber stress at end diastole and end systole: a target for in silico design of heart failure treatments. *Journal of applied physiology*, 117(2):142–152, 2014.

[12] Serdar Göktepe, SNS Acharya, et al. Computational modeling of passive myocardium. *International Journal for Numerical Methods in Biomedical Engineering*, 27(1):1–12, 2011.

[13] Mark Golob, Richard L Moss, et al. Cardiac tissue structure, properties, and performance: a materials science perspective. *Annals of biomedical engineering*, 42(10):2003–2013, 2014.

[14] William Grossman. Cardiac hypertrophy: useful adaptation or pathologic process? *The American journal of medicine*, 69(4):576–584, 1980.

[15] Fr´ed´eric Guyon and Rodolphe Le Riche. Least squares parameter estimation and the levenberg-marquardt algorithm: Deterministic analysis, sensitivities and numerical experiments. Technical Report 041/99, Institut National des Sciences Appliqu´ees, 2000.

[16] John E Hall and AC Guyton. Role of the kidneys in long-term control of arterial pressure and in hypertension: the integrated system for arterial pressure regulation. *Guyton and Hall Textbook of Medical Physiology (13th ed.), edited by Hall JE and Guyton AC. Philadelphia, PA: Elsevier*, pages 227–243, 2016.

[17] Patrick A Helm, Laurent Younes, et al. Evidence of structural remodeling in the dyssynchronous failing heart. *Circulation research*, 98(1):125–132, 2006.

[18] Gerhard A Holzapfel, Thomas C Gasser, et al. A new constitutive framework for arterial wall mechanics and a comparative study of material models. *Journal of elasticity and the physical science of solids*, 61:1–48, 2000.

[19] Roelof M Huisman, Gijs Elzinga, et al. Measurement of left ventricular wall stress. *Cardiovascular research*, 14(3):142–153, 1980.

[20] Jay D Humphrey. Continuum biomechanics of soft biological tissues. *Proceedings of the Royal Society of London. Series A: Mathematical, Physical and Engineering Sciences*, 459(2029):3–46, 2003.

[21] JD Humphrey and FC Yin. A new constitutive formulation for characterizing the mechanical behavior of soft tissues. *Biophysical journal*, 52(4):563–570, 1987.

[22] Kung-Ming Jan. Distribution of myocardial stress and its influence on coronary blood flow. *Journal of biomechanics*, 18(11):815–820, 1985.

[23] Stefan Klotz, Ilan Hay, et al. Single-beat estimation of end-diastolic pressure-volume relationship: a novel method with potential for noninvasive application. *American Journal of Physiology-Heart and Circulatory Physiology*, 291(1):H403–H412, 2006.

[24] Ekaterina Kovacheva, Lukas Baron, et al. Optimization framework to identify constitutive law parameters of the human heart. *Current Directions in Biomedical Engineering*, 6(3):95–98, 2020.

[25] Adarsh Krishnamurthy, Christopher T Villongco, et al. Patient-specific models of cardiac biomechanics. *Journal of computational physics*, 244:4–21, 2013.

[26] Lik Chuan Lee, Jonathan F Wenk, et al. Analysis of patient-specific surgical ventricular restoration: importance of an ellipsoidal left ventricular geometry for diastolic and systolic function. *Journal of applied physiology*, 115(1):136–144, 2013.

[27] D. Legner, S. Skatulla, et al. Studying the influence of hydrogel injections into the infarcted left ventricle using the element-free galerkin method. *International Journal for Numerical Methods in Biomedical Engineering*, 30:416–429, 2014.

[28] Ian J LeGrice, BH Smaill, et al. Laminar structure of the heart: ventricular myocyte arrangement and connective tissue architecture in the dog. *American Journal of Physiology-Heart and Circulatory Physiology*, 269(2):H571–H582, 1995.

[29] Carolin Lim, Edyta Blaszczyk, et al. Quantification of myocardial strain assessed by cardiovascular magnetic resonance feature tracking in healthy subjects—influence of segmentation and analysis software. *European radiology*, 31:3962–3972, 2021.

[30] Eloi Marijon, Mariana Mirabel, et al. Rheumatic heart disease. *The Lancet*, 379(9819):953–964, 2012.

[31] Eloi Marijon, Phalla Ou, et al. Prevalence of rheumatic heart disease detected by echocardiographic screening. *New England Journal of Medicine*, 357(5):470–476, 2007.

[32] Martin S Maron and Barry J Maron. Clinical impact of contemporary cardiovascular magnetic resonance imaging in hypertrophic cardiomyopathy. *Circulation*, 132(4):292–298, 2015.

[33] Daniele Muser, Simon A Castro, et al. Clinical applications of feature-tracking cardiac magnetic resonance imaging. *World journal of cardiology*, 10(11):210, 2018.

[34] Amir Nikou, Shauna M Dorsey, et al. Computational modeling of healthy myocardium in diastole. *Annals of biomedical engineering*, 44:980–992, 2016.

[35] Arnab Palit, Sunil K Bhudia, et al. Computational modelling of left-ventricular diastolic mechanics: Effect of fibre orientation and right-ventricle topology. *Journal of biomechanics*, 48(4):604–612, 2015.

[36] Niels Rahe-Meyer, Christian Weilbach, et al. In vivo myograph measurement of muscle contraction at optimal length. *BioMedical Engineering OnLine*, 6:1–10, 2007.

[37] Ritesh R Rama and Sebastian Skatulla, et al. Towards real-time cardiac mechanics modelling with patient-specific heart anatomies. *Computer Methods in Applied Mechanics and Engineering*, 328:47–74, 2018.

[38] Ritesh R Rama, Sebastian Skatulla, et al. Real-time modelling of diastolic filling of the heart using the proper orthogonal decomposition with interpolation. *International Journal of Solids and Structures*, 96:409–422, 2016.

[39] Ritesh Rao Rama and Sebastian Skatulla. Real-time nonlinear solid mechanics computations for fast inverse material parameter optimization in cardiac mechanics. *Journal of Engineering Mechanics*, 145(4):04019020, 2019.

[40] RR Rama and S Skatulla. Towards real-time modelling of passive and active behaviour of the human heart using podi-based model reduction. *Computers & Structures*, 232:105897, 2020.

[41] Michael S Sacks. Biaxial mechanical evaluation of planar biological materials. *Journal of elasticity and the physical science of solids*, 61:199–246, 2000.

[42] Stanley J Sarnoff, E Braunwald, et al. Hemodynamic determinants of oxygen consumption of the heart with special reference to the tension-time index. *American Journal of Physiology-Legacy Content*, 192(1):148–156, 1957.

[43] Andreas Schuster, Kan N Hor, et al. Cardiovascular magnetic resonance myocardial feature tracking: concepts and clinical applications. *Circulation: Cardiovascular Imaging*, 9(4):e004077, 2016.

[44] Dianne Sika-Paotonu, Andrea Beaton, et al. Acute rheumatic fever and rheumatic heart disease. In *Streptococcus pyogenes: Basic Biology to Clinical Manifestations [Internet]*. University of Oklahoma Health Sciences Center, 2017.

[45] S Skatulla and C Sansour. On a path-following method for non-linear solid mechanics with applications to structural and cardiac mechanics subject to arbitrary loading scenarios. *International Journal of Solids and Structures*, 96:181–191, 2016.

[46] Sebastian Skatulla, Carlo Sansour, et al. Non-invasive in silico determination of ventricular wall pre-straining and characteristic cavity pressures. *arXiv preprint arXiv:2308.00461*, 2023. http://arxiv.org/abs/2308.00461.

[47] Karen Sliwa and Peter Zilla. Rheumatic heart disease: the tip of the iceberg, 2012.

[48] Gerhard Sommer, Andreas J Schriefl, et al. Biomechanical properties and microstructure of human ventricular myocardium. *Acta biomaterialia*, 24:172–192, 2015.

[49] TP Usyk, R Mazhari, et al. Effect of laminar orthotropic myofiber architecture on regional stress and strain in the canine left ventricle. *Journal of elasticity and the physical science of solids*, 61(1-3):143–164, 2000.

[50] HM Wang, H Gao, et al. Structure-based finite strain modelling of the human left ventricle in diastole. *International journal for numerical methods in biomedical engineering*, 29(1):83–103, 2013.

[51] Jonathan F Wenk, Doron Klepach, et al. First evidence of depressed contractility in the border zone of a human myocardial infarction. *The Annals of thoracic surgery*, 93(4):1188–1193, 2012.

[52] Jing Xu, Wenjing Yang, et al. State-of-the-art myocardial strain by cmr feature tracking: Clinical applications and future perspectives. *European Radiology*, 32(8):5424–5435, 2022.

[53] FC Yin. Ventricular wall stress. *Circulation research*, 49(4):829–842, 1981.

[54] Liesl J Zühlke, Mark E Engel, et al. Incidence, prevalence and outcome of rheumatic heart disease in south africa: a systematic review of contemporary studies. *International journal of cardiology*, 199:375–383, 2015.

13th South African Conference on Computational and Applied Mechanics (SACAM2024)
22 – 23 January, 2024, Stellenbosch, South Africa

DEVELOPMENT OF A SPILLWAY AND SLUICE GATE DEMONSTRATION SYSTEM

Niël Gertenbach[1*], Jan-Hendrik Kruger[2†]

[*] Presenting author: nielgertenbach@gmail.com
[†]Corresponding author: jandhendrik.kruger@nwu.ac.za
[1,2]North-West University, School of Mechanical Engineering
Private Bag X6001, Potchefstroom, 2520

Abstract: Hydraulic control structures form vital components in modern infrastructure and engineering, from hydroelectric power generation to water storage and canal locks, and they need to be designed appropriately. Mechanical engineering students will benefit from practical exposure to hydraulic control structures supplementing their studies of fluid mechanics. This paper covers the analysis, simulation, and experimental results of a test bench to perform open channel flow demonstrations for engineering education. The flow phenomena incorporated into the designed apparatus were focused on the development of hydraulic jumps in the open channel and the transition from free to submerged flow at a sluice gate opening. CFD simulations using the Volume-Of-Fluids methodology in Ansys Fluent were validated with an analytical model, as both indicated the formation of a hydraulic jump at specific gate openings and flow rates. The effects of three variables were then further characterised during subsequent simulations and experiments, namely the height of the gate opening, the flow rate, and the stop log depth. The testing and construction of the test bench was completed at North West University and demonstrated that the test bench was successful and could be used to investigate other hydraulic structures such as culverts, radial gates, and weirs.

Key words: Hydraulic control structures, sluice gate, computational fluid dynamics, volume-of-fluids

1 Introduction

What is the best way to teach fluid mechanics to undergrad students? Arguably, it is an approach that introduces students to the fundamental fluid flow equations, but in a strong context of practical application [1]. As such, the School of Mechanical Engineering identified the need for a test bench that would allow the practical demonstration of fluid mechanics concepts and open channel flow phenomena, which will add to the theory taught in class.

Therefore, the focus of this study was the development of a test bench that demonstrates the effect of a sluice gate on downstream channel flow. As proof of concept, the formation of a hydraulic jump at specified gate openings, flow rates, and downstream depths was investigated using CFD simulations and experimentally validated.

2 Theory

This section discusses the main equations related to the hydraulic jump and its transition, as used in the current study for analytical validation.

2.1 Hydraulic Jump

A hydraulic jump is the rapid transition in the flow regime from supercritical to subcritical. For example, a spillway gate creates a constriction in the flow path, which causes supercritical flow at the gate opening, which then transitions to subcritical downstream due to flow disturbances. Hydraulic jumps are useful for dissipating the energy of supercritical flow. [2]

Hydraulic jumps can be predicted by investigating the Froude number of a flow region. The Froude number, Fr, is a dimensionless variable that describes how fast a typical wave moves relative to the flow field. When the Froude number is smaller than 1, the flow will be subcritical and supercritical for a Froude number larger than 1. When the Froude number is equal to 1, the flow is critical [2].

The Froude number can be calculated according to [2] as shown in Eqn. (1) where V is the flow velocity, g the gravitational acceleration and y the flow depth.

$$Fr = \frac{V}{\sqrt{gy}} \tag{1}$$

If the Froude number indicates supercritical flow, a hydraulic jump could occur when there is an obstacle that causes a disturbance in flow and a change in the flow regime. With subcritical flows, the depth of the fluid will only change gradually when disturbances or obstacles are encountered. Table 1 shows the hydraulic jump classification for a range of Froude numbers. [2]

Table 1: Hydraulic Jump Classification [2]

Froude Number	Jump Classification
<1	Impossible
1 to 1.7	Standing wave or undulant
1.7 to 2.5	Weak jump
2.5 to 4.5	Oscillating
4.5 to 9	Stable and steady
>9	Rough, strong

When water in the supercritical flow regime encounters, for example, a stop log at the end of a flume, the water will lose velocity and the depth will increase. This region of low-velocity water will cause a hydraulic jump where the two flow regimes meet.

The specific energy equation is used to evaluate open channel flow energy considerations and is shown in Eqn. (2) [2]. Specific energy is especially helpful in investigating the possibility of a hydraulic jump and in predicting the flow depths.

$$E = y + z + \frac{V^2}{2g} \tag{2}$$

Assuming the elevation, z to be zero, Fig. 1 illustrates the relationship between specific energy and flow depth at a constant flow rate of 30 l/min. It shows a sudden turning point at a depth of 0.0173 m and a specific energy of 0.026 m. This turning point corresponds to the critical depth at which the Froude number is one. The arrow shows the determination of a potential hydraulic jump from a depth of 0.006 m to 0.08 m.

According to [3], the critical depth, which is the point at which the specific energy is at its minimum, can be derived from Eqn. (2). This is shown in Eqn. (3) where q is the flow rate per channel width.

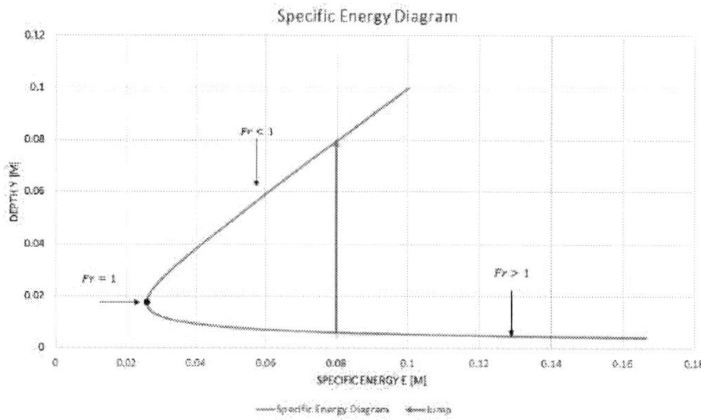

Figure 1: Specific energy as function of flow depth at a constant flow rate

$$y_c = \sqrt[3]{\frac{q^2}{g}} \tag{3}$$

Using Eqn. (3) and a flow rate of 30 l/min, a critical depth of 0.0173 m can be calculated, similar to what was determined from the specific energy diagram.

To calculate the depth of a hydraulic jump, Eqn. (4) can be used, which is derived from the momentum equation when assuming uniform and steady flow, see [2].

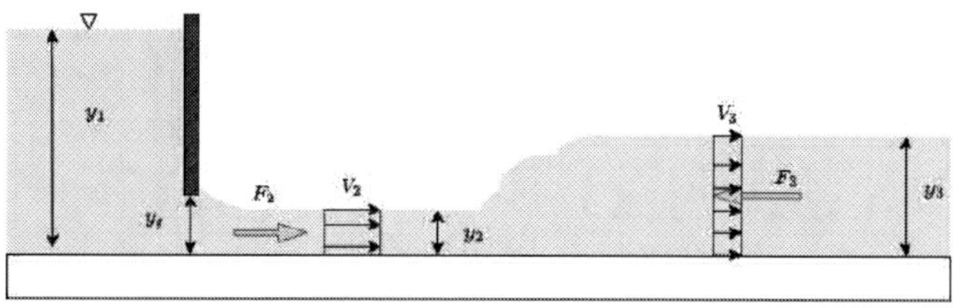

Figure 2: Conservation of momentum over a hydraulic jump. Adapted from [2].

$$\frac{y_3}{y_g} = \frac{1}{2}(-1 + \sqrt{1 + 8Fr_g^2}) \tag{4}$$

2.2 Free or submerged flow transition

The discharge from a spillway gate can be classified as free-flow or submerged-flow. According to [4], free flow will occur when the upstream depth is significantly larger than the gate opening. Submerged flow occurs when the ratio of upstream depth to gate opening is relatively low and will develop when the following conditions are met:

$$y_3 < y_1 < 0.81 y_3 (\frac{y_3}{y_g})^{0.72} \tag{5}$$

2.3 Discharge Coefficient

The discharge of a sluice gate can be determined using Eqn. (6), where C_d is the discharge coefficient. This equation can also be used to determine the upstream depth (y_1) for a given flow rate and gate opening.

$$Q = C_d y_g b \sqrt{2gy_1} \tag{6}$$

The discharge coefficient for the free flow from the sluice gate can be calculated from the various flow depths with Eqn. (7) [4]:

$$C_d = 0.611\left(\frac{y_1 - y_g}{y_1 + 15y_g}\right)^{0.072} \tag{7}$$

3 Development of test bench

The goal was to develop experimental equipment to demonstrate how theoretical knowledge and principles are applied in real life. This was done by developing an open channel flow test bench with a sluice gate to allow the formation of a hydraulic jump. Water is circulated with a centrifugal pump connected to a sump at the inlet and the channel at the outlet. After flowing through the channel, the water drops back into the sump. Fig. 3 shows the construction of the test bench and some of its components. Fig. 4 shows the actual test bench with a hydraulic jump.

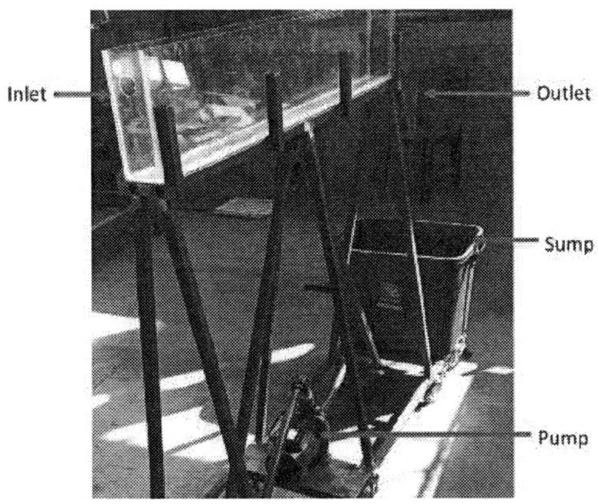

Figure 3: Test Bench Construction

Figure 4: Test bench showing a hydraulic jump in process

4 Experimental methodology

Stop logs are beams that are usually added incrementally to a spillway or channel to seal it and allow maintenance of the spillway or gate. In this experiment, stop logs were used to regulate the depth of the downstream flow. Eight 5 *mm* plexiglass rods were used as stop logs and dropped into a slot at the end of the channel. The water then flowed over the stop logs, and the barrier raised the flow depth.

The following experimental procedure should be followed:

- Switch on the pump and open the valve.
- Allow the flow to stabilise.
- Adjust the opening of the gate (y_g) and the height of the stop logs.
- Allow the flow to stabilise.
- Measure the depth of the hydraulic jump (y_3) and the depth upstream of the gate (y_1).
- Repeat the measurements three times to obtain a dataset.

The hydraulic jump depth was determined analytically using Eqn. (4), and adding the height of the stop logs. The depth upstream of the gate was analytically determined using Eqn. (6).

The values of the analytical equations were then compared to the actual measurements as part of the validation process.

5 Simulation methodology

CFD simulations were conducted to serve as an additional tool along with analytical calculations to verify the design and operation of the test bench. Simulations were performed using Ansys Fluent, release 2023R1, with multiphase modelling.

The simulations were conducted in two dimensions since no elements changed along the width of the channel and to save computational time. The simulation operating conditions were atmospheric and steady state. The standard multiphase volume-of-fluid (VOF) model was used to represent the interaction between water and air and to provide a solution for the interface between these, in other words, to calculate the movement of the surface of the water.

The VOF method uses a continuum indicator function (α) to represent the various fluids contained in the model domain. It typically has a value of zero in regions completely occupied by the first fluid and a value of one for regions occupied by the second fluid. In the interface region between the two fluids, the indicator function varies between one and zero, that is, $0 < \alpha < 1$. Using the indicator function as a weighting factor, the fluid properties of the mixture can then be determined, e.g. the density function shown in Eqn. (8).

$$\rho = \alpha \cdot \rho_1 + (1 - \alpha) \cdot \rho_2 \tag{8}$$

The momentum and energy conservation equations are then solved according to the standard SIMPLE-C algorithm, but using the weighted properties of the fluid mixture instead of only one fluid [5]. Second-Order Upwind discretization was used to achieve second-order accuracy, and a least squares cell-based interpolation method (gradient) was used as it is less computational intensive, but the same accuracy as the more computational intensive Green-Gauss Node-Based interpolation method. .

The standard k-ε turbulence model was chosen and is suitable for turbulent flows that occurred in the channel. The mesh dependency was not fully investigated, as the mesh resolution was already at the finest that the available computer could handle. The maximum iterations were set at 1000, but convergence often occurred earlier.

Fig. 5 shows the Spaceclaim model used in the CFD simulations, and the important elements of the model are also indicated. The flow straightener was added to impart uniform flow, and a buffer plate was used to reduce the incoming velocity.

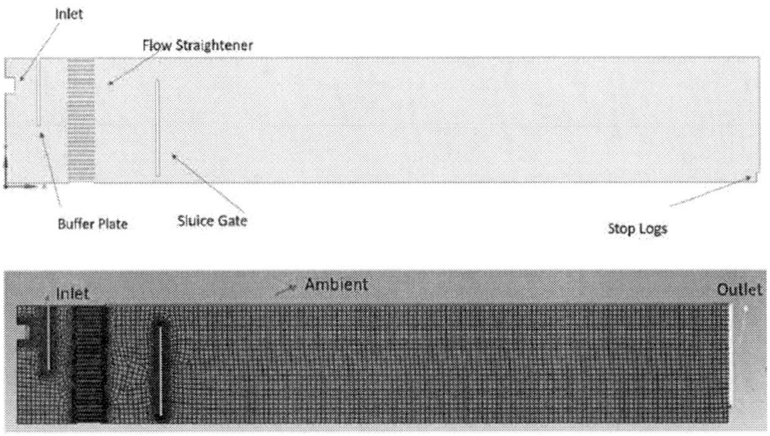

Figure 5: Simulation model and mesh. **Top:** Spaceclaim CAD model. **Bottom:** Simulation mesh showing local refinement and boundary conditions.

The CFD mesh is shown at the bottom of Fig. 5 and indicates the different boundary conditions of the channel. The top section from the inlet to the end of the flow straightener was closed to keep water from splashing out, but it was open to the atmosphere at the "Ambient" edge. Local mesh refinement was employed around the buffer plate, flow straightener, and sluice gate components to capture fine details of the highly complex flow in those regions.

6 Results

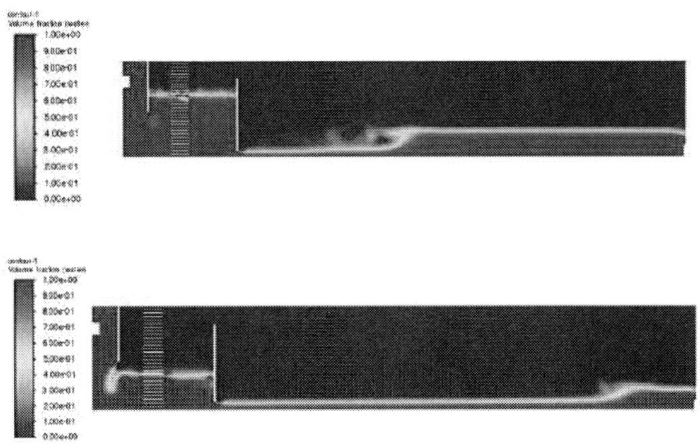

Figure 6: Hydraulic jump formation under different conditions. **Top:** Gate opening of 12 *mm* with 15 *mm* stop logs. **Bottom:** Gate opening of 15 *mm* with 10 *mm* stop logs.

Fig. 6 show the results of the simulation carried out at a flow rate of 50 l/min, with 12 and 15 *mm* gate openings, and stop log heights of 15 and 10 *mm*, respectively. The blue region represents air, and the red region represents water. The sudden increase in flow depth indicates a hydraulic jump. It can be seen that with fewer stop logs (therefore, lower flow resistance), the hydraulic jump occurs later in the stream.

Fig. 7 shows the influence of the height of the stop logs on the depth of the hydraulic jump (y_3) at a flow rate of 73 l/min and a gate opening (y_g) of 20 *mm*. The height of the stop logs was incrementally increased from 0 to 40 *mm*. The inconsistency at the lower gate openings was due to the hydraulic jump

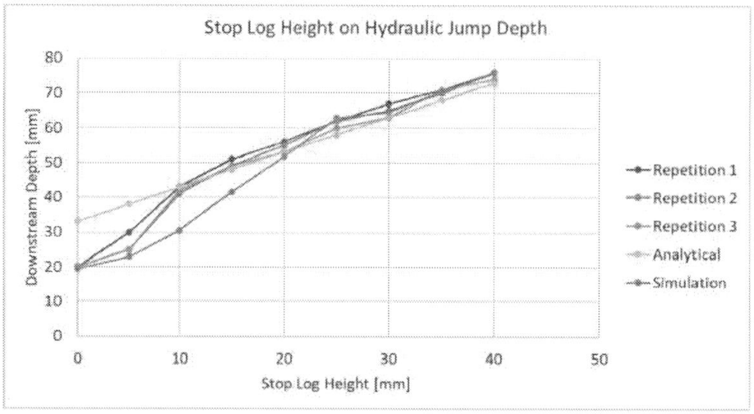

Figure 7: Hydraulic jump depth as function of stop log height. (Gate opening of 20 *mm* and flow rate of 73 l/min.)

not developing within the channel length. With the increase in the height of the stop logs, the hydraulic jump was pushed upstream. Fig. 7 also shows that the simulation results were initially lower. This is possibly due to the 2D nature of the simulations, where wall friction cannot be accounted for. With lower frictional resistance in the channel, the disturbance required to trigger a hydraulic jump increased. Hydraulic jumps therefore often occurred at slightly higher stop log heights in the simulations than during the experiments.

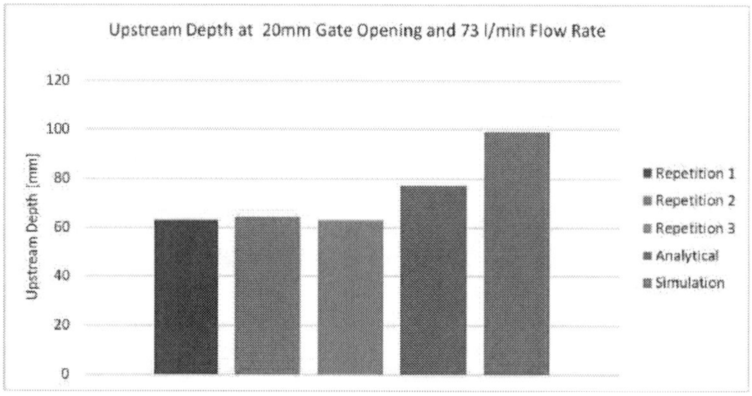

Figure 8: Results for depth upstream of the gate. (Gate opening of xx *mm* and flow rate of xx l/min.)

Fig. 8 shows a comparison of computational, simulation, and analytical results for flow depth (y_1) upstream of the gate. Analytical results were obtained using Eqn. (6). It shows that the test bench delivered repeatable results, but 18% lower than the analytical results. This deviation is suspected to be due to small leaks present between the gate and the channel walls.

The results of the simulation were comparably higher than the analytical and experimental values. The higher depth predicted by the simulations was possibly due to (1) the absence of leakage flow effects in the CFD model and (2) how the flow straightener, used to impart uniform flow, was modelled in 2D, which was not an exact representation of the actual 3D device on the test bench.

Whether the discharge from the sluice gate was submerged or free also influenced the formation of the hydraulic jump. The submerged flow lowers the velocity, which subsequently lowers the Froude number of the flow. If the velocity is lowered enough, the discharged flow will be sub-critical, making a hydraulic jump impossible.

Submerged flow was experienced when the stop logs were too high and pushed the hydraulic jump upstream up to the gate, and at low flow rates, which required an extremely small gate opening to achieve supercritical flow.

An experiment was conducted to investigate the transition from free flow to submerged flow as a result of the increase in the stop log height. It was carried out with a gate opening of 20 *mm*, a flow rate of 116 l/min, and an incremental increase in the height of the stop log. Fig. 9 shows the upstream depth and the right-hand side (RHS) of Eqn. (5). The intersection between the upstream depth and the RHS is the point where the transition occurred. When the RHS is greater than the upstream depth, the flow is submerged. As expected, the depth upstream (y_1) of the gate also increased as the flow became submerged.

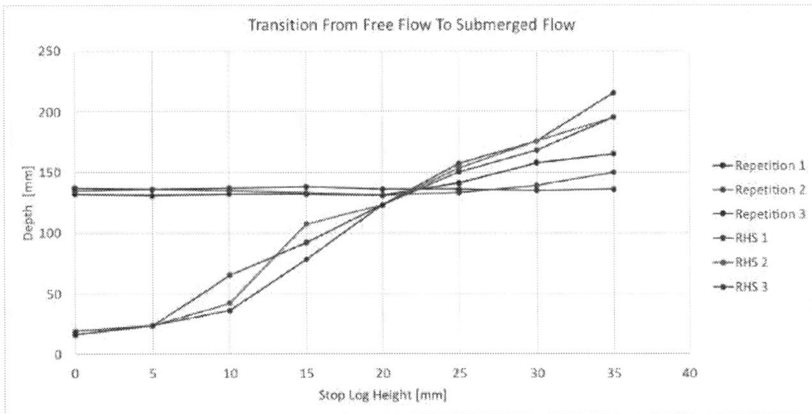

Figure 9: Experimental results showing transition from free to submerged flow, related to variations in the stop log height. (Gate opening of 20 *mm* and flow rate of 116 l/min.)

7 Conclusion

During the experiments, it was observed that lower gate openings lead to a stronger hydraulic jump as a result of the higher discharge velocity and also moved the jumps downstream. More stop logs moved the jump upstream, and higher flow rates moved the jump downstream, and they also had stronger jumps.

The test bench was successful in demonstrating open-channel flow geometries and regimes, the development of a hydraulic jump, similarity to analytical calculations, the transition from free to submerged flow, and the upstream and downstream shifting of a hydraulic jump.

The test bench does have a few practical limitations, which include not having a real time flow measurement device, unmeasured friction due to the silicon used to seal the channel, manual gate operation, slight leaking between the gate and the channel walls, and lastly the small operating range due to the short length of the channel.

The simulations showed encouraging agreement with experimental results, but may be improved by developing a full 3D model to fully incorporate wall friction effects and the 3D effects of the flow straightener. A more extensive numerical investigation may also be performed to explore the effects of the interface capture scheme used in the VOF method to determine whether it has a significant effect on the prediction of the height and location of the hydraulic jump.

Nomenclature

α	Density indicator function	–
γ	Specific weight of water	$\frac{N}{m^3}$

ρ	Density	$\frac{kg}{m^3}$
C	Discharge coefficient	–
E	Energy head	m
Fr	Froude number	–
H	Water head	m
Q	Volume flow rate	$\frac{l}{min}$
q	Specific flow rate	–
R_h	Hydraulic radius	m
Re	Reynolds number	–
V	Velocity	m/s

Acknowledgement

I wish to thank the Lord for blessing me with my skills, for guidance and discipline. I wish to also thank Dr. Kruger for his guidance on this project, as well as my parents who funded my studies and exposed me to the wonderful world of engineering. I also acknowledge Marinique for her support and motivation.

References

[1] Ashwin Vaidya. "Contributions to the Teaching and Learning of Fluid Mechanics". In: *Fluids* 6 (July 2021), pp. 269–269. DOI: 10.3390/fluids6080269. URL: https://www.mdpi.com/2311-5521/6/8/269/htm (visited on 12/10/2023).

[2] B.R. Munson et al. *Fluid Mechanics*. 7th. Wiley & Sons, 2013.

[3] M. Hanif Chaudhry. *Open-Channel Flow*. 2nd ed. Springer, 2008. URL: https://link.springer.com/book/10.1007/978-0-387-68648-6 (visited on 08/15/2023).

[4] Prabhata K. Swamee. "Sluice-Gate Discharge Equations". In: *Journal of Irrigation and Drainage Engineering* 118 (1992), pp. 56–60. URL: https://www.academia.edu/45391822/Sluice_Gate_Discharge_Equations (visited on 08/13/2023).

[5] ANSYS Incorporated. *Introduction to ANSYS FLUENT*. Dec. 2010. URL: https://imechanica.org/files/fluent_13.0_lecture05-solver-settings.pdf.

Numerical Modelling of Blood Flow During Syringing

J. Goedhals[1,2*], A. de Villiers [1], G.J.F. Smit[1]
[1] Division of Applied Mathematics, Department of Mathematics, Stellenbosch University,* jaimegoedhals@gmail.com
[2] DSI-NRF Centre of Excellence in Mathematical and Statistical Sciences (CoE-MaSS), South Africa

Abstract: This study investigates modelling blood flow through a syringe and the use of numerical models to calculate resulting hemolysis. Hemolysis, defined as the breaking open of red blood cells, can occur due to a range of factors which can be physical, chemical, or biological in nature. Physical damage can manifest in scenarios such as emergency rapid blood transfusions conducted through syringing, a practice which may be employed in resuscitation procedures, particularly in rural hospitals. It has been demonstrated that such syringe-based transfusions result in noteworthy hemolysis, significantly surpassing the impact of pressure bag usage. This heightened hemolysis carries potential negative implications for the recipient of the transfusion. The core objective of this project revolves around emulating the blood flow through a sudden contraction, simulating a syringe, to emulate and quantify the underlying mechanisms driving hemolysis. Employing numerical simulations alongside a variety of hemolysis models, the goal is to identify the specific flow conditions that are most conducive to hemolysis occurrence. The modelling process commences with the depiction of blood flow traversing a sudden contraction, representing the syringe's action, assuming both Newtonian and non-Newtonian fluid models. The scope extends further to encompass a three-dimensional setup, incorporating a representative syringe, and hypodermic needle. The time history of shear stress along streamlines is then extracted to facilitate hemolysis analysis and comparisons are made to published experimental results.

Keywords: Hemolysis, Blood flow, Fluid modelling, Numerical simulation

1. Introduction

In emergency medical situations, a rapid blood transfusion may need to be administered. These rapid blood transfusions are typically administered by increasing the pressure gradient over the intravenous (IV) cannula using a pressure bag. In low resource settings, such as rural hospitals, the transfusion may be alternatively expedited through a process known as syringing, which involves drawing blood up and forcefully injecting it into the patient. However, this process can cause physical damage to the red blood cells (RBCs) and cause hemolysis (or breaking open of RBCs), which can have negative effects on the recipient of the transfusion, such as hemolytic anemia, and the inactivation of nitric oxide which may result in thrombosis, and smooth muscle dystonias [1], [2]. It has been shown that syringing causes significantly higher levels of hemolysis than the use of a pressure bag [3].

To investigate hemolysis resulting from blood flow through a syringe, this study aims to model blood flow and use numerical models to calculate hemolysis. Under high shear rates, in large vessels such as the aorta, blood has been modelled as a Newtonian fluid [4]. However, blood has also been shown to have non-Newtonian properties. For example, blood viscosity has been shown to vary with shear rate [5] and has shown dependence on the size of the vessel it is flowing through [6]. Thus, some authors model blood using non-Newtonian models, particularly when investigating flow through small vessels. Examples of non-Newtonian models which have been used include viscoelastic models, such as those within the Oldroyd family [7], the power law model, the Casson model, and the Herschel-Bulkley model [8–10].

Various numerical models have been used in combination with computational fluid dynamics (CFD) simulations to calculate hemolysis, which are described as being either stress-based or strain-based. Stress-based models tend to be empirical models which relate hemolysis to the shear rate, whereas strain-based models include various parameters, such as red blood cell deformation, sometimes in combination with each other to calculate hemolysis. Approaches to modelling hemolysis can be divided into those using a Lagrangian particle tracking approach and those that make use of Eulerian transport equations. Lagrangian models calculate haemolysis along pathlines, or along streamlines for steady-state simulations, while Eulerian models calculate hemolysis concurrently to solving the continuity and Navier-Stokes equations as the relevant transport equations are implemented in a CFD solver [11], [12]. Eulerian methods generally require more computational memory and time than Lagrangian methods which can be implemented in a post-processing step. However, Lagrangian methods tend to be less accurate than Eulerian methods, since the results obtained from Lagrangian methods are dependent on how well the pathlines capture the domain. Thus, Lagrangian methods for modelling hemolysis are best suited for simple flow problems where there is minimal re-circulation [11].

2. Methodology

Fluid simulations were carried out using both Ansys Fluent and the open source software deal.II [16]. All simulations assumed laminar, steady flow with zero body forces, and blood was assumed to be an incompressible fluid with constant density, ρ.

The Navier-Stokes equations implemented in deal.II are given by

$$\begin{aligned} \rho(\mathbf{u}\cdot\nabla)\mathbf{u} - \nabla\cdot(2\eta\nabla^s\mathbf{u}) + \nabla p &= \mathbf{0} && \text{in } \Omega,\\ \nabla\cdot\mathbf{u} &= 0 && \text{in } \Omega,\\ \eta_1 + \eta_2[\kappa + (\lambda\dot{\gamma})^a]^{\frac{n-1}{a}} &= \eta && \text{in } \Omega,\\ \mathbf{u} &= \mathbf{u}_0 && \text{on } \Gamma_D, \end{aligned} \tag{1}$$

where $\nabla^s\mathbf{u} = \frac{1}{2}[\nabla\mathbf{u} + (\nabla\mathbf{u})^T]$, is the symmetric part of the velocity gradient, $\dot{\gamma} = 2\sqrt{\frac{1}{2}\nabla^s\mathbf{u}:\nabla^s\mathbf{u}}$, is the shear rate and η is the viscosity. Equation $(1)_c$ is a generic viscosity equation used by Pacheco et al. [17], which allows for the implementation of different viscosity models by varying the parameters η_1, η_2, κ, λ, a, and n respectively. For example, when $\eta_1 = \kappa = 0$, Equation $(1)_c$ represents the power-law model, and the Newtonian model can be implemented by setting $\eta_2 = 0$.

The weak form of Equations $(1)_a$, $(1)_b$, and $(1)_c$ was derived and the finite element formulation was implemented in deal.II. The Newton-Raphson method was used to find solutions for the velocity, pressure, and viscosity fields, $\mathbf{u}, p$, and η.

Fluid simulations were also carried out in Ansys Fluent using the pressure-based solver together with the second order upwind scheme.

The percentage hemolysis values were calculated in a post-processing step in MATLAB [18] using velocity gradient data exported along streamlines generated in ParaView [19] and CFD-Post using the Runge-Kutta method for the deal.II and Ansys Fluent simulations, respectively. The shear stress values are calculated using the velocity gradients using the following equation:

$$\tau = 2\eta(\dot{\gamma})\sqrt{\frac{1}{2}\nabla^s\mathbf{u}:\nabla^s\mathbf{u}}, \tag{2}$$

where the apparent viscosity, $\eta(\dot{\gamma})$, is calculated using Equation $(1)_c$.

Following Chen et al. [20] the hemolysis is calculated by integrating a hemolysis power-law (HPL) model over the length of a streamline,

$$H(\tau, t) = \sum_j C\tau_{i,j-1}^{\alpha}\left(t_{i,j} - t_{i,j-1}\right)^{\beta}, \tag{2}$$

where H represents the percentage ratio of free hemoglobin present to the total hemoglobin $(\frac{\mathrm{Hb_{free}}}{\mathrm{Hb_{total}}} \times 100)$, $\tau_{i,j-1}$ is the stress corresponding to step $j - 1$ on streamline i, and $\left(t_{i,j} - t_{i,j-1}\right)$ is the exposure time for which the blood is subjected to the stress, $\tau_{i,j-1}$.

Three different sets of values for the parameters C, α, and β were experimentally derived by Giersiepen et al. [13], Heuser and Opitz [14], and Zhang et al. [15], respectively. The Giersiepen et al. [13] parameter set has been shown to overpredict hemolysis. However, both the Heuser and Opitz [14] and the Zhang et al. [15] parameter sets have produced results comparable to experimental results in relevant literature [11].

For this study, Equation (3) was used to calculate percentage hemolysis in two different ways. The first assumed that any amount of shear stress would cause hemolysis, thus, hemolysis is calculated over the entirety of every streamline ($\tau_s = 0$). The second, involves checking whether the sum of the preceding shear stresses exceeds the specified threshold shear stress, τ_s, resulting in haemolysis only being calculated once the cumulative shear stress along a streamline exceeds the threshold shear stress ($\sum_{k=1}^{j-1} \tau_{i,k} > \tau_s$). The threshold shear stress was set to 250 Pa [11].

3. Results and discussion

Blood flow through a two-dimensional contraction was investigated first to test the simulations on a simple domain. The dimensions, which can be seen in Figure 1, were chosen to approximately represent a 20 ml syringe and 18-gauge needle. Different meshes were used for the deal.II and Ansys Fluent simulations based on the computational costs, such as runtime, and desired accuracy for deal.II and Ansys Fluent respectively. Due to post-processing difficulties in deal.II, the two-dimensional power law simulation and the three-dimensional simulations were only carried out in Ansys Fluent.

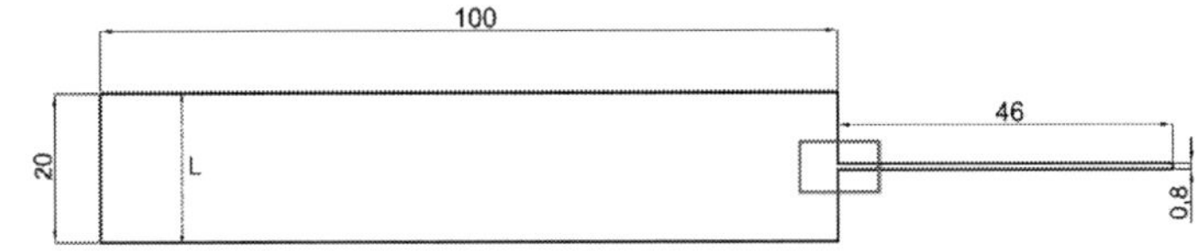

Figure 1 A simple, two-dimensional representation of a 20 ml syringe and 18-gauge needle with dimensions shown in ml. The block indicates the area of focus for future figures.

An inlet velocity of 0.03 m/s was implemented, which corresponds to a Reynolds number of 159 for the Newtonian simulation and 17.57 for the power law simulation. The Reynolds number for this study was calculated using the following formula:

$$\mathrm{Re} = \frac{\rho u L}{\left[\eta_1 + \eta_2\left(\kappa + \left(\lambda \frac{u}{L}\right)^{a}\right)^{\frac{(n-1)}{a}}\right]}. \tag{4}$$

Ten thousand streamlines were generated, in a post-processing step in ParaView [19] and CFD-Post, from the inlet of the domain over which the percentage hemolysis was calculated.

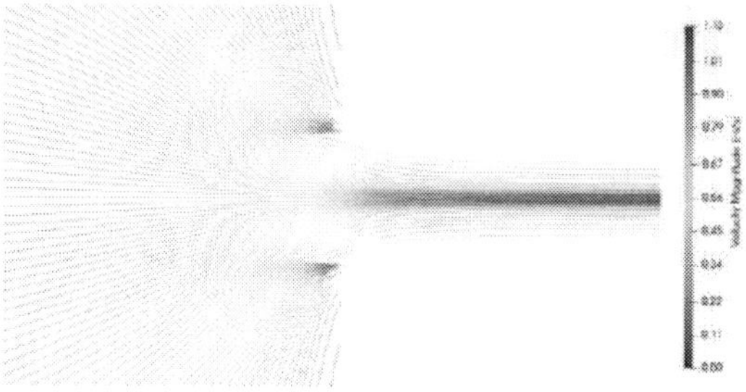

Figure 2 Representative streamlines, coloured by velocity magnitude, which were generated in ParaView from the Newtonian deal.II solution.

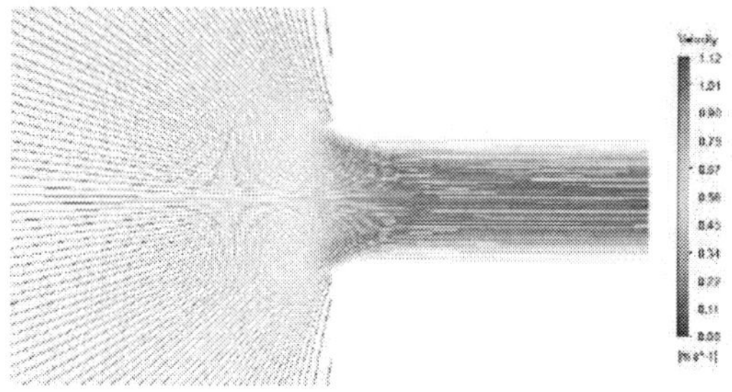

Figure 3 Representative streamlines, coloured by velocity magnitude, which were generated in CFD-Post from the Newtonian Ansys Fluent solution.

Representative streamlines from both the deal.II and the Ansys Fluent simulations are shown in Figures 2 and 3, respectively. The Newtonian and power law results are summarised in Table 1.

Table 1 Maximum percentage hemolysis results obtained from the two-dimensional sudden contraction simulations carried out in deal.II and Ansys Fluent.

			Giersiepen et al.	Heuser and Opitz	Zhang et al.
deal.II	Newtonian $\tau_{max} = 69.75$ Pa	$\tau_s = 0$	0.01750%	0.0002878%	0.004490%
		$\sum_{k=1}^{j-1} \tau_{i,k} > \tau_s$	0.01202%	0.0002227%	0.003611%
Ansys Fluent	Newtonian $\tau_{max} = 54.26$ Pa	$\tau_s = 0$	0.06173%	0.0009586%	0.01309%
		$\sum_{k=1}^{j-1} \tau_{i,k} > \tau_s$	0.06145%	0.0009537%	0.01301%
	Power law $\tau_{max} = 16.46$ Pa	$\tau_s = 0$	0.007958%	0.0001826%	0.002431%
		$\sum_{k=1}^{j-1} \tau_{i,k} > \tau_s$	0.007490%	0.0001722%	0.002279%

Despite the deal.II simulation producing a higher τ_{max} than the Ansys Fluent simulation, the percentage hemolysis values calculated from the Ansys Fluent simulation were consistently larger than the deal.II percentage hemolysis values. This may be a result of gradient smoothing arising due to a coarser mesh being implemented in deal.II.

The effect of the cumulative threshold shear stress method on the results can be seen in Table 1. The largest difference in results between the two methods can be seen in the deal.II results, however, this could also have been influenced by gradient smoothing. Comparing the Newtonian and power law Ansys Fluent results indicate that the difference in results between the two methods will be smaller the larger the shear stresses are.

A second two-dimensional simulation was carried out with an inlet velocity of 0.1 m/s, however, the streamlines generated from the deal.II simulation failed to sufficiently capture the domain which negatively impacts the accuracy of the percentage hemolysis results. Implementing a Eulerian-based hemolysis model would avoid the dependence of the hemolysis results on the streamlines and could be more suitable for a finite element simulation but was outside the scope of this study.

The two-dimensional domain was extended to a three-dimensional representative syringe through which flow was simulated in Ansys Fluent using both the Newtonian and the power law model. An inlet velocity of 0.1 m/s was implemented, and the number of streamlines generated was increased to 20 000, with the results listed in Table 2.

Table 2 Maximum percentage hemolysis results obtained from the three-dimensional representative syringe simulations carried out in Ansys Fluent.

		Giersiepen et al.	Heuser and Opitz	Zhang et al.
Newtonian $\tau_{max} = 694.0$ Pa	$\tau_s = 0$	70.19%	0.2508%	3.529%
	$\sum_{k=1}^{j-1} \tau_{i,k} > \tau_s$	70.19%	0.2508%	3.529%
Power law $\tau_{max} = 82.41$ Pa	$\tau_s = 0$	0.2643%	0.002512%	0.03769%
	$\sum_{k=1}^{j-1} \tau_{i,k} > \tau_s$	0.2642%	0.002511%	0.03767%

The two methods used to calculate the percentage hemolysis produced the same results for the Newtonian simulation, however, this is expected at high shear rates, since only small shear rates are excluded by the threshold shear stress method.

De Villiers et al. [3] reported a median percentage hemolysis of 1.28%, calculated by Hemolysis % $= \frac{(100-\text{Hct})\text{Hb}_{\text{free}}}{\text{Hb}_{\text{total}}}$, from a syringing experiment involving a 20 ml syringe and an 18-gauge IV cannula. Therefore, the results in Table 2 were similarly scaled by the average hematocrit reported by De Villiers et al. [3] (Hct = 52.07%) to compare the results. These results are referred to as the hematocrit scaled percentage hemolysis, or H_{Hct}.

Table 3 The maximum hematocrit scaled percentage hemolysis values from the three-dimensional Newtonian simulation, as well as the median percentage hemolysis result reported by De Villiers et al. [3].

De Villiers et al. (Median %)	Giersiepen et al.	Heuser and Opitz	Zhang et al.
1.28%	33.64%	0.1202%	1.692%

The hematocrit scaled power law results are significantly smaller than those reported by De Villiers et al. [3] and have been omitted for brevity since they are not comparable.

The Giersiepen et al. [13] parameter set overpredicts the H_{Hct} values, while the Heuser and Opitz [14] parameter set underpredicts the H_{Hct} values when compared to the results reported by De Villiers et al. [3]. Most notably, the Zhang et al. [15] parameter set shows the best agreement with De Villiers et al.'s [3] results.

Lastly, blood flow through a 16-gauge hypodermic needle was investigated. The setup used is similar to that of Chen et al. [20] and mimics blood drawn up through a needle. The needle has a total length of 45.19 mm, an inner diameter of 1.219 mm, and an outer diameter of 1.651 mm. The point of the needle is formed by a primary bevel of 14° and two secondary bevels of 45° and was surrounded by a rectangular prism domain, with dimensions 3 mm × 9 mm × 3 mm, of stationary fluid. A symmetry plane down the centre of the needle and the box was used for the CFD simulation. The meshed setup is shown in Figure 4. To compare results with Chen et al. [20], input values similar to theirs were used where possible. Specifically, a pressure drop of 44817 Pa was imposed across the needle by setting pressure inlet conditions (Γ_{p}) on the faces of the fluid box (excluding the symmetry plane) and assigning a pressure boundary condition to the outlet of the needle, and blood was modelled as a Newtonian fluid with viscosity $\mu = 0.00351$ Pa·s and density $\rho = 1050$ kg/m^3.

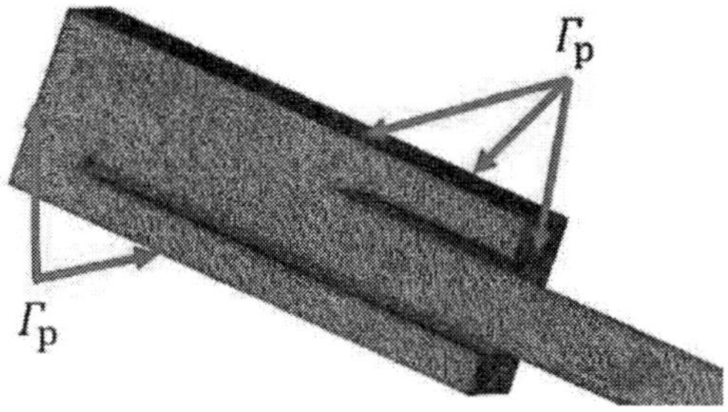

Figure 4 Meshed setup of a 16-gauge needle submerged in a fluid box.

Figure 5 Representative streamlines near the point of the needle and in the fluid box, coloured by shear rate, from the Newtonian simulation.

The percentage hemolysis results, as well as the relevant results from Chen et al. [20], are listed in Table 4. The results from the Newtonian cumulative threshold method were omitted as the results were the same as from the standard HPL method.

Table 4 Maximum percentage hemolysis results from the Newtonian and power law simulations of flow through a hypodermic needle, as well as pertinent results from Chen et al. [20].

		Experimental Result	Giersiepen et al.	Heuser and Opitz	Zhang et al.
	Chen et al.	0.029%	21.1%	0.1405%	-
Newtonian $\tau_{max} = 559.8$ Pa	$\tau_s = 0$	-	13.60%	0.1041%	1.861%
Power law $\tau_{max} = 72.98$ Pa	$\tau_s = 0$	-	0.07155%	0.0009015%	0.01538%
	$\sum_{k=1}^{j-1} \tau_{i,k} > \tau_s$	-	0.07147%	0.0008658%	0.01535%

The Newtonian simulation produced results representative of the simulation results reported by Chen et al. [20]. However, the power law simulation together with the Zhang et al. [15] parameter set produced the percentage hemolysis value which was closest to the experimental result reported by Chen et al. [20].

4. Conclusion

For all simulation cases, the Giersiepen et al. [13] parameter set produced the largest values of percent hemolysis. This is consistent with the existing literature. In comparison, the Zhang et al. [15] parameter set produced percentage hemolysis values comparable to existing experimental results both for a Newtonian simulation and for a power law simulation.

The cumulative threshold shear stress method implemented for this study did allow for 0% hemolysis to be calculated over streamlines with sufficiently low shear rates, however, the method consistently produced results that were either the same or only slightly smaller than those obtained from the standard HPL method for the cases investigated. Alternative methods which aim to capture the accumulation of damage to an RBC could capture this phenomenon better and could be explored further.

While the Lagrangian-based HPL model successfully produced results comparable to experimental results in two cases, the accuracy of one set of results obtained was negatively impacted by poor coverage of the streamlines. Implementing a Eulerian-based hemolysis model could produce more accurate results and would avoid the problems caused by streamlines failing to capture the domain sufficiently.

This study only implemented one type of hemolysis model, however implementing Eulerian models, as mentioned previously, or a more complex strain-based hemolysis model could result in more accurate results. Furthermore, investigation into how different non-Newtonian models, such as a viscoelastic model or a yield stress generalised Newtonian model, affect the resulting percentage hemolysis could also be done.

Nomenclature

Roman

a	Fitting parameter	p	Pressure (Pa)
C	Empirical constant	t	Exposure time
H	Percentage hemolysis (%)	$\mathbf{u}, u$	Velocity, speed (m/s)
H_{Hct}	Hematocrit scaled percentage hemolysis (%)	V	Volume
K	Consistency index	x, y	Cartesian coordinates
n	Behaviour index		

Greek

α, β	Empirical constants	μ	Viscosity (Pa·s)
Γ	Boundary of the domain	ρ	Density (kg/m^3)
$\dot{\gamma}$	Shear rate (1/s)	τ	Magnitude of the viscous shear stress tensor (Pa)
η	Apparent viscosity (Pa·s)	τ_s	Threshold shear stress (Pa)
η_1, η_2, κ	Fitting parameters	Ω	Domain
λ	Characteristic (relaxation) time (s)		

Subscripts

D	Dirichlet	max	Maximum value
free	Amount of free or unbound hemoglobin	min	Minimum value
i, j, k	Indices	s	Threshold
in	Inlet	total	Total amount

Abbreviations

CFD	Computational fluid dynamics	IV	Intravenous
Hb	Hemoglobin	RBC	Red blood cell
Hct	Hematocrit	Re	Reynolds number
HPL	Hemolysis power-law		

Acknowledgement

This work is based on the research supported in part by the DSI-NRF Centre of Excellence in Mathematical and Statistical Sciences (CoE-MaSS), South Africa. Opinions expressed and conclusions arrived at, are those of the author and are not necessarily to be attributed to the CoE-MaSS.

References

[1] M. Nakamura, S. Bessho *et al.*, 'Analysis of Red Blood Cell Deformation under Fast Shear Flow for Better Estimation of Hemolysis', *International Journal for Numerical Methods in Biomedical Engineering*, vol. 30, no. 1, pp. 42–54, 2014, doi: 10.1002/cnm.2587.

[2] R. P. Rother, L. Bell *et al.*, 'The Clinical Sequelae of Intravascular Hemolysis and Extracellular Plasma Hemoglobin: A Novel Mechanism of Human Disease', *JAMA*, vol. 293, no. 13, pp. 1653–1662, 2005, doi: 10.1001/JAMA.293.13.1653.

[3] W. L. De Villiers, A. A. Murray *et al.*, 'Expediting red blood cell transfusions by syringing causes significant hemolysis', *Transfusion (Paris)*, vol. 57, no. 11, pp. 2747–2751, 2017, doi: 10.1111/trf.14283.

[4] S. S. Gopalakrishnan, B. Pier *et al.*, 'Dynamics of pulsatile flow through model abdominal aortic aneurysms', *J Fluid Mech*, vol. 758, pp. 150–179, 2014, doi: 10.1017/jfm.2014.535.

[5] S. Chien, 'Shear dependence of effective cell volume as a determinant of blood viscosity', *Science*, vol. 168, no. 3934, pp. 977–979, 1970, doi: 10.1126/SCIENCE.168.3934.977.

[6] R. Fåhræus and T. Lindqvist, 'The viscosity of the blood in narrow capillary tubes', *https://doi.org/10.1152/ajplegacy.1931.96.3.562*, vol. 96, no. 3, pp. 562–568, 1931.

[7] I. G. Donev and B. D. Reddy, 'Time-dependent finite element simulations of a shear-thinning viscoelastic fluid with application to blood flow', *Int J Numer Methods Fluids*, vol. 75, no. 9, pp. 668–686, 2014, doi: 10.1002/fld.3914.

[8] E. Lakzian and P. Akbarzadeh, 'Numerical investigation of unsteady pulsatile Newtonian/non-Newtonian blood flow through curved stenosed arteries', *Biomed Mater Eng*, vol. 30, no. 5–6, pp. 525–540, 2020, doi: 10.3233/BME-191072.

[9] M. Trias, A. Arbona *et al.*, 'FDA's nozzle numerical simulation challenge: Non-Newtonian fluid effects and blood damage', *PLoS One*, vol. 9, no. 3, 2014, doi: 10.1371/journal.pone.0092638.

[10] A. H. Golshirazi, S. G. Etemad *et al.*, 'Three-Dimensional Numerical Investigation of Steady State and Physiologically Realistic Pulsatile Flow through the Left Coronary Curved Artery with Stenosis', *Theoretical Foundations of Chemical Engineering*, vol. 54, no. 3, pp. 489–499, 2020, doi: 10.1134/S0040579520030045/FIGURES/10.

[11] H. Yu, S. Engel *et al.*, 'A Review of Hemolysis Prediction Models for Computational Fluid Dynamics', *Artif Organs*, vol. 41, no. 7, pp. 603–621, 2017, doi: 10.1111/AOR.12871.

[12] M. E. Taskin, K. H. Fraser *et al.*, 'Evaluation of Eulerian and Lagrangian models for hemolysis estimation', *ASAIO Journal*, vol. 58, no. 4, pp. 363–372, 2012, doi: 10.1097/MAT.0B013E318254833B.

[13] M. Giersiepen, L. J. Wurzinger *et al.*, 'Estimation of Shear Stress-related Blood Damage in Heart Valve Prostheses - in Vitro Comparison of 25 Aortic Valves', *Int J Artif Organs*, vol. 13, no. 5, pp. 300–306, 1990, doi: 10.1177/039139889001300507.

[14] G. Heuser and R. Opitz, 'A Couette viscometer for short time shearing of blood', *Biorheology*, vol. 17, no. 1–2, pp. 17–24, Jan. 1980, doi: 10.3233/BIR-1980-171-205.

[15] T. Zhang, M. E. Taskin *et al.*, 'Study of Flow-Induced Hemolysis Using Novel Couette-Type Blood-Shearing Devices', *Artif Organs*, vol. 35, no. 12, pp. 1180–1186, 2011, doi: 10.1111/J.1525-1594.2011.01243.X.

[16] D. Arndt, W. Bangerth *et al.*, 'The deal.II Library, Version 9.4', *Journal of Numerical Mathematics*, vol. 30, no. 3, pp. 231–246, 2022.

[17] D. R. Q. Pacheco, R. Schussnig *et al.*, 'An efficient split-step framework for non-Newtonian incompressible flow problems with consistent pressure boundary conditions', *Comput Methods Appl Mech Eng*, vol. 382, p. 113888, 2021, doi: 10.1016/J.CMA.2021.113888.

[18] The MathWorks Inc., 'MATLAB version 9.10.0 (R2021a)'. https://www.mathworks.com, 2021. [Accessed: 7 September 2022].

[19] J. Ahrens, B. Geveci *et al.*, 'ParaView: An End-User Tool for Large Data Visualization', in *Visualization Handbook*, Elsevier, 2005, ISBN-13: 9780123875822.

[20] Y. Chen, T. L. Kent *et al.*, 'Testing of Models of Flow-Induced Hemolysis in Blood Flow Through Hypodermic Needles', *Artif Organs*, vol. 37, no. 3, pp. 256–266, 2013, doi: 10.1111/J.1525-1594.2012.01569.X.

COST EFFECTIVE BALLISTIC PROTECTION FOR VEHICLES

John D. Jones[1,*], J. David Reinecke[2] and Lwazi Pambuka[1]
[1]School of Mechanical, Industrial and Aeronautical Engineering, University of the Witwatersrand, South West Engineering Building, East Campus, Braamfontein, Johannesburg
john.jones@wits.ac.za
[2]Landward Sciences Impact Area, Defence and Security Cluster, CSIR
Building 45. CSIR Campus, Meiring Naude Drive, Brummeria
DReinecke@csir.co.za

Abstract
In the design of protected vehicles there is a constant trade-off between mobility, protection, and cost. To protect against increasing threat levels, designers are usually required to use armour materials with increased mass and thickness. However, this has a negative effect on the vehicle's mobility. Reducing the mass of armour plates for the same level of protection usually requires the use of more expensive materials, thus increasing the cost. The aim of this project was to investigate how the areal density of armour plates, used for vehicle protection against a NATO Level 3 ballistic threat, can be reduced whilst still maintaining the required level of threat protection, and optimising cost by exploring varying material layers in a composite armour plate assembly. This work used computational modelling to evaluate protection capabilities of various combinations of lower cost materials that were then manufactured and tested. The test plate combination initially selected were based on the published computational work of Rahman et al., [2]. These proposed, multilayered, plates computationally provided a reduction in aerial density of 12% compared to equivalent homogeneous amour steel plate. Additional plate combinations, using Strenx 700E Al-7075-T6 with Kevlar and Dyneema layers, were proposed and computationally evaluated and assessed. These multilayered plates were then manufactured and subjected to ballistic tests against NATO level 3 (7.62 x 51 mm Tungsten Carbide (WC)) armour piecing rounds. None of the proposed and computationally verified plates provided the required ballistic protection. The main reason for this is ascribed to the application and use of only published material parameters and the implementation of the failure model.

Keywords: Ballistic limit, 7.62 mm bullets, Johnson-Cook, Recht-Ipson

1. Introduction

In recent times, there has been an increased interest in using multilayered plates for the protection of structures from ballistic and explosive threats [1–12]. The advantages of using multilayered plates are that the protection can be optimised to utilise the strengths of the different materials to prevent perforation, damage, and shock, whilst potentially reducing the density and mass of the overall plate. There may also be significant benefits in the manufacture and repair [11] of multilayered plates as opposed to their monolithic counterparts, especially as the monolithic thick armour plates are not always manufactured to the specified nominal dimensions [3,8]. There is also a continual drive to reduce the cost of protection provided by armour plates as high quality, monolithic armour plate is significantly more expensive than structural steels. This has specific relevance in the design of protected vehicles where there is a constant competing trade-off between mobility, protection, and cost.

Several researchers have investigated the performance of single monolithic plates versus layered plates. However, the literature shows that there is still some debate as to whether a layered plate configuration can offer equal, or better, protection to an equivalent thickness monolithic plate, noting that this would be comparing the plates performance based on a specific monolithic plate thickness to provide protection against a specific threat level as detailed in [13,14].

Dey *et al.*, [6] investigated the ballistic performance of single and double layer Weldox 700E plates impacted by blunt and ogival projectiles at sub ordnance velocities. Their findings showed that for a total target thickness of 12 mm, the double layer target plate performed better than the single layer monolithic target plate when impacted with a blunt projectile. However, for an ogival projectile, the performance of the layered plates was slightly worse than its monolithic counterpart. This research highlighted that the type of projectile can affect the protection level offered by a specific target plate(s) as well as the fact that the failure mechanism can change, depending on the target plate configuration. These findings have been numerically verified for double layered plates of the same metal in [11] as well as for layered plates of different materials in [9].

In stark contradiction, Wei *et al.* [5] showed that the ballistic performance of a monolithic plate target was better than its layered counterpart when impacted with a blunt projectile which is consistent with the findings in [7,10] for sub-ordinance velocities. As such, there is still a question as to whether using layered plates, of the same or differing material types, offers any benefit in terms of ballistic performance compared to an equivalent thickness monolithic plate, particularly for impact velocities above 500 m/s. It is evident, however, that using layered plates, and the order in which the plates are layered, can affect the failure mechanism of the plates [5,9] and the ballistic performance.

As such, layered armour plates, specifically using different materials, are currently an interesting area of research as they can offer significant weight and / or cost savings. Rahman *et al.* [2] numerically investigated the penetration resistance of different plate configurations against a 7.62 APM2 round at various velocities, to determine whether any of the plate configurations could meet the ballistic limit of 930 ± 20 m/s specified in [13,14] for a level 3 threat, whilst reducing the overall areal density of the plate configurations, when compared to a monolithic armoured plate of 25 mm thickness. The authors investigated using high strength steel and Al7075 T6 layered plate configurations to stop the specified projectile. These computational results by Rahman *et al.* [2] showed that the areal density could be reduced by 20% and still meet the ballistic limit requirements, with triple layered plate configurations performing better than the equivalent areal density double-layered plates.

The work reported in this paper aimed to build on the computational work of Rahman *et al.* [2] and determine whether double and triple layered plates, constructed from lower cost and lower density materials, can be used to meet the NATO Level 3 requirements [13,14] for both a 7.62 APM2 and 7.62 x 51 mm APWC bullet. Both experimental and computational analysis have been used, with the computational models based solely on material data available in published literature. The experimental method applied was to validate the computational threat penetration model, design and test selected plate combinations based on computational results and additional post-test computational modelling with the WC ballistic threat.

2. Validation of Numerical Modelling

To validate the ballistic model of a 7.62 APM2 round impacting a target plate, literature data [3,4,12] was used to simulate a 7.62 APM2 round impacting a monolithic 12 mm Weldox 700E target and a monolithic 20 mm Al7075 T651 target, at various velocities, however bullet rotational velocity was not considered following literature [3,4,12].

A finite element model (FEM) of the 7.62 APM2 projectile was created using quarter symmetry in Ansys Explicit Dynamics, as shown in Figure 1, with a mesh size of 0.4 mm being used for both the projectile and the region of interest on the target plates. As far as possible, only 8 node hexahedral elements were used for the meshing. The mesh size was determined by running convergence studies on both the plate and the projectile.

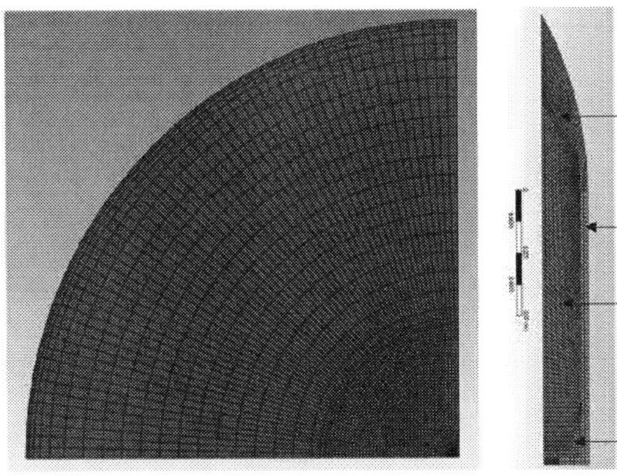

Figure 1: Quarter symmetry model of 7.62 APM2 bullet and target plate.

The Johnson-Cook (JC) constitutive model [15] presented in Equation (1) was used to model the material behaviour of both the bullet and the target plates.

$$\sigma = [A + B\varepsilon^n][1 + C ln\dot{\varepsilon}^*][1 - T^{*m}] \quad (1)$$

In Equation (1), σ is the von Mises flow stress, A is the yield stress, B and n represent the effects of strain hardening, ε is the equivalent plastic strain, $\dot{\varepsilon}^*$ is the dimensionless plastic strain at a specific strain rate $\dot{\varepsilon_0}$, T^* is homologous temperature and C, m are material constants. The material properties used in the validation models are shown in Table A1, Appendix A.

For the target plates, the JC failure model [16] in Equation (2) was used to model the material failure of both the bullet and the target plates.

$$\varepsilon^f = [D_1 + D_2 e^{(D_3\sigma^*)}][1 + D_4 ln\dot{\varepsilon}^*][1 + D_5 T^*] \quad (2)$$

In Equation (2), ε^f is the fracture strain, D_1 - D_5 are material constants and σ^* is the dimensionless pressure-stress ratio. The material constants used for the failure model are shown in Table A2, Appendix A. The material and failure parameters used for the Weldox 700E plate were taken from [12,17] whereas the properties for Al7075 T651 were taken from [18] as the JC failure parameters were not available in the original papers [3,4,12] as a different failure criterion was used.

In ballistic testing and simulation, the ballistic limit (v_{bl}) is important as it gives an indication of the maximum velocity that a protective plate can stop for a specified bullet type. A common way of predicting the v_{bl}, based on experimental work, is using the Recht-Ipson model [19] in Equation (3).

$$V_r = a(V_i^P - V_{bl}^P)^{\frac{1}{P}} \tag{3}$$

In Equation (3), V_r is the residual velocity, a is constant which is set to 1 when no plugging occurs during penetration, V_i is the impact velocity and P is a fitting constant.

Using in Equation (3) for determining the v_{bl}, simulations were run by impacting both the 12 mm Weldox 700E plate and the 20 mm Al7075 T651 plate with a 7.62 APM2 bullet at increasing velocities until the v_{bl} could be determined. These results were then compared to the experimental and Recht-Ipson values from literature [3,4,12], and presented in Figure 2. For the Weldox 700E target plate results shown in Figure 2A, the difference in predicted and experimental v_{bl} value was 2.1%. Note also the small difference between the P value for the Recht-Ipson fits to the experimental data from literature [3] and the simulation. The Aluminium 7075 T651 target plate results in Figure 2B show a larger disparity with a predicted v_{bl} 8.3% higher than the v_{bl} determined using the experimental data from [4]. This disparity is also reflected in the P values. Given that previous results showed an overestimation of 4% for the 12 mm Weldox 700E Plate and 11% for the 20 mm Al 7075 T651 plate [3], these results were considered acceptable validation of the combined bullet and target model.

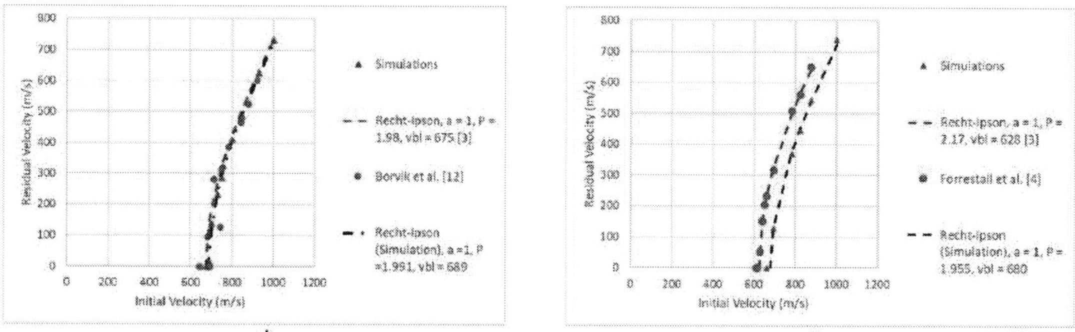

A B

Figure 2: Comparison of experimental results [3,4,12] and simulation results for a 7.62 APM2 projectile impacting (A) a 12 mm Weldox 700E plate and (B) a 20 mm Al 7075 T651 plate.

3. Design of Target Plates

Using the results of the work in [2] and the authors validated 7.62 APM2 and plate models, the eight different panel configurations shown in Table 1 were simulated at an impact velocity of 930 m/s from a 7.62 APM2 bullet. The protection plates comprised a combination of lower cost armoured steel Strenx 700F, Aluminium Al7075 T6 and Dyneema® resin matrix panels. The predicted potential reduction in areal density is shown in Table 1 and is based on the specified areal density of 196.5 kg/m^2 for a 25 mm monolithic plates specified in [2]. These plate configurations offered significant weight and cost savings as lower cost steel and less dense materials were chosen. For these targets, the material properties and the failure constants used in the simulations were taken from literature, as experimental data was not available at the time of writing, and are shown in Table A2 and Table A3 noting that the aluminium material and related failure model parameters differed from the validation model.

The image results of these simulations for all combinations listed in Table1 showing the final position of the projectile are shown in Figure 3. Only plate combinations B4 and C1 did not stop the bullet at 930 m/s, and full penetration occurred. For plate combination A2, A3 and B3, partial penetration occurred but the bullet was still effectively stopped. Based on these results, and despite the predicted perforations it was decided to physically manufacture and test all these panel configurations.

Table 1: Description of target plates showing the predicted reduction in areal density for each plate configuration.

Code	Description	Layers	Bonded	Areal density (kg/m^2)	Reduction in areal density
A1	16 mm Strenx 700E-10 mm Al7075 T6	2	No	145.14	26.1%
A2	8 mm Strenx 700E-10 mm Al7075 T6-8 mm Strenx 700E	3	No	146.44	25.5%
A3	8 mm Strenx 700E-10 mm Al7075 T6-8 mm Strenx 700E	3	Yes	146.44	25.5%
B1	16 mm Strenx 700E-9 mm Dyneema	2	Yes	134.40	31.6%
B2	8 mm Strenx 700E-9 mm Dyneema-8 mm Strenx 700E	3	Yes	134.40	31.6%
B3	16 mm Strenx 700E-9 mm Aramid	2	Yes	136.67	30.4%
B4	8 mm Strenx 700E-9 mm Aramid-8 mm Strenx 700E	3	Yes	136.67	30.4%
C1	8 mm S355JR-10 mm Al7075 T6-8mm S355JR	3	No	150.90	23.2%

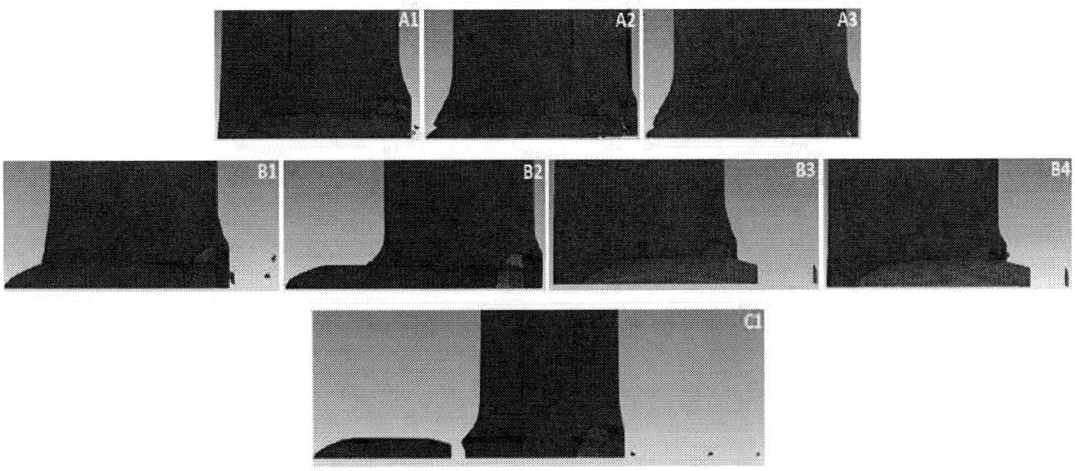

Figure 3: Images of simulations of designed target plates, Table 3, versus 7.62 APM2 bullet showing if the bullet was stopped (A1, B3) at 930 m/s or if the plates were partially penetrated (A2, A3, B1) or fully penetrated (B2, C4).

4. Ballistic Testing

All the plate configurations shown in Table 1 were manufactured and then tested by ARMSCOR. During testing, only the 7.62 x 51 mm Armour Piercing Tungsten Carbide (APWC) bullet was available, and only one velocity was used, namely 930 ± 20 m/s. The target plates were manufactured to dimensions of 300 mm x 300 mm and a maximum of 3 shots were allowed per target plate. The target plates were placed in a frame, 30 m from the barrel. During testing, the V_i and V_r were determined by using Tema Pro T2023a-64 software to analyse the test videos recorded using a Photron SA4 high-speed camera. The average results of the ballistics tests carried out are shown in Table 2.

As shown in Table 2, at a nominal impact velocity of 930 ± 20 m/s not one of the panel configurations was able to stop the 7.62 x 51 mm APWC round. Interestingly, for target plates A1, A2 and A3, the ballistic performance of the triple layered plates was better than that of the double layered plate, with the bonded triple layered plate having the lowest average residual velocity value. In contrast, when composite materials were used instead of the Al7075 T6, the double layered plates performed better than the triple layered plates.

Table 2: Average results of ballistic testing.

Target Plate	Average Impact Velocity (m/s)	Average Residual Velocity (m/s)	Number of Shots
A1	912.9	461.0	6
A2	928.4	452.1	6
A3	915.3	428.8	3
B1	917.9	565.6	3
B2	915.7	596.0	3
B3	929.5	585.1	3
B4	923.0	635.0	3
C1	917.0	493.0	3

Inspection of the plates post-impact revealed that the Al7075 T6 material exhibited a significantly different failure behaviour when sandwiched between two Strenx 700E (A2, A3) plates to being placed at the unsupported rear layer of the target plate (A1). The exit material failure appears brittle on target plate A1 as opposed to the more ductile failure features evident on the sandwiched A2 and A3 configurations. These exit hole material failures are shown in Figure 4.

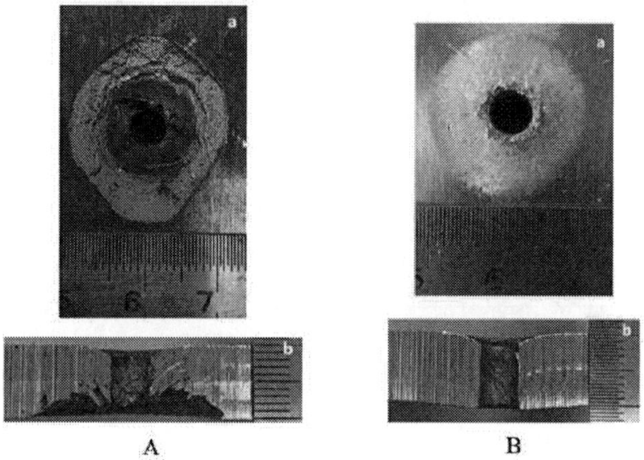

Figure 4: Aluminium plate exit hole images from target plate (A) A1 and (B) A2 plate configuration showing the (a) plate surface and (b) plate section.

5. 7.62 x 51 mm APWC simulations

On completion of the experimental testing, it was clear that the modelled 7.62 APM2 bullet penetration performance at 930 m/s impact velocity is not similar to the 7.62 x 51 mm APWC bullet at the same impact velocity. Further simulations were conducted to determine whether numerical modelling could be used to predict the experimental results for the 7.62 x 51 mm APWC bullet. The same plate material properties shown in Tables A1, A2 and A3 were used, however the bullet needed to be re-modelled. A 7.62 x 51 mm bullet was weighed and sectioned and the actual measured geometry of the bullet was used, as far as possible, to create the new bullet model, see Figure 5. As with the 7.62 APM2 model, a mesh size of 0.4 mm was used for the projectile and plates and, as far as possible, only 8 node hexahedral elements were used. The JC material properties used for the WC projectile were taken from [23] and are shown in Table A4, noting that the density was reduced slightly to match the measured mass of the projectile. A standard Ansys Explicit Dynamics material model was used for the aluminium sabot. As before, rotational velocity of the bullet was omitted.

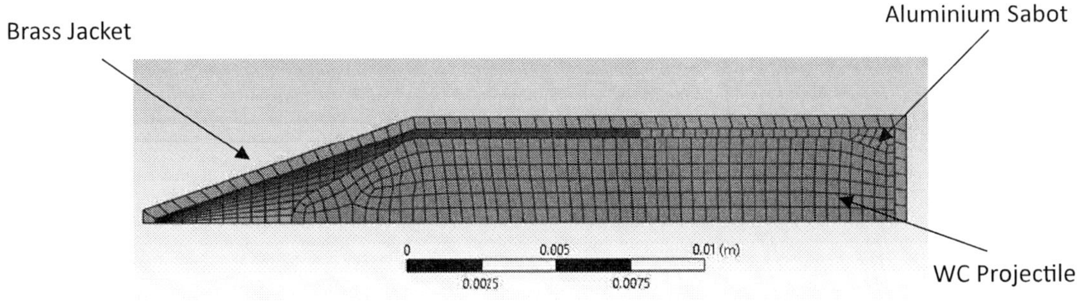

Figure 5: Meshed quarter symmetry model of 7.62 x 51 mm bullet

Simulations were run on each plate configurations listed in Table 1, at different impact velocities to enable the determination of the v_{bl} using the Recht-Ipson formulation [19]. Once the Recht-Ipson parameters had been determined using the Matlab® Curve Fitting Toolbox [24], these parameters were used to estimate the v_{bl} of the designed plate configurations using the actual impact velocities. The results of this approach are shown in Figure 6 for the A1, A2 and A3 and B4 plate configurations. The difference between the simulated v_{bl} and estimated experimental v_{bl} values for these plate configurations were 14.4%, 5%, 8% and 76.9% respectively. A summary of the Richt-Ipson curves for the test and experimental results for all the plate configurations tested are shown in Table 3. For plates A2 and similarly for A3, the simulations emulated the more ductile deformation and failure the test plates underwent, whereas for plate A1, the simulation was unable to capture the brittle nature of the plate exit failure, as shown in Figure 7 and Figure 8.

Table 3: Summary of simulated and estimated v_{bl} values based on 7.62 x 51 mm APWC bullet impact simulations.

Code	Simulated v_{bl}	Estimated v_{bl}	% Difference
A1	835	730	14.4%
A2	800	760	5.3%
A3	810	750	8.0%
B1	830	650	27.7%
B2	800	618	29.4%
B3	1048	620	69.0%
B4	920	630	76.9%
C1	750	690	8.7%

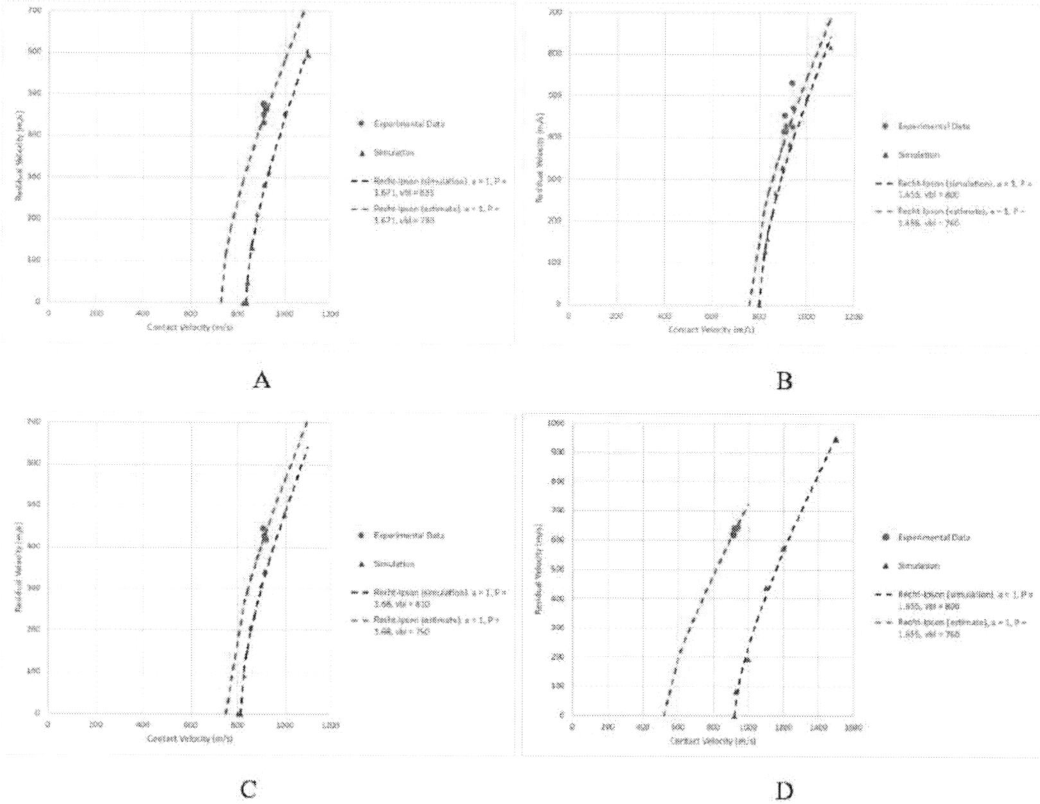

Figure 6: Richt-Ipson plots of the simulations for plates (A) A1, (B) A2, (C) A3 and (D) B1 showing the simulation data, the fitted Recht-Ipson model as well as the experimental data and estimated v_{bl} of the plate configuration.

Figure 7: Comparison of plate damage and deformation evident on sectioned plate (A2) versus simulation.

Figure 8: Comparison of plate damage and deformation evident on sectioned plate (A1) versus simulation.

6. Discussion

It is evident from the comparison of the literature, experimental and simulated results, presented as Richt-Ipson plots shown in Figure 6, that the computational models for the Weldox 700E plate were significantly more accurate than those for the Al7075 T651. This is largely attributed to the use of literature data for the JC constitutive material properties as well as the JC failure model for Al7075 T651 that were not the same as the experimental results reported in [24]. There is also a significant question about the suitability of the JC failure model for brittle materials, particularly as it is more commonly used for ductile materials [25]. This is particularly relevant for the Al7075 T651 plate which may be significantly more brittle than the Weldox 700E plate. The results also raise the issue of whether the computational models accurately represent tensile failures at the free surfaces under impact loading. The accuracy of these models may also be improved if a failure model were used on the projectile material. Here again, when efforts were made to use the JC failure criterion on the WC projectile, the simulations generally ended with the projectile squashing into the target rather than penetrating it which once again brings into question the suitably of this failure model for these types of hard brittle materials.

The simulations of the 7.62 APM2 bullet impacting the designed plate configurations were consistent with the results shown in [5, 9] as the majority of the plate configurations stopped the projectile at 930 m/s with little or no penetration. The question over whether a double plate is better than a triple plate is still open for further investigation as the experimental results shown in this paper indicate that it largely depends on the materials used, the specific layering sequence and the failure mechanisms that occur in the individual plates. In this regard, these findings are consistent with the literature [23–25]. Evidently further research is required to fully understand how the failure mechanism of a plate changes depending on its position in the overall plate configuration and how this would affect the protection offered by different plate configurations.

What is evident by comparison of the simulation results for the 7.62 APM2 and 7.62 x 51 mm APWC is that the WC bullet offers a significantly greater penetration threat than its APM2 counterpart. This can be seen on the APM2 simulations that showed that at 930 m/s the plate configurations would be affective in stopping the bullet whereas, for the WC simulations and experimental results, at 930 m/s, all the plate configurations were easily penetrated.

Comparison of the 7.62 x 51 mm APWC simulations with the experimental results showed that all the simulations significantly underestimated the protection offered by the designed plate configurations. This is largely attributed to the material and failure models used in the computational model for the plate materials as well as the WC projectile. Specifically for the plates, it is evident that the failure behaviour of the aluminium plate changes depending on where in the plate configuration it is placed and, as has been shown, this behaviour has not been captured by the material and failure models used in the simulations.

With regard to the WC projectile, as has been reported in the literature, measuring the high strain rate properties and characterising the WC projectile materials, and then creating a verified material and failure model that captures its behaviour adequately, is an area of ongoing research and requires more investigation [23,25,26].

The simulation results for the metal-metal layered plates were useful for design purposes and can be used for future plate configuration developments as, particularly for the triple layered plates, the difference between the experimental and simulated results is acceptable. However, the plate configurations used in this work were ineffective at stopping the WC bullet. The results bring into question whether structural steel pates should be used on the impact face of ballistic protection and the suitability of Aluminium for this type of bullet. Similarly, using Dyneema and Kevlar panels in these configurations proved particularly unsuitable for ballistic protection against this type of bullet. In addition, it is evident from the simulation results for the Dyneema and Kevlar plates that the material model used in this work for these materials were not suitable.

7. Conclusion

This work has gone a long way to improving our understanding as to how, metal plates respond to bullet impact and provide ballistic protection. The use of computational simulations is clearly a viable option for future plate designs by providing protection improvement trends with multilayer panels, however, it is evident that the material and failure models need to be improved, and that, ideally, the properties of the bullets and the plate models should be experimentally extracted.

Based on this initial computational and test work it is believed that multi-layered plates can be used to reduce the areal density of protection plate configurations, however, it is questionable as to whether structural materials will be suitable, which will have a significant cost implication, especially when trying to stop the WC bullet. As ever, the challenge remains to reduce the weight and cost of the plates, whilst still meeting the required protection level.

Nomenclature

ε	equivalent plastic strain
ε^{f}	fracture strain
$\dot{\varepsilon}^{*}$	dimensionless plastic strain at a specific strain rate
$\dot{\varepsilon}_0$	strain rate
σ	von Mises flow stress
σ^{*}	dimensionless pressure-stress ratio
a	constant
A	yield stress
B	strain hardening parameter
C	material constant
D_1 - D_5	material constants
P	fitting constant.
m	material constant
n	strain hardening parameter
T^{*}	homologous temperature
v_{bl}	ballistic limit
V_r	residual velocity
V_i	impact velocity

Abbreviations

Al	Aluminium
JC	Johnson-Cook
WC	Tungsten Carbide

References

[1] K. Senthil and M. A. Iqbal, "Prediction of superior target layer configuration of armour steel, mild steel and aluminium 7075-T651 alloy against 7.62 AP projectile," *Structures*, vol. 29, pp. 2106–2119, Feb. 2021.

[2] N. A. Rahman, S. Abdullah, W. F. H. Zamri, M. F. Abdullah, M. Z. Omar, and Z. Sajuri, "Ballistic Limit of High-Strength Steel and Al7075-T6 Multi-Layered Plates Under 7.62-mm Armour Piercing Projectile Impact," *Latin American Journal of Solids and Structures*, vol. 13, no. 9, pp. 1658–1676, Sep. 2016.

[3] E. A. Flores-Johnson, M. Saleh, and L. Edwards, "Ballistic performance of multi-layered metallic plates impacted by a 7.62-mm APM2 projectile," *International Journal of Impact Engineering*, vol. 38, no. 12, pp. 1022–1032, Dec. 2011.

[4] M. J. Forrestal, T. Børvik, and T. L. Warren, "Perforation of 7075-T651 Aluminum Armor Plates with 7.62 mm APM2 Bullets," *Experimental Mechanics*, vol. 50, p. 1245, 2010.

[5] Z. Wei, D. Yunfei, C. Z. Sheng, and W. Gang, "Experimental investigation on the ballistic performance of monolithic and layered metal plates subjected to impact by blunt rigid projectiles," *International Journal of Impact Engineering*, vol. 49, pp. 115–129, Nov. 2012.

[6] S. Dey, T. Børvik, X. Teng, T. Wierzbicki, and O. S. Hopperstad, "On the ballistic resistance of double-layered steel plates: An experimental and numerical investigation," *International Journal of Solids and Structures*, vol. 44, no. 20, pp. 6701–6723, Oct. 2007.

[7] N. K. Gupta, M. A. Iqbal, and G. S. Sekhon, "Effect of projectile nose shape, impact velocity and target thickness on the deformation behavior of layered plates," *Int. J. Impact Eng.*, vol. 35, no. 1, pp. 37–60, Jan. 2008.

[8] J. A. Zukas and D. R. Scheffer, "Impact effects in multilayered plates," *International Journal of Solid Structures*, vol. 38, pp. 3321–3328, 2001.

[9] X. Teng, T. Wierzbicki, and M. Huang, "Ballistic resistance of double-layered armor plates," *International Journal of Impact Engineering*, vol. 35, no. 8, pp. 870–884, Aug. 2008.

[10] J. Radin and W. Goldsmith, "Normal Projectile Penetration and Perforation of Layered Targets," *International Journal of Impact Engineering*, vol. 7, no. 2, pp. 229–259, 1988.

[11] X. Teng, S. Dey, T. Børvik, and T. Wierzbicki, "Protection performance of double-layered metal shields against projectile impact," *Journal of Mechanics of Materials and Structures*, vol. 2, no. 7, pp. 1309–1329, Sep. 2007.

[12] T. Børvik, S. Dey, and A. H. Clausen, "Perforation resistance of five different high-strength steel plates subjected to small-arms projectiles," *International Journal of Impact Engineering*, vol. 36, no. 7, pp. 948–964, Jul. 2009.

[13] Nato Standardization Agency, "Procedures for evaluating the protection level of armoured vehicles-kinetic energy and artillery threat," NATO AEP-55-vol 1 (Ed2), 2011.

[14] Nato Standardization Agency, "Stanag 4569 (Edition 2): Protection Levels for Occupants of Armoured Vehicles," AC/225-D(2011)0012 (PFP), 2011.

[15] G. R. Johnson and W. H. Cook, "A constitutive model and data for metals subjected to large strains, high strain rates, and high temperatures," presented at the Proceedings 7th International Symposium on Ballistics, The Hague, 1983.

[16] G. R. Johnson and W. H. Cook, "Fracture characteristics of three metals subjected to various strains, strain rates, temperatures and pressures," *Eng. Fract. Mech.*, vol. 21, no. 1, pp. 31–48, Jan. 1985.

[17] S. Dey, T. Børvik, O. S. Hopperstad, and M. Langseth, "On the influence of fracture criterion in projectile impact of steel plates," *Computational Materials Science*, vol. 38, no. 1, pp. 176–191, Nov. 2006.

[18] K. Senthil, M. A. Iqbal, P. S. Chandel, and N. K. Gupta, "Study of the constitutive behavior of 7075-T651 aluminum alloy," *International Journal of Impact Engineering*, vol. 108, pp. 171–190, Oct. 2017.

[19] R. F. Recht and T. W. Ipson, "Ballistic Perforation Dynamics," *Journal of Applied Mechanics*, vol. 30, no. 3, pp. 384–390, Sep. 1963.

[20] T. J. Holmquist, G. R. Johnson, and W. A. Gooch, “Modeling the 14.5 mm BS41 projectile for ballistic impact computations,” in *Projectile Impact*, WIT Press, 2014, pp. 73–86.
[21] The MathWorks Inc., *Curve Fitting Toolbox: 3.5.13 (R2021a)*. Natick, Massachusetts, United States: The MathWorks Inc., 2023.
[22] J. F. Moxnes, J. A. Teland, S. Skriudalen, S. M. Bergsrud, L. Sundem-Eriksen, and H. Fykse, “Development of material models for semi-brittle materials like tungsten carbide,” Norwegian Defence Research Establishment (FFI), FFI-rapport 2010/02225, 2010.
[23] C. Francart, “Mechanical characterization and modeling of the stress flow of a cemented tungsten carbide over wide ranges of temperatures and strain rates,” *hal-04165509*, 2023.
[24] D.-N. Zhang, Q.-Q. Shangguan, C.-J. Xie, and F. Liu, “A modified Johnson–Cook model of dynamic tensile behaviors for 7075-T6 aluminum alloy,” *Journal of Alloys and Compounds*, vol. 619, pp. 186–194, Jan. 2015.
[25] L. M. Bresciani, A. Manes, A. Ruggiero, G. Iannitti, and M. Giglio, “Experimental tests and numerical modelling of ballistic impacts against Kevlar 29 plain-woven fabrics with an epoxy matrix: Macro-homogeneous and Meso-heterogeneous approaches,” *Composites Part B*, vol. 88, pp. 114–130, Mar. 2016.
[26] M. K. Hazzard, R. S. Trask, U. Heisserer, M. Van Der Kamp, and S. R. Hallett, “Finite element modelling of Dyneema® composites: From quasi-static rates to ballistic impact,” *Composites. Part A, Applied Science and Manufacturing*, vol. 115, pp. 31–45, Dec. 2018.

Appendix A: Material Properties used in simulations.

Table A1: Johnson Cook material parameters used for validating the 7.62 APM2 model and the target plate design.

Property	Material					
	Weldox / Strenx 700E [12]	Aluminium 7075 T651 [18]	Aluminium 7075 T6 [20]	Brass Jacket [3]	Lead Tip [3]	Steel Projectile [3]
Density (kg/m^3)	7850	2700	2700	8520	10600	7850
Modulus (GPa)	210	64	64	115	1	210
Poisson's Ratio	0.3	0.31	0.31	0.31	0.42	0.33
A (MPa)	819	448.45	448.45	206	24	1200
B (MPa)	308	475.808	475.808	505	3.00E+02	50000
n	0.64	0.3948	0.3948	0.42	1.00E+00	1
Reference Strain Rate (s^{-1})	0.0005	0.0001	0.0001	0.01	0.0005	0.0005
C	0.0098	0.0012	0.0012	0.0005	0.1	0.1
Tm (Degree C)	1527	620	620	916	487	1527
m	1	1.29	1.29	1.68	1	1

Table A2: Johnson Cook damage model parameters used for validating the 7.62 APM2 model and the target plate design.

Constant	Material		
	Strenx 700E [17]	Aluminium 7075 T651 [18]	Aluminium 7075 T6 [20]
D1	0.361	-0.428	0.3714
D2	4.768	0.7566	-0.1233
D3	-5.107	-3.4078	-1.9354
D4	-0.0013	-0.003	0.0101
D5	1.333	24.93	0

Table A3: Orthotropic material parameters for Kevlar® 29 and Dyneema® HB26

Property	Kevlar® 29 [21]	Dyneema® HB26 [22]
Modulus, X (MPA)	18500	34257
Modulus, Y (MPA)	18500	34257
Modulus, Z (MPA)	6000	3260
Poisson's Ratio, XY	0.25	0
Poisson's Ratio, YZ	0.33	0.013
Poisson's Ratio, XZ	0.33	0.013
Shear Modulus, XY (MPA)	770	173.8
Shear Modulus, YZ (MPA)	5430	547.8
Shear Modulus, XZ (MPA)	5430	547.8

Table A4: JC parameters of WC projectile taken from [23].

Property	WC projectile
Density (kg/m^3)	13805
Modulus (GPa)	620
Poisson's Ratio	0.215
A (MPa)	3000
B (MPa)	89000
n	0.65
Reference Strain Rate (s^{-1})	1
C	0
Tm (Degree C)	1527
m	1

13th South African Conference on Computational and Applied Mechanics (SACAM2024)
22 – 23 January, 2024, Stellenbosch, South Africa

AERODYNAMIC STABILITY OF THE TILTING PV WING OF THE NWU SOLAR CAR

Etienne Kruger[1*], Jan-Hendrik Kruger[2†], Johan Bosman[3]

[*] Presenting author
[†]Corresponding author: jandhendrik.kruger@nwu.ac.za
[1,2,3]North-West University, School of Mechanical Engineering
Private Bag X6001, Potchefstroom, 2520

Abstract:
As seen in the NWU Naleid2.0 solar car, a tiltable photovoltaic wing that dynamically aligns with the sun is an effective method of optimising solar radiation collection. However, this introduces a subsequent uncertainty regarding the lateral stability of the car, especially in crosswinds. The tilted wing increases the area exposed to crosswinds and, therefore, its negative effect on stability by an uncertain amount. To characterise the detrimental influence of these crosswinds on the tilted wing, CFD simulations were performed using Ansys Fluent. The simulation model used for crosswind simulations was validated first using Xfoil software, and secondly, through an experimental procedure on the Naledi2.0 car by comparing results with acceptable uncertainty. To account for the presence of both laminar and turbulent flow, a boundary layer transition model was used during all simulations. A range of wing angles was simulated for a maximum wind speed found along the typical race path of Naledi2.0. From the results of these simulations, Naledi2.0 was determined to be stable against roll and slide for all wing angles at the maximum wind speed. The effect of the crosswind was determined to be directly proportional to the wing angle for all cases except in the roll condition with a downward-orientated wing.

Key words: Solar car, tilting wing, lateral vehicle stability, crosswind, roll-over

1 Introduction

Naledi 2.0 is the latest attempt of North West University (NWU) in South Africa at producing a competitive solar car. One of the features that makes Naledi 2.0 so efficient is its tilting photovoltaic (PV) wing, improving its solar energy harnessing efficiency. However, this wing also heavily affects the stability of the vehicle - especially in crosswind conditions - to an extent that has not been quantified. This paper uses a validated simulation model to characterise the stability of Naledi 2.0 under crosswind conditions for the condition of slide and roll.

1.1 Problem Statement

Under crosswind conditions, the stability characteristics of the NWU solar car are unknown. This negatively affects the competitive capabilities of Naledi 2.0 in cross-country racing conditions.

1.2 Aim and Objectives

This report aims to investigate and characterise the drag and stability of Naledi 2.0 as influenced by its tilting PV wing at different angles under crosswind conditions through CFD simulations. Short-term objectives in reaching this aim include a literature study, validation of a simulation model, and lastly performing crosswind simulations using this validated simulation model.

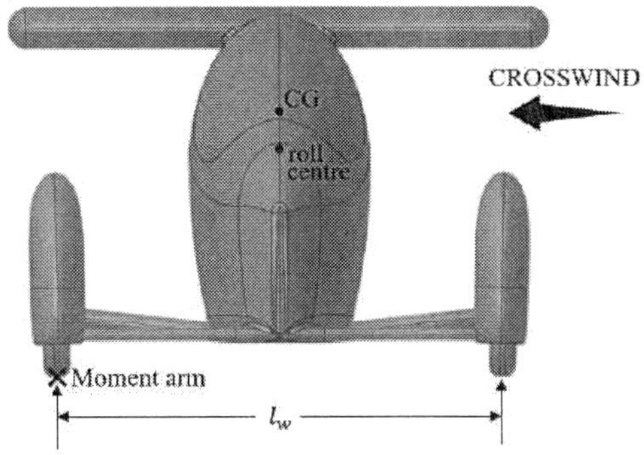

Figure 1: Moment balance for roll instability

2 Literature Study

To simulate representative wind speed conditions, the decision was made to simulate maximum wind speeds that occur during cross-country racing conditions. The NWU solar car typically participates in the Sasol Solar Challenge. The maximum wind speed experienced by the solar car along the path during this race was estimated to be $10[m/s]$ [1]. The simulation of the crosswind condition can be described as flow over a submerged body. Therefore, from the boundary layer theory, laminar, transitional and turbulent flow is expected [2]. Thus, from Morgado et al. [3], the $k-kl-\omega$ transition model is the most adequate simulation model for the expected transition flow. This model uses the specific dissipation rate equation, the laminar kinetic energy equation, and the turbulent kinetic energy equation for simulation [3].

To quantify the stabilising moment against roll that the solar car can generate under its own weight and the weight of the driver, Rajamani [4] added the moments around the indicated wheel in Fig. 1, with the result given in Eqn. (1). Assuming that the vehicle acceleration is zero and taking the moment arm as half of the wheel span, this equation is simplified to Eqn. (2). For the derivation of this formula, the angle between the centre of gravity (CG) and the roll centre of the car is assumed to be zero [4]. To quantify the stabilising force, the simple static friction equation Eqn. (3) is used. The vertical force for all simulations was determined to be negligible in this regard, therefore, the normal force in this equation was taken as the weight of the car and driver.

$$ma_yh + F_{zl}l_w - mg\left(\frac{l_w}{2}\right) = 0 \tag{1}$$

$$M_{stable} = \frac{mg \cdot l_w}{4} \tag{2}$$

$$F_{stable} = \mu_s \cdot F_N \tag{3}$$

The ratio of the length and width of an object over which fluid is flowing, as shown in Fig. 2, has a large effect on its drag characteristics. It was determined from the literature that as the l/D ratio increases, the object is classified as more streamlined and less blunt. For a streamlined body with a larger l/D ratio, the drag will be primarily the result of friction [2]. On the contrary, for a blunt object with a smaller l/D ratio, the pressure drag is more dominant [2]. Therefore, it is determined that the bluntness of the tilted wing will affect the drag characteristics of the car during crosswinds. Furthermore, the drag force is used to calculate the extent to which an object is affected by fluid that flows over it using Eqn. (4) [2].

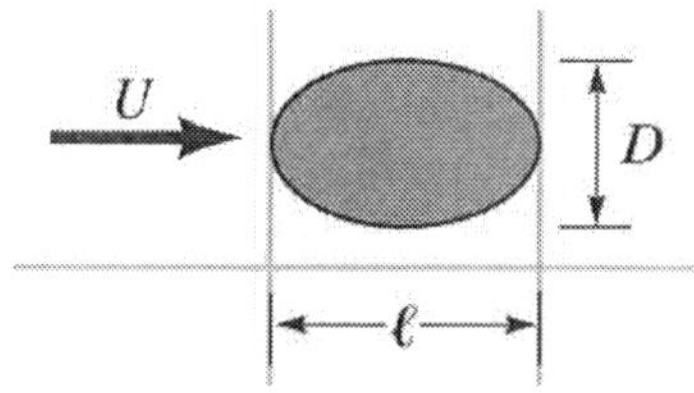

Figure 2: l/D ratio of an object inside a moving fluid

$$F_D = \frac{1}{2}\rho A C_D V^2 \tag{4}$$

From this equation, it is seen that the drag force on an object is directly proportional to the cross-sectional area exposed to fluid flow. Therefore, the initial assumption is made that, for any wing orientation with a larger cross-sectional area exposed to flow, the effect of the crosswind would be directly proportional.

3 Methods

Simulations quantifying the behaviour of airflow over the solar car geometry, as well as the effect thereof on the stability of the car, were performed using Ansys Fluent. The simulation model used to perform crosswind simulations was validated using two different methods. First, by simulating a simple aerofoil geometry on Ansys Fluent [5] and Xfoil [6] that delivered an error percentage of 14.8%. Second, a drag experiment was performed on the NWU solar car and the results were compared to a Fluent simulation for the same conditions as the experiment. This second validation method obtained an error of 8.56%. Subsequently, the simulation model was deemed valid for use for cross-wind simulations. This validated simulation method involved three main steps, namely representation, meshing, and setup before simulation could commence. These steps were followed during the validation of the simulation method and subsequently for all crosswind simulations.

Representing the Naledi2.0 geometry in Ansys SpaceClaim was the first step in this process. During this phase, the car was modelled to give the PV wing the ability to rotate along its longitudinal axis and then be trimmed flush with the body of the car again to form a watertight geometry. Ansys SpaceClaim was used in conjunction with Siemens NX [7] to do the simulation modelling. A flow domain was modelled around this geometry shown to the left in Fig. 3. While modelling the flow domain around the car geometry, the intersection of the car wheels with the road was given a rounded fillet using the Pull tool shown to the right in Fig. 3. This allowed the wheel intersection point to be meshed much easier using a mesh size slightly finer than the size of the fillet.

Next, this representation was used to apply mesh to the simulated geometry using Ansys Fluent Meshing. During this step, it was important to consider the trade-off between a fine mesh, which would deliver more accurate numerical solutions, and the longer simulation times that accompanies such a fine mesh. This was considered to ensure that the simulation results were mesh-independent. A body of influence (BOI) was added, as can be seen in pink in Fig. 3 to contribute to reducing the computing time. The maximum size used for the BOI was 0.03 $[mm]$, with a growth rate of 1.15. Smooth transition inflation boundary layers were added to the whole car geometry, which included five layers with a transition ratio of 0.272 and a growth rate of 1.15. Mesh refinements in more intricate portions of the geometry were used to define these areas more accurately while saving computing power and time. The parameters used to apply these refinements, as well as a surface mesh, are shown in Table 1. Also shown in this table are the parameters used to finally apply a volume mesh to the geometry. For both the surface mesh and the volume mesh, the respective *Improve Mesh* tasks within Fluent were used to obtain a mesh of reasonable quality and skewness.

Lastly, after successfully applying a reasonable mesh to the flow domain, the simulation setup followed.

Table 1: Meshing Parameters

Name	Target	Type	Min. Size [mm]	Max. Size [mm]	Growth Rate
Local Sizing	Whole car	Curvature	0.70	15.0	1.15
Local Sizing	Small Faces	Proximity	0.10	0.70	1.15
Surface Mesh	Whole Car	Curvature	0.75	600	1.15
Name	**Target**	**Type**	**Min. Length [mm]**	**Peel Layers**	**Quality Lim.**
Volume Mesh	Whole Car	Poly-Hexcore	0.1	1	0.3

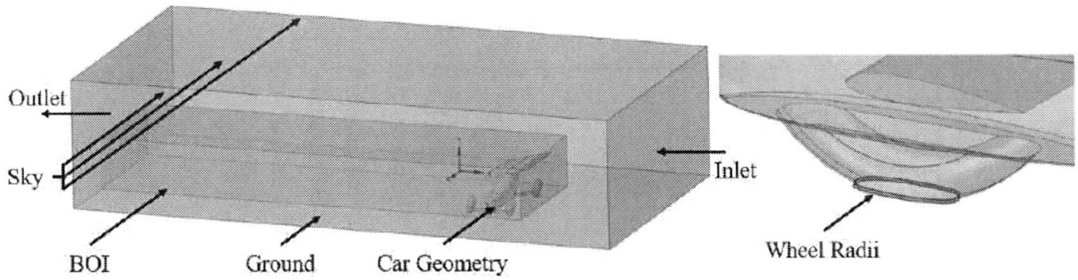

Figure 3: Crosswind simulation flow domain in Ansys Fluent

The $k-kl-\omega$ model was used for the simulation in the current study, as described in the Literature Study. The fluid used for the simulation was air, with a standard density of 1.225 $[kg/m]$ and a viscosity of $1.7894e-05$ $[kg/m \cdot s]$. For each simulated wing angle, the projected area was determined to be used as a reference value for the simulation setup. This was done using a simple plane in SpaceClaim, and the values ranged from 3.221 $[m^2]$, for the flat wing, to 4.584 $[m^2]$, for the maximum wing tilt. Shown in Fig. 3 are the named selections used to apply boundary conditions for simulation. The boundary conditions applied to these selections are specified in Table 2.

The stability of the solar car was analysed for two different scenarios. In the first scenario, it was evaluated whether the car would slide laterally due to the lateral force of the crosswind on the car. In the second scenario, it was evaluated whether the car would roll over due to a moment caused by the crosswind on the car. For each of these scenarios, the maximum stabilising friction force and moment that the car could generate under its own weight were calculated. The results obtained from simulations are categorised similarly and then compared to the stabilising value for both categories. If in any case the force or moment obtained from the simulations due to the crosswind on the car exceeded the stabilising force or moment, the car would be deemed unstable. Eqn. (3) and Eqn. (2) were used to calculate the maximum stabilising force and moment, respectively, for the specific geometry, shown in Fig. 1, and the weight of the NWU solar car. Fig. 4 indicates the notation that is used to refer to the orientation of the wing to the crosswind. This notation is used throughout the simulation results and discussion.

Table 2: Boundary Conditions for Simulation

Name	Type	Value
Inlet	Velocity inlet	10 $[m/s]$
Outlet	Pressure Outlet (gauge)	0 $[kPa]$
Car Geometry	Stationary no-slip wall	-
Ground	Stationary no-slip wall	-
Sky	Moving Wall	10 [m/s]

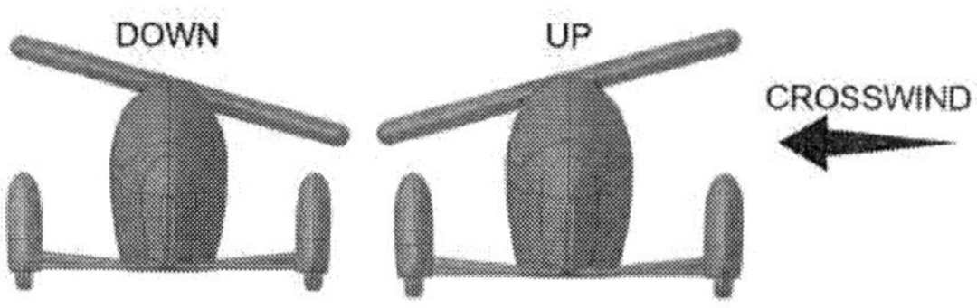

Figure 4: Syntax used to indicate wing orientation

To determine the values of the lateral force and the roll moment on the car due to pressure and friction drag, the built-in *Results* tab inside Ansys Fluent was used. The lateral force was determined for the whole car geometry as the total force in the negative-x direction. The moment was determined around the z axis, with the point of origin being the middle of the opposing wheel, as indicated in Fig. 1. Therefore, the moment axis was $(0,0,-1)$ and the moment centre at the wheel was $(1.045, 1.045, 1.792)$ $[m]$.

4 Results

The results obtained are processed as follows to draw meaningful conclusions. Table 3 and Table 4 show the results obtained from calculations and simulation for the case of slide and roll, respectively. The stabilising force and moment are shown in their own row first, after which the values obtained for different wing angles are shown.

Table 3: Lateral Sliding Force Results

Force Name	**Force Value** $[N]$
Maximum Stabilising Force	2163.1
Lateral Force at $[\theta = 0°]$ (FLAT)	220.73
Lateral Force at $[\theta = 15°]$ (UP)	330.16
Lateral Force at $[\theta = 25°]$ (UP)	400.00
Lateral Force at $[\theta = 15°]$ (DOWN)	260.08
Lateral Force at $[\theta = 25°]$ (DOWN)	282.20

Fig. 5 shows the colour-coded velocity distribution of airflow over the solar car, obtained from two different wing orientations under crosswind conditions. These figures give a clear comparison of the behaviour of airflow between the up- and down-orientated wing cases.

5 Discussion

The tables and figures shown in the previous section yield interesting results. Table 3 and Table 4 both show that for all wing angles at maximum wind speed, the force and moment exerted on the car by the crosswind are smaller than the stabilising force and moment, respectively. This makes the car stable against slide and roll for all scenarios simulated. For both scenarios evaluated, the influence of the

Table 4: Roll Moment Results

Moment Name	**Moment Value** $[N \cdot m]$
Maximum Stabilising Moment	929.53
Roll Moment at $[\theta = 0°]$ (FLAT)	276.44
Roll Moment at $[\theta = 15°]$ (UP)	527.44
Roll Moment at $[\theta = 25°]$ (UP)	606.01
Roll Moment at $[\theta = 15°]$ (DOWN)	126.81
Roll Moment at $[\theta = 25°]$ (DOWN)	26.14

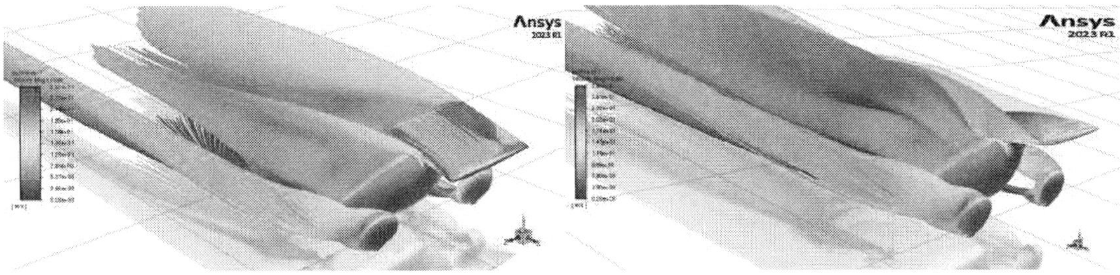

Figure 5: Velocity distribution of airflow over the down (left) and up (right) wing orientations

crosswind was expected to be directly proportional to the angle at which the wing is tilted. Therefore, it was expected that the most severe effect of the crosswind on the car would be at the maximum angle of wing tilt.

Visually inspecting Fig. 5, along with the values in the tables, the results can be further discussed. The results in Table 3 indicate that the magnitude of the lateral force on the car due to crosswinds is directly proportional to the tilt angle of the wing. This result corresponds to expectations, since a wing tilted at a larger angle increases the cross-sectional area that is exposed to flow and subsequently the effect of a fluid moving over it. Additionally, it is seen that for the up-orientated wing, a larger lateral force is exerted on the car compared to the same wing angle for a down-orientated wing. This result can be substantiated by visually inspecting Fig. 5 and comparing the behaviour of the cross flow on the down- and up-orientated wings. For the up-orientated wing to the right, the flow over the car is seen to be much more disturbed. The up-orientated wing acts like a drag-intensifying device, catching more fluid particles and, therefore, increasing the force exerted on the car. For the down-orientated wing to the left, on the other hand, the flow is seen to be more streamlined, resulting in a smaller lateral force being exerted on the car.

The results in Table 4 indicate that the magnitude of the roll moment on the car due to crosswinds is directly proportional to the tilt angle of the wing for the up-orientation, but indirectly proportional for the down-orientation. Therefore, the result for the up-orientated wing is as expected, whereas the results for the down-oriented wing are contrary to initial expectations. However, this phenomenon observed for the down-orientated wing under crosswinds can be logically justified. The crosswind over the down-orientated wing generates a lateral force, contributing to the roll moment, as well as a down force. This downforce generated a moment on the car in the same direction as the stabilising moment. As the wing tilt angle increases, the lateral force component decreases, and the down force component increases. Therefore, as the wing angle increases, the moment in the same direction as the stabilising moment increases and the net roll moment decreases. This justifies the indirect proportionality of the roll moment generated by the crosswind to the wing tilt angle for the down-orientated wing.

6 Conclusion

In conclusion, an Ansys Fluent simulation method was validated and used to obtain accurate cross-wind simulation results that were used to characterise the stability of the NWU solar car during crosswinds. The first validation of this method obtained an error below 15% and the second an error below 10%. From the cross-wind simulations conducted, for the slide condition, the wing angle was seen to be directly proportional to the extent to which the crosswind affected the car. Furthermore, it was found that the car was stable against roll for all wing orientations at maximum wind speed. For the roll condition, the angle of the wing was seen to be directly proportional to the extent to which the crosswind affected the car for the up-orientated wing, but indirectly proportional for the down-orientated wing.

These results conclude that the car would be stable against roll and slide under naturally occurring cross-winds under typical cross-country racing conditions - regardless of the wing orientation. For crosswind scenarios that do not occur naturally, such as passing larger motor vehicles, it is concluded that tilting the wing downward with respect to oncoming traffic would leave the car with a greater chance of remaining stable. It is recommended that in future work, a scenario such as this is explicitly simulated to further the abilities of Naledi2.0 on the solar racing stage.

Nomenclature

Acronyms and Abbreviations:

BOI	Body of Influence	
CFD	Computational Fluid Dynamics	
PV	Photovoltaic	

List of Symbols:

μ	Coefficient of friction	$N \cdot s/m^2$
θ	Angle of the PV wing	$^\circ$
F	Force	N
l	Length	m
M	Moment	$N \cdot m$
m	Mass	kg

List of Subscripts:

D	Aerodynamic Drag	
N	Normal value, eg. Normal Force	
s	Static value, eg. Static Friction	
w	Lateral wheel span of the NWU solar car	
$stable$	A value that can be generated by the car purely from its physical properties to keep the car in a stable condition. When this value is exceeded, the car is deemed unstable.	

Acknowledgements

My gratitude goes out to Dr Jan-Hendrik Kruger for his patience with me and unwavering support. The encouragement and technical support I received played a pivotal role in pushing me beyond my perceived limits and achieving the milestone of completion. I thank my parents and my friends for their words of encouragement and fervent prayers. A special thanks to my peer this year working on Naledi, Düter Struwig for the friendship and teamwork that we brewed in this short year. I thank my big brother, Stephan, for his proofreading and logical contributions. Lastly, I thank my beautiful wife, Adri, for being present supportive for every single day of this project.

References

[1] Sawea.org.za. 2021. URL: `https://sawea.org.za/south-african-wind-atlas/`.

[2] A.L. Gerhart, J.I. Hochstein, and P.M. Gerhart. *Munson, Young and Okiishi's Fundamentals of Fluid Mechanics*. Wiley, 2020. ISBN: 9781119597308. URL: `https://books.google.co.za/books?id=F6ALEAAAQBAJ`.

[3] J. Morgado et al. "XFOIL vs CFD performance predictions for high lift low Reynolds number airfoils". In: *Aerospace Science and Technology* 52 (2016), pp. 207–214. ISSN: 1270-9638. DOI: `https://doi.org/10.1016/j.ast.2016.02.031`. URL: `https://www.sciencedirect.com/science/article/pii/S1270963816300839`.

[4] Rajesh Rajamani. *Vehicle Dynamics and Control (2nd Edition)*. 2nd. Springer, 2012. 498 pp. ISBN: 978-1-4614-1432-2.

[5] 2021. URL: https://www.ansys.com/.

[6] 2024. URL: https://web.mit.edu/drela/Public/web/xfoil/.

[7] 2024. URL: https://plm.sw.siemens.com/en-US/nx/.

13th South African Conference on Computational and Applied Mechanics (SACAM2024)
22 – 23 January, 2024, Stellenbosch, South Africa

EXPERIMENTAL AND NUMERICAL ANALYSIS OF DILUTE GAS-PARTICLE FLOW IN A PIPE SECTION

Jan-Hendrik Kruger[1*], Louis le Grange[2], Michael Roberts[3]

[*] Presenter and corresponding author: janhendrik.kruger@nwu.ac.za
[1,2,3]North-West University, School of Mechanical Engineering
Private Bag X6001, Potchefstroom, 2520

Abstract: Coal-fired power plants depend on the milling and conveying of Pulverised Coal (PC), where the coal is milled down and then distributed through a network of pipes, using pneumatic transport, to individual burners that feed the boilers. These systems can be plagued by various operational problems due to varying flow conditions, which can even lead to unplanned plant breakdowns and illegal emission of pollutants.

The quality of PC gas-particle mixtures and flow stability can be improved by online measurement of particle mass flow and size distributions. A Particle Flow Test Loop (PFTL) was available that includes a blower fan providing gas flow for the system, a solid particle feeder to mix particles with the gas, a number of long pipe sections, and finally, a cyclone separator and gas cleaning system.

To enable particle flow measurements, a new pipe test section for the PFTL was designed and analysed using computational fluid dynamics (CFD) simulations. The multiphase software code Neptune_CFD by EDF was used in conjunction with Ansys Fluent and Siemens Star-CCM+. The final design was manufactured and fitted to the PFTL and can accommodate measurement devices such as an offline isokinetic sampler as well as online laser-based devices. Two-phase CFD modelling was used to predict cross-sectional solid concentrations for comparison with experimental measurements of particle velocity and particle size distribution. In this paper, selected results from the gas-particle flow simulations are presented, together with comparisons to experimental measurements taken inside the PFTL test section.

Key words: Gas-particle flow measurement, two-phase CFD, pulverised coal, pneumatic transport

1 Introduction

The current economic and energy situation in South Africa requires Eskom to continue to use coal-fired power plants for the foreseeable future [1]. Eskom is under continuous pressure to improve plant performance, reduce emissions, and, through regular maintenance, improve general plant availability.

Many coal-fired power plants experience challenges related to the milling and conveying of pulverised coal (PC). The coal is milled to a typical average particle size of 75 μm and then distributed through a network of pipes, using pneumatic transport, to individual burners that feed the boilers. These systems can be plagued by problems such as the variable fineness of the coal, the settling, and the choking of the PC pipes and the variation of the mass flow as the PC enters the burners. These varying flow conditions can cause costly operational problems such as high wear and fouling of the boiler tubes, leading to unplanned plant breakdowns, as well as the emission of toxic pollutants due to poor combustion conditions [2].

A first step towards an improved PC milled quality and gas-particle flow stability is accurate and on-line measurement of particle mass flow and particle size distribution. Due to extremely challenging

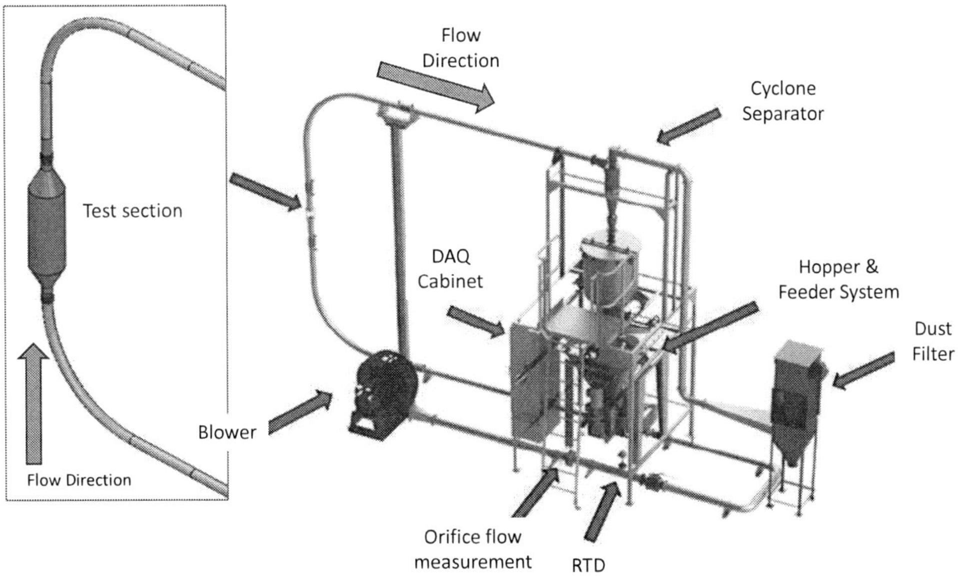

Figure 1: PFTL Test rig

operational conditions in an actual power plant, obtaining experimental measurements is fraught with difficulty, especially when high accuracy and repeatability is required for comparison with research simulations.

A Particle Flow Test Loop (PFTL) was developed by Colin F. du Sart [3] for the original purpose of investigating the settling velocities of PC in pipes. The PFTL shown in Fig. 1 includes a blower fan that provides gas flow to the system, a solid particle feeder to mix the particles with the gas, several long sections of pipe, and finally a cyclone separator and a gas cleaning system. The facility was conceived and developed at the University of Cape Town in 2016 and transferred to the North-West University in 2020 for further research focussing on the evaluation of different gas-particle flow measurement techniques.

2 Problem statement

The original PFTL test section was developed to investigate the velocity of particle settling and therefore had the same diameter as the feeder pipe sections. A new and larger test section was required that would induce the required flow conditions and could be used with online and offline particle flow measurement instruments (see red frame detail in Fig. 1).

3 Aim and Objectives

The aim of the study was to develop an operational test facility for particle flow measurements, satisfying the following objectives:

- Commission gas-particle test loop facility that produces gas-particle flow.
- Develop a test section for industrial-sized measurement devices.
- Perform measurements using a laser-based measuring probe (Parsum IPP70, [4]).
- Validate CFD simulations with measured results.
- Enable demonstrations and training for academic staff and students.

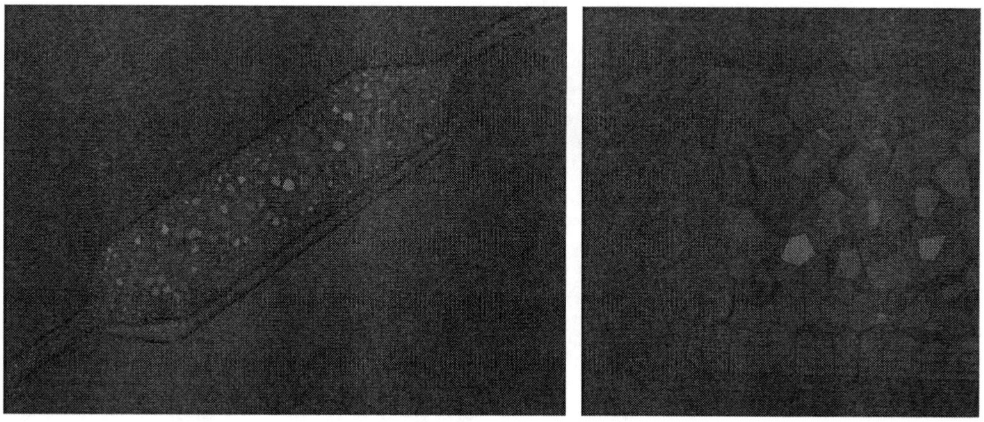

Figure 2: Mesh setup

4 Materials and Methods

The general methodology was to perform detailed computational fluid dynamics (CFD) simulations of gas-particle flow in the test section using three different approaches, and compare those results with experimental measurements and ascertain if validation is possible.

4.1 Simulation methodology

Three approaches were used during simulation development: first, clean air simulations were performed using Ansys Fluent [5] to get a general idea of flow patterns within the test section. Following that, two commonly used gas-particle flow modelling methods were used to incorporate gas-particle interaction, namely an Euler-Lagrangian gas-particle model (using Siemens Star-CCM+, [6]) and an Euler-Euler two-fluid model (using EDF Neptune_CFD, [7]).

Software packages from different vendors were specifically chosen to facilitate verification between model predictions and implementation strategies.

4.1.1 Clean air simulations

Clean air flow simulations were set up in Ansys Fluent to obtain a baseline flow pattern understanding [8]. The predicted flow velocity profiles were compared with the air velocity measurements taken by a pitot tube that traversed the test section.

The numerical domain encompassed the test section, as well as the inlet and outlet bend sections (see Fig. 1) so that the effects of the flow momentum were adequately captured in the test section. An example of the mesh in the test section is shown in Fig. 2, where the inner polyhedral cells with inflation layers can be seen.

Notable mesh parameters included a local target mesh size of 0.01 m and growth rate of 1.2, with a minimum surface mesh size of 0.02 m and maximum of 0.03 m. The domain boundaries consisted of a velocity inlet and pressure outlet, with adiabatic walls for the structure. The internal boundary layers (a.k.a. prism or inflation layers) were defined as three layers with a smooth transition growth rate of 1.2 and transition ratio of 0.27. The polyhedral volume mesh was defined with a maximum cell length of 0.07 m and growth rate of 1.3.

The air flow in the PFTL is controlled by setting the fan motor frequency and for the case shown in Fig. 3 the control frequency was set at 50 Hz, which corresponds to an average flow rate of 0.134 kg/s with an air temperature of 49.4°C.

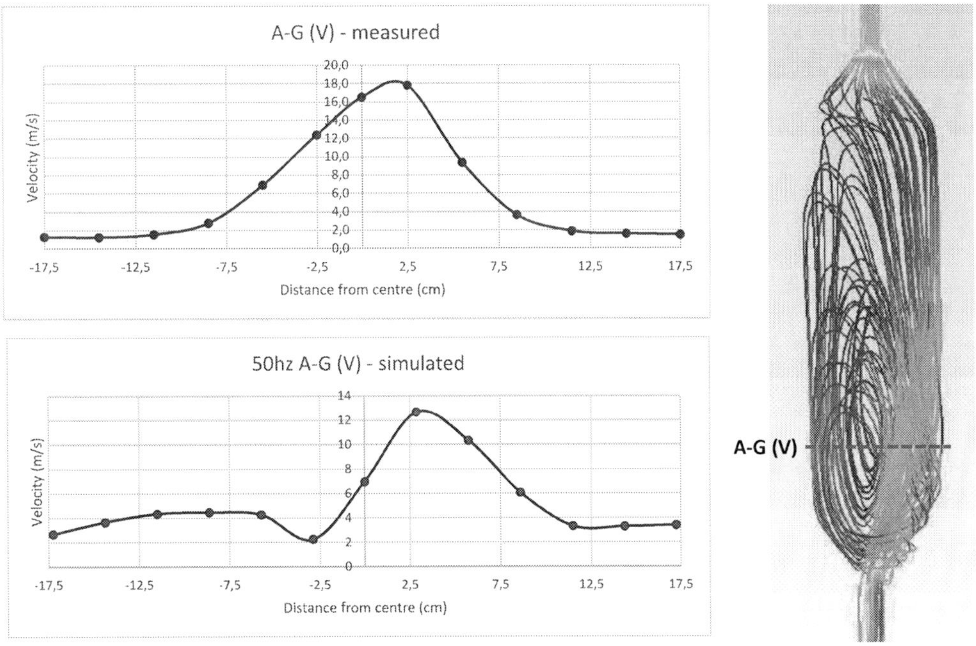

Figure 3: Clean air simulations

The streamlines in Fig. 3 clearly indicate the high-speed air flow stream moving to the outer region inside the test section. Measurements of velocity profiles in the A-G (V) plane support the general trend of the velocity profile, as predicted by the simulations. The discrepancy of about 6 m/s can be attributed in part to the highly directional measurement characteristic of the pitot tube, while the flow inside the test section actually experiences a high degree of recirculation that complicates any measurement of flow properties.

4.1.2 Euler-Lagrangian Simulations

The Euler-Lagrangian approach solves the gas flow as an Eulerian fluid using a continuum framework [9]. Particle motion is calculated by calculating the dynamic fluid forces on individual particles (or groups of particles) and tracking their movement through the numerical domain. Newton's second law with discrete time steps is used as the basis of the modelling approach, and a strength of the method is that a high-resolution solution of particle movement can be obtained, but with a strong one-way effect of gas flow on particles. However, the method breaks down when large amounts of particles need to be considered (such as in dense particle-laden flows with a strong interactive coupling between phases).

In Fig. 4 a snapshot of the particle movement for 40 μm particles is shown as calculated by Star-CCM+ [6], which illustrates the flow pattern in which the particles flow to the outside of the inlet bend and then continue to the outside of the test section, before recirculating. The simulation parameters are shown in the included table, where it must be noted that the air velocity indicated is the velocity inside the diameter of the small pipe and that the average flow rate corresponds to the other examples.

The volume fraction of the particles as plotted in Fig. 5 gives additional insight into the behaviour of the particles in the test section. The three plotting planes are indicated in their positions relative to the test section, and the formation of a core of high velocity particles can be seen on plane A (also with the position offset to the outside of the bend); the particle concentration then disperses as the flow moves upwards, as shown on planes B and C. It must be remembered that these plots only show the scalar volume fraction and are not an indication of particle velocity or flow direction.

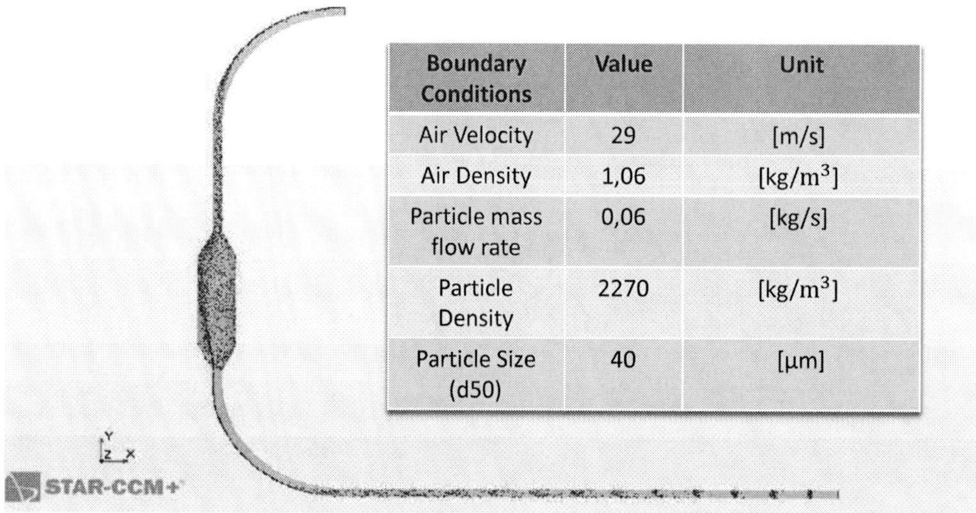

Boundary Conditions	Value	Unit
Air Velocity	29	[m/s]
Air Density	1,06	[kg/m^3]
Particle mass flow rate	0,06	[kg/s]
Particle Density	2270	[kg/m^3]
Particle Size (d50)	40	[µm]

Figure 4: Discrete particle flow movement

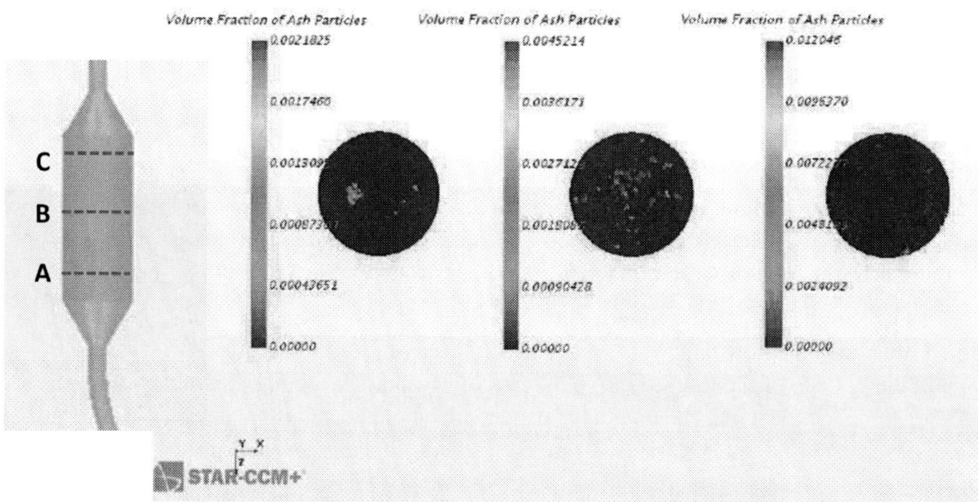

Figure 5: Volume fraction of particle phase - Euler Lagrangian

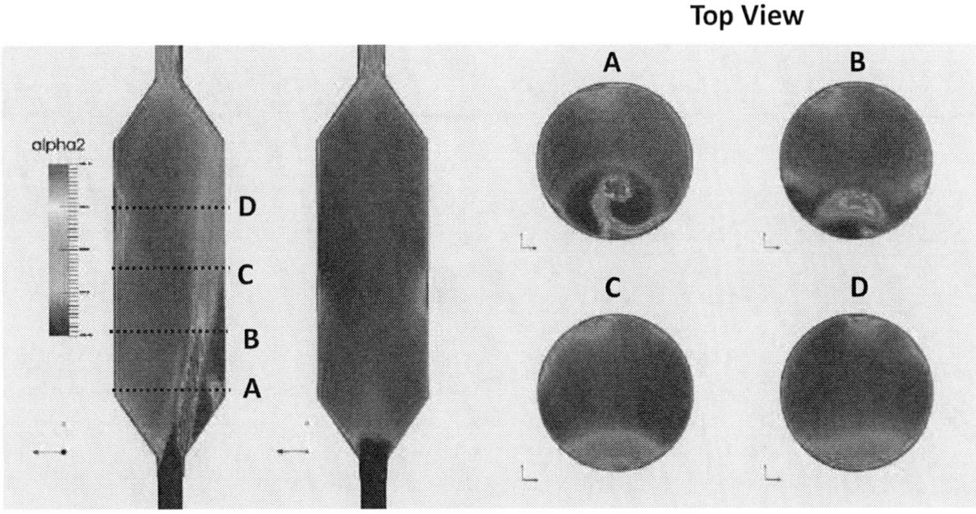

Figure 6: Volume fraction of particle phase - Euler Euler

4.1.3 Euler-Euler Simulations

The Euler-Euler approach (a.k.a. the Eulerian or two-fluid approach) considers both particles and gas as interpenetrating fluids [9]. A strength of the method is that a strong interaction between gas flow and the particle phase can be modelled, making it well suited for densely laden gas-particle flow situations. Compared to an Eulerian-Lagrangian approach, it is also much more computationally efficient for dense flow, especially when transient flow is modelled.

In Fig. 6 the volume fraction of the particle phase is shown, as calculated by using Neptune_CFD from EDF [7], for the same operating conditions as in the other examples. Four cut planes were defined from A to D, and the particle concentration plume can be seen where it appears strongest in plane A and then disperses from plane B through to D. On plane D a slightly higher concentration of particles can also be seen on the opposite side of the main plume, which is due to the onset of recirculation.

4.2 Experimental methodology

4.2.1 Features of test section

The test section illustrated in Fig. 7 had to satisfy several design requirements, while still achieving the main objectives. These design features were the most important:

- Due to the existing pipe work of the test rig, there was a limitation on the length of the test section.
- The diameter of the test section was restricted by the length of the Parsum and Gravimat probes, thus also prescribing a certain aspect ratio for the test vessel.
- The requirement of being able to use various instruments with their own, but different, inlet ports necessitated a rotating test section, so that cross-sectional flow profiles can still be measured using single-line traversals.
- The test section should induce repeatable, quasi-steady-state gas-particle flow profiles.
- Recirculating gas-particle flow should be induced to evaluate the instrumentation under adverse measurement conditions.

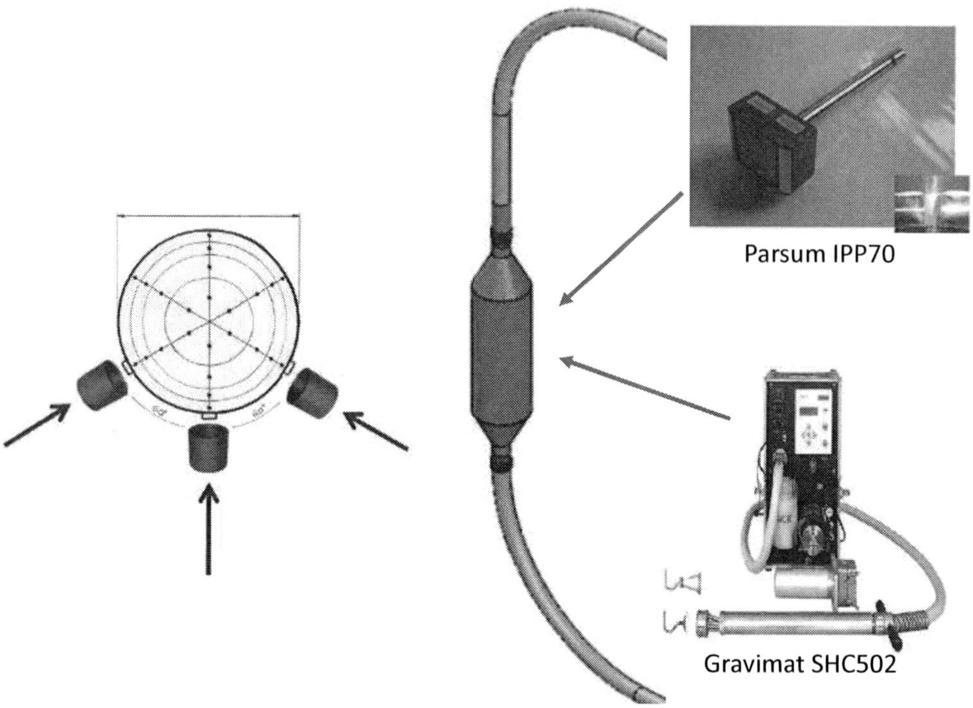

Figure 7: Features of test section

4.2.2 Experimental Measurements

The test rig with the Parsum IPP70 probe installed is shown in Fig. 8. The probe required its own digitiser and software to enable measurements to be taken. After the rig was turned on, the fan motor frequency was selected as required to generate a specific average flow velocity. Once the air flow temperature stabilised, the particle feeder mechanism was activated to feed the particles into the loop so that measurements can be taken.

In Fig. 9 a diagram of the measurement plane with two lines of traversal locations can be seen. The horisontal line with fewer measuring points is for the Gravimat SHC502 isokinetic sampler [10], while the vertical line contains the laser probe locations. Keep in mind that the plane as shown will be rotated to enable the measurement of a complete cross-sectional profile.

After initial tests with the isokinetic sampler, it became clear that the offline nature of this method would be completely unsuited to the objectives of the project, and further work focused on the online laser probe.

For a certain angle of rotation of the test section, the Parsum IPP70 probe [4] was then traversed by hand through the test section, stopping precisely at the locations indicated in Fig. 9. The controlling software measured and recorded the particle size distribution, as well as the particle velocity distribution. When a specific traversal line was completed, the test section was rotated to the next angle, and the process was repeated again, until all the angles of rotation were covered.

5 Results and Discussion

The laser probe measured the particle velocity, and gave the particle count for a specific situation. Of these two quantities, only the particle velocity measurement can be directly compared with the simulated values, while the particle count only gives an indication of possible trends.

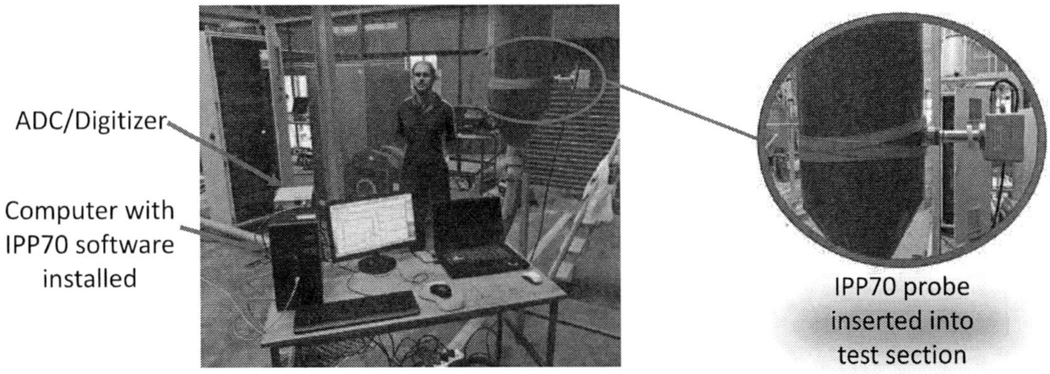

Figure 8: Test rig photo

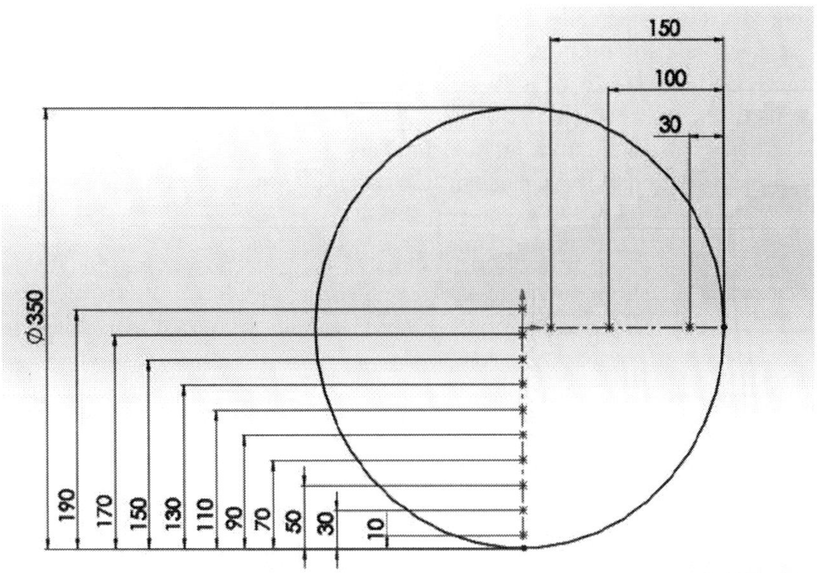

Figure 9: Probe traversing locations

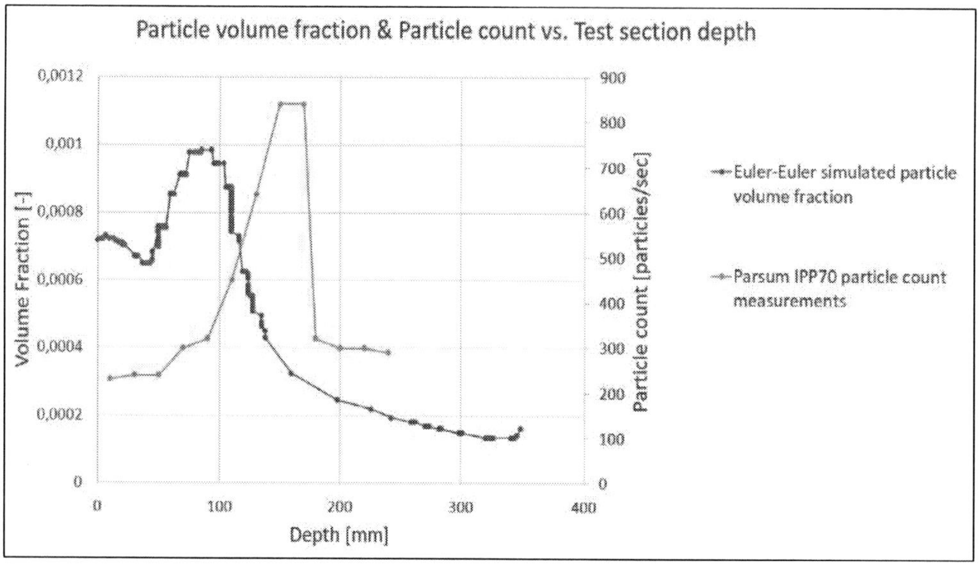

Figure 10: Particle volume fraction vs test section depth

When considering the changes in volume fraction and particle count over the depth of the test section, as shown in Fig. 10, interesting conclusions may be drawn (recall that the Euler-Euler simulations do not determine individual particle properties, unlike the probe which does count individual particles). The simulated volume fraction shows the maximum concentration about halfway between the inlet and the centre of the section.The probe, however, shows the maximum particle count measurements closer to the centre of the test section.

It is encouraging that there is actually some semblance between the trends of the two quantities, even though these are fundamentally different quantities and are not required to exhibit the same behaviour. The particle count can, for example, be influenced by particle recirculation, which can cause particles to be counted several times when located in the (what would be predicted by the simulation) same volume fraction region. There is also a complex relationship between die particle size distribution of the actual particles used in the test rig and the manner in which the two-fluid simulation handles such particle concentrations.

The velocity profiles shown in Fig. 11 exhibit a strong correlation between the simulated air and particle velocities, underscoring the strong effect of the gas phase on particles in dilute phase mixtures. The IPP70 measured particle velocities again show a profile peak that is located more towards the centre of the test section, however, it is at least of a comparable magnitude (within about 3 m/s of the simulated values).

An important point to note with relation to the Parsum IPP70 probe is that due to the design of the probe, it also has a strict directional measurement bias (similar to a pitot tube) and that measurement errors may be introduced when the directions of particle movement and the probe's measurement slot do not align closely.

6 Conclusion

In conclusion, most of the project objectives were completed. The gas-particle test loop facility was successfully commissioned and produced gas-particle flow that could be controlled and measured. The test section was constructed for the industrial-sized measurement devices, and both the isokinetic sampler and the laser probe could be used to complete basic exploratory measurements. In addition, this level of

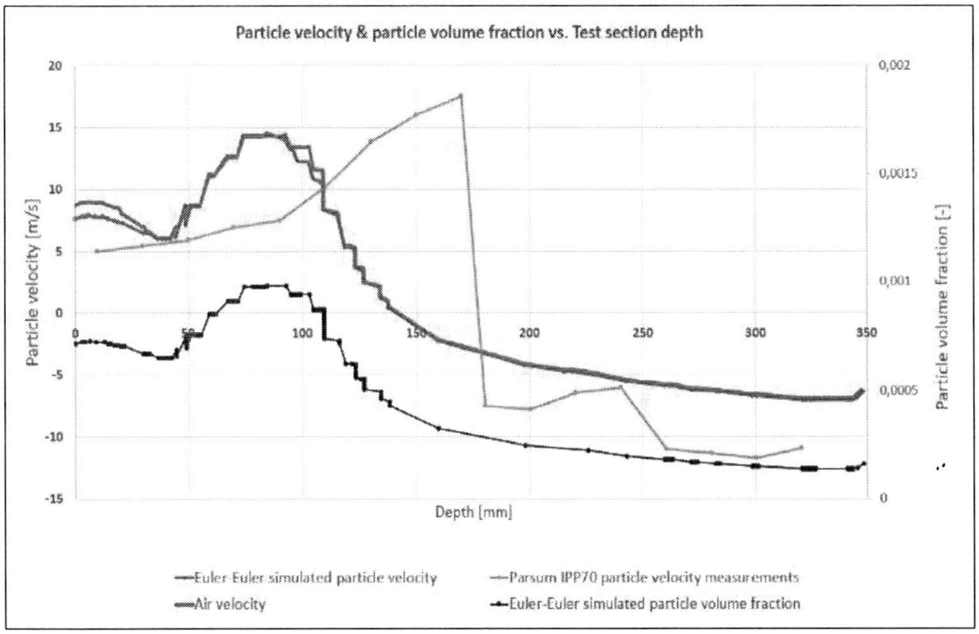

Figure 11: Particle velocity vs test section depth

functionality is already useful for demonstration purposes and training of staff and students.

Subsequent investigations were performed with the Parsum probe, where the particle count and particle velocity profiles were determined for a variety of test cases. However, comparisons with the simulated profiles for velocity and particle concentrations were inconclusive, even though there was encouraging agreement in terms of the large-scale trends.

Recommendations for future work include investigating the use of a more uniform particle medium to reduce the effect of the PSD on measurements. The viability of flow simulations that capture the details of the probe geometry and its measurement slot can be evaluated to obtain more appropriate comparisons between experimental measurements and simulated quantities.

Nomenclature

CFD Computational Fluid Dynamics

PC Pulverised Coal

PFTL Particle Flow Test Loop

References

[1] Department of Mineral Resources and Energy. “Integrated Resource Plan 2023”. In: *Republic of South Africa Government Gazette* 49974 (2024).

[2] Hamresin Archary. “Condition Monitoring and Performance Optimisation of Pulverised Fuel Vertical Spindle Type Mills”. Master of Science in Engineering. Johannesburg: University of the Witwatersrand, 2015. 225 pp.

[3] Colin F. du Sart. “Design and prototyping of a test facility to investigate the transport properties of dilute phase gas-particle flows applicable to coal-fired power plants”. Masters of Science degree in Mechanical Engineering. Cape Town: University of Cape Town, 2017. 355 pp.

[4] Dieter Petrak, Stefan Dietrich, Günter Eckardt and Michael Köhler. “In-line particle sizing for real-time process control by fibre-optical spatial filtering technique (SFT)”. In: *Advanced Powder*

Technology, International Journal of Sciences & Technology of Powders & Particulate Materials 22.2 (2011), p. 18.

[5] Ansys, Inc. *Ansys Engineering Simulation Software*. ANSYS, Inc., 2023. URL: https://www.ansys.com/.

[6] Siemens Digital Industries Software. *Simcenter STAR-CCM+, version 2020.1*. Version 2020.1. 2020.

[7] Lavieville, J., Quemerais, E., Mimouni, S., Boucker, M. and Mechitoua, N. *NEPTUNE CFD V1. 0 Theory Manual*. Rapport interne EDF H-I81-2006-04377-EN Rapport NEPTUNE Nept_2004_L, 1(3). EDF, 2006.

[8] Rayden Peyper. "Identification of flow patterns and losses in the particle flow test loop's test section using CFD". Final Year Project Report. Potchefstroom: North-West University, 2022. 70 pp.

[9] Andrew L Gerhart, John I Hochstein, Philip M. Gerhart. *Munson, Young and Okiishi's Fundamentals of Fluid Mechanics*. 9th ed. Wiley, 2021.

[10] SICK AG. *Operating Instructions Gravimat SHC502 (Gravimetric Dust Concentration Measuring System)*. 8007941. SICK Engineering GmbH, 2014. URL: https://cdn.sick.com/media/docs/6/16/316/operating_instructions_shc502_gravimat_en_im0011316.pdf.

Thirteenth South African Conference on Computational and Applied Mechanics
SACAM24
Stellenbosch, 22—23 January 2024

USING EXPERIMENTAL FLOW VISUALISATION TECHNIQUES TO INTERPRET COMPUTATIONAL FLUID DYNAMICS DATA

Craig Law[*,1], **Jan Nordström**[†,‡,2]

[*]School of Mechanical, Industrial and Aeronautical Engineering
University of the Witwatersrand, PO Wits, Johannesburg, 2050, South Africa,
[1]craig.law@wits.ac.za.

[†]Department of Mathematics
Linköping University, SE-581 83, Linköping, Sweden, [2]jan.nordstrom@liu.se

[‡]Department of Mathematics and Applied Mathematics
University of Johannesburg, P.O. Box 524, Auckland Park 2006, South Africa.

Keywords: Compressible flow, Computational Fluid Dynamics, Flow Visualisation

Abstract

A short overview of the application of schlieren and shadowgraph image construction has been presented with a focus on its use in interpreting compressible CFD data. Numerical flow visualisation methods have been used for validation of numerical models of real flow fields but are also a valuable tool for identifying and interpreting flow field features and provide a useful means for extracting information from the numerical data that may not be readily apparent.

1 Introduction

Most of the post-processing techniques used to present computational data, have a flow visualisation analogue of one form or another. Examples include the form of ribbon traces in CFD data that are comparable to dye or smoke trail injection into a flow field; or the use of thermochromic paint to produce the experimental equivalent of a colour flood plot of temperature variation on a surface due to flow as demonstrated by Atkins et al. [1]. The application of simulated shadowgram and schlieren[1] images to interpret Computational Fluid Dynamics (CFD) data is becoming a generally well accepted method. Recent articles by Kumar et al. [2] (2D) and Luthman et al. [3] (3D) demonstrate the interpretation of CFD results using this. Tecplot has now added a shadowgraph feature to its data analysis suite for post processing of numerical data.

Kumar et al. [2] demonstrated the application of numerical shadowgraph and schlieren imaging of a 2D simulation in OpenFOAM of a Mach 3 flow over a forward facing step. While the images are of poor quality, this has more to do with the implementation method and resolution of the simulation. The results of Kumar et al. [2] clearly show that the qualitative information required to describe the flow features are more readily apparent in the shadowgram and schlieren images than compared to a flood plot of density gradient.

[1]It is a lower case 's' as it is not a persons name but the German word for shadow.

While the work of Kumar et al. [2] and Luthman et al.[3] are two of many recent examples, the technique has been around for some time and is used extensively within the engineering and shock wave research communities, as demonstrated by the work of Law and Skews [4].

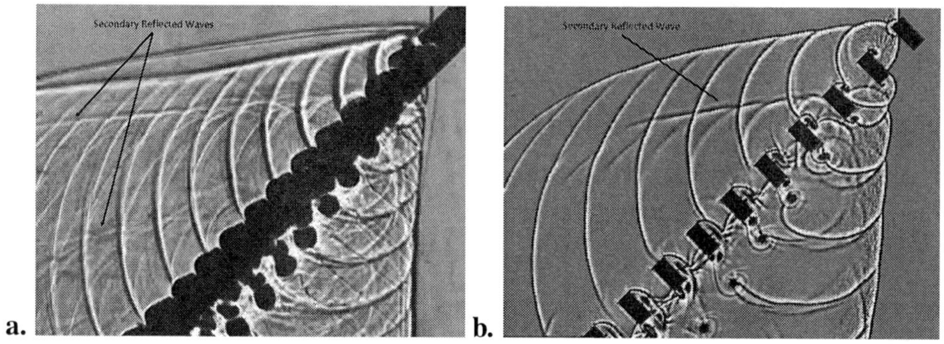

Figure 1: Shadowgrams of a Mach 1.353 shock impacting a grid at 45.0° with a blockage ratio of 33%. **a.** experimental, **b.** numerical. [4]

As CFD has became an increasingly important investigational tool for compressible flow research, it was a logical step to begin applying experimental flow visualisation techniques to CFD data. This was done primarily as a means to qualitatively validate the numerical models by comparing them to experimental images as these were often the primary type of experimental data. An example of this is shown in Fig.1 where a side by side comparison of experimental and numerical results are presented for a shock interacting with a rigid grid. The numerical result gives insight into the flow in the gaps, which would ordinarily not be possible given the requirement for structural support of the grid in the shock tube. Since there is good qualitative agreement between experiment and numerical image for the flow field below the grid, it was assumed that the flow field in the gaps had been adequately captured by the numerical simulation [4]. It was then possible to describe the flow field in the gaps and infer the behaviour of the flow field from this descriptive model.

Luthman et al. [3] provides a mathematical description of the calculation of the planar light intensity distributions that would be typical of a 3D flow field, using a ray tracing technique. They go on to apply the method to Large Eddy Simulations of a jet in supersonic cross flow. Their experimental and numerical 3D flow field images compare well. An alternative to this approach is the line integral approach as originally presented by Merzkirch [5].

This study will provide a brief overview of the use of flow visualisation methods as a way of interpreting and interrogating CFD data drawing largely on a small sample of the experimental and numerical studies undertaken since 2002 by the Flow Research Unit (FRU) at the University of the Witwatersrand. The objective is to present a case for using flow visualisation methods as a standard approach to interrogating numerical data.

2 Contour and vector plots

While contour and contour flood plots give a good representation of quantitative variations in a flow field, it can be very difficult to identify the position of features in the flow field or indeed the type of phenomena which make up a flow field. As an example, a shear layer or slip stream in a flow field represents a discontinuity in velocity, temperature or density but will not appear on a pressure contour plot. Depending on the range of density variation difference across the slip stream, it can be difficult to identify or locate such a feature on a contour plot. This is illustrated in Fig. 2 and Fig. 3.

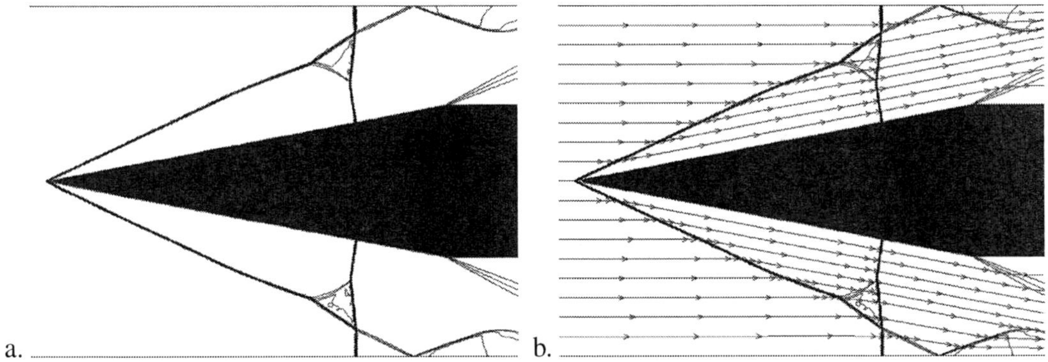

Figure 2: a.) Density contour plots and b.) density contour plots with overlayed streaklines, of a Mach 1.581 shock propagating to the right and impinging on a wedge traveling at Mach 3.664 in the opposite direction.

The Mach number of the wedge and shock represent a unique case in shock-wave moving body interactions where the velocity of the flow relative to the wedge changes as the impinging shock passes over it, but the Mach number of the shock remains unchanged. This is counterintuitive but is evident from the bow shock angle which is the same either side of the impinging shock (refer to Law [6] for more detail). The slipstream generated at the four shock interaction point in Fig. 2 is unclear. A contour is a line of constant value, so that a density contour is a line of constant density, but such a line can be distorted by the small variations in density due to the numerical noise inherent to any simulated flow field. This noise can create contours in steady regions of flow highlighting the slight variations in the flow field that are typical in a numerical simulation. Usually these noise contours can be reduced by reassigning the contours by either increasing or decreasing the number and spacing of contours. This can be seen by comparing the streamlines around the four shock interaction points in the upper and lower halves of the images.

Vector plots and their associated streak- and streamline plots are a good indicator of flow direction and stability, but can be difficult to interpret and are not recommended for locating flow phenomena. The plots are also highly sensitive to the reference frame in which they are made. The normal convention for these types of plots is a laboratory reference frame in which the body of interest is stationery. The frame of reference used for the numerical simulations shown in Fig. 2 and 3 was centred on the origin of the domain with the wedge and shock moving in oposite directions through the domain to ensure precise control of the Mach numbers of the shock and wedge. This means that the reference frame of the data had to be changed in order to generate Fig. 2b, where the stream/streak lines are in the reference frame of the wedge. A detailed investigation of the shock-wave/moving-body interactions illustrated in Fig. 2 and Fig. 3 can be found in [7], [8] and [9].

In experiments, flow features such as slip surfaces and shock waves usually show up clearly in shadow and schlieren images, where it is the rate at which density changes occur that is

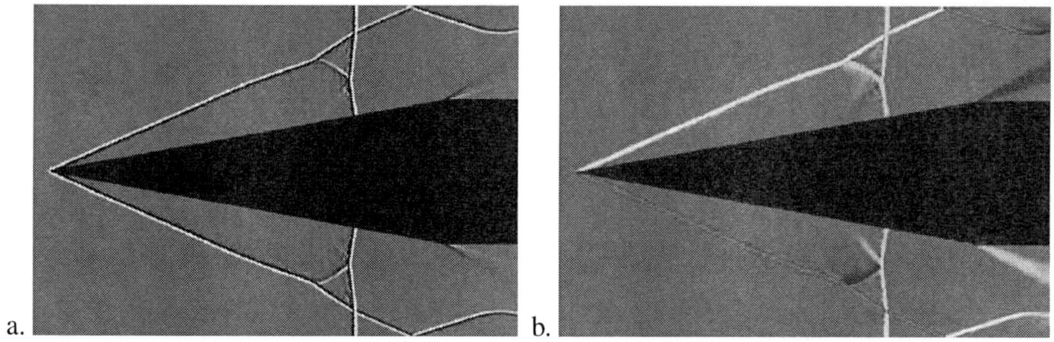

Figure 3: Constructed Shadowgraph (a.) and Schlieren (b.) Images of the results presented in Fig.2

important. This is illustrated in Fig. 3 which contains numerical shadowgraph and three-colour schlieren images of the same flow field shown in Fig. 2. While the weak shear layer eminating from the four shock interaction is not visible in the shadowgraph, it is clearly visible in the schlieren image.

3 Line integral approach to image construction

In both shadowgraph and schlieren images, parallel light is passed through a flow field containing density variations. A typical light bench setup for a schlieren system is illustrated in Fig. 4b. The difference between the shadow and schlieren imaging techniques is the knife edge which has the effect of polarising the light before it reaches the imaging plane. The density variations in the compressible flow field, diffract the light passing through it creating areas of varying light intensity on an imaging plane as illustrated in Fig. 4a. The deflection of the light path is cumulative through the area of interest so that the wider a test section is, the more pronounced the variation in light intensity is. Conversely a test section that has no thickness, would not create any light intensity variations. The image formed on the imaging plane allows phenomena such as shocks and contact surfaces to be clearly visualised since the rate at which flow density varies across such phenomena is large. Theoretically, the density gradient across such phenomena would be infinite, but in real flow fields they have finite but small thickness.

In a gas, the refractive index, n, of the gas is directly proportional to the density of the gas and is given by:

$$n - 1 = K\rho \tag{1}$$

where K is the Gladstone-Dale constant for the gas. K is a function of the wavelength of the light, the temperature of the gas and its pressure. Merzkirch [5] shows that variations in K are slight if the temperature and pressure variations are moderate i.e. real gas effects such as dissociation and ionisation have not occurred in the gas. Since a perfect gas model is often used in numerical modelling, K can be assumed to be independent of pressure and temperature variations in the majority of flow field simulations.

The image construction techniques demonstrated in this paper emphasise 2D flow fields but the technique can be readily extended to 3D flow fields by integrating through the test section using a ray tracing technique such as demonstrated by Luthman et al. [3], or by treating the 3D flow

a.
b.
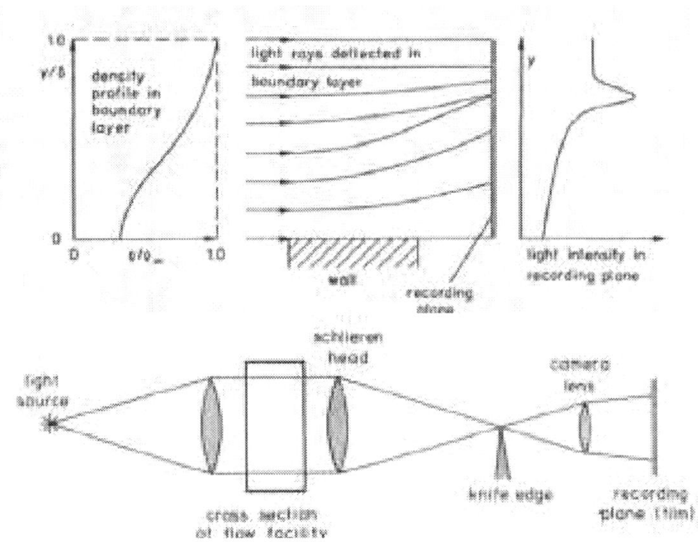

Figure 4: Principle of operation of shadow and schlieren potography (a.) Effect of density variation on light intensity distribution (fig7.5 of [10]); (b.) Toepler schlieren system light path showing parallel light passing through the test section. (fig8.1 of [10])

field as a sequence of 2D fields, calculating the change in intensity distributions and summing them. The discrete nature of the numerical model makes the density gradient a function of cell size which places an upper limit on the image resolution. This is seen in Fig. 1 where all flow discontinuities are thicker than could be obtained experimentally.

The image generation was originally implemented using the Matlab image processing toolbox and a set of custom scripts. This is still used at times if the dataset is small enough, but it is possible to generate shadowgrams and schlieren images directly in Tecplot, including three-colour schlieren. Three-colour schlieren is very memory intensive as an intensity map is required for each of the three colours.

3.1 Shadowgraphs

Shadowgraph images have a long association with compressible gas flow research. A typical light bench setup for a schlieren system is illustrated in Fig. 4b however, by removing the knife edge, the system then becomes a shadowgraph system. Merzkirch [5] derives an expression for the light intensity distribution within a shadowgraph image, by analysing the path of a ray of light through a shadowgraph system and shows that:

$$\frac{\triangle I}{I} = l \int_{z0}^{z1} \left(\frac{\partial^2}{\partial x^2} + \frac{\partial^2}{\partial y^2} \right) \ln{(n)}\, dz \tag{2}$$

In Eqn. 2, the z direction is parallel to the undisturbed light beam path and $\frac{\triangle I}{I}$ is a normalised ratio of the variation in light intensity through the test section along a ray. Given that the numerical model is two dimensional, the test section width, l, as it appears in Eqn. 2 would be zero. Instead a pseudo-test section of fixed width l is assumed and the image constructed accordingly. Since the model is two-dimensional and the variations in density are small compared to the rate at which density varies within a compressible flow field, it can be shown that:

$$\frac{\Delta I}{I} = F\left(\frac{\partial^2 \rho}{\partial x^2} + \frac{\partial^2 \rho}{\partial y^2}\right) \tag{3}$$

where $F = c\frac{Kl}{\lambda}$ is a constant with λ representing the dominant wave length of the light source, and c representing an intensity factor used to control the contrast of the image – analogous to adjusting the light source intensity in an optical system. Experience has shown that an exact value of F is not strictly necessary and a value of $F = 16$ was found to work well for most of the images presented here that are drawn from earlier work such as Fig.3.

The image can then be constructed by:

1. Calculating $\frac{\partial^2 \rho}{\partial x^2} + \frac{\partial^2 \rho}{\partial y^2}$ across the grid using a first order finite differencing scheme.
2. Thresholding the data by a factor of F. This is done by multiplying the data generated in the previous step by F and then setting all terms in the dataset greater than 1 equal to 1 and all negative terms equal to zero.
3. Generating gray scale flood plots of the intensity distribution generated in the previous step.

Depending on how the method is applied, it may be necessary to apply an interpolation to the two-dimensional light intensity distribution, to convert it from the original CFD data and mesh structure to a structured rectangular grid with the number of grid elements in each direction being directly proportional to the resolution of the image in each direction. This step is typically required when using the image processing toolbox in Matlab to convert the intensity maps into images. Where the computational dataset is large, it may be preferable to perform this interpolation before calculating the image intensity maps. Adjusting F allows for the contrast of the image to be adjusted.

In the case of most post processing software, the calculation of the intensity maps would be done using the original dataset. This may require either the use of a user defined function to calculate the derivatives of the density distributions or the calculation of the density gradients using a post processing package such as Tecplot. Using Tecplot as an example, a shadowgraph image is obtained by calculating the intensity distribution using Eq. 3 and then generating a grayscale flood plot of the light intensity distribution.

3.2 Schlieren imaging

As with shadowgraph imaging, Merzkirch [5] derives an expression for the light intensity distribution within a schlieren image by analysing the path of a ray of light through a schlieren system and shows that:

$$\frac{\Delta I}{I} = \frac{Kf_2}{a}\int_{z0}^{z1} \frac{\partial \rho}{\partial y} dz \tag{4}$$

where a is the height of the light source image perpendicular to the light source mask on the imaging plane and f_2 is the focal length of the schlieren head. From this it can be shown that:

$$\frac{\triangle I}{I} = F\frac{\partial \rho}{\partial \zeta} \tag{5}$$

where $F = c\frac{Kf_2l}{a}$ is a constant similar to that of Eqn. 3 and ζ is a direction perpendicular to the knife edge of the light source. A conventional schlieren system is restricted to showing variations in density gradient in one direction only which is perpendicular to the knife edge. Three-colour schlieren techniques were developed to provide an experimental imaging system which is sensitive to density variations in all directions in the flow field. A typical example of such as system was that developed by Seitz, Scott and Skews [11], using three coloured light sources inclined at 120° to each other. This three-colour system was applied to numerical images in the method described below.

Given that the three light sources are each masked in a different direction, the intensity matrix for each colour needs to be generated separately. The schlieren images can then be generated by:

1. Calculating $\frac{\partial \rho}{\partial x}$ and $\frac{\partial \rho}{\partial y}$ using an appropriate finite difference scheme.
2. For a gray scale schlieren image; skip step 3 and proceed to the next.
3. Calculating $\frac{\partial \rho}{\partial \zeta}$ for each of the colours using:

   $$\left(\frac{\partial \rho}{\partial \zeta}\right)_{col} = \frac{\partial \rho}{\partial x}\cos\left(\theta_{col}\right) + \frac{\partial \rho}{\partial y}\sin\left(\theta_{col}\right) \tag{6}$$

   where θ_{col} is the angle of incidence of the colour specific knife edge under consideration. These are $\theta_{red} = 0°$ for the red light source, $\theta_{green} = 120°$ for green and $\theta_{blue} = 240°$ for blue, as applied by [11].
4. Thresholding each density gradient map by F using the same method as before. Changing F will change the sensitivity of the schlieren image, so that finer changes in density can be detected by increasing F. It is important to note that a value for F which is too large, will make the image will appear overexposed and too small will make the image appear underexposed. The image can also be biassed toward one of the colours by using a larger threshold value for that colour. This is analogous to allowing more light through for one of the colours and is what is typically done in 3 colour systems to ensure that there is enough light in the test section to illuminate the model.
5. For a grey scale schlieren image, continue as per step 3 of the shadowgraph procedure. For a three-colour image each of the 3 datasets needs to be turned into a gray scale flood plot which then needs to be tinted to the colour associated with the dataset (red, green or blue). These are then overlayed and set to 50% transparency to create the final colour image.

The final step is much simpler to accomplish in Matlab as the three-colour intensity maps can be turned into an image directly. The same comment applies as before regarding interpolation between an unstructured mesh and a structured grid however, along with the associated loss of image quality.

While shadowgraphs are sensitive to large changes in density such as would occur across a shock or in a vortex, schlieren images are sensitive to small changes in density typical associated with expansion fans. Often strong features will blur in an experimental schlieren image and details of shock position and shape are better determined using shadowgraphs. The bluring can be attributed to a large extent to the image being more sensitive to the three dimensional nature of the flow field. Shadowgraph images tend not to be as sensitive to the weaker density gradients that are introduced by a 3D flow field.

3.3 Interpreting the data

It is possible to perform interferometric image construction on the numerical data as well and this has been used in the past by the Flow Research Unit. The experimental set up usually requires a laser light source and the cost and complexity does not warant the level of detail which today can be obtained from a well validated simulation.

None of the flow visualisation or data presentation techniques described here can be said to be the ideal data presentation method. Each has its own strengths and weaknesses so that the ideal form of numerical data interpretation lies in the use of a combination of the techniques. Since the imaging algorithms use the same numerical data, it is possible to generate synchronous colour schlieren and shadowgraph images of the numerical flow field. These images can then be combined with contour and other plots, to provide a detailed analysis of the numerical flow field. An approach similar to that applied in analysing experimental images can then be applied to the numerical results. The combined information from all the plots and images provide a much richer picture that can be used to understand subtlties in the flow field that are not readily apparent from just viewing contour or flood plots, or the images on their own.

Imaging lends itself far more readily to geometric analysis of flow fields, while vector and contour plots give a better idea of the size and type of flow property change. By using all of the techniques described, a clearer understanding of the numerical results, and comparisons with analytical modelling can be made with greater certainty.

4 Application to High Order Accurate DNS calculations of a shock diffracting over a wedge

Law et al. [12] used a hybrid $6^{th}/3^{rd}$ order summation by parts (SBP) direct numerical simulation (DNS) solver with a weak slip boundary formulation to simulate the diffraction of a Mach 1.5 shock over a complex, convex wall. Details of the numerical solver were discussed by Eriksson et al. [13] with a focus on how the order of the SBP solver was lowered around the discontinuities in the flow field in order to preserve numerical stability. Law et al. [12] looked at the effect of viscosity on the flow field while the geometry and Mach number were kept constant. A detailed demonstration of the convergence and validation of the models was presented and a series of schlieren images were used to demonstrate the change in flow features with Reynolds number. The computational results for a Mach 1.5 shock at a Reynolds number of 200×10^3 have been reprocessed here to demonstrate the use of flow visualisation imaging in interpreting CFD data. Fig. 5 contains traditional colour flood plots of density and Mach number for the flow around the first corner.

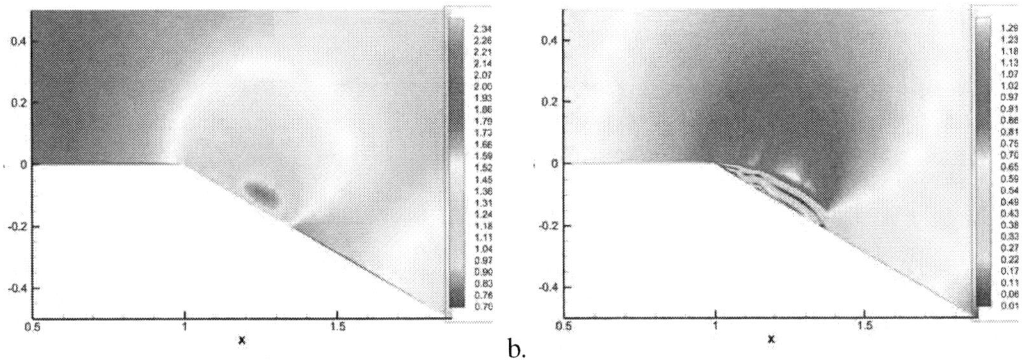

a. b.

Figure 5: Mach 1.5 Shock diffracting over a 30° backward facing wedge (a.) Density flood plot (b.) Mach number flood plot

In Fig. 5 it is possible to identify a series of shocks in the flow field and a small oval region of low density. By using the Mach contour and density plots togther, it is possible to clearly identify a shear layer propagating from the corner and the shocks can be seen to form a lambda shock pattern that is typical of this type of flow field. It is much easier to identify these features when you expect them to be there and know what to look for. This should be compared to Fig. 7 where the strong features (shocks and shear layer) are clearly identifiable.

Figure 6: Mach 1.5 Shock diffracting over a 30° backward facing wedge. 3 Colour schlieren reconstruction of the numerical data.

Something that was not apparent from Fig. 5 is that the shear layer rolls up into a vortex resulting in the oval low density region. What can also be seen from the Fig. 6 is that there are a series of supersonic flow regions, weak shocks and interactions with the shock waves that have propagated through the shear layer. In addition, a contact surface can be near the right edge of the image. At the time, computational resource constraints meant that it was not possible at the time to re-run the simulations with greater mesh refinement in the region of the shear layer and vortex.

While Fig. 6 clearly shows the many shocks and the shear layer clearly, it is possible to identify these features by examining various flood plots and streamline overlays in combination. This would typically require an examination of a number of images in combination. The schlieren image makes it possible to rapidly identify the many interesting features in the flow.

The images in Fig. 7 combine the flood plots of Fig. 5 with a 50% transparent grey scale schlieren image sensitive to vertical variations in density gradient.

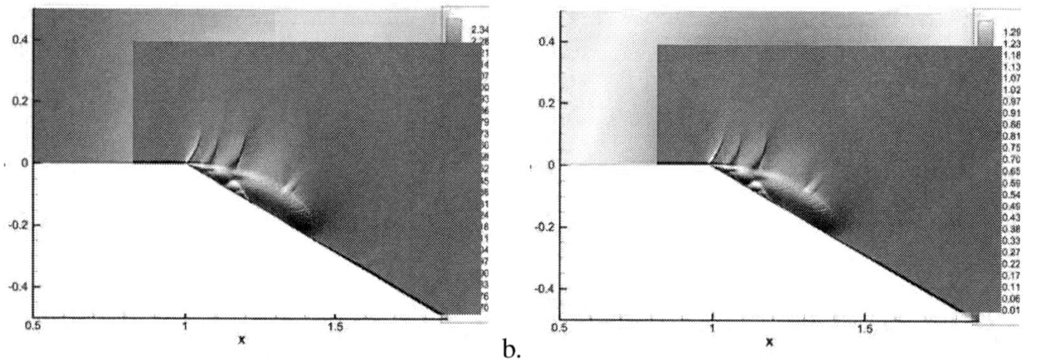

a. b.

Figure 7: Mach 1.5 Shock diffracting over a 30° backward facing wedge (a.) Density flood plot (b.) Mach number flood plot. Both plots have an overlay of a greyscale schlieren reconstruction

By overlaying the two types of images, it is possible to readily correlate features that were identified in the schlieren image with density variations and changes in Mach number as illustrated in Fig. 7. Note that the contact surface shown in Fig. 6 is very difficult to identify in Fig. 5a. and not at all visible in Fig. 5b. This contact surface represents the signal propagation front that separates the flow that is induced by the passage of the diffracting shock front only from the disturbances initiated at the corner.

5 Conclusion

A brief description of how to convert CFD data into shadowgraph and schlieren images has been provided. This is accompanied by examples of where these flow visualisation reconstructions have been used to identify flow features that were captured in the numerical data. Numerical flow visualisation provides a powerful means of quickly doing a qualitative comparison between experimental and numerical data. It can also be used to identify weak features in the flow field such as contact surfaces which are not readily apparent in most forms of CFD data presentation. The combined information from traditional flood plots can be combined with flow visualisation reconstruction images to provide a much richer picture that can be used to understand subtlties in the flow field that are not readily apparent when using traditional CFD post-processing on its own.

References

[1] MD Atkins, FW Kienhöfer, TJ Lu, and T Kim. Local heat transfer distributions within a rotating pin-finned brake disk. *ASME. J. Heat Transfer*, 142(11), 2020.

[2] A Kumar, P Dubey, and BT Kannan. Numerical flow visualization of supersonic flow using OpenFOAM. In *First International Conference on Mathematical Techniques and Applications*, 2020.

[3] E Luthman, N Cymbalist, D Lang, G Candler, and P Dimotakis. Simulating schlieren and shadowgraph images from LES data. *Experiments in Fluids*, 60(134), 2019.

[4] C Law and B Skews. Numerical modelling of shock porous body interactions. In *Proceedings of the 7th South African Conference on Computational and Applied Mechanics SACAM10*, 2010.

[5] W Merzkirch. *Flow Visualization*. Academic Press, 1987.

[6] Craig Law. *Two dimensional shock-supersonic body interactions*. PhD thesis, University of the Witwatersrand, 2002.

[7] C Law and B Skews. Flow effects resulting from relative motion between shocks and supersonic bodies. *Shock Waves*, 13(4):271 – 281, 2003.

[8] C Law, LT Felthun, and B Skews. Two-dimensional numerical study of planar shock-wave/moving-body interaction part i. *Shock Waves*, 13(5):381 – 394, 2003.

[9] C Law and B Skews. Two-dimensional numerical study of planar shock-wave/moving-body interaction part ii. *Shock Waves*, 13(5):395 – 408, 2003.

[10] W Merzkirch. *AGARD-AG-302:TECHNIQUES OF FLOW VISUALIZATION*. NATO Science and Technology Organisation, 1987.

[11] MW Seitz, DM Scott, and B Skews. Colour schleiren photographic system. *Optical Engineering*, vol. 33(no. 9):2907–2910, September 1994.

[12] C Law, Q Abbas, J Nordström, and B Skews. The effect of Reynolds number in high order accurate calculations with shock diffraction. In *Proceedings of the 7th South African Conference on Computational and Applied Mechanics SACAM10*, 2010.

[13] S Eriksson, C Law, J Gong, and J Nordström. Shock calculations using a very high order accurate Euler and Navier-Stokes solver. In *Proceedings of the 6th South African Conference on Computational and Applied Mechanics SACAM08*, 2008.

ON THE COMPLEXITY OF INTERNAL CHAMBERS IN SOFT PNEUMATIC ACTUATORS

Philip Ligthart[1,*], Martin Venter[2]

[1,2]Mechanical and Mechatronic Department of Stellenbosch University, [*]*pfligthart04@gmail.com*

Abstract: As a result of the complex material behaviour of soft robots, many factors need to be considered during the design process. In this paper, the design of the internal chambers of soft pneumatic actuators is considered. Manual building and testing can be expensive and time-consuming and the simulation alternative is computationally expensive. To design soft pneumatic actuators more efficiently the complexity the internal chamber in a soft pneumatic actuator needs to be considered. Previous works have shown that complex soft robot behaviour can be achieved through a combination of simple primitive units. In the context of simple primitives the output deformation of a unit is analyzed based on the internal chamber geometry. Regular polygons are used to perform comprehensive factorial design analysis and the results are interpreted with a bespoke deformation measure. t-SNE embedding is used to reveal the underlying structure in the high-dimensional data created from the factorial designs. The analysis concludes with a recommendation on the types of shapes to use in soft pneumatic actuators for maximizing the potential of diverse output deformation while minimizing the overall parameter count required to define the chamber shape.

Keywords: Soft Robot; t-SNE; Finite Element Method

1. Introduction

Soft robotics is an interesting field that is vastly different from conventional robotics and has gained significant traction in recent years. Unfortunately, the field remains largely interesting as opposed to useful as a result of the difficulty in designing soft robots. This is evident by the large number of interesting yet currently unusable anecdotes published in recent literature. A handful of such anecdotes are included here; the first entirely soft autonomous robot [1], an uncontrolled mase escaping soft robot [2], a magnetically actuated fiber-based soft robot [3] and, a geko-inspired climbing soft robot [4]. Despite the large and complex design domain trial and error remains the defacto design methodology [5]. Attempts have been made to improve the design process by articulating a 'concrete path to the solution' [6]. Although this is helpful for new designers in the field, it does not address the fundamental problem of exploring the large and complex design domain more efficiently. Computational tools such as finite element analysis and regression techniques have been leveraged in design frameworks, such as those presented in [7] and [8]. These frameworks attempt to use both designer intuition and computational resources to more efficiently design soft robots.

Among the computational tools available to aid in the design of soft robots are physics engines. [9] created a soft robot benchmark, Evogym, in which different types of discrete units are used to design soft robots using a set of algorithms including, genetic algorithms, Bayesian optimization, and compositional pattern-producing networks. The benchmark uses co-evolution of the morphology and the control of the robots using proximity policy optimization. The design of a soft robot takes place on a grid, such as a 5×5 grid. Each grid unit can be occupied by one of five units, hard, soft, horizontal actuation, vertical actuation, or empty. An example of such a soft robot is shown in fig. 1. From this relatively simple encoding, soft robots with complex behaviors can be created. For example, robots that perform carrying, climbing, and catching are designed. In the benchmark, actuation is enforced and no 'real' actuation takes place. This paper is concerned with the chamber design of such units which can be pneumatically actuated.

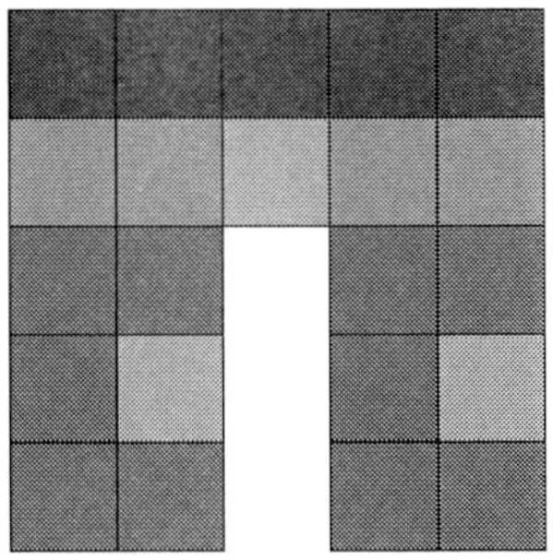

Figure 1: A 5×5 array of square primitives soft robot. Figure adapted from [9]. In their study, black represents hard material, gray represents soft material, and blue and orange represent the two different types of actuators.

Since pneumatic actuation works by applying pressure to a chamber embedded within soft material the design of the soft unit is the design of the internal chambers of the soft unit. In order for the chamber to be designed the shape needs to be defined. In a two-dimensional case, a chamber may be defined by a set of connected points. Such a shape definition allows for a large variety of both convex and non-convex polygons but has the disadvantage of having a large number of design variables. For example, the case when the shape is defined by 7 connected points, see fig. 2a. This shape will have 14 design variables, the x and y coordinates of each of the points. This can be compared to a simple polygon, such as the triangle in fig. 2b, which can be completely defined by 4 design variables, the center point coordinates, side length, and rotation. In terms of an optimization problem, it is advantageous to reduce the number of dimensions in the problem. This is all the more important when more than a single cavity is added to the design problem.

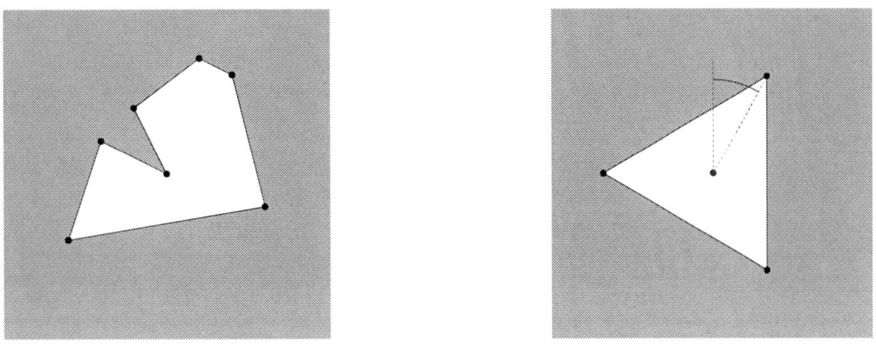

(a) (b)

Figure 2: Example of **(a)** non-convex seven-point chamber shape and **(b)** triangle chamber shape

A common theme in pneumatically actuated soft robots is the design of bending actuators. Within the many publications concerned with this target, there are mainly two approaches, firstly, a network of connected rectangular chambers is used. The second alternative is where a freeform chamber design is used. The freeform designs are achieved in different ways such as using topology optimization [10]. The free-form method is often computationally prohibitive as a result of the large number of variables. To overcome this, the rectangular chamber design has become the de facto chamber design. Often the spacing between the chambers is varied to achieve different effects. The rectangular design was introduced by [11] in their design of the novel fast pneumatic network in 2014. This design has persisted and is still used in many recent publications, for example in the geko-inspired soft robot [4]. Interestingly, the chamber design obtained by a freeform design method using a bespoke topology optimization method results in almost rectangular chambers. This paper explores the effect of different internal chamber shapes and complexity on the output deformation in pneumatically actuated units.

2. Methods and Materials

A factorial design was set up to explore the design space on a single chamber unit of silicone. Figure 3 shows an example finite element simulation of a $25\,\mathrm{mm} \times 25\,\mathrm{mm}$ square of Ecoflex 0030 silicone with a rectangular chamber. The factorial design considered three regular polygons namely, circles, squares, and triangles. When considering the number of corners of each shape these shapes represent the extrema, with a triangle having the least possible corners and a circle having an infinite amount of corners. Additionally, the factorial design considered the area, pressure, rotation, and center point position of the shapes. Due to the resulting chambers having different shapes and sizes, the center point could not effectively be included in the factorial design in the conventional factorial manner. If a fixed set of center points were prescribed it would result in small chambers not being placed as close to the outside of the shape as larger chambers. To overcome this limitation, 50 uniformly distributed random center points for each factorial design entry were simulated. The factorial design parameters are presented in table 1 and produce 18 000 data points.

Table 1: Factorial design of the parameters. The rotation is measured in the percentage of a full rotation, the amount of rotation changes between the different shapes. A square can rotate between $\pm 90°$ before it is back at the same orientation, whereas a triangle can rotate between $\pm 120°$. The rotation factorial levels are selected in increments of $15\,\%$ from $0\,\%$ upwards and $100\,\%$ downwards. The $45\,\%$ and $55\,\%$ entries in the middle are replaced with a $50\,\%$ entry. The $100\,\%$ entry is removed as it is the same as the $0\,\%$ entry. A circle has no rotation. For each combination in the factorial design, a set of 50 centre points were generated with a uniform distribution

Parameters	Samples	Unit
Shape	Square, Circle, Triangle	—
Area	2.0, 9.5, 17.0, 24.5, 32.0	mm^2
Pressure	5, 15, 25, 35	kPa
Rotation	0, 15, 30, 50, 70, 85	%
Center Point	50 samples $(x, y) \sim U\,([4, 21] \times [4, 21])$	mm

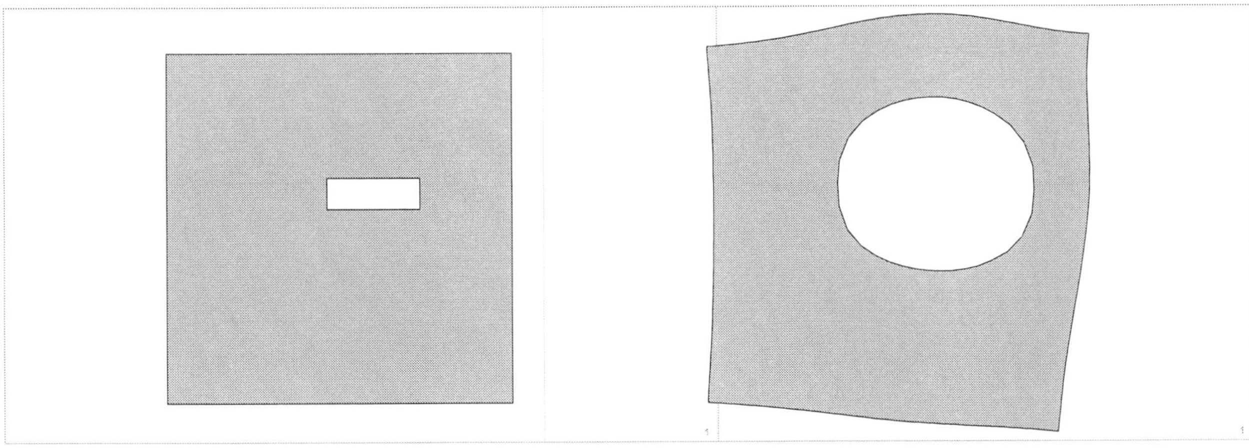

(a) (b)

Figure 3: Deformation of a unit square with a single chamber. **(a)** The undeformed unit square with a single chamber. **(b)** The resulting deformed unit square with pressurized chamber.

From the created data set, assessing the deformed silicone units and comparing the deformations created from the different internal chamber shapes was done using a bespoke deformation measure in the absence of a useful definition of shape. Each side of the deformed unit is discretized along the mesh nodes. The four sides are then separated and rotated to start at the origin and extend in the positive x direction. A polynomial can then be fit to each of the sets of points. The process is depicted in fig. 4. The order of the polynomial fit to the data points can then be used as an indication of shape uniqueness. Polynomials are fit in increasing order from a first-order polynomial up to a maximum of fifteenth order. Measures such as the Akieke information criterion (AIC), the Bayesian information criterion (BIC), and the adjusted coefficient of determination (R^2_{adj}) were used to determine the best-fit polynomial, however, these measures resulted in a very low variance output with all curve fits occupying only two or three polynomial orders. To allow for a richer representation of the output deformation with more variance two approaches were followed. Firstly, an absolute error measure was used to determine the best-fit polynomial. A subset of the data was inspected and an absolute mean squared error (MSE) was set. Polynomials of increasing order were fit to the data and the MSE was calculated. The polynomial order was increased until the MSE was below the set threshold. This was done for all four sides of the unit and the resulting polynomial orders were summed. The second process followed was selecting a single polynomial order to fit all samples. If a polynomial of too high an order is fit the unnecessary explanatory variables are pushed to zero. A sixth-order polynomial was able to fit all samples in the dataset with an acceptable error, given as:

$$y = a_0 + a_1x + a_2x^2 + a_3x^3 + a_4x^4 + a_5x^5 + a_6x^6 \quad (1)$$

The 7 coefficients, a_0 to a_6 of the polynomials then give insight into the data with a large variance. However, this results in 4 sides $\times 7$ coefficients $= 28$ different values to be interpreted. t-distributed Stochastic Neighbour Embedding (t-SNE) [12] was used to inspect the higher dimensional structure in two dimensions. The input data for the embedding was the factorial design values, as in table 1, and the seven coefficients for all four sides of each unit.

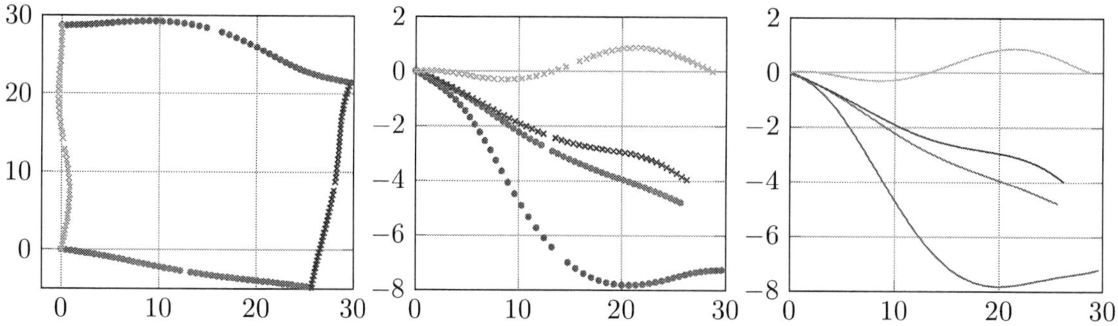

(a) (b) (c)

Figure 4: Process followed to determine the best-fit polynomial order for each side of the deformed unit **(a)** The deformed shape is discretised into 50 points per side **(b)** Each of the sides shifted and rotated to start at the origin and extend in the positive x direction **(c)** The polynomial fits to each side of the shape

In the factorial design discussed above the regular polygons had a fixed aspect ratio. To consider the aspect ratio a similar factorial design was performed as in table 1 but with the addition of the aspect ratio as a parameter. In order to define aspect ratio in the accepted sense as height divided by width, is not general to all shapes, and as such a term called radial distance ratio is used instead. The radial distance ratio is defined as the ratio between the maximum and minimum distance from the center of a regular polygon to its side. The radial distance ratio is depicted in fig. 5. A rectangle was used for this factorial design. The radial distance ratio was varied between 2 and 8. The second factorial design analysis is shown in table 2.

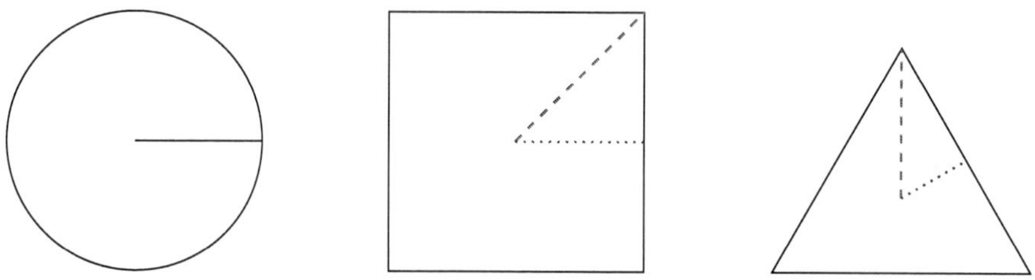

Figure 5: Radial distance ratio definition. The minimum dimension is shown with a blue dotted line and the maximum dimension is shown with a red dashed line. For the circle, these are the same and shown in black.

Table 2: Factorial design of the parameters for the secondary hypothesis. For each combination in the factorial design, a set of 25 centre points were generated with a uniform distribution

Parameters	Samples	Unit
Shape	Rectangle	—
Area	17.0, 24.5, 32.0	mm^2
Pressure	15, 25, 35	kPa
Rotation	0, 27, 54, 90, 126, 153	$^\circ$
Center Point	25 samples $(x, y) \sim U([4, 21] \times [4, 21])$	mm
Radial distance ratio	2, 4, 6, 8	—

3. Results

The first factorial design, as presented in table 1, results in the polynomial order sum for each shape presented in fig. 6. The histograms for the circle and square are very similar. The histogram for the triangle, although of similar form, is distinct from the other two regular polygons. From these results, it is proposed that the triangle chamber design has more unique output deformations when compared to the other two-chamber designs. When calculating the radial distance ratio of all three of the shapes the triangle has the highest radial distance ratio, followed by the square and then the circle, with values of, 2, 1.4 and 1 respectively.

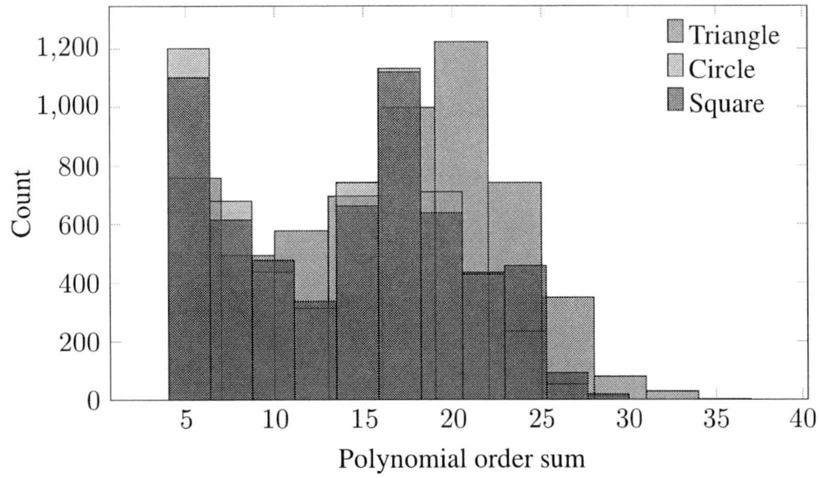

Figure 6: Distribution of the sum of the polynomial orders for each shape. The polynomial orders are calculated for each side of the shape. The sum of the polynomial orders is used as a measure of the complexity of the deformation.

When considering the right-hand side of each of the distributions in fig. 6 the tail lengthens by the radial distance ratio of the three shapes. The longest tail is found on the triangle chamber with a radial distance ratio of 2. To further explore the data t-SNE dimensionality reduction was performed. Figure 8 shows the resulting two-dimensional t-SNE embedding. Clear clusters are formed in the embedding and represent the area of the chambers. This result is expected as larger chamber area will result in more deformation than smaller chamber areas which provide diversity in output deformation and hence yield clusters.

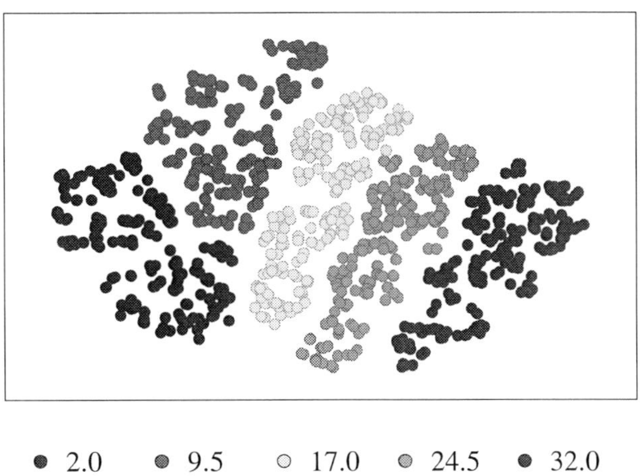

Figure 7: t-SNE representation coloured by area (mm^2). A random sample of 1000 data points is shown

t-SNE was used to further investigate the effect of radial distance ratio on the second factorial design described in table 2. The resulting two-dimensional embeddings also show area clusters being formed, however, in contrast the the previous analysis where the radial distance ratio was not considered in the factorial design, sub-clusters are formed which represent rotation. These results show that the radial distance ratio has a positive effect on the uniqueness of the deformation which enhances the rotation effect of the shape. This result is further supported by the fact that the triangle has the highest radial distance ratio and the longest tail in the polynomial order sum distribution. Additionally, when considering the case of a circle, the radial distance ratio is one and rotation has no effect.

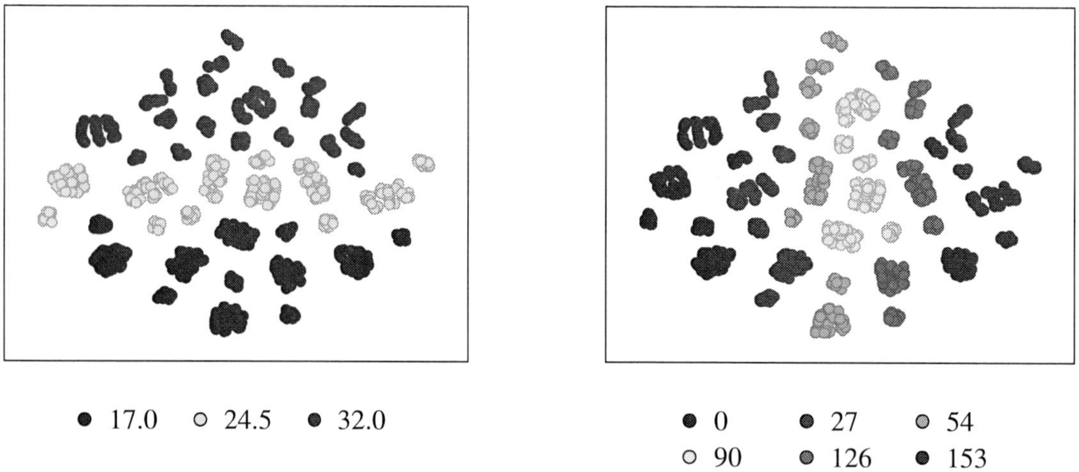

(a) Colored by Area (mm^2)

(b) Colored by Rotation ($^\circ$)

Figure 8: 2D t-SNE representation of the factorial design with aspect ratio. In all plots a random sample of 1000 data points is shown

4. Discussion

When considering the numerous soft pneumatic bending actuators presented in literature and the design of their chambers, the findings of this work are in good agreement. The majority of the designs of bending actuators have elongated rectangular chambers. This work suggests that the chamber shape that allows for larger range of radial distance ratios is more desirable as it allows for more diverse output deformations which is clear from the distinct clusters which form in the two-dimensional t-SNE embedding. Other factors are also at play, such as the practicality of simulating and realizing chamber shapes, and should be considered depending on the specific use case. The design of a soft robot with more sophisticated deformations requires more variables than only chamber shape, however, choosing a chamber shape that has a low variable count but has a versatile deformation output is advantageous. This is all the more important when multiple primitive units are combined to create capable soft robots. To this end, this paper recommends the use of shapes that allow a range of aspect ratios, such as triangles and rectangles.

Acknowledgment

Gerrit Visser for his help with the finite element simulations.

References

[1] Wehner, Michael and Truby, Ryan L., et al. *An integrated design and fabrication strategy for entirely soft, autonomous robots*. Nature, 536(7617), 2016, pages 451-455.

[2] Zhao, Yao and Hong, Yaoye, et al. *Physically intelligent autonomous soft robotic maze escaper*. Science Advances, 9(36), 2023.

[3] Lee, Youngbin and Koehler, Florian, et al. *Magnetically Actuated Fiber-Based Soft Robots*. Advanced Materials, 35(38), 2023.

[4] Schiller, Lars and Seibel, Arthur, et al. *Toward a gecko-inspired, climbing soft robot*. Frontiers in Neurorobotics, 13, 2019.

[5] Pinskier, Joshua and Howard, David. *From Bioinspiration to Computer Generation: Developments in Autonomous Soft Robot Design*. Advanced Intelligent Systems, 4(1), 2022.

[6] Seibel, Arthur and Schiller, Lars. *Systematic engineering design helps creating new soft machines*. Robotics and Biomimetics, 5, 2018.

[7] Runge, Gundula and Peters, Jan, et al. *Design optimization of soft pneumatic actuators using genetic algorithms*. 2017 IEEE International Conference on Robotics and Biomimetics, ROBIO 2017, 2018-January, 2017, pages 393 - 400.

[8] Ligthart, Philip Frederik and Venter, Martin Philip. *A Hierarchical Design Framework for the Design of Soft Robots*. Mathematical and Computational Applications, 28(2), 2023, Article Number 47.

[9] Bhatia, Jagdeep Singh and Jackson, Holly, et al. *Evolution Gym: A Large-Scale Benchmark for Evolving Soft Robots*. Advances in Neural Information Processing Systems, Volume 3, 2021, pages 2201 - 2214.

[10] Chen, Yongyu and Xia, Zeyang, et al. *Optimal Design of Soft Pneumatic Bending Actuators Subjected to Design-Dependent Pressure Loads*. IEEE/ASME Transactions on Mechatronics, 24(6), 2019, pages 2873 - 2884.

[11] Mosadegh, Bobak and Polygerinos, Panagiotis, et al. *Pneumatic networks for soft robotics that actuate rapidly*. Advanced Functional Materials, 24(15), 2014, pages 2163 - 2170.

[12] Van Der Maaten, Laurens and Hinton, Geoffrey. *Visualizing data using t-SNE*. Journal of Machine Learning Research, 9, 2008, pages 2579 - 2625.

MEASUREMENT OF LAMB WAVE PHASE VELOCITY USING LOW-COST EQUIPMENT

Philip W. Loveday[1,*], Paul Fromme[2]

[1]School of Mechanical, Industrial and Aeronautical Engineering, University of the Witwatersrand, Johannesburg, South Africa

[*]Philip.Loveday@wits.ac.za

[2]Department of Mechanical Engineering, University College London (UCL), London, UK

p.fromme@ucl.ac.uk

Abstract: Lamb waves are a family of guided ultrasonic waves that propagate in plate structures. The phase velocity is an important characteristic for material characterisation and locating defects in non-destructive testing applications. The phase velocity may be measured by exciting the Lamb wave at one location and then measuring the response at a set of equally spaced points extending radially from the source. These measurements are often performed using a fixed piezoelectric excitation transducer and a non-contact laser vibrometer to measure the response at several points. This research aims to use low-cost equipment to perform these measurements by replacing the laser vibrometer with a moving piezoelectric transducer or electromagnetic acoustic transducer (EMAT). It was previously found that the piezoelectric transducer had limited repeatability, and this is investigated further. An EMAT was developed and compared to the piezoelectric transducer. The EMAT has poor signal to noise ratio and a high number of averages was required. Measurements at a single location showed repeatability similar to the piezoelectric transducer. When the EMAT was moved to a series of measurement positions the performance was not better than that of the piezoelectric transducer and development of a more accurate positioning technique is required.

Keywords: guided wave ultrasound, low-cost measurement system

1. Introduction

Guided ultrasonic waves can propagate in structures where the boundaries of the structure constrain the propagation of the elastic waves [1]. Lamb waves, propagating in plate structures, are guided waves of great importance in non-destructive testing and monitoring applications because of the potential to inspect a large area from a few transducer locations [2]. The use of guided waves is complicated by the existence of numerous modes of propagation and dispersion, which causes distortion of the wave packet with propagation [1]. Research in this area generally requires both simulation and experimental measurement. While expensive measurement equipment is available in some laboratories, the development and use of low-cost equipment, costing less than USD 1000, would make this field of research accessible to more researchers. Measurements of the phase velocity of the A_0 Lamb wave mode in plates was performed using two custom made piezoelectric transducers and low-cost instrumentation and these were compared to measurements using a state-of-the-art laser vibrometer based system [3]. One piezoelectric transducer was permanently glued to the plate and was used to transmit the waves. The transducer used to measure the response must be positioned at a series of measurement points and therefore was not bonded. It was observed that measurements were not completely repeatable, and it was speculated that the coupling of the piezoelectric transducer to the plate using ultrasonic coupling gel was less consistent than required. In this paper an alternative sensor, an electromagnetic acoustic transducer (EMAT) is developed and compared to the piezoelectric sensor. The EMAT consists of a copper wire coil and permanent magnets and has the potential to be a very low-cost sensor. The sensors do not require mechanical coupling to the structure and can operate as noncontact sensors. The signal to noise ratio (SNR) of EMATs is known to be significantly lower than that of piezoelectric sensors. However, the relative repeatability of the EMAT measurements was not known and this is investigated.

2. Measurement Techniques and Equipment

The phase velocity in plates is often measured to verify theoretical predictions for composite structures or anisotropic materials. When there are multiple modes of propagation, measurements are performed at a series of equally spaced measurement points arranged in the direction of wave propagation. The two-dimensional Fourier transform was used to estimate the dispersion curves from these measurements in [4] and this was extended by adding the matrix pencil method in [5] where 488 equally spaced points were measured using a laser vibrometer. When only a single mode of propagation is present it is possible to estimate the phase velocity from measurements at two points and this is illustrated in Figure 1. Figure 1a shows two response signals measured at two positions, which were 5 mm apart. The first wave packet is the excitation signal electrically coupled in the oscilloscope and this is independent of the receive transducer position. The wave packet arriving at 10 ms is the arrival of the A_0 mode (fundamental flexural/bending mode of propagation), which is of interest and the slight delay between the two signals is visible. The wave packet at the end of the time trace is a wave reflected from a boundary of the plate. The phase delay between the two signals is estimated by firstly time gating the signals to retain only the parts of the signals containing the arrival of the A_0 Lamb wave mode as shown in the figure. The fast Fourier transform (FFT) of the time gated signals is then computed and one frequency domain signal is divided by the other to yield a ratio. The phase angle of this ratio is then computed, and this is plotted in Figure 1b. At the frequency of 101169 Hz the phase difference between the two signals is 1.72236 rad. The wavelength may be calculated using the distance between the two measurement points of 5 mm as 18.24 mm. The phase velocity is then calculated by multiplying the wavelength by the frequency (1845 m/s). Note that the spacing between the two measurement points was less than one wavelength and this is required.

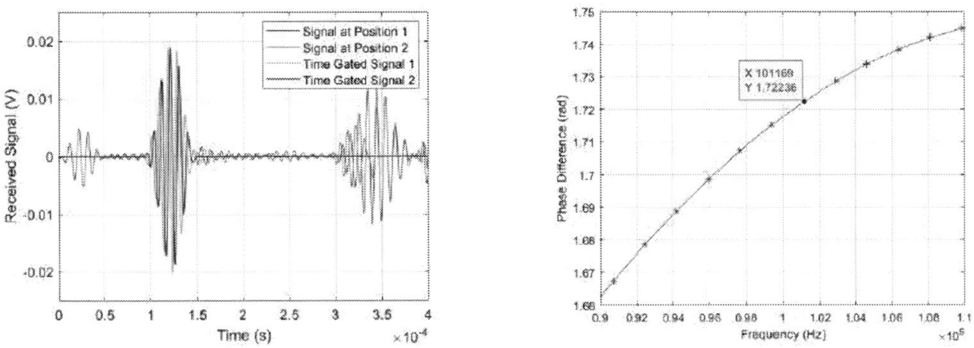

(a) (b)

Figure 1. (a) Measured signals at two positions 5 mm apart and (b) estimation of the phase difference.

While it is possible to obtain an estimate of the phase velocity from only two measurement points it is expected that a better estimate will be obtained by using a larger number of points spanning a greater length on the plate. In our previous work [3], 100 measurement points were used in the laser vibrometer measurements, while 21 measurements were performed when manually scanning the piezoelectric receive transducer.

The equipment used in the low-cost measurement setup used in this paper is listed in Table 1 along with price estimates. The cost of the equipment is less than the target of USD 1000, believed to be acceptable for most researchers. The focus of this paper is not on further cost reduction but rather on investigating whether the EMAT transducer can provide better performance than the piezoelectric transducer as a replacement for the laser vibrometer.

Table 1: Low-cost measurement equipment.

Equipment	Make and Model	Cost Estimate Piezo System (USD)	Cost Estimate EMAT System (USD)
USB Oscilloscope	Digilent Analogue Discovery 2 Pro Bundle	430	430
Power Amplifier	Juntek DPA-2698	70	70
Power Transformer	EPCOS N87 RM14 ferrite core	-	20
Piezoelectric Transducers	Custom made - estimate for material purchases of USD 50 per transducer	100	50
EMAT Transducer	N42 permanent magnets	-	20
Pre-amplifier	Custom made – estimate for electronic components.	-	30
Total Cost		**600**	**620**

A piezoelectric transducer was glued to the plate using a gel superglue and this transducer was used to excite the plate. The transducer was a custom-made sandwich transducer with stainless steel front and back sections and a single piezoelectric ceramic ring (Ferroperm Pz29 material). A second identical piezoelectric transducer and a custom made EMAT were used to measure the response of the plate. The changes in position of the receive transducers were measured using a vernier calliper attached to the plate. The transducers are shown in Figure 2.

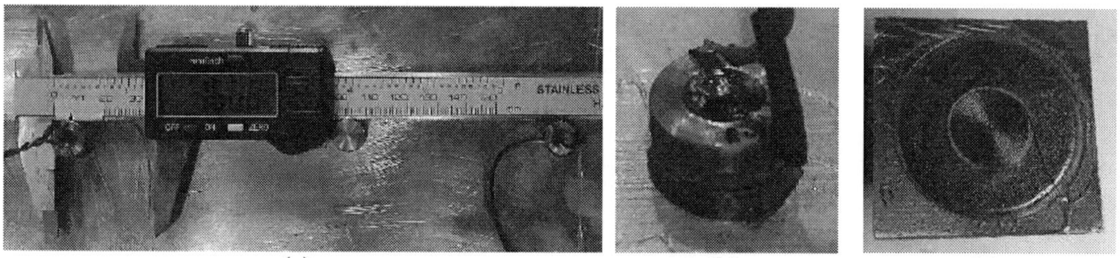

(a) (b) (c)

Figure 2. Transducers and positioning (a) piezoelectric transducer on the right glued to plate, (b) moveable piezoelectric transducer, (c) EMAT showing coil which would be facing the plate.

The transmit transducer was driven with a peak-to-peak amplitude of 53 Vpp by using a Juntek DPA-2698 amplifier and a transformer. The signal supplied to the amplifier was a 10 cycle Hanning windowed tone burst generated by a Digilent Analog Discovery 2 USB oscilloscope. The USB oscilloscope was used to measure the response from the transducers. When the piezoelectric transducer was used for measuring the response, the transducer was connected directly to the USB oscilloscope. The EMAT produced a very small current and a custom-made preamplifier was used to amplify the signal. This preamplifier used a transimpedance amplifier as the first stage to convert the current to a voltage and then two subsequent gain stages to increase the amplitude of the voltage signal. Averaging and filtering were performed by the USB oscilloscope and the data was saved in files for later processing.

3. Electromagnetic Transducer Development

EMATs may be used when waves in conductive materials are to be excited or sensed. The EMAT comprises a magnet (usually a permanent magnet) and a coil. The transduction mechanisms can include Lorentz force, magnetisation force, and magnetostriction in magnetic metals. When the structure is aluminium the only mechanism of transduction is the Lorentz force. The Lorentz force arises when current flows within a magnetic field and the force generated is directed perpendicularly to the current and magnetic field. It should be noted that in an EMAT a conductor is positioned near the structure and induces a current in the skin of the conductive structure, therefore the Lorentz force acts near the surface of the structure. This means that EMATs can operate without physical contact with the plate. The same mechanism may be used for sensing the velocity of the metallic surface. In this case the velocity of the surface normal to the magnetic flux gives rise to a current flowing in the surface of the structure and this current induces a current in the nearby conductor, which can be measured. A quantitative analysis of the excitation and reception of Lamb waves in plates was provided by Thompson [6].

The wavelength of the A_0 mode in the plate at 100 kHz is approximately 20 mm and the active area of the transducer should be no more than half of the wavelength. Ideally the receive transducer should be omnidirectional, i.e., have the same sensitivity for waves approaching from any direction so that the angular orientation of the sensor is not critical. It was decided to use a spiral coil and axisymmetric magnets in the EMAT design. When a cylinder is positioned vertically above a horizontal plate the magnetic flux density is perpendicular to the plate along the axis of the cylinder. At points on the plate away from the axis of symmetry there will be a component of the magnetic flux density in the radial direction and this component is largest near the outer radius of the cylinder. The two components of magnetic flux density will cause radial shear stresses and normal stresses and can excite and sense radially polarized shear waves and longitudinal waves in a dual-mode EMAT [7]. In this paper the radial component of the magnetic flux is used with the spiral coil to detect the motion of the surface in the direction normal to the surface. Instead of using a small cylindrical magnet it was thought that it would be possible to use a larger ring magnet and use the radial component of the magnetic flux density near the inner radius of the magnet. The magnetic flux densities in the plate were investigated using finite element modelling.

The software FEMM 4.2 was used to analyse the magnetic field produced in the surface of an aluminium plate by a permanent magnet positioned above the plate. As the magnet geometries are cylinders and rings it was possible to use axisymmetric models. The FEMM software was driven from Matlab 2020b using the OctaveFEMM interface. This meant that parametric studies could be rapidly conducted. Only results for two available magnet sizes are presented in this paper although other sizes were considered. The dimensions of the cylinder magnet were 3 mm diameter and 4 mm length, while the ring had 25 mm outside diameter, 7 mm inside diameter and 5 mm thickness. The material of the neodymium magnets was N42. Four models were prepared: a cylindrical magnet above an aluminium plate, a ring magnet above the plate, a combination of the concentric cylinder and ring magnets above the plate and a combination of the concentric cylinder and ring magnets above the plate but with the polarisation of the cylinder magnet reversed. The plate was aluminium and had a thickness of 5 mm. The gap between the magnets and the plate was 0.5 mm in all models. Figure 3 shows the model of the cylinder and ring magnets and the computed magnetic flux density distribution for this model. The magnetic flux density a small distance inside the surface of the plate was extracted from the models. The component of the magnetic flux density tangent to the plate (radial component) for the different configurations is plotted in Figure 3c as a function of the radial position. These results show that for the cylinder alone the maximum of the tangential magnetic flux density occurs near the outer radius of the cylinder (1.5 mm) and for the ring alone there is a maximum near the inner radius (3.5 mm) and a minimum near the outer radius (12.5 mm). The combination of the cylinder and the ring produced greater magnetic flux densities close to centre of the model as expected. When the cylinder was inverted (polarisation direction reversed) the magnetic flux density was initially negative peaking near the outer radius of the cylinder and then becoming positive and peaking near the inner radius of the ring. This configuration would be expected to be significantly less sensitive than the original cylinder – ring configuration, which is predicted to be the most sensitive configuration. Due to the wavelength of the Lamb wave to be measured the diameter of the coil was limited to 10 mm, therefore we are interested in the magnetic flux density from the axis of symmetry to a radius of 5 mm. The flux density was numerically integrated from a radius of 0 mm to 5 mm to provide a single measure of sensitivity for each configuration, which could be compared to the measured sensitivity of each configuration.

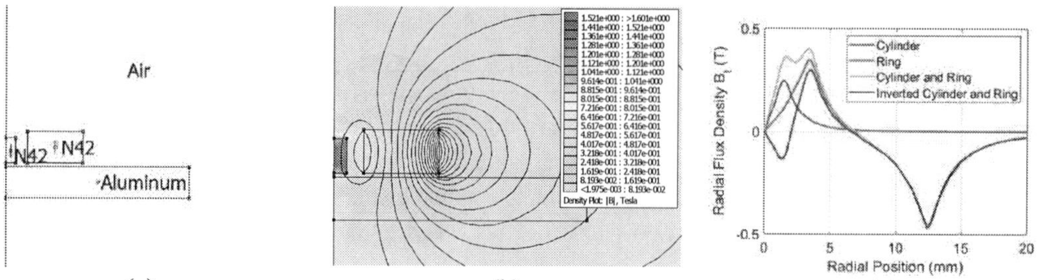

(a) (b) (c)

Figure 3. Magnetic flux modelling (a) model of cylinder and ring, (b) computed magnetic flux density, (c) radial component of magnetic flux density in plate surface for the four models.

A spiral coil of approximately 10 mm diameter was wound by hand using 0.18 mm enamel coated copper wire. Different magnet configurations were placed on top of the coil to determine the sensitivity of each configuration.

The measurement setup described in section 2 was used and the response at approximately 260 mm from the excitation was measured for each magnet configuration for the first arrival of the ultrasonic waves. The results of these measurements are listed in Table 2 where they are compared to the integrals of the magnetic flux densities from the numerical modelling. The measured and computed values were normalised by dividing by the value for the cylinder-ring configuration. The normalized results show very good correlation between the measured and predicted sensitivities for the three less sensitive configurations. It is believed that the numerical model could be used to optimise the dimensions of the magnets if non-standard sizes are to be considered in future.

Table 2: Measured sensitivities compared to predicted magnetic flux densities.

	Measured (Vpp)	**FEM Integral (T.mm)**	**Measured Normalised**	**FEM Integral Normalised**
Cylinder	0.55	3.21	0.32	0.37
Ring	1.15	5.57	0.66	0.64
Cylinder & Ring	1.73	8.72	1	1
Inverted Cylinder & Ring	0.48	2.41	0.28	0.28

The cylinder-ring configuration was selected for further development and testing. The cylinder magnet was glued inside the ring magnet and the coil was glued to the magnets. An epoxy layer was applied to the coil side of the EMAT. To allow for accurate orientation of the EMAT a rectangular piece of printed circuit board with a hole in the middle was glued to the EMAT, as shown in Figure 2c.

4. Measurement Results

When performing the phase velocity measurement there are various sources of error. Noise in the signals, changes in the contact conditions between the receive transducer and plate and inaccuracies in the measurement of the position of the transducer will cause errors. Changes in the transducer (either transmitting transducer or receiving transducer) and changes in the phase velocity in the plate (possibly due to temperature changes) will appear as phase velocity errors. Measurements were performed without moving the transducers to determine how noise in the signals contributes to phase errors. Measurements were then performed with removal of the transducer and replacement in the same position (against a mechanical guide) between measurements to determine if this had an influence on the measured phase of the signals. Finally, measurements were performed with the receive transducers positioned at a series of measurement points for estimating the phase velocity.

A series of measurements was performed with the EMAT and a piezoelectric receive transducers to compare the performance of the two transducers without repositioning. The random noise in the signals may be reduced by averaging and the noise decreases in proportion to the square root of the number of averages. One hundred measurements were performed with 1, 10, 100 and 1000 averages each. The decrease in noise that can be obtained by averaging when using the EMAT is shown in Figure 4 where each plot contains 100 measurements performed with different numbers of averages. The measurements performed with the piezoelectric receive transducer do not show such a large improvement with averaging because the SNR is already good without averaging.

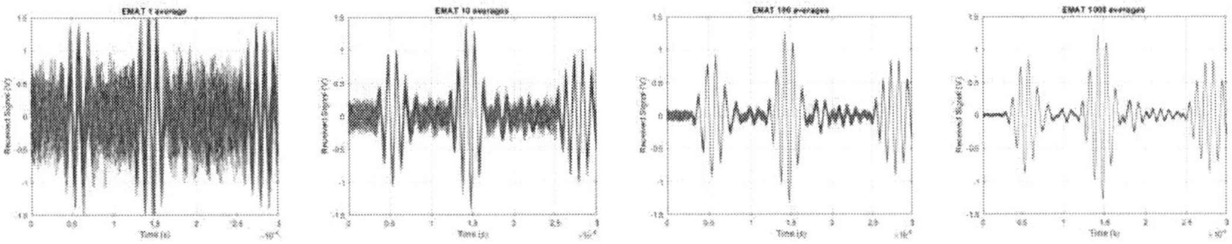

Figure 4. Decrease in EMAT measurement noise by increasing number of averages (1, 10, 100, 1000).

Each of the 100 signals captured for each of the four numbers of averages was processed to extract the phase of the signal at 100 kHz and this is plotted in Figure 5 for the two different receive transducers. The average phase and the standard deviation of the 100 measurements is listed in Table 3. It is evident that the variation in the phase measurements performed with the EMAT decreases with the number of averages. The standard deviation improved slightly for the piezoelectric receiver measurements with 10 or 100 averages but was worse for the 1000 average measurements. The 1000 average measurements took approximately 90 minutes and there appears to be some gradual change in the average phase during this time, which could be due to temperature changes. From these results it was decided that 1000 averages should be used when measuring with the EMAT while 10 or 100 averages is adequate when measuring with the piezoelectric receive transducer.

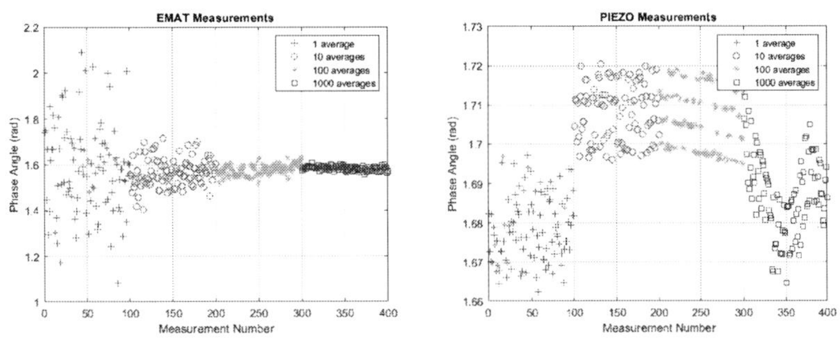

Figure 5. Variations in phase measurement for different numbers of averages.

Table 3: Influence of averaging on the repeatability of phase measurements.

No. of Averages	EMAT average phase (rad)	EMAT standard deviation (rad)	Piezo average phase (rad)	Piezo standard deviation (rad)
1	1.5854	0.1982	1.6788	0.0086
10	1.5653	0.0615	1.7079	0.0072
100	1.5763	0.0237	1.7071	0.0074
1000	1.5821	0.0099	1.6871	0.0101

Next the effect of removing the transducers and replacing them between measurements was considered. A position guide was glued to the plate and used to ensure repeatable repositioning of the transducers between measurements.

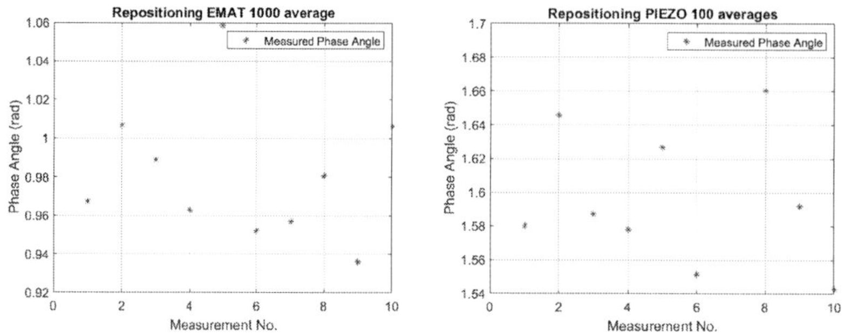

Figure 6. Phase measurements with repositioning of the transducers between measurements.

When the piezoelectric receive transducer was removed the ultrasonic coupling gel was wiped off the transducer and the plate and new gel was applied before the transducer was repositioned. The phase measurement results are shown in Figure 6. The standard deviation of the EMAT phase measurements was 0.0354 rad while that of the piezoelectric transducer measurements was 0.0481 rad. These values are similar but larger than those recorded without repositioning in Table 3, especially for the piezoelectric transducer. The repositioning of the transducers has therefore caused a decrease in the consistency of the phase measurements.

Finally, the measurements were performed along a line from the transmit transducer at 21 positions with 5 mm steps between measurement locations. The phase was estimated at each of the locations and is plotted in Figure 7a for the EMAT measurements and Figure 7b for the piezoelectric receiver measurements. A linear curve fit was performed. The slope of this line can be used to calculate the phase velocity. It is noticeable that the measured points lie closer to the line for the piezoelectric transducer measurements than for the EMAT measurements. The distances between the measurement points and the lines (the residuals) are plotted in Figures 7c and 7d.

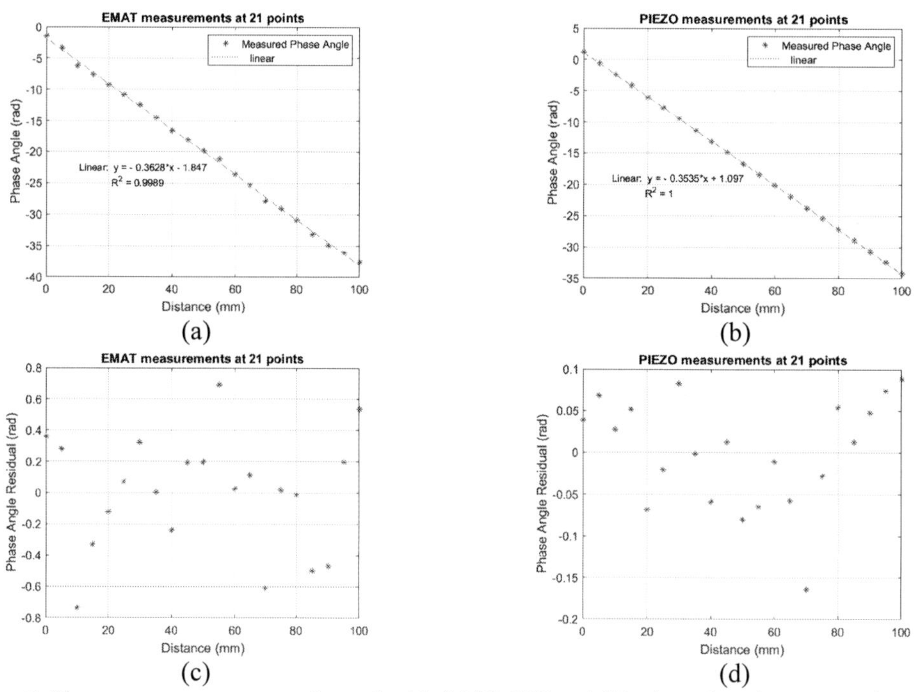

Figure 7. Phase measurements performed with (a) EMAT and (b) piezoelectric sensor, phase angle residuals from linear regression of (c) EMAT measurement and (d) piezoelectric sensor measurement.

The standard deviation of the residuals was 0.3716 rad and 0.0654 rad for the EMAT and piezoelectric transducer measurements respectively. The standard deviation of the piezoelectric transducer measurements was only slightly worse than that of the measurements performed with repositioning of the transducer, while for the EMAT the standard deviation was significantly worse. During the measurements it was difficult to position the EMAT and obtain signals with consistent amplitude and it is suspected that the positioning jig was lifting the EMAT at times. Further work is required to improve the positioning of the EMAT without disturbing the orientation and distance from the plate surface.

5. Conclusions and Recommendations

The EMAT developed for this research produced high levels of noise and averaging of 1000 repeated measurements was required to reduce the influence of the noise. Improving the SNR could be investigated in future by trying better electrical shielding of the EMAT and higher drive levels on the piezoelectric transmit transducer. Removing and replacing the transducers in nominally the same position caused increased variations in the phase of the received signals, especially when the piezoelectric receive transducer was used. A more repeatable coupling method could be explored for this transducer. When the EMAT transducer was positioned at different locations there appeared to be positioning errors and the method of positioning should be improved. If this can be achieved it is expected that the EMAT could produce phase velocity measurements with similar accuracy as those achieved with the piezoelectric receive sensor. Significant further improvements would be required to achieve the performance of the laser vibrometer system.

Acknowledgements

The authors acknowledge funding from the Wits-UCL Research/Teaching Collaborative Activity Seed Fund 2022/23.

References

[1] J. L. Rose, *Ultrasonic Guided Waves in Solid Media*. Cambridge University Press, 2014.

[2] R. P. Dalton, P. Cawley et al. "The Potential of Guided Waves for Monitoring Large Areas of Metallic Aircraft Fuselage Structure," *J Nondestr Eval*, vol. 20, no. 1, pp. 29–46, 2001.

[3] P. W. Loveday and P. Fromme, "Measurement of Ultrasonic Guided Waves in Plates using Low-Cost Equipment," in *50th Annual Review of Progress in Quantitative Nondestructive Evaluation*, American Society of Mechanical Engineers, Jul. 2023.

[4] D. N. Alleyne and P. Cawley, "A two-dimensional Fourier transform method for the measurement of propagating multimode signals," *J Acoust Soc Am*, vol. 89, no. 3, pp. 1159–1168, 1991.

[5] F. Schöpfer, F. Binder *et al.*, "Accurate determination of dispersion curves of guided waves in plates by applying the matrix pencil method to laser vibrometer measurement data," *CEAS Aeronaut J*, vol. 4, no. 1, pp. 61–68, Apr. 2013.

[6] R. B. Thompson, "A model for the electromagnetic generation of ultrasonic guided waves in ferromagnetic metal polycrystals," *IEEE Trans Sonics Ultrason*, vol. 25, no. 1, pp. 7–15, 1978.

[7] M. Hirao and H. Ogi, *Electromagnetic Acoustic Transducers*. Tokyo: Springer Japan, 2017.

APPROPRIATE MODE OF STORAGE OF PORCINE HEART MYOCARDIA WITHOUT TEMPERING WITH ITS PASSIVE MECHANICAL PROPERTIES

I. Mabuda[1], T. Pandelani[1*], H. Ngwangwa[1], F. Nemavhola[2]
[1]Unisa Biomedical Engineering Research Group, Department of Mechanical Engineering, School of Engineering, College of Science Engineering and Technology, University of South Africa, Science Campus, Florida, 1710.
[2]Faculty of Engineering and the Built Environment, Durban University of Technology, Durban, South Africa
*Correspondence: epandet@unisa.ac.za; Tel.: (011) 471-2137

Abstract

One of the important aspects of mechanics of soft tissues is perseveration during and before mechanical testing. This study addresses the critical aspect of preserving soft tissue integrity for mechanical testing, particularly focusing on the myocardium of the porcine heart. The primary objective of the study was to assess the impact of different preservation methods on the mechanical properties of this tissue.

A series of experiments were conducted. Porcine heart myocardium was subjected to tensile testing using a uniaxial machine as a means of evaluating the traditional preservation methods employed in the field. To simulate real-world scenarios, harvested myocardium samples were initially preserved by immersion in either fresh water or a NaCl solution. These samples were then stored under varying temperature conditions, including freezing at -2°C, -10°C, and -20°C, or refrigeration at 2°C-3°C. The preserved myocardium underwent a thawing process before undergoing mechanical testing.

The study specifically focused on the right ventricle of the porcine heart, known for its significant role in cardiac function. Notably, frozen right ventricular myocardium exhibited a slight increase in tensile strength, particularly in the circumferential direction, when compared to fresh tissue.

Overall, this research sheds light on the potential impact of preservation methods and storage temperatures on myocardial mechanical properties, emphasizing the importance of considering these factors in biomechanical studies of cardiac tissues.

Keywords: Soft tissue, heart myocardium, tensile testing, preservation

1. Introduction

The mechanical properties of soft tissues, especially those of cardiac muscle, are of paramount importance in the field of biomedical engineering and cardiology. Myocardial tissue, being a vital component of the heart, requires rigorous investigation and preservation methods to ensure accurate assessment of its passive mechanical properties. This literature review explores existing research related to the preservation of porcine heart myocardium and its impact on passive mechanical properties, laying the foundation for understanding the significance of appropriate storage modes.

The passive mechanical properties of myocardium, characterized by parameters like stiffness, elasticity, and tensile strength, play a crucial role in cardiac function and pathology [1-4]. Studies have shown that myocardial tissue exhibits anisotropic behavior, meaning that its mechanical properties vary depending on the direction of deformation [4]. The understanding of these properties is essential for the development of therapies and interventions in cardiac diseases [5-6].

Preservation of myocardial tissue is necessary to maintain its structural integrity and mechanical properties during experimentation. Commonly used preservation methods include freezing, refrigeration, and immersion in preservative solutions. While these methods are prevalent, the literature suggests variations in their effectiveness in preserving the mechanical properties of myocardium [7-10].

Various studies have investigated the impact of preservation methods on myocardial mechanical properties. The effects of freezing and refrigeration on myocardial stiffness were investigated [7]. They reported that refrigeration at 4°C better maintained the tissue's passive mechanical properties, including elasticity, compared to freezing at -20°C.

The temperature at which myocardium is stored is a critical factor affecting tissue preservation. Research by [8] demonstrated that freezing myocardium at -80°C leads to a significant reduction in tensile strength, highlighting the importance of considering storage temperature.

Studies have explored the use of preservation solutions to mitigate the detrimental effects of freezing on myocardium. Another study investigated the impact of cryoprotectants in combination with freezing and found improvements in tissue preservation, suggesting the potential for optimized preservation protocols [9].

While the literature primarily focuses on porcine myocardium, it is essential to acknowledge that species differences can influence tissue properties and preservation requirements. Another research emphasised the importance of selecting appropriate animal models for specific studies [10].

The preservation of porcine heart myocardium is a crucial aspect of research into its passive mechanical properties. Existing literature highlights the influence of storage methods, temperatures, and preservation solutions on these properties. While various studies have contributed valuable insights, there remains a need for further research to establish standardized protocols for preserving myocardial tissue without altering its passive mechanical properties. This topic holds great potential for advancements in cardiac biomechanics and the development of therapeutic strategies for cardiovascular diseases.

2. Material Preparation

a. Sample Collection

The study involved the collection of 150 hearts from South African landrace pigs, carefully selected within the age range of 6 to 8 months and weighing 200kg and above. This selection criterion aimed to ensure a representative sample set [4]. The hearts were sourced from an abattoir based on their availability, aligning with the practical constraints of the research context.

b. Preservation and Transportation

Maintaining tissue condition as close to its physiological state as possible was a paramount concern in this study. To achieve this, the entire porcine hearts were meticulously packaged and transported from the abattoir to the testing laboratory at the University of South Africa. This transportation was executed with precision, utilising a cooler box to regulate temperatures within the range of -2°C to -4°C. This approach to preservation during transportation was adopted to mitigate tissue degradation and maintain sample integrity during transit, ensuring that the samples remained suitable for subsequent mechanical testing [8].

c. Storage Conditions

Upon arrival at the laboratory, the hearts were subjected to various freezing temperatures, with specifics detailed in the research protocol. These hearts were stored at different freezing temperatures, ranging from below freezing to moderately cold, for a maximum duration of one week. This predetermined storage period was implemented to simulate practical conditions commonly encountered in research scenarios [9].

d. Dissection and Specimen Preparation

Post-storage, the hearts underwent meticulous dissection procedures. The septum, left ventricle, and the upper section of the heart, including both atria and their associated valves, were carefully dissected away. This process left only the right ventricle wall available for testing. The dissection was performed with precision to ensure that the right ventricle wall was cleanly separated, and the myocardial tissue specimen was extracted from the mid-section of the wall. By selecting the thickest section of the wall for specimen extraction, the study aimed to minimize potential variations in the samples [1,5].

All excess tissue, including the pericardium and endocardium, was diligently removed using surgical knives and scissors, ensuring that the myocardial specimens remained free of extraneous materials. Two identically sized specimens, each measuring 40 mm in length, 20 mm in width, and 10 mm in thickness, were consistently cut from each wall. This standardization in specimen dimensions was essential for maintaining uniformity throughout the study. Additionally, before testing, samples were further prepared by carefully separating the endocardium and pericardium from the myocardium, further ensuring that the specimens were pure myocardial tissue [7].

The thickness of the myocardial samples was meticulously maintained using a digital vernier, a precise measurement instrument, to adhere to established research standards for tissue preparation [4, 9].

3. Experimental testing

Each tissue specimen underwent mechanical testing utilizing the CellScale UStretch system, specifically designed for the evaluation of soft tissue mechanical properties [3,4]. This system, renowned for its precision and applicability to soft tissue mechanics, was chosen to ensure accurate and reliable data collection.

During cyclic testing, a saw-tooth function was employed, a common approach in soft tissue mechanical testing. The testing regimen consistently utilized a total testing length of 25 mm, thereby leaving clamping lengths of 7.5 mm at each end. This standardization in testing length contributes to the reproducibility and comparability of test results [4].

To ascertain the accurate dimensions of each sample, including length, width, and thickness, a Vernier caliper was utilized. Four measurements were taken for each dimension to enhance precision, and the average dimensions were employed for subsequent data processing [4, 5].

Before collecting mechanical data, a normal preload of 0.1 N was applied to remove any tissue slackness, ensuring that subsequent test results accurately represented the tissue's response to cyclic loading [11].

The tissues underwent a total of three cycles during testing, each characterized by a maximum strain of 40%. This specific strain magnitude was chosen to mimic the physiological conditions of the mammalian heart. Each cycle had a duration of 6 seconds, contributing to the simulation of realistic cardiac mechanical behavior [6].

To maintain tissue hydration and replicate physiological conditions, a 0.91% w/v NaCl saline solution was placed in the testing bath and maintained throughout the testing duration. This practice aligns with established procedures in cardiac biomechanics research [5].

4. Experimental results

Figure 1 shows the porcine right ventricle uniaxial stress and strain graphs three cycles (stretching and relaxing), A (Circumferential) and B (Longitudinal), immersed in water during storage.

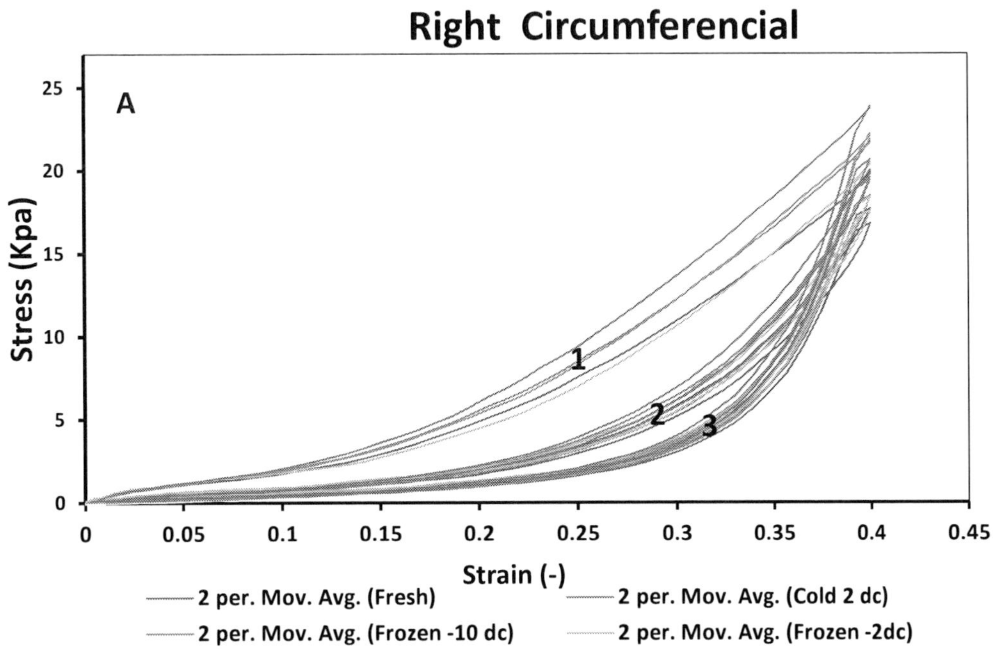

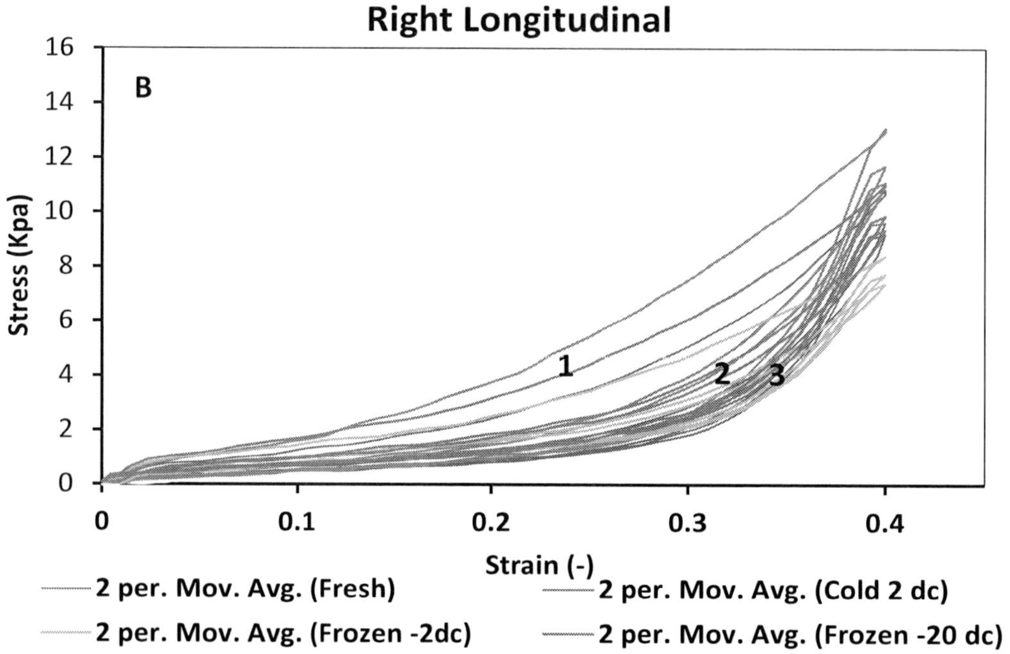

Figure 1: Porcine right ventricle uniaxial stress and strain graphs three cycles (stretching and relaxing), A (Circumferential) and B (Longitudinal), immersed in water during storage.

Figure 2 describe the porcine right ventricle uniaxial stress and strain graphs three cycles (stretching and relaxing), A (Circumferential) and B (Longitudinal), immersed in NaCl solution during storage.

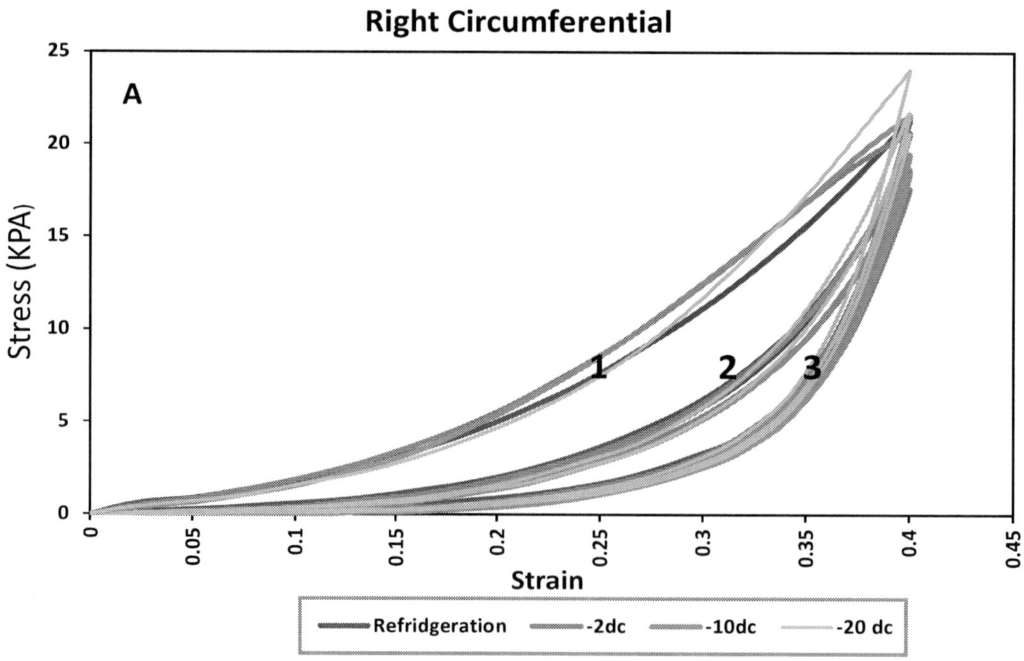

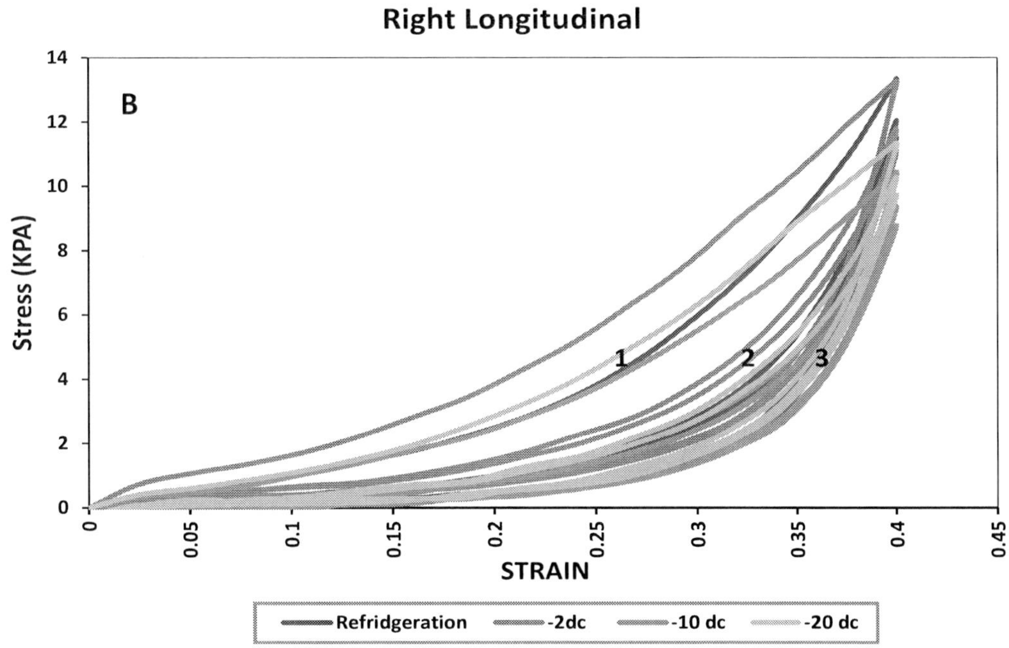

Figure 2: Porcine right ventricle uniaxial stress and strain graphs three cycles (stretching and relaxing), A (Circumferential) and B (Longitudinal), immersed in NaCl solution during storage.

Figure 3 shows the uniaxial stretched circumferential stress-strain curve (A-C) and Uniaxial stretched longitudinal stress- strain curve (D-F) for porcine right ventricle at 6.67 s-1 strain rate and all previously immersed in water.

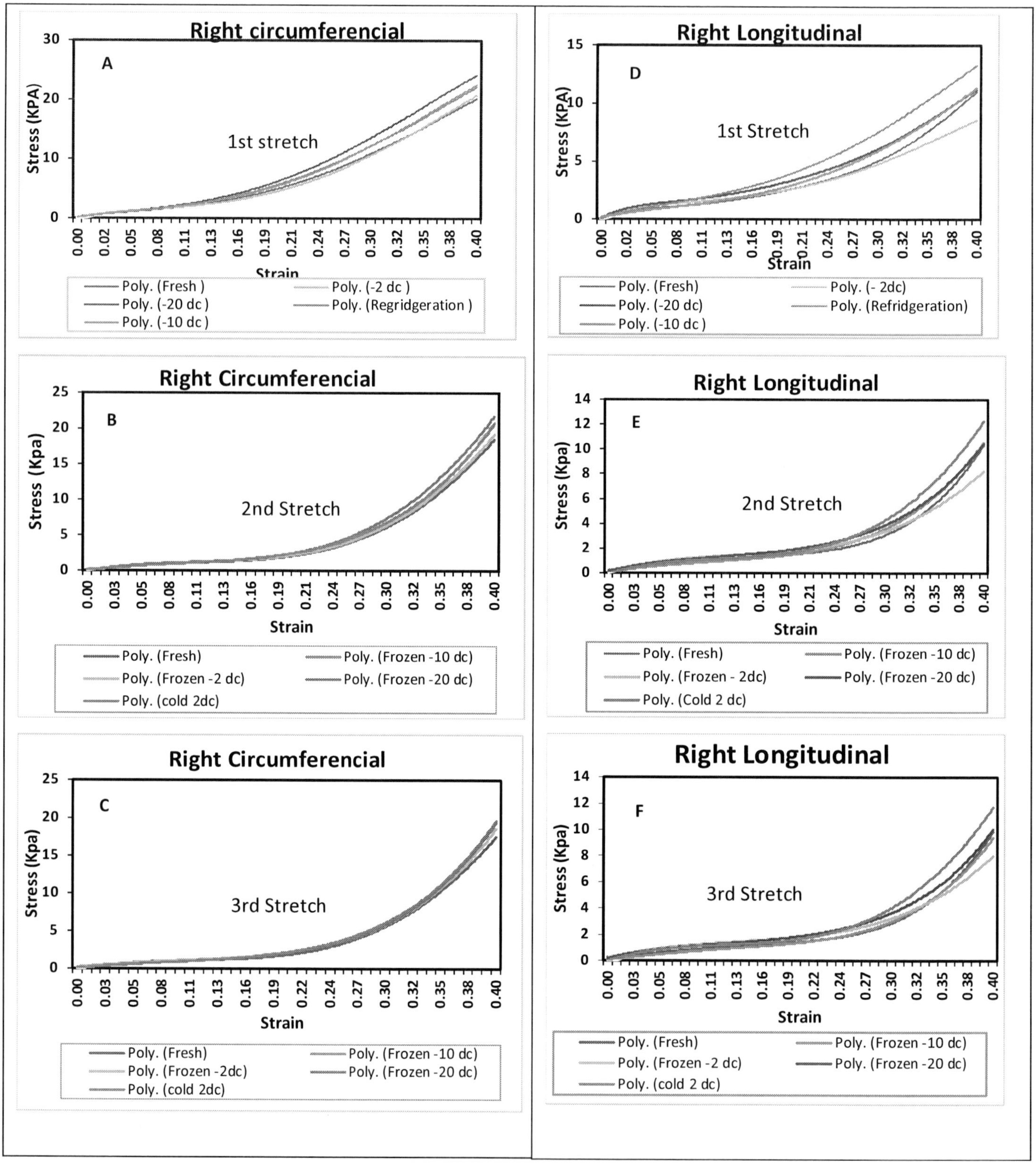

Figure 3: Uniaxial stretched circumferential stress-strain curve (A-C) and uniaxial stretched longitudinal stress- strain curve (D-F) for porcine right ventricle at 6.67 s-1 strain rate and all previously immersed in water.

5. Discussion and concussions

The present study aimed to investigate the influence of various preservation methods and storage temperatures on the mechanical properties of porcine heart myocardium, focusing primarily on the right ventricle. The findings provide valuable insights into the potential impact of preservation techniques on the mechanical behavior of cardiac tissue.

The utilisation of different preservation methods, such as immersion in still water and a NaCl solution, allowed for a comprehensive assessment of their impact on myocardial mechanical properties. While these preliminary findings indicate differences in tensile strength, further analysis is required to elucidate the specific mechanisms underlying these variations. Existing research has shown that preservation solutions can affect tissue osmolarity and hydration, potentially influencing tissue mechanical behavior [9].

The storage temperatures tested, ranging from -2°C to -20°C in a freezer and 2°C to 3°C in a refrigeration unit, revealed that frozen right ventricular myocardium exhibited a slight increase in tensile strength compared to fresh tissue, particularly in the circumferential direction. This observation is consistent with previous studies that have reported alterations in tissue properties with changes in temperature [8]. The mechanisms behind these temperature-related effects warrant further investigation.

The decision to focus on the right ventricle for mechanical testing is scientifically justified by its vital role in cardiac function. Understanding the mechanical properties of the right ventricle, especially in relation to the left ventricle, can provide valuable insights into cardiac biomechanics and pathophysiology [1, 5, 6].

When comparing these preliminary findings to existing literature, several important considerations arise. Previous studies have investigated the effects of preservation methods and storage temperatures on soft tissues, including myocardium [7-8]. However, the specific impact of these factors on right ventricular myocardium and their implications for cardiac biomechanics remain relatively unexplored.

This study presents initial findings suggesting that preservation methods and storage temperatures may influence the mechanical properties of right ventricular myocardium. These observations raise intriguing questions about the underlying mechanisms and the potential clinical significance of such alterations. Future research should delve deeper into the effects of specific preservation techniques and temperature conditions on cardiac tissue, aiming to enhance our understanding of myocardial biomechanics and its relevance to cardiovascular health.

References

[1] Sirry, M.S.; Butler, J.R.; Patnaik, S.S.; Brazile, B.; Bertucci, R.; Claude, A.; McLaughlin, R.; Davies, N.H.; Liao, J.; Franz, T. Characterisation of the mechanical properties of infarcted myocardium in the rat under biaxial tension and uniaxial compression. J. Mech. Behav. Biomed. Mater. 2016, 63, 252–264.

[2] Martonová, D.; Alkassar, M.; Seufert, J.; Holz, D.; Dương, M.T.; Reischl, B.; Friedrich, O.; Leyendecker, S. Passive mechanical properties in healthy and infarcted rat left ventricle characterised via a mixture model. J. Mech. Behav. Biomed. Mater. 2021, 119, 104430.

[3] Nemavhola, F., Biaxial quantification of passive porcine myocardium elastic properties by region. Engineering Solid Mechanics, 2017: p. 155-166.

[4] Ngwangwa, H., et al., Determination of Cross-Directional and Cross-Wall Variations of Passive Biaxial Mechanical Properties of Rat Myocardia. Processes, 2022. 10(4): p. 629.

[5] Sacks, M. S., & Yoganathan, A. P. (2007). Heart valve function: a biomechanical perspective. Philosophical Transactions of the Royal Society B: Biological Sciences, 362(1484), 1369-1391.

[6] Nemavhola, Fulufhelo, Harry M. Ngwangwa, and Thanyani Pandelani. "An investigation of uniaxial mechanical properties of excised sheep heart muscle fibre–fitting of different hyperelastic constitutive models." (2021).

[7] He, Z., Wan, F. H., Xie, L. Y., & Fan, Y. (2019). Effects of preservation on the mechanical properties of porcine myocardium. Journal of Surgical Research, 233, 381-386.

[8] Yin, J., Yang, L., Xie, Z., & Song, X. (2016). The effect of cryopreservation on mechanical properties of myocardium tissue. Cryobiology, 72(3), 258-262.

[9] Phan, L., Bordes, P., & Macchi, L. (2020). Effect of cryoprotectants on the mechanical properties of myocardium tissue. Cryobiology, 95, 139-143.

[10] Bell, S. P., Nyland, L., & Tischler, M. D. (2014). Species specificity in the biomechanical response of preserved myocardium. Journal of Biomechanical Engineering, 136(12), 121003.

[11] Calvo-Gallego, Jose L., Jaime Domínguez, Tomás Gómez Cía, Gorka Gómez Ciriza, and Javier Martínez-Reina. "Comparison of different constitutive models to characterize the viscoelastic properties of human abdominal adipose tissue. A pilot study." Journal of the mechanical behavior of biomedical materials 80 (2018): 293-302.

Studying transvalvular aortic haemodynamics with simplified fluid-structure interaction and experiment: Lessons Learnt

Eugene Marais [1,*], Martin P Venter [1], Ryno Laubscher [1] and Philip Herbst [2]
[1]Department of Mechanical and Mechatronic Engineering, Stellenbosch University, Stellenbosch, South Africa
[2]Division of Cardiology, Faculty of Medicine and Health Sciences, Stellenbosch University, Cape Town, South Africa
*21659087@sun.ac.za; *eclaudemarais@gmail.com

Abstract: Cardiovascular diseases (CVDs) contribute to 32% of global deaths, with Sub-Saharan Africa (SSA) facing a disproportionate burden, where 75% of CVD-related deaths occur. Unlike high-income countries, SSA has a high prevalence of rheumatic heart disease affecting young and middle-aged adults due to an untreated strep throat, leading to aortic valve damage and eventually death. Non-invasive personalised treatments are crucial, and current computational precision medicine involves time-consuming fluid-structure interaction (FSI) simulations, focusing on fluid flow and leaflet deformations. This paper discusses the experimental validation of finite element simulations for aortic heart valves, emphasising healthy and moderately stenosed valves, with a focus on the valve opening area. The structural simulation forms part of a simplified FSI which uses a lumped parameter model (LPM) to describe haemodynamics. Discrepancies were found between the experimental and simulated data, prompting further investigation. Issues were traced back to the manufacturing method which resulted in varying thicknesses of the test valve leaflets. These differences resulted in a non-symmetrical opening behaviour in the test valves. Future work will involve adopting a new manufacturing approach and repeating experimental tests to improve the accuracy.

Keywords: aortic valve; aortic stenosis; fluid-structure interaction; lumped parameter model, polymeric heart valve

1. Introduction

Globally, cardiovascular disease is a leading cause of death, resulting in 19 million deaths annually [1], [2]. Although both first- and third-world countries experience the social and healthcare burden of CVDs, third-world countries, which are common in sub-Saharan Africa, suffer from large economic knocks as the primary workforce aged 20 – 50 years suffer from CVDs [3], [4]. Along with this knock to their economy, many children also fall victim to CVDs in these regions [5], commonly suffering from the later stages of rheumatic fever, mainly due to a lack of medical infrastructure.

Computational precision medicine can be integrated into the current medical infrastructure in low-income countries. This would allow researchers and clinicians alike to reproduce the effects of lesions and obstructions caused by stenosis resulting from either from calcification due to age related health complications or rheumatic heart disease and produce less invasive personalised treatments for patients at a far lower cost than some of the currently available treatment procedures [6].

The current work experimentally validates the finite element structural simulation of a simplified FSI. The simplified FSI is comprised of a lumped parameter model, in which the valve model has been replaced with a 3 dimensional (3D) structural model. However, the experimental area openings proved to differ significantly compared to the structural simulation-predicted area openings. An investigation was then conducted to determine the cause of these faults. Through this many lessons were learnt that can be further utilised by others to accelerate their own work.

2. Methods and Materials

2.1 Coupling

The coupling used in this study can be described as a simplified fluid-structure interaction (FSI) simulation. "Simplified" is used to describe the coupling, as the fluid side of the FSI is made up of a LPM rather than a computational fluid dynamic model. The LPM used was that of Laubscher et al. [7], which accounts for the Reynolds number-dependent pressure losses and valve motion.

Many LMPs describe heart valves as diodes that are either open or closed due to pressure gradients, or as beams that have opening and closed positions relative to pressure gradients, both of which do not capture the true effect of leaflet flexure on haemodynamics. To capture this behaviour to its fullest extent, the beam model in the utilised LMP was replaced with a 3D finite element (FE) model of a generic bioprosthetic heart valve. Compared to fully coupled FSIs, making use of only a 3D FE model is much less computationally expensive and allows for the observation of leaflet flexure on the haemodynamics of the body. The valve design was based on Abbott's "Trifecta GT" bioprosthetic heart valve, as shown in Figure 1.

Figure 1 Generic aortic valve made based on the design of the Abbotts "Trifecta GT"

2.2 Simulation boundary conditions

In the case of the original LPM vs the simplified FSI (LPM-FE FSI) the boundary conditions were as follows:

The LPM models the heart valve as two rigid beams which opening position is determined by the pressure difference over the heart valve. A FE analysis of the valve in the simplified FSI was performed using LS-Dyna. The simulation was quasi-static. This was initially chosen over dynamic, as the opening and closing action of the valve occurs over a few milliseconds, leaving the valve in a static-like state for the majority of the systole phase [8]. Only one leaflet of the valve was simulated, as the geometry of the valve was trisymmetric. A shell element mesh composed of first-order quadrilateral elements using 5391 elements was created to represent the 3D solid geometry of a single leaflet. The thicknesses of the elements were kept consistent to simplify the model, and the thickness of the leaflet was determined from the measurement taken from the manufactured valves using a using a 0-25 mm Oxford precision micrometer. The convergence of the mesh was determined using the valve's area ratio (Ar), (measured area opening)/(theoretical full area opening), as the criterion. Convergence was achieved using a element size of 0.5 mm with a relative error of 0.93 percent.

The Ogden (N=3) hyperelastic model was used to model the non-linear stress-strain of the polymeric material used for the valves. The Ogden N=3 model was selected as it can characterize hyperelastic materials strains from 0% - failure unlike other models which only capture regions of elongation accurately [9] working well for membranes which undergo large distortions. Smoothsil 950 was the chosen silicone to model the valve. The Ogden (N=3) material parameters for Smoothsil 950 is chosen below in Table 1 [10].

Table 1 Material Properties of SmoothSil 950

Material Property	Value	Unit
Density	1211	kg/m^3
Poisson ratio	0.49	N/A
Bulk Modulus	679	kPa
Shear Modulus	50.59	MPa
μ_1*	-0.3062	MPa
μ_2*	0.0283	MPa
μ_3*	$6.596e^{-9}$	MPa
α_1*	-3.059	N/A
α_2*	4.597	N/A
α_3*	17.69	N/A

In the dynamic simulation which was setup in Star-ccm+ the following boundary conditions were utilized:

A 3D solid model of the aortic valve was used and the outer face of the valve was fully clamped. The pressure and leaflet thicknesses were set to the same as the quasistatic simulation. To model the stenosed condition, sections of the leaflets were fused together. The material used was the same as the quasistatic model.

2.3 Experimental Setup

Figure 2 shows the experimental setup [11] used to perform the tests and to measure the valve area openings. The system was set to run at 60 beats/min with a cardiac output of 6.6 l/min. The temperature was set to 36 °C. The resistances and compliances of this system were fine-tuned such that the end-resulting pressures for the aorta and ventricle matched the physiological pressures. To simulate blood a blood analogue was prepared using a ratio of water to glycerol 60/40 [11]. Altogether, six valves were tested for two different scenarios: three healthy valves and three moderately stenosed valves.

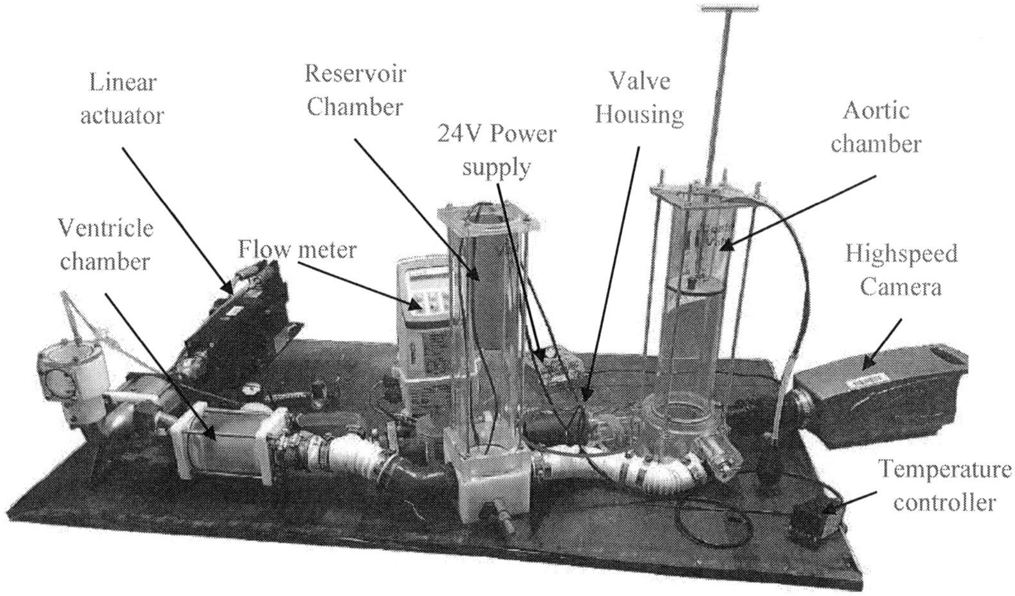

Figure 2 Experimental setup used to test valves

To capture the opening of the valve, a high-speed camera, Olympus I3 speed, along with a Nikon macro lens combined with a two-times-teleconverter, was used. As shown in Figure 2, the high-speed camera was positioned next to the aortic chamber with the lens looking at the bottom of the aortic chamber. The camera lens looks through a viewing port which allows the user to see the valve while the system is operating. Figure 3 illustrates a simplified diagram of the camera positioned next to the aortic chamber capturing the motion of the aortic valve.

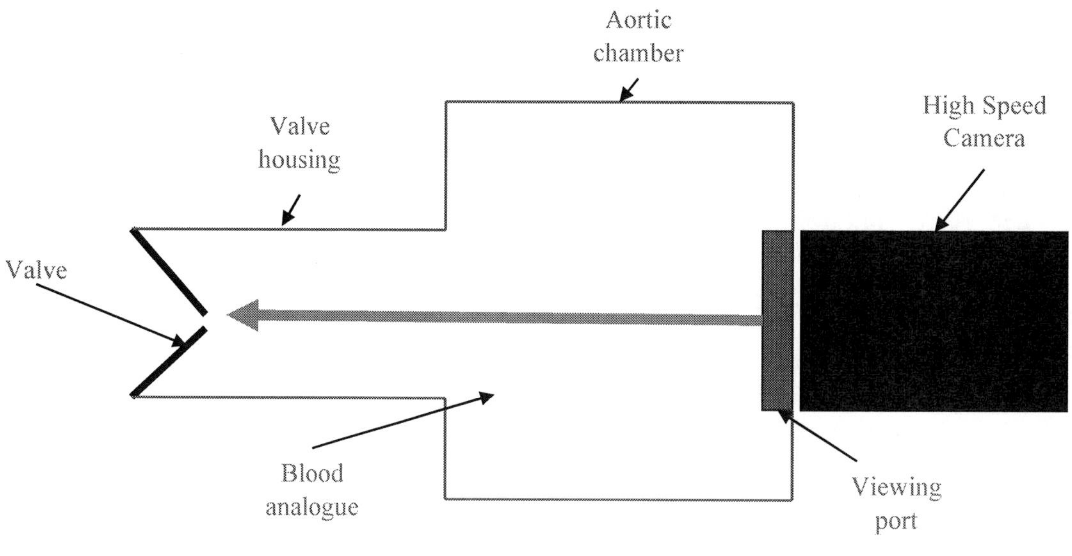

Figure 3 Valve and camera setup

2.4 Test Valves

Six valves were tested in this study. These valves were produced using a two-part core and cavity resin mould produced using a Formlabs Form 3 resin printer. Because the valve was made entirely of silicone, it was completely compliant, resulting in the outer edge needing reinforcement with a rigid material. This reinforcement cage was attached to the valve using specialised silicone glue called Sil-poxy. This can be observed in Figure 4 which shows the top views of the healthy and stenosed valves in their open positions. This was done so that the valve would not collapse in the experimental setup, and further reinforcement was added by adding surgical stitching which tied the soft valve to the outer reinforcement. The additional reinforcement is illustrated in Figure 4.

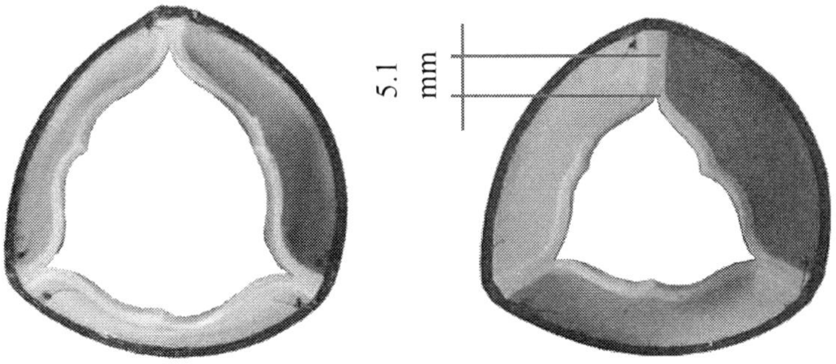

Figure 4 Top view of the tested valves. Left is the healthy case study, and right is the moderately stenosed case study.

The valves were cast in a closed position, and an indentation was made in the mould to allow the user to see where to cut with the surgical scalpel to open the valve. Moderate stenosis occurs when the heart valve opens only one-third of its previous area opening [12]. Through a set of trial-and-error simulations, the correct area opening for moderate stenosis was found, and it was determined that the valves needed to be 5.1 mm closed compared to the healthy valves. This is shown in Figure 4. The trail-and-error simulations consisted of placing seed nodes on the opening edge of the aortic valve, at a certain distance from the outer edge and then constraining that section of the opening edge, resulting in a smaller opening. This was done till the max opening area of the stenosed valve was a third of the healthy valve, resulting in the fixed section of the opening edge to be 5.1 mm.

3. Experimental Results

3.1 Coupling

The simplified FSI coupling (LPM-FE FSI) was used to simulate various other haemodynamic parameters, such as volumetric flow rate, transvascular pressure, and general pressure waveforms. However, as shown in Figures 5 and 6, the area openings predicted by the simplified FSI (LPM-FE FSI) for a healthy heart valve and moderately stenosed heart valve, respectively.

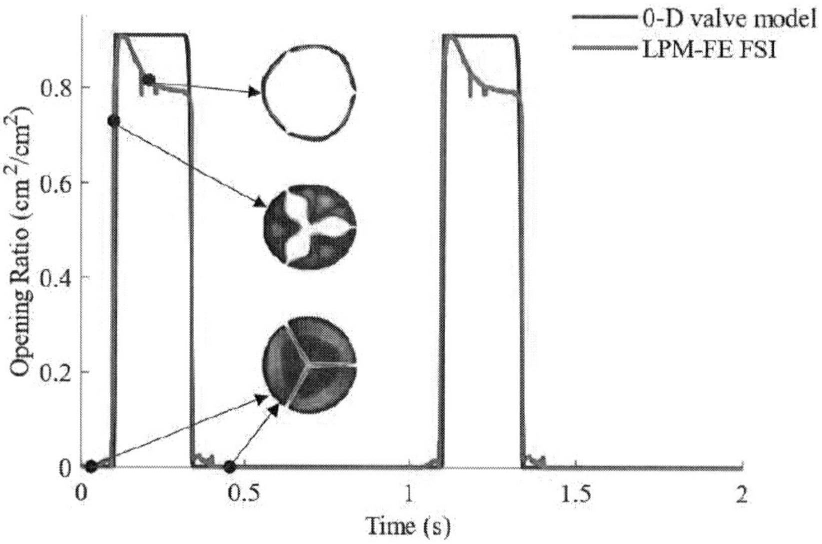

Figure 5 Healthy area opening of 0-D valve model and simplified coupling (LPM-FE FSI)

In both scenarios, the 0-D valve model predicted that the heart valves should open immediately, and the simplified FSI (LPM-FE FSI) predicted different behaviours correlating more to what would be seen in actual hard valves [8]. In the case of a healthy heart valve, there is a peak opening area that occurs in the early systole phase, leading to a slow closing but still fully open plateau, after which the valve fully closes [8]. The moderately stenosed opening is not immediate but takes a bit longer owing to the smaller outlet area of the valve. The resulting full maximum opening does not occur at peak systolic pressure, but rather a bit later, forming a parabolic peak.

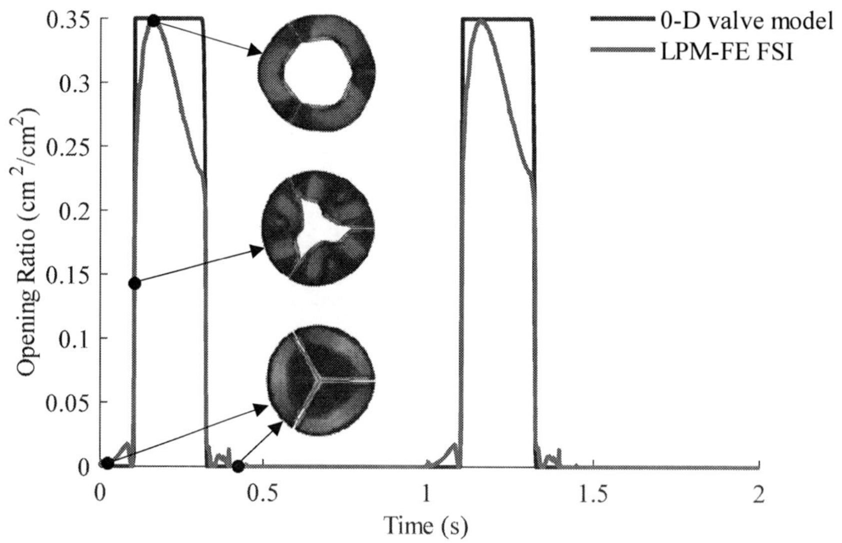

Figure 6 Moderately stenosed area opening of 0-D valve model and simplified coupling (LPM-FE FSI)

3.2 Experimental valve openings

Figure six shows the measured area opening for both healthy and sick cases for the singular valves. An immediate standout that can be picked up is that both the sick and healthy cases have the same form, where the peak opening is reached in the centre rather than an immediate opening. Furthermore, visible notches can be seen in both figures. These notches indicate nonlinear opening of the three leaflets of the valve. This opening pattern is due to the varying thicknesses in the leaflets which were measured using a 0-25 mm Oxford precision micrometer.

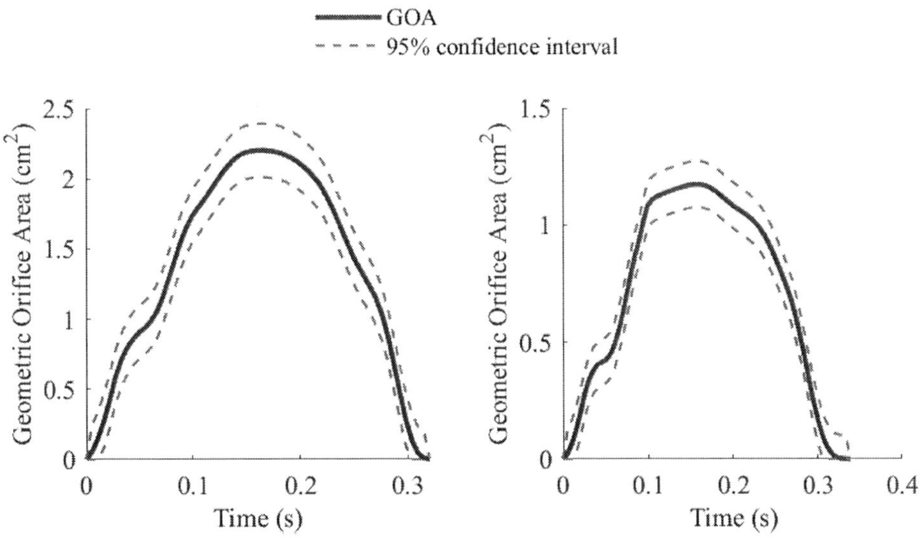

Figure 7 Area opening of one healthy test valve (left) and one moderately stenosed valve (right).

3.3 FE Validation

The experimental and simulation results were compared to validate the accuracy of the FE model. To achieve this, the average measured transvalvular pressure wave was extracted for each experimental test and then applied to the FE simulation. To capture the effects of the varying thicknesses, that were measured, the valve models leaflets thicknesses were set to measured the maximum and minimum thicknesses. This was done for each valve. Dynamic FE simulations were performed using Star CCM+ to further verify the accuracy of the experimental tests. This is illustrated in Figure 7.

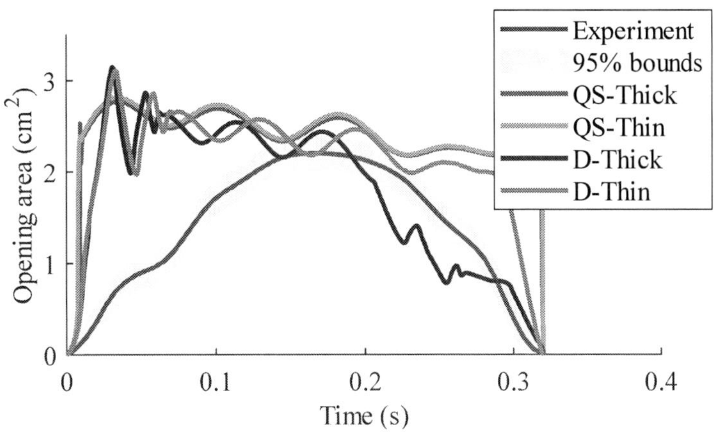

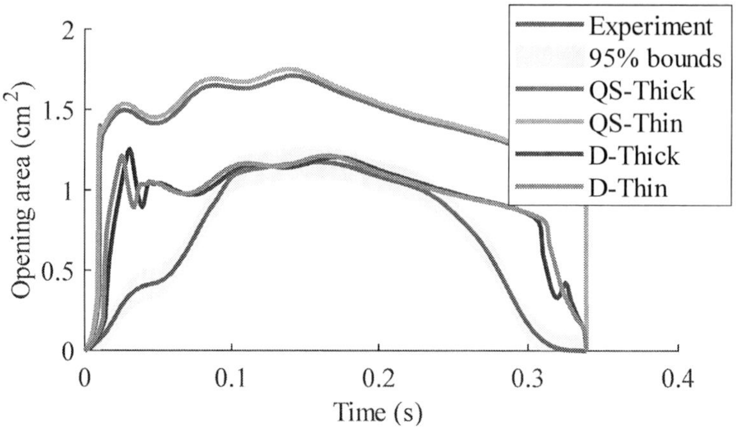

Figure 8 Area openings of a single healthy valve (top) and a single moderately stenosed valve (bottom). QS denotes quasi-static and D denotes dynamic

Ultimately, it was found that in all cases of the experiment, the maximum area was far less than that predicted by the simulations, and all the valves were stenosed based on the way they opened. When comparing the two FE simulations, the dynamic simulation agrees with the experiment in more places than the quasi-static simulation; however, the experimental results still differ greatly from those of both FE simulations. The reasons for these differences, and why the experimental results differ significantly from the simulation results, are discussed in the following section.

4. Lessons Learnt

Owing to the great difference between the experimental and simulation results, an investigation was conducted to determine why the two results differ from one another. It was found that the main factors resulting in these differences could be due to manufacturing choices. The effects of manufacturing choices on the heart valve are shown in Figure 8.

In both instances, for the healthy and stenosed valves, when the valve is in the diastole state (closed), a twisting action can be seen. This phenomenon is known as pinwheeling, which occurs because of the inconsistent thickness of leaflets [13]. In Figure 8 (b), for both case studies, it can be observed that as the valve starts, the leaflets open in a nonlinear fashion owing to the varying thicknesses. This type of behaviour is often observed in patients' heart valves which suffer from the early stages of stenosis, and as it progressively worsens. Leaflets begin to thicken, causing them to lag behind thinner leaflets [14]. This lagging in the opening causes the fluid pressure profiles in the experiment, which act on the leaflets, to change and take the direction of the easiest flow, resulting in the notches seen earlier in Figure 6. These types of dynamic pressure profiles would prove difficult to measure and simulate in an FE simulation. Furthermore, the FE simulation would not be able to capture these movements even if the leaflets had different thicknesses because the pressure was constant and did not change, as in the experiment. One way to simulate this would be with a fully coupled FSI simulation; however, this is outside the scope of the current work and is possible in future work.

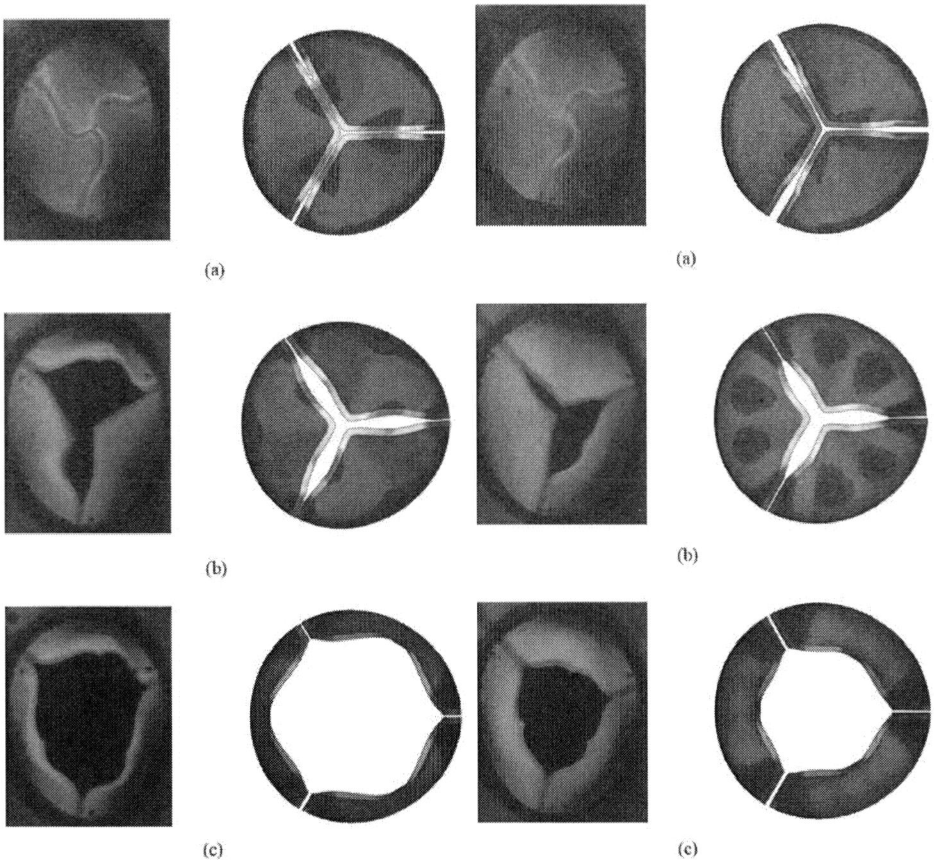

Figure 9 Left: Opening stages of a healthy aortic test valve. Right: Opening stages of the moderately stenosed valve. (a) The valve is closed in the diastole stage, (b) the valve starts to open in early systole, and (c) the valve is fully open at peak systole. The right side of each experimental image is the predicted simulated opening area at the same time. Experimental images appear elliptical this was caused by the lens and working fluid distortion. This distortion was corrected in the area opening calculations.

As mentioned earlier, the main reason for all the previously stated problems was the valve manufacturing process. Wu-Chiao Shih et al. [15] investigated the fabrication and hemodynamic performance especially the measured opening area of polymeric valves which were manufactured in different ways. They explored two methods for manufacturing these valves. The first was an assembly method which consisted of making compliant leaflets and attaching them either mechanically or chemically to a rigid body. The second method used a two-part mould and cast the valve as one, resulting in the entire valve being compliant. This was performed in the current study. Wu-Chiao Shih found that the assembled valve area opening matched that of their non-coupled FSI, while the single-piece valve only behaved like the FSI early systole while the valve was opening. As the peak systole was reached, the one-piece valve started to behave like a valve with thicker leaflets, causing the valve to not be fully open, unlike the assembled valve. They determined that these faults were due to the manufacturing methods. They found that when casting a single-piece valve, any error in the male or female piece of the mould would result in thickness differences. Sometimes, the casting medium does not flow into the cavity correctly, resulting in inconsistencies within the valve. Another problem they found was that the entire structure was compliant with varying thicknesses within the outer edge, which is rigid, and the assembled valve would result in slower valve opening and pinwheeling. These effects were observed in the current work with thickness difference of $\pm$ 0.05 mm in the leaflets.

When they compared the area opening of the valves, it was found that the assembled valve had an average area opening of 3 cm^2 and stable outputs for the area openings at different flow rates. For the one-piece valve, the average opening area varied with increasing flow, resulting in unstable outputs. This was also observed in the experimental opening area; as the flow rate was changed in these valves, the opening area was greater.

5. Conclusion

After determining the reasons why the results differ significantly from the simulations and experimental work, future work will entail producing a new test valve which will only have compliant leaflets, and the rest of its structure will be rigid. The experiments and simulations contribute significantly to the existing body of work in the field, enhancing research and fostering a deeper understanding of heart valve manufacturing and modelling within the University of Stellenbosch's Institution of Biomedical Engineering (IBE) Cardiovascular Department. While the results may not be groundbreaking, they lay the groundwork for further advancements and improvements in future studies. This work marks a significant milestone and a step forward for the department, as it re-establishes its presence in the field of cardiovascular research.

Nomenclature

Abbreviations:

Cardiovascular diseases (CVDs) Sub-Saharan Africa (SSA) fluid-structure interaction (FSI) a lumped parameter model (LPM) 3 dimensional (3D) finite element (FE)
Institution of Biomedical Engineering (IBE)

References

[1] C. W. Tsao *et al.*, 'Heart Disease and Stroke Statistics—2022 Update: A Report From the American Heart Association', *Circulation*, vol. 145, no. 8, Feb. 2022, doi: 10.1161/CIR.0000000000001052.
[2] 'Aortic Stenosis Overview', www.heart.org. Accessed: Nov. 22, 2022. [Online]. Available: https://www.heart.org/en/health-topics/heart-valve-problems-and-disease/heart-valve-problems-and-causes/problem-aortic-valve-stenosis
[3] V. T. Nkomo, 'Epidemiology and prevention of valvular heart diseases and infective endocarditis in Africa', *Heart*, vol. 93, no. 12, pp. 1510–1519, Dec. 2007, doi: 10.1136/hrt.2007.118810.
[4] O. S. Ogah, A. Adebiyi, K. Sliwa, O. S. Ogah, A. Adebiyi, and K. Sliwa, *Heart Failure in Sub-Saharan Africa*. IntechOpen, 2019. doi: 10.5772/intechopen.82416.
[5] S. G. M. Jivanji, S. Lubega, B. Reel, and S. A. Qureshi, 'Congenital Heart Disease in East Africa', *Front Pediatr*, vol. 7, p. 250, Jun. 2019, doi: 10.3389/fped.2019.00250.
[6] L. Garber, S. Khodaei, and Z. Keshavarz-Motamed, 'The Critical Role of Lumped Parameter Models in Patient-Specific Cardiovascular Simulations', *Arch Computat Methods Eng*, vol. 29, no. 5, pp. 2977–3000, Aug. 2022, doi: 10.1007/s11831-021-09685-5.
[7] R. Laubscher, J. van der Merwe, J. Liebenberg, and P. Herbst, 'Dynamic simulation of aortic valve stenosis using a lumped parameter cardiovascular system model with flow regime dependent valve pressure loss characteristics', *Medical Engineering & Physics*, vol. 106, p. 103838, Aug. 2022, doi: 10.1016/j.medengphy.2022.103838.
[8] M. Thubrikar, *The Aortic Valve*, 1st ed. CRC Press, 1989.
[9] A. Harish, 'How to Choose a Hyperelastic Material Model for your FEA?', SimScale. Accessed: Jun. 16, 2023. [Online]. Available: https://www.simscale.com/blog/how-to-choose-hyperelastic-material/

[10] D. R. Ellis, M. Venter, and G. Venter, 'Generative Design Procedure for Embedding Complex Behaviour in Pneumatic Soft Robots', Doctor, Stellenbosch University, Faculty of Mechanical and Mechatronic Engineering, 2020.
[11] R. A. Rodriguez, J. H. Müller, and K. H. Dellimore, 'Redesign and Performance Evaluation of a Cardiac Pulse Duplicator', Masters, Stellenbosch University, South Africa, 2017.
[12] R. Rajami, J. Hancock, and J. B. Chambers, 'The art of assessing aortic stenosis', pp. 14–22.
[13] V. Stanová *et al.*, 'Effects of hemodynamic conditions and valve sizing on leaflet bending stress in self-expanding transcatheter aortic valve: An in vitro study', *Artificial Organs*, vol. 44, no. 7, pp. E277–E287, 2020, doi: 10.1111/aor.13654.
[14] F. Bakhtiary *et al.*, 'Opening and closing kinematics of fresh and calcified aortic valve prostheses: an in vitro study', *J Thorac Cardiovasc Surg*, vol. 134, no. 3, pp. 657–662, Sep. 2007, doi: 10.1016/j.jtcvs.2007.02.050.
[15] Wu-Chiao Shih, Ha-Hai Vu, Cheung-Hwa Hsu, Kai-Wen Yang, and Yaw-Hong Kang, 'Effective Orifice Area of a Newly Developed Polyurethane Aortic Valve: Manufacturing and Evaluation', *Journal of Marine Science and Technology*, vol. 22, no. 6, Jan. 2014, doi: 10.6119/JMST-014-0321-5.

The effect of optimization model on the accuracy of constitutive model.

Lebogang Mathebela[1]*. Rudzani Sigwadi[2]. Tetelo Lekwana[1]. Harry Ngwangwa[1]. Thanyani Pandelani[1]. Fulufhelo Nemavhola[3].
[1]Unisa Biomechanics Research Lab. Department of Mechanical Engineering. School of Engineering. College of Science Engineering and Technology. University of South Africa. Pretoria. 0001. South Africa.
[2]Department of Chemical Engineering. School of Engineering. College of Science Engineering and Technology. University of South Africa. Pretoria. 0001. South Africa
[3]Department of Mechanical Engineering. Faculty of Engineering and the Built Environment. Durban University of Technology. Durban. 4000. South Africa
*Correspondence: lebokmathebela@gmail.com; Tel.: +27 11 471 2137

Abstract

Selecting the best constitutive model from the curve fitting software requires numerous trial runs from the software in attempt to find the model that fits the experimental data optimally when developing material parameter of curly hair. Additionally, selecting the optimization method could also add to the length of time taken to process the experimental data. The single curly hair fiber material parameters are mainly not present on the literature. The purpose of the study is to assess the effect of the optimizing methods on the accuracy of the constitutive model. The average stress-strain curve data from N=10 specimens of 20 mm in lengths processes from Uniaxial 44 N Load cell with the accuracy of 0.2 % of the rated full-scale load were used for curve fitting on Hyperfit software. The hyperelastic constitutive model Mooney-Rivlin model selected and their material parameter were used to assess the effect of optimizing model when doing trial runs for different constrained optimizing methods. The coefficient of variation was also utilized to find the data variances per material parameters. The material parameter C_{10} and C_{01} satisfy material stability criteria for the model on all the optimizing methods. However, the Levenberg-maruardt, Nelder-Mead Powell, CG &COBYLA methods yield the similar and better results than Gradient decent and Nelder Mead on all the material parameter. The developed of material parameters from this study is essential for accurate development of computational models for detailed studying of hair fiber behaviour.

Keywords: Curly hair fibers. strain rate. tensile testing. hyperelastic constitutive model

1. Introduction

Characterizing the single fibers under tensile loading requires the large number of sampling and it is time consuming [1]. Taking into considerations the complicating aspects include load frame compliance, additional grip effects, and slack in the specimen or slipping/seating in the grips [2]. However, the slack and grip effects are covered by ASTM D3822, "Standard Test Method for Tensile Properties of Single Textile Fibres," which is designed for "natural and man-made" fibres [3]. Additionally the ASTM D3822 does not take into the account the geometric profile of the hair fiber along the fiber. This is because the curly fibers possess the cross-section area with twist and bends with narrowed cross-sectional area along the fiber [4]. This can create the weak spots because when excessive force is applied ,it can create high stress point over the small area of the fiber. To reduce uncertainties in terms of mechanical properties and geometric profile, accurate material parameters need to be developed on the curve fitting software. The study also assessed the effect constrained optimizing methods on the accuracy of the material parameters. Namely, C_{10}. C_{01}. R^2. R. NRMSE. NE. SS & SA. The hair care sector can benefit from the application of hyperelastic constitutive modelling in the areas of research and development, time-saving strategies for product launch and cost saving option.

2. Methods and materials

2.1 Specimen acquisition and preparation

To prevent the variance in mechanical properties driven by age, gender, and ethnicity, the N = 10 specimens were obtained from one African female in her early 30s. The neck or occipital portions of the scalp were used because they are less susceptible to sunlight or other external environmental aggressors [5].There was no cosmetic treatment done on the hair for the last six months, and only daily washing in water and drying was done on the hair prior to sample harvest.

2.3 Mechanical tensile testing

A Cellscale UStretch uniaxial testing machine, with a load cell of 44 N was used for the tensile testing of the hair fibers. The UStretch can perform both dry and wet conditions and it can be operated in both vertical and horizontal configuration with precision of 0.2 % of the rated full-scale load. It has a high-performance actuator with an inline load cell and an integrated temperature-controlled media bath. The bath can only be used when operating the equipment in a horizontal configuration. In this study, the uniaxial force was applied to single curly hair (N = 10) with the gauge length of 20 mm at strain rate of $10^{-2}.s^{-1}$. The paper frame technique proposed by Adusumalli et al. [6] for the clamping of single hair fibers Figure 2 [A] was then modified by using the A4 papers instead of the cardboard tabs. The fabric glue was used to glue the specimen onto the paper tabs. The lags on the specimen were removed by hand when gluing the specimen onto the paper. The 20 mm specimens were tested under room temperature (24 °C) and stretched up to 60 % elongation. Only the specimens that broke in the middle of the gauge length were considers for calculations of this this work and the specimens that snapped closed to the glue. This was to eliminate the influence of the mechanical stress where the specimens were fixed and the influence of the glue to the specimens.

2.4 Constitutive modelling

Hyperelastic models have been utilised in understanding the behaviour of materials under mechanical loadings [7-9]. The material parameters derived from the hyperelastic models were utilised in developing the computational models [10-12]. In this study, the stress-strain behaviour of curly hair fibers was modelled as hyperelastic incompressible isotropic material using the Hyperfit software. The mechanical behaviour of the specimens were investigated using the Mooney-Rivlin Model. The non-linear optimisation method selected with a constrained optimisation linear approximation namely, Gradient decent. Nelder Mead. Levenberg-maruardt. the: Nelder-Mead. Powell. CG and COBYLA algorithms implemented on Hyperfit software. In this study, the algorithm minimises a scalar error function between the model calculated stress and experimentally measured stress and returns the models' material parameters that yields optimal errors.

The uniaxial tension or compression test data protocols from the incompressible form were used when fitting the tensile data onto Hyperfit. The new strain was generated, and model parameters were also generated by the software to be used for mathematical characterisation of the curly fiber.

Mooney-Rivlin model (MRM)

Mooney first proposed the MRM hyperelastic isotropic material model, which is used to explain the stress-strain characteristics of solids that resemble rubber [13]. Incompressible and Nearly Incompressible (NI) formulations of this model are implemented in Hyperfit.

Given below is the strain energy function for incompressible material:

$$\Psi = C_{10}(I_1 - 3) + C_{01}(I_2 - 3) \quad (3)$$

Volumetric strain energy function for Nearly Incompressible formulation is:

$$\Psi_{vol} = C_{10}(I_1 - 3) + C_{01}(I_1 - 3) + \Psi_{vol} \quad (4)$$

The above incompressible formulation is considered nearly incompressible (NI), and can be given by appropriate extension analogous to the 2-parameter model described above, where C, is the right Cauchy-Green deformation tensor, and I_1 and I_2 are the first and the second invariants of C. The C_{10} and C_{01} are the stress-like parameters defining the stiffness of the material. Ψ_{vol} is a volumetric change contribution to the strain energy function.

2.4 Data analysis

The average cross-section of each specimen that was used in a prior investigation and assumed to have an elliptic shape [14], was used to calibrate the stress-strain behaviour of curly hair fibers. The performance of each model was assessed using the evaluation index per strain-rate, based on the model that best captured its behaviour in terms of R^2 The study further assessed the two parameter Mooney-Rivlin model to satisfy the restriction of the model to reproduce the real material parameters of the model and it is expressed as follows [15]:

- $C_{10} + C_{01} > 0$ for the Mooney-Rivlin model with two parameters.

The adopted system for model fitting has been used across the literature to investigate constitutive formulation that will provide the best characterisation of biological material mechanics [16-[18]

This study is intended to assess the material parameter of the of the curly fiber for the purpose of getting the best calibrated behaviour of the fibers to be used for simulation. After fitting the experimental data on the Hyperfit, the material parameters for each test of all six algorithms are given by Table 1. The representation of the best optimization was represented by the curve fitting that yields the R^2 closed to 1.

3 Results

Stress strain behavior of all optimizing methods namey. Gradient decent. Nelder Mead. Levenberg-maruardt. Nelder-Mead (3rd party: Scipy). Powell (3rd party: Scipy). CG (3rd party: Scipy) & COBYLA (3rd party: Scipy) and experimental data are presented in this section. According to Figure

1 , the experimental stress-strain resembles the behaviour of the wool and other keratinized material [19] and it also evident that all the Money-Rivlin fit well with the experiment in all constrained optimising Methods.

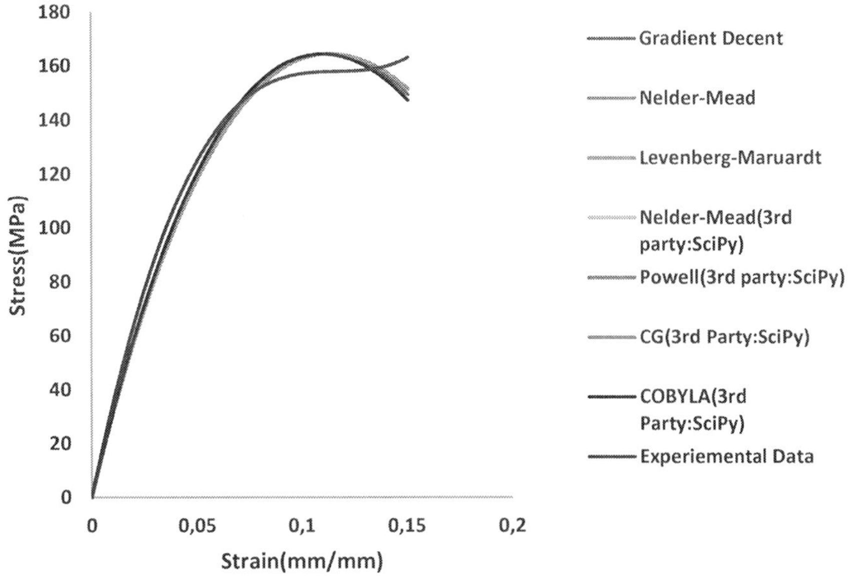

Figure 1; The average stress strain data from the experiement showing the stress-strain curve after fitting the Mooney-Rivlin model using different constrained optimizing methods namely, Gradient decent. Nelder Mead. Levenberg-maruardt. Nelder-Mead (3rd party: Scipy). Powell (3rd party: Scipy). CG (3rd party: Scipy) & COBYLA (3rd party: Scipy)

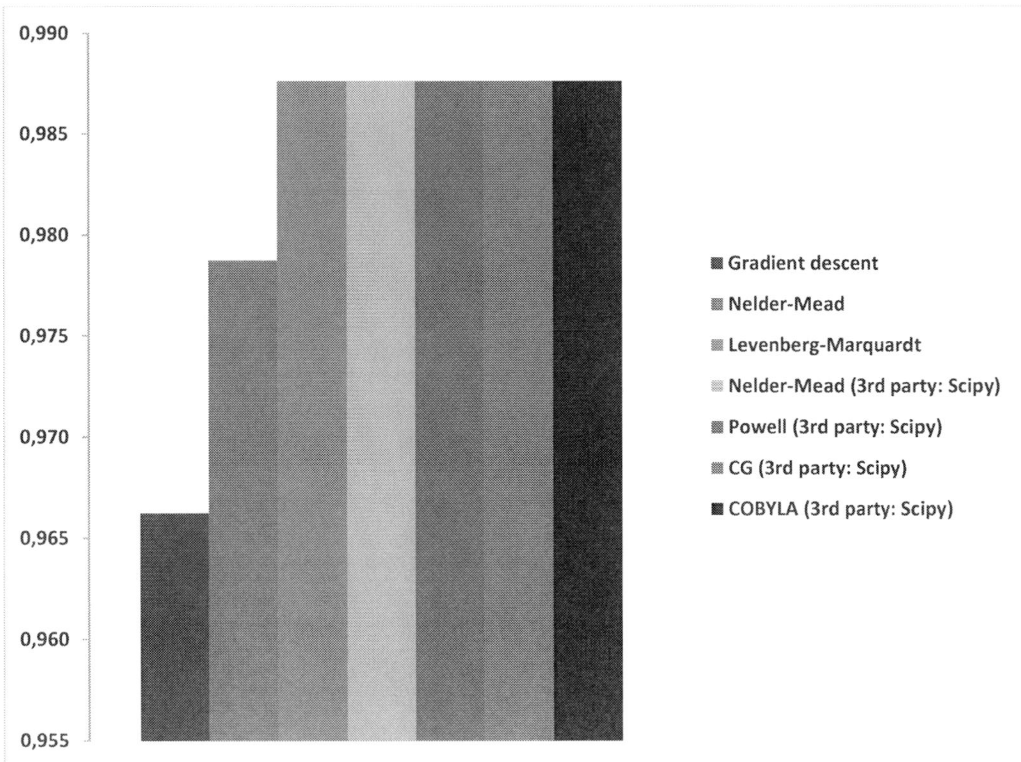

Figure 2: The comparism of the R^2 parameter for curly hair fiber pulled @ 10-2.s-1 on Mooney-Rivlin Model using constrained optimizing methods namely, Gradient decent. Nelder Mead. Levenberg-maruardt. Nelder-Mead (3rd party: Scipy). Powell (3rd party: Scipy). CG (3rd party: Scipy) & COBYLA (3rd party: Scipy).

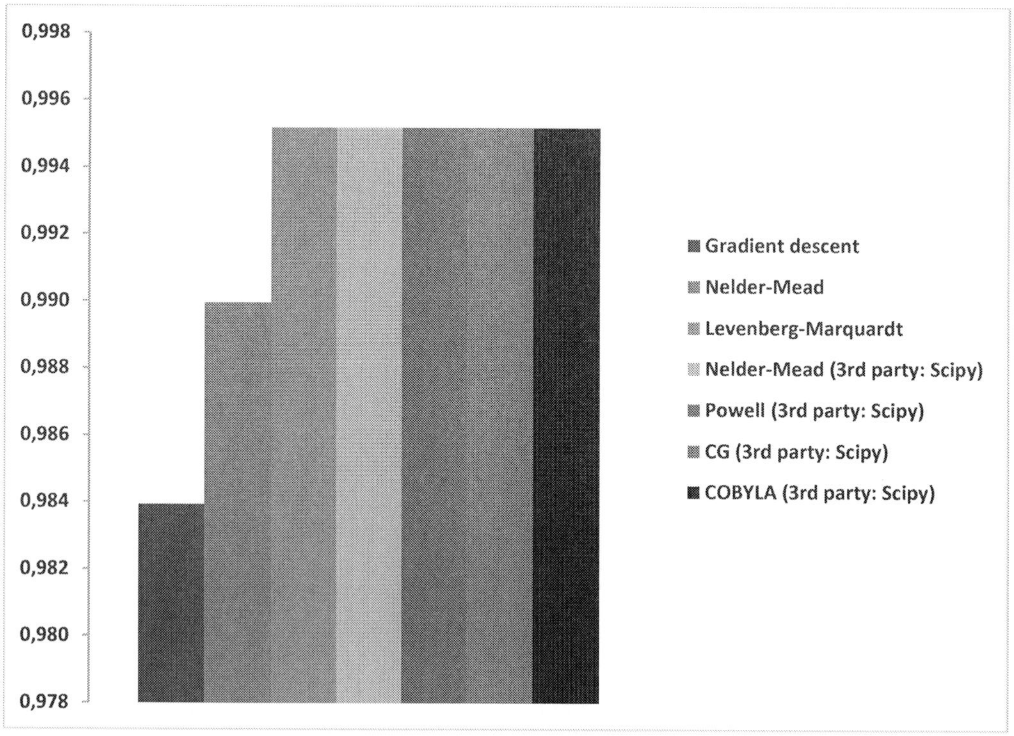

Figure 3; The comparism of the R parameter for curly hair fiber pulled @ $10^{-2}.s^{-1}$ on Mooney-Rivlin Model using constrained optimizing methods namely, Gradient decent. Nelder Mead. Levenberg-maruardt. Nelder-Mead (3rd party: Scipy). Powell (3rd party: Scipy). CG (3rd party: Scipy) & COBYLA (3rd party: Scipy)

From Figure 2 & 3, the value of R^2 and R parameter results for all the models shows a correlation between the axial loading stress-strain results and constitutive relation with the highest value of above 0.9 R^2 and R. it is also interesting to notice that Levenberg-maruardt. Nelder-Mead (3rd party: Scipy). Powell (3rd party: Scipy). CG (3rd party: Scipy) & COBYLA (3rd party: Scipy) yield the same results. This then indicate that their implications on the nature of the parameter can be regarded as the optimising methods that does not influence the outcome of the results when selecting the best optimizing methods for curve fitting. On the other hand Gradient decent and Nelder Mead should be considered the list when selecting optimising methods for the experiments.

Table: 1 two parameter Mooney-Rivlin Model assessing the stability of the material where √= satisfy the material restrictions, -=not applicable and X=unsatisfactory

	GD	NM	LB	NM(3RD)	Pw	CG
$C_{10} + C_{01} > 0$	√	√	√	√	√	√

4 Discussion

The stress-strain curve presented by this study resembles that of the wool and keratinized fibers [20]. It is also interesting to note that the models selected for this study to assess the material parameter of the single curly fibers can fit well with the experiments on both linear and non-linear curves. The performance of the models indicates that they yield better correlation and R^2 values within the linear elastic region than the non-linear regions of the stress-strain curve.

Interns of the statistical dispersion in assessing the material parameter by the three models, the Polynomial model has the least variation over the five different tests for each condition of between 0.00 and 0.08.This followed by Mooney Rivlin Model at 0.00 and 0.09.On the other hand the Yeoh model has the largest variation of between 0.00 and 0.018 % which is still within the acceptable limit.

The material stability for the model was further done on the two parameters Mooney-Rivlin model to verify if the parameters are within the restriction in order to reproduce the real behaviour material. It was discovered that the two parameter Mooney-Rivlin model satisfy the material restriction for all the conditions where the sum of material like stiffness $C_{10} + C_{01} > 0$. This will intern minimise the errors if the materials for curly fiber can be produced from Mooney-Rivlin model and directly used for Finite element analysis.

Table 2: The hyperelastic constitutive material parameter of curly fiber for Mooney Rivlin model of optimizing method including , C_{10}, C_{01},NE, NRMS, SS & SA. The average value for the parameters and standard deviation.

Optimizing Methods	C_{01}	C_{10}	NRMSE	NE	SA	SS
GD	-2238,09	2790,20	0,08	0,07	1331,72	17489,91
NM	-2169,52	2714,10	0,07	0,06	1154,94	14645,88
LB	-2246,32	2799,41	0,07	0,06	1134,35	14357,45
NM(3RD)	-2246,32	2799,41	0,07	0,06	1134,35	14357,45
Pw	-2790,92	3404,11	0,11	0,09	1739,80	28677,48
CG	-2246,33	2799,41	0,07	0,06	1134,35	14357,45
CB	-2222,04	2772,43	0,07	0,06	1142,68	14480,93
Average	-2308,51	2868,44	0,08	0,07	1253,17	16909,51
Std	198,59	220,52	0,01	0,01	209,49	4919,10
CV	-8,60	7,69	17,36	15,15	16,72	29,09

5. Conclusion

The optimizing methods' parameters satisfy the material stability criteria of the Mooney-Rivlin hyperelastic model.

The CV from this study can help you avoid employing optimising approaches that can result in parameter duplication and ensure that you choose the most optimised methods that could produce the best results for your data.

All five optimizing methods yielded good parametric values on R, R^2, NRMSE, NE, SS and SE. The value R^2 was used to compare the performance of the algorithm method. The results presented here could be utilised for the development of computational models in studying how curly hair behaves under mechanical loadings. However careful consideration should be made when deciding on the optimizing method that will yield the minimum errors and by assessing the stability of the materials.

Data availability

All data is available within the manuscript.

Conflict of interest

It is stated by the authors that they have no competing interests.

Acknowledgment.

Grant Number (129380) from the National Research Foundation (NRF) is gratefully acknowledged. The Department of Mechanical Engineering, School of Engineering, College of Science, Engineering, and Technology was able to purchase the Uniaxial testing equipment thanks to the Unisa CAPEX Program.

3 References

[1] A. R. Bunsell, S. Joannès, and A. Marcellan, *Testing and characterization of fibers*. 2018. doi: 10.1016/B978-0-08-101272-7.00002-X.

[2] A. E. Engelbrecht-Wiggans and A. L. Forster, "Analysis of strain correction procedures for single fiber tensile testing," *Compos Part A Appl Sci Manuf*, vol. 167, p. 107411, 2023, doi: https://doi.org/10.1016/j.compositesa.2022.107411.

[3] ASTM D3822 Standard test method for tensile properties of single textile fibres https://doi.org/10.1520/D3822.

[4] T. Lasisi *et al.*, "High-throughput phenotyping methods for quantifying hair fiber morphology," *Sci Rep*, vol. 11, no. 1, 2021, doi: 10.1038/s41598-021-90409-x.

[5] T. R. Ekmekci, D. Sakiz, and A. Koslu, "Occipital involvement in female pattern hair loss: Histopathological evidences," *Journal of the European Academy of Dermatology and Venereology*, vol. 24, no. 3, pp. 299–301, 2010, doi: 10.1111/j.1468-3083.2009.03411.x.

[6] R. B. Adusumalli, K. C. Venkateshan, C. Kunchi, and S. R. Vadlamani, "Tensile testing of single fibres," *Procedia Structural Integrity*, vol. 14, no. 2018, pp. 150–157, 2019, doi: 10.1016/j.prostr.2019.05.020.

[7] A. E. Kiszewski, M. Bevilaqua, and L. B. De Abreu, "Alopecia Areata : A New Therapeutic," *Int J Trichology*, vol. 9, no. 1, pp. 50–53, 2018, doi: 10.4103/ijt.ijt.

[8] Z. Ndlovu, D. Desai, T. Pandelani, and H. Ngwangwa, "Biaxial Estimation of Biomechanical Constitutive Parameters of Passive Porcine Sclera Soft Tissue," vol. 2022, 2022.

[9] H. Ngwangwa, F. Nemavhola, T. Pandelani, M. Msibi, and I. Mabuda, "Determination of Cross-Directional and Cross-Wall Variations of Passive Biaxial Mechanical Properties of Rat Myocardia," no. March, 2022, doi: 10.3390/pr10040629.

[10] F. Nemavhola, "Fibrotic infarction on the LV free wall may alter the mechanics of healthy septal wall during passive filling," *Biomed Mater Eng*, vol. 28, pp. 579–599, 2017, doi: 10.3233/BME-171698.

[11] F. Nemavhola, "Mechanics of the septal wall may be affected by the presence of fibrotic infarct in the free wall," vol. 11, no. 3, pp. 205–225, 2019.

[12] Z. Ndlovu, F. Nemavhola, and D. Desai, "Biaxial Mechanical Characterization and constitutive modelling of sheep sclera soft tissue," vol. 24, pp. 84–96, May 2020, doi: 10.15593/RJBiomech/2020.1.09.

[13] M. Mooney, "A theory of large elastic deformation," *J Appl Phys*, vol. 11, no. 9, pp. 582–592, 1940, doi: 10.1063/1.1712836.

[14] A. Franbourg, P. Hallegot, F. Baltenneck, C. Toutain, and F. Leroy, "Current research on ethnic hair," *J Am Acad Dermatol*, vol. 48, no. 6 SUPPL., 2003, doi: 10.1067/mjd.2003.277.

[15] D. Vlasceanu, H. Gheorghiu, and S. D. Pastrama, "Experimental determination of the Mooney-Rivlin parameters for hyperelastic materials like rubber," *28th Danubia - Adria - Symposium on Advances in Experimental Mechanics, DAS 2011*, pp. 281–282, 2011.

[16] P. Taylor, D. A. Morrow, T. H. Donahue, G. M. Odegard, and K. R. Kaufman, "A method for assessing the fit of a constitutive material model to experimental stress – strain data," vol. 13, No. 2, no. April 2010, pp. 247–256, 2010, doi: 10.1080/10255840903170686.

[17] M. Kohandel, S. Sivaloganathan, G. Tenti, and J. M. Drake, "The constitutive properties of the brain parenchyma Part 1 . Strain energy approach," vol. 28, pp. 449–454, 2006, doi: 10.1016/j.medengphy.2005.01.005.

[18] R. L. C. Akkermans and P. B. Warren, "Multiscale modelling of human hair," *Philosophical Transactions of the Royal Society A: Mathematical, Physical and Engineering Sciences*, vol. 362, no. 1821, pp. 1783–1793, 2004, doi: 10.1098/rsta.2004.1395.

[19] B. Erik, H. Havitcioglu, S. Aktan, and N. Karakus, "Biomechanical properties of human hair with different parameters," *Skin Research and Technology*, vol. 14, no. 2, pp. 147–151, 2008, doi: 10.1111/j.1600-0846.2007.00268.x.

[20] I. P. Seshadri and B. Bhushan, "In situ tensile deformation characterization of human hair with atomic force microscopy," *Acta Mater*, vol. 56, no. 4, pp. 774–781, 2008, doi: 10.1016/j.actamat.2007.10.033.

OPTIMIZING DRIVING BEHAVIOUR FOR UNDERGROUND ELECTRIC VEHICLES UNDER VARYING LOADS

R.F. Meeser[1], D.N.Wilke[2], N.J. Theron[3]
[1] University of Pretoria, riaan.meeser@up.ac.za
[2] University of Pretoria
[3] University of Pretoria (retired)

Abstract:

Reducing energy consumption is essential in moving towards a sustainable future. This study aims to reduce the overall energy consumption of an underground electric vehicle by providing the controller with an optimal velocity profile. Energy consumption is mainly a function of the vehicle type, route travelled, gross mass and the velocity profile required to meet logistical constraints. This project is performed in three phases: route identification, mass estimation and velocity profile optimization, which are performed in real-time to facilitate real-world usability. Underground route identification is performed by evaluating patterns in heading and barometric altitude changes compared to data recorded on previous traversing of routes in a computationally simple method without external communication, like GPS. Mass estimation is achieved by applying Newton's second law to a vehicle in motion, rearranging the equation to solve for the vehicle mass. A strategy is developed that provides an optimal target velocity profile based on the route topography and logistical time constraints. A total consumed energy reduction of 21% in real-time is seen over a normal driver velocity profile for the 10 km above-ground test route evaluated, proving that improved velocity profiles significantly reduce the energy required to operate a vehicle.

Keywords: Underground route identification, Real-time mass estimation, real-time optimization

1. Introduction

Reducing natural resource consumption is essential in moving towards a sustainable society[1]. One such application field is the mining sector, where electric locomotives haul tons of ore for many kilometres underground. Reducing the overall energy consumption of these vehicles will result in higher machine utilization, increased profitability, reduced overall cost of ownership and reduced environmental damage. To optimize the energy consumption of such a vehicle in real-time it is of great importance to know the energy requirements for the vehicle while it is covering its route.

There are several contributors to the energy consumed by a vehicle covering a route. 1. The type of vehicle, such as a car, light delivery vehicle, truck, or even rail-bound vehicle, along with their respective wheel types. 2. The route being travelled, which includes the topography of the route as well as the length and surface. 3. The gross combined mass of the vehicle, its loads, and potentially loaded trailers. 4. The velocity profile required to complete the route in the logistical time constraints imposed. Regenerative braking, various powertrain efficiencies and environmental conditions also significantly affect the overall energy consumed on a route. The focus of this study is to develop a robust method of finding the best velocity profile to cover a route by, in real-time, looking at factors external to the vehicle's internal efficiencies to provide the vehicle controller with an optimal target velocity profile throughout the route. The internal efficiencies can be considered in a separate study, using the target velocity values from this study as the input. This study is not concerned with internal vehicle efficiencies, which allows the findings to be easily adapted to optimization applications for various vehicle types, environments, and applications.

A route identification strategy is developed to identify an underground route being travelled in real-time using a simple database of routes without requiring external communication systems, such as GPS. The route is identified by patterns in magnetic heading and altitude change data and compared to a database of known routes in a computationally simple method in real-time. If a route is not identified, the recorded data is stored as a new route and can be used in future traversing of the route[2].

A real-time mass estimation strategy is developed that uses a simple torsional load cell on the driving axle of the vehicle. This method simplifies mass estimation as it doesn't require multiple load cells on the vehicle and its trailers. Newton's second law of motion for the vehicle is applied, with the equations rearranged to solve for the unknown mass of the vehicle rig. Test results show that the vehicle's gross combined mass can be estimated to be within5%of the true value.

The main controllable variable in energy consumption is the vehicle velocity profile. A strategy is developed that proposes an optimal velocity profile for a vehicle in real-time such that the overall energy consumed is reduced [3]. The results show an energy reduction from 8.5 MJ to 6.7 MJ (21%) is obtained over a normal driver velocity profile for a 10 km test route simulated in real-time.

2. Route identification in real-time

Successful optimization of the energy usage of a vehicle relies on knowing not only the current state of the onboard energy systems, such as the total mass, fuel level (battery state of charge for an electric vehicle), etc., but also the future energy demands that the vehicle will encounter on its route[4], [5]. The topography of a route directly affects the instantaneous energy requirements, as a flat route can have a fuel consumption of 15 to 20 % less than that of hilly terrain [6], [7]. The advantage of having a hybrid or electric power system is more emphasized by hilly terrain, as energy going downhill can be regenerated using sophisticated electric braking and energy storage systems [8]–[10]. It can thus be of great benefit if the route being travelled can be identified in real-time making the topography known to facilitate finding an optimal velocity profile that reduces the energy consumed.

Vehicle location above ground can easily and accurately be pinpointed through a global positioning system (GPS), and route identification is easily performed, but these systems do not work underground. This study aims to present a method by which routes may be easily and cheaply identified underground with sufficient accuracy, using easily obtainable environmental data and not requiring any transmitters or beacons to be placed on the routes. To prove the functionality of this strategy, the initial testing is done above ground with a road vehicle, after which underground testing is performed to prove the availability of usable data for underground route identification.

The novelty of this method lies in the use of magnetic heading and barometric altitude data, which are easily measurable environmental attributes that do not change as a function of driving style or vehicle used. A 3-axis magnetic field strength sensor and a digital barometric pressure sensor are used with a wheel rotation encoder to record heading and altitude-change data patterns at 100-metre intervals. The heading and altitude data for three passes of a test route are shown in Figures 1a and b, respectively. From this, there is very good repeatability in the data for the three passes. Jumps in heading are for values close to north, where approaching from the positive side yields headings close to zero and from the negative yields headings close to 360°. Due to the affect of weather on barometric altitude data, which was found to vary by as much as 60 metres in a day, the change in altitude over a set distance increment is used instead of the absolute value. An elementary statistical comparison model is implemented that compares the values measured in real-time to those of saved routes. Patterns in the results are evaluated to perform route identification. The method uses the normal probability density function (PDF) to evaluate the correlation between each new data point and the points in a library of routes point by point. The MATLAB function used for this is *normpdf*(X,μ,σ), with X being the new data point's numeric value, μ the saved data point and σ a preset standard deviation used to set the sensitivity of the system. If the value of a new data point lies close to one of the stored values, the equation will yield a high correlation value. Otherwise, it will be very low due to the nature of the *normpdf* function. These correlation values for the real-time data vs. the saved data are stored in a matrix for both the heading and altitude change data sets. Jumps in heading data are accounted for by using the highest correlation of three heading point values, the first being the original data point and the other two being the original data point plus and minus 360°. The resultant correlation matrix for heading and the correlation matrix for altitude change are multiplied point for point to yield a new matrix in which an element only has a high value if a high correlation exists simultaneously in heading and altitude change. This is repeated for all the routes in the library. One axis of the matrix represents the saved data and the other the real-time stream, so if these correspond, there should be a high diagonal. The resultant matrix for a route compared to itself (left) and another route (right) is shown in Figures 2a and b, respectively. One disadvantage of this method is that the data needs to line up exactly, meaning that slight distance discrepancies will result in correlation values alternating from one diagonal band to the next, which makes real-world application difficult as these slight discrepancies are expected to occur. A smoothing filter is applied to the data to allow misaligning correlation values to still contribute to the correlation. A threshold is applied with values above the threshold evaluating to 1 (successful correlation) and 0 otherwise (See Figures 2c and d).The mean of the diagonal bands for this matrix is now evaluated and a route is identified if this value is above a preset threshold. If no route can be identified, the system is set up to save this data as a new route for use in future passes of the route.

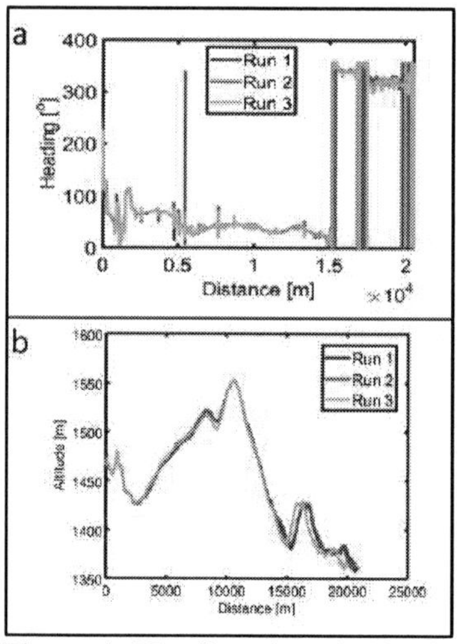

Figure 1: Heading a) and altitude b) data for a test route

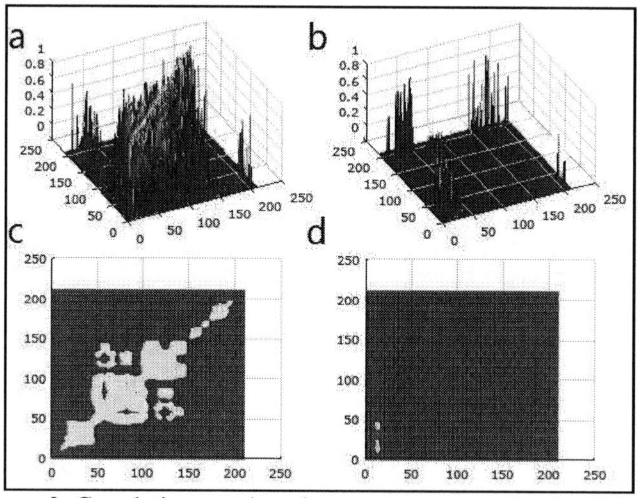

Figure 2: Correlation matrices for a route compared to itself (left) and a different route (right)

Eight routes were evaluated during testing of the system. These routes are shown in Figure 3a. It is noted that routes 1, 3, and 6 have the same initial strip of road, meaning that incorrect route identification needs to be tolerated initially, as there is no way to know where the vehicle will turn off. Figure 3b shows the correlation plotted while travelling on route R1 from A to B. As no information is initially available, there is no way to identify a route at its start. As more data is streamed, the correlation increases until a route is identified. A high correlation exists for routes 1, 3, and 6 up to the point where routes 3 and 6 turn off, which is expected to occur.

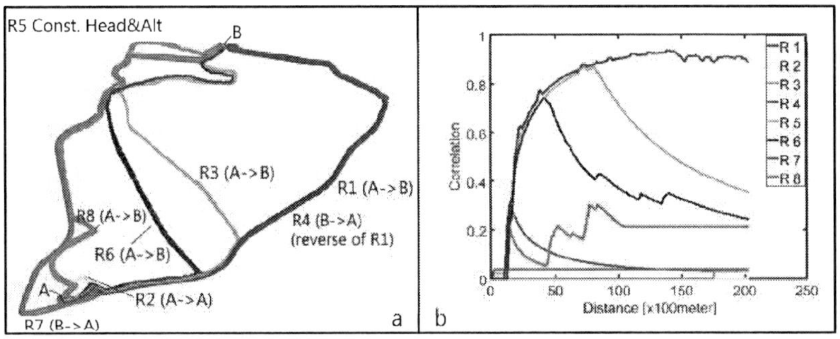

Figure 3: a) Above-ground routes evaluated, b) route correlation vs. distance

The Route Identification Strategy (RIS) proposed is first evaluated above ground using a normal road vehicle and eight different routes. The results showed that it can correctly identify a route, usually within less than 40 data points, with no prior knowledge of the route. Heading and barometric altitude readings were taken underground in a basement parking lot and a decommissioned gold mine. Those proved capable of yielding the same quality in route identification as the above-ground tests [2].

3. Mass estimation in real-time

Hellström et al. [11] proposed an optimised control strategy to reduce the fuel used in a heavy diesel truck. This strategy utilises information about the road ahead, specifically looking at the route's topography. They minimise the fuel mass used while considering the maximum route time. They state that mass is the most important parameter in their generic analysis of a heavy vehicle. Vahidi et al. [12] reported that the mass of a heavy-duty vehicle can vary by as much as 400%. Wilhelm et al. [13] found that the mass of a vehicle can be estimated sufficiently accurately using Newton's second law of motion if sensor data is accurately gathered. They found that the best accuracy (<3% error) is obtained during wide-open throttle (full power). During normal driving, the model accuracy deteriorates to an error of around 10%. Real-time mass estimation can significantly improve the required energy and distance-to-empty estimations.

Newton's second law is applied to a vehicle in motion, as is shown by the free body diagram in Figure 4. Friction forces include aerodynamic and rolling resistance, always opposing the direction of motion. Incline force, propulsion force and acceleration force can have signs in both directions. The propulsion force is determined by measuring the rear axle's driving torque and dividing it by the effective wheel radius. In the case of an electric vehicle, the torque can be accurately estimated by measuring the electric motor current. This study used a specially designed load cell mounted to the vehicle's driving axle [3]. Summing all forces and rearranging to solve for the unknown mass of the vehicle results in equation 1.

This equation does, however, pose two risks:1. If the propulsion force equals the aerodynamic drag force, the numerator in equation 1 approaches zero, yielding zero mass calculated. 2. If the sum of the acceleration, rolling resistance and incline terms approach zero, the denominator causes the mass value to approach ± infinity. Equation 1 still holds during these cases, but the inaccuracy of the sensor data causes massive errors in mass estimation. This needs to be carefully accounted for during the mass estimation calculations. By evaluating the mass result only in specific cases of minimum amounts of power being applied, excluding braking events and excluding estimations expected to yield unrealistically high mass values, allows the mass of the entire vehicle rig (vehicle and trailer) to be estimated by taking the average of several mass estimation values, as is shown in Figure 5. The blue line represents the real-time mass estimations and the red line is the average.

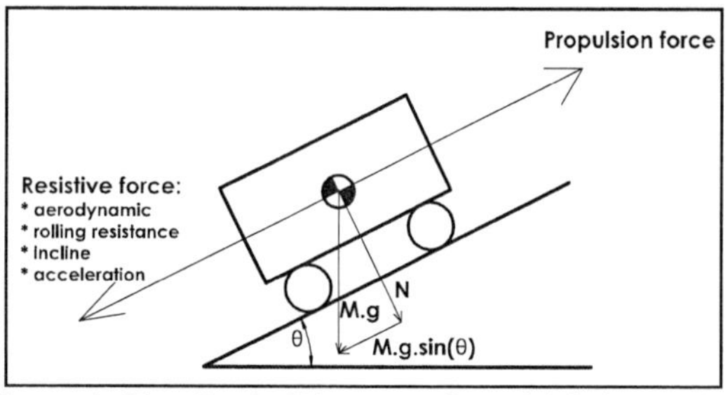

Figure 4: Free Body Diagram of a vehicle in motion on an incline

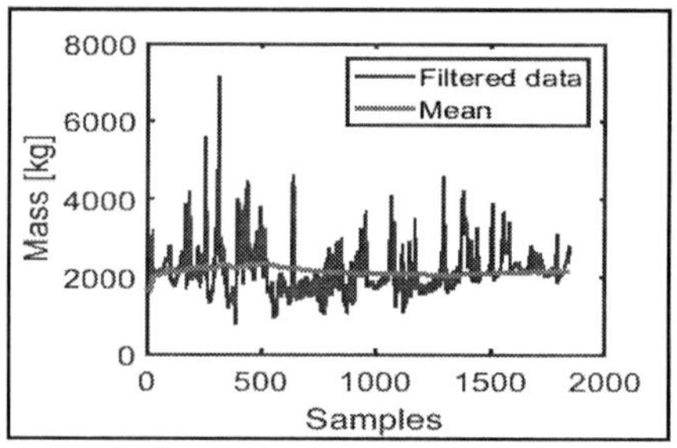

Figure 5: Real-time mass estimation. Blue: Mass estimates, Red: Mean

$$MASS = \frac{\left(\frac{T_{axle}}{wheelradius} - \frac{1}{2}\rho C_d A_f V^2\right)}{\left(a + C_{rr}.g + g.\sin(\theta)\right)} \qquad \text{Equation 1}$$

A variety of loading cases were evaluated, which included an inside-vehicle payload variance of 330kg (≈15% change) from fully laden to empty, as well as a small or large trailer with various loads, attached to the vehicle, equating to a Gross Combined Mass (GCM) variance of up to ≈52%. The standard deviation for the error in estimating the mass from 27 tests performed is found to be 5.2%. This mass is used to estimate the energy required for a vehicle to complete its route.

4. Velocity profile optimization in real-time

Back et al.[4] mention that the route can't be optimized for topography and loads as many times the optimal plan will be inhibited by external influences, like traffic. They mention that considering all the internal and external factors yields a complicated optimization problem that can't be solved in real-time. A compromise is therefore required between optimality and realistic real-time execution ability amongst uncertainties. The approach followed is to provide an optimal target velocity profile that the vehicle controller can try to follow which considers the route topography, maximum allowable completion time, which considers the route topography, maximum allowable completion time, and the regeneration efficiency of the vehicle drivetrain. If the optimal velocity profile is able to update in real-time it will allow compensation for the external influences as the route is being covered, resulting in reduced overall energy consumption. Zhou et al. [14] report that aggressive and inappropriate driving styles can increase fuel consumption by 68% for urban and 47% for rural roads, emphasizing the effect of improper velocity on route energy. A vehicle fitted with a powertrain capable of storing energy can significantly reduce effective energy consumption [6], [8]–[10]. Johannesson et al. [8] propose that speeding up when going downhill is preferred to regenerative braking due to less wear on the battery and hybrid system and better efficiency in exchanging between kinetic and potential energies.

A first principles model is used to model a vehicle's energy consumption while travelling a known route accurately. It accounts for all forces mentioned in Figure 4, including the ability to adjust the regeneration efficiency of the vehicle. The regeneration efficiency was assumed to be constant for this study. The MATLAB gradient-based solver (*fmincon*) was used to find a vehicle's optimal velocity profile to complete the route in the maximum allowable time and consume the least energy. From this, the optimal velocity profile had a shape similar to an inverted version of the route topography. Figure 6 illustrates the optimal route velocity profile for various test routes. Figure 6a is for routes with constant inclines and declines ranging from flat to 10%, showing the energy required to complete the route in the legend of the figure. It is noted that a 2% decline results in almost no energy required to complete the route. For routes simply being a constant uphill, the optimal is simply constant. For routes with varying inclines, the optimal profiles approach an inverted version of the topographic profiles, as seen in Figure 6b for test routes with 7% up-down-up-down and 7% down-up-down-up topographic profiles.

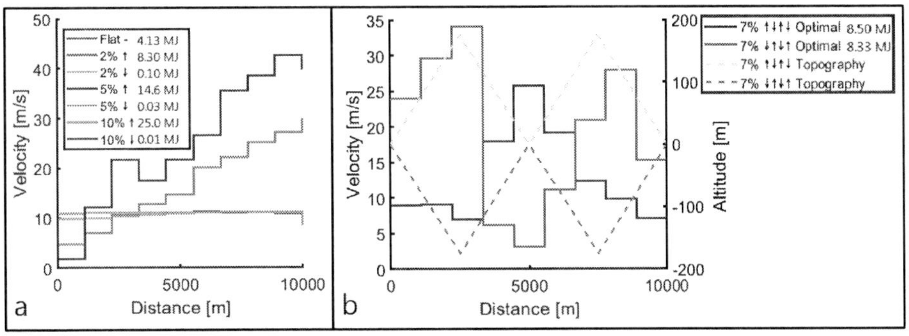

Figure 6: Optimal velocity profiles obtained for test route profiles. a) constant, b) zig-zag

The effect of mass on the optimal route profile is studied to see whether it influences the optimal velocity profile. Figure 7a shows the optimal velocity profiles obtained for a vehicle travelling a real-world test route with the mass varied from 2 up to 4 tons. This shows that the optimal velocity profile is relatively insensitive to the vehicle's mass, with slight differences attributed to noise in the optimizers' convergence.

Then, the effect of regeneration efficiency is investigated by sweeping the constant value from zero (a normal vehicle without regeneration ability) up to 100%. Figure 7b shows that the total route energy consumed reduces if it is possible to store energy in times of excess, such as when driving down a hill. It is also seen that the need to speed up when going downhill reduces as the regeneration efficiency increases, up to a point where it is possible to recover all the energies and travel at a minimum constant velocity throughout (100% regeneration efficiency). However, a realistic limitation to this is capped at around 40% for currently available regeneration systems [15].

The simulation results presented above take over 3 hours each, making their use in a real-time application impractical. To create an approximation to these optimal profiles that can run in real-time, a low-fidelity optimization problem is set up that linearly inverts, scales and shifts the topographic profile to yield an approximation to the optimal velocity profile. This approximation is based on the route, maximum completion time and regeneration efficiency. The scale and shift parameters are obtained by sweeping a range of realistic regeneration efficiencies and maximum completion times for each route. They are stored in a lookup table for fast retrieval, allowing a real-time estimate of the optimal velocity profile. The error in route final energy when using the approximated values compared to a full optimization is less than 14%. An approximated profile is obtained in 0.6 milliseconds. Figure 8 shows a real-world route driven by an uninformed driver (red line), along with the real-time updating optimal velocity profile (blue line), with the original target optimal profile shown in faint black. When driving below the proposed optimal velocity, the remainder of the route's velocity profile increases, and vice versa. The real-world completion of the route consumes 8.5 MJ of energy, whereas the real-time optimized profile, if followed from the start, would have consumed 6.7 MJ, and the complete optimal solution required only 5.8 MJ. Thus, the complete optimal can reduce the required energy by 32%, whereas the real-time optimal could reduce the required energy by 21% for a vehicle without regeneration ability. This difference between the real-time optimal and complete optimization solution reduces as regeneration efficiency increases. For an achievable regeneration efficiency of 40%, the real-world velocity profile consumes 7.6 MJ, the complete optimal 5.3 MJ and the real-time optimal 5.8 MJ. Resulting in a reduction of 30% for the complete optimal solution and 24% for the real-time solution.

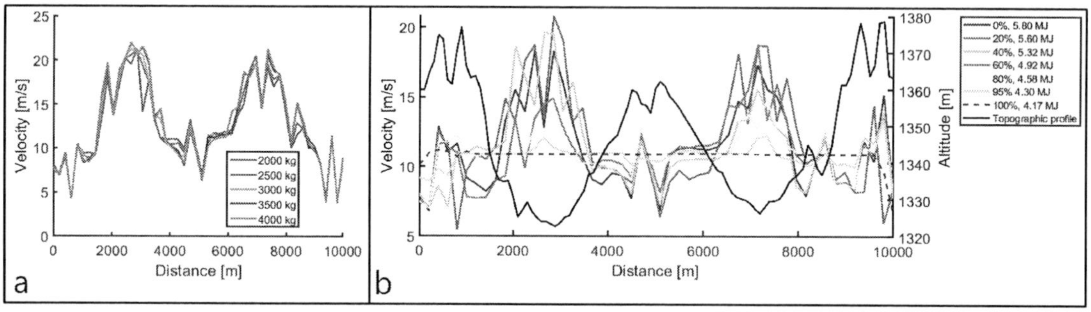

Figure 7: Optimal velocity profiles for varying a) mass and b) regeneration efficiency

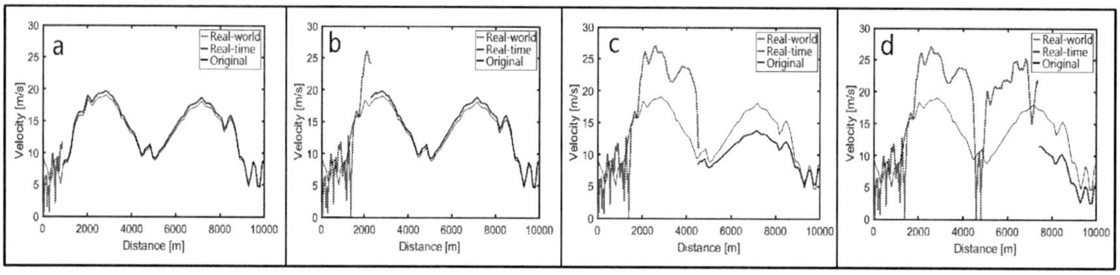

Figure 8: Real-time updating of the optimal velocity profile as the route is being completed

5. Conclusion

This project set out to reduce the energy consumption of an underground vehicle. The method incorporates a strategy to identify an underground route by evaluating patterns in heading and altitude change data in a simplistic manner. Having the topography available yields accurate real-time incline estimates used to estimate the vehicle mass. Driving a vehicle over hilly terrain can increase fuel consumption significantly, which motivates finding improved velocity profiles that use the topography to its advantage. The velocity profile is insensitive to mass, though the energy consumed still is, influencing estimations on remaining distance with the available amount of energy. A method is developed that can use an adapted version of the route topography to estimate the optimal velocity profile that a vehicle should travel the route by in real-time. This method shows capable of providing an improvement of 21% for the real-time travelling of the test route evaluated.

Nomenclature

X = In route identification, this is the streamed-in data point's numeric value
μ= In route identification, this is the saved route data point numeric value
σ= In route identification, this is a preset standard deviation used to set the sensitivity of the system
T_{axle} = Torque on axle measured by strain gauges, Nm
ρ = Air density determined from barometric pressure, kg/m^3
C_d= Drag coefficient, 0.4 (from vehicle datasheet)
A_f= Frontal area of vehicle, 2.33m^2 (from vehicle datasheet)
V = Vehicle velocity from odometer sensor, m/s
C_{rr} = Coefficient of rolling resistance, typically 0.012 - 0.015 for radial rubber tyres on tarmac
θ = Incline angle, radians

a = Acceleration, m/s^2
F = Force, N

References

[1] A. Taghavipour, M. Vajedi, N. L. Azad, and J. McPhee, "A Comparative Analysis of Route-Based Energy Management Systems for Phevs: Analysis of Route-Based Energy Management Systems for PHEVs," *Asian Journal of Control*, vol. 18, no. 1, pp. 29–39, Jan. 2016, doi: 10.1002/asjc.1191.
[2] R. F. Meeser and N. Theron, "Real-time underground route identification and route progress using simple on-board sensing and processing," *J. S. Afr. Inst. Min. Metall.*, vol. 123, no. 1, pp. 41–52, Feb. 2023, doi: 10.17159/2411-9717/1893/2023.
[3] R. Meeser, "A system and method for estimating a mass of a vehicle," PCT/IB2022/061330, 2022
[4] M. Back, M. Simons, F. Kirschaum, and V. Krebs, "PREDICTIVE CONTROL OF DRIVETRAINS," *IFAC Proceedings Volumes*, vol. 35, no. 1, pp. 241–246, 2002, doi: 10.3182/20020721-6-ES-1901.01508.
[5] Y. Yokoi *et al.*, "Driving pattern prediction for an energy management system of hybrid electric vehicles in a specific driving course," in *30th Annual Conference of IEEE Industrial Electronics Society, 2004. IECON 2004*, Busan, South Korea: IEEE, 2004, pp. 1727–1732. doi: 10.1109/IECON.2004.1431842.
[6] H. Liu, H. Li, M. O. Rodgers, and R. Guensler, "Development of road grade data using the United States geological survey digital elevation model," *Transportation Research Part C: Emerging Technologies*, vol. 92, pp. 243–257, Jul. 2018, doi: 10.1016/j.trc.2018.05.004.
[7] K. Boriboonsomsin and M. Barth, "Impacts of Road Grade on Fuel Consumption and Carbon Dioxide Emissions Evidenced by Use of Advanced Navigation Systems," *Transportation Research Record: Journal of the Transportation Research Board*, vol. 2139, no. 1, pp. 21–30, Jan. 2009, doi: 10.3141/2139-03.
[8] L. Johannesson, N. Murgovski, E. Jonasson, J. Hellgren, and B. Egardt, "Predictive energy management of hybrid long-haul trucks," *Control Engineering Practice*, vol. 41, pp. 83–97, Aug. 2015, doi: 10.1016/j.conengprac.2015.04.014.
[9] A. Lajunen, "Fuel economy analysis of conventional and hybrid heavy vehicle combinations over real-world operating routes," *Transportation Research Part D: Transport and Environment*, vol. 31, pp. 70–84, Aug. 2014, doi: 10.1016/j.trd.2014.05.023.
[10] R. Wang and S. M. Lukic, "Review of driving conditions prediction and driving style recognition based control algorithms for hybrid electric vehicles," in *2011 IEEE Vehicle Power and Propulsion Conference*, Chicago, IL, USA: IEEE, Sep. 2011, pp. 1–7. doi: 10.1109/VPPC.2011.6043061.
[11] E. Hellström, J. Åslund, and L. Nielsen, "Design of an efficient algorithm for fuel-optimal look-ahead control," *Control Engineering Practice*, vol. 18, no. 11, pp. 1318–1327, Nov. 2010, doi: 10.1016/j.conengprac.2009.12.008.
[12] A. Vahidi, A. Stefanopoulou, and H. Peng, "Recursive least squares with forgetting for online estimation of vehicle mass and road grade: theory and experiments," *Vehicle System Dynamics*, vol. 43, no. 1, pp. 31–55, Jan. 2005, doi: 10.1080/00423110412331290446.
[13] E. Wilhelm, L. Rodgers, and R. Bornatico, "Real-time electric vehicle mass identification," vol. 6, p. 6, 2013.
[14] Y. Zhou, A. Ravey, and M.-C. Péra, "A survey on driving prediction techniques for predictive energy management of plug-in hybrid electric vehicles," *Journal of Power Sources*, vol. 412, pp. 480–495, Feb. 2019, doi: 10.1016/j.jpowsour.2018.11.085.
[15] A. Boretti, "Improvements of Vehicle Fuel Economy Using Mechanical Regenerative Braking," SAE International, 2010. doi: 10.4271/2010-01-1683.

13th South African Conference on Computational and Applied Mechanics (SACAM2024)
22 – 23 January, 2024, Stellenbosch, South Africa

LOW-SPEED WIND TUNNEL TESTS OF TWO AIRFOILS FOR STANDARD CLASS SAILPLANE APPLICATION

Tristan Melman*, Johan Bosman†, Attie Jonker

* Presenting author

† Corresponding author: johan.bosman@nwu.ac.za

North-West University, School of Mechanical Engineering

Private Bag X6001, Potchefstroom. 2520

ABSTRACT

A significant portion of aircraft drag is attributed to the aerodynamic performance of the wings, which is largely dependent on the geometric shape of the airfoils. For a new standard-class (no flaps) sailplane, a novel airfoil (OPT110) was developed through a genetic optimization algorithm that utilizes XFOIL (vortex panel method) as the aerodynamic calculation engine. When compared to a baseline case, the sailplane with improved airfoils exhibited remarkable improvements in calculated average cross-country speeds.

Although 2D airfoil research is generally considered resolved by most aerodynamic academics, discrepancies between numerical analysis methods were observed. While using CFD to compare the differences in aerodynamic performance between the baseline and the new airfoil, the improvement is not as substantial as previously indicated by XFOIL. Numerous CFD simulations were conducted, incorporating the best-known transition models and methodologies, which included far-field drag calculations in an attempt to achieve comparable results with XFOIL, but without success.

When designing new aircraft, confidence in performance prediction is essential, and significant differences between the results of airfoil analysis methods create uncertainties that, in turn, impede aircraft development. It is for this reason that both the baseline and new airfoil sections were tested in a low wind-speed wind tunnel facility provided by the CSIR to determine their true performance characteristics. Wind tunnel models for both airfoil sections were manufactured using CNC-milled molds, incorporating pressure taps built into the skins, which are connected to a 64-channel miniature Scanivalve analogue pressure scanner. Drag was measured with a wake rake, and the test rig also integrates loadcells to mechanically measure lift and drag.

These experimental results are expected to offer insights into the ability of XFOIL and CFD to analyze aerodynamic characteristics of 2D airfoil sections designed for low-speed range sailplanes.

Key words: Airfoil design, Low speed aerodynamics, CFD transition models, Sailplanes, Wind tunnel experiments

1. Introduction

Modern Sailplanes operate in the Reynolds number range where extensive laminar flow regions on the exterior surface is possible. This has been exploited for many years with airfoil sections which have been designed for extensive laminar flow over the top and bottom surfaces. The design of glider airfoils was in the past a difficult problem that was solved with extensive wind tunnel testing. This was simplified with the arrival of airfoil design codes such as XFOIL [1] and the Eppler Airfoil code [2].

These codes have been used with great success in the past for the development of airfoil sections for modern gliders. XFOIL has been exclusively used for the design of the airfoil family for the JS1, JS2 and JS3 gliders. Currently the JS family of gliders are considered to be some of the best performing gliders in the world as is shown by recent World Gliding Championship results [3].

The JS family of gliders all uses flapped airfoil sections and the accuracy of XFOIL in the design of flapped airfoil sections have been proved through wind tunnel testing as well comparative flight testing between full-scale gliders.

In the development of the JS4 glider, which is a standard class glider and uses an unflapped airfoil section, XFOIL was again used in the development of this airfoil section. The new airfoil (OPT110) was developed through a genetic optimization algorithm that utilizes XFOIL (vortex panel method) as the aerodynamic properties calculation engine.

The XFOIL calculations of this new airfoil showed far superior performance over a baseline model and suspicion arose as to the validity of these results. It was decided to analyze the airfoil using CFD methods as a verification of the XFOIL results. Even though best practice CFD methods with verified transition models and far-field drag calculations was used, the results differed considerably from the XFOIL data. It was thus impossible to tell which code gave the correct answer and the only way out is through wind tunnel testing [4].

2. Problem definition

It is well known that numerical methods often give inaccurate absolute results, but are quite useful to determine performance differences between geometries. This can be applied to airfoil design as well. The results of most airfoil analysis codes are not necessarily correct in relation to real-world answers, but the trends and difference in performance they show between airfoil geometries are assumed to be correct.

XFOIL has therefore been used with great success in the past efforts to design flapped airfoil sections by just looking at the difference in performance between different design iterations. This means that if one geometry gives a better performance than a previous one, the improvement was assumed to be correct.

During the design of the flapped airfoil section, OPT110, this approach was also used. An optimization routine was used with XFOIL to search for the best unflapped airfoil section based on performance specific criteria. The optimized airfoil geometry that was obtained is shown in Figure 1 and was numbered OPT110.

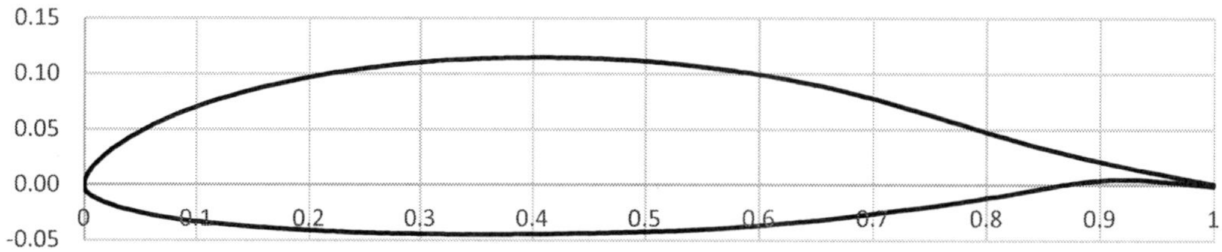

Figure 1: OPT110 Airfoil geometry

OPT110 was developed from a baseline airfoil section called ST1, which is an approximate of airfoil sections currently used on production standard class gliders. The performance difference between OPT110 and ST1 as predicted by XFOIL is shown in Figure 2. The XFOIL results show that the ST1 has lower drag between Cl = 0.3 and 0.43, but the OPT110 airfoil has lower drag over the rest of the polar. A significant improvement over the baseline is shown for higher lift coefficients.

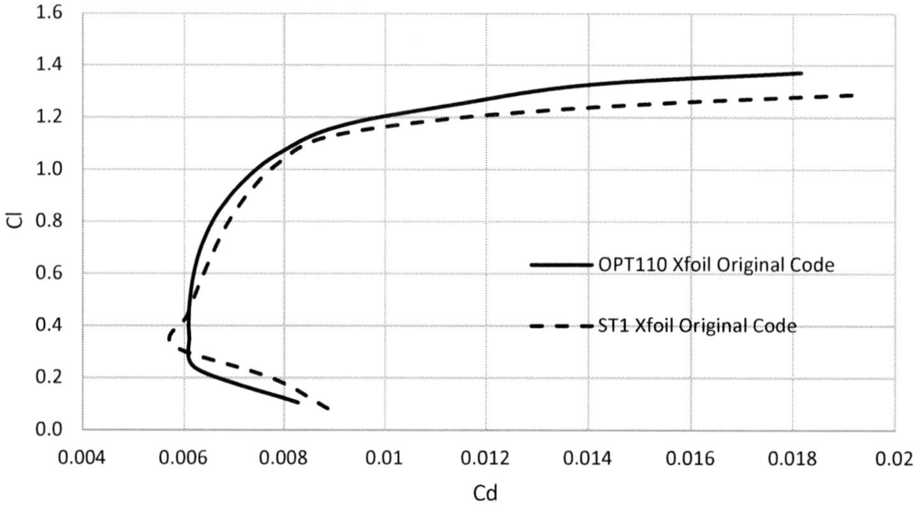

Figure 2: XFOIL Cd, Cl Results

The two airfoils were also compared using CFD analysis. In this case STAR-CCM+ was used with transition models and mesh parameters as shown in Table 1. The calculated performances for OPT110 and ST1 are shown in Figure 3. The CFD results do not show the same results than that of XFOIL. Here ST1 has a lower drag over the complete speed range (above Cl = 0.2), with a significantly lower drag shown in the Cl = 0.2 to Cl =0.6 range in comparison to the XFOIL results.

Table 1: STAR-CCM+ Continua and mesh values

Continua:		
$\gamma - Re_{\theta_t}$ Transition		
Density	0.998	kg/m^3
Dynamic Viscosity	1.842E-5	Pa-s
K-Omega Turbulence [5]		
Segregated Flow Model		
Turbulence Intensity	0.199	
Velocity	32.8	m/s

Mesh:		
Base Size	0.4	m
Target Surface Size	0.8	%
Minimum Surface Size	0.1	%
Surface Growth Rate	1.06	
Number of Prism Layers	96	
Prism Layer Stretching	1.05	
Prism Layer Total Thickness	0.015	
Wake Distance	18	m
Wake Isentropic Size	2.0	%
Wake Refinement	1.2	

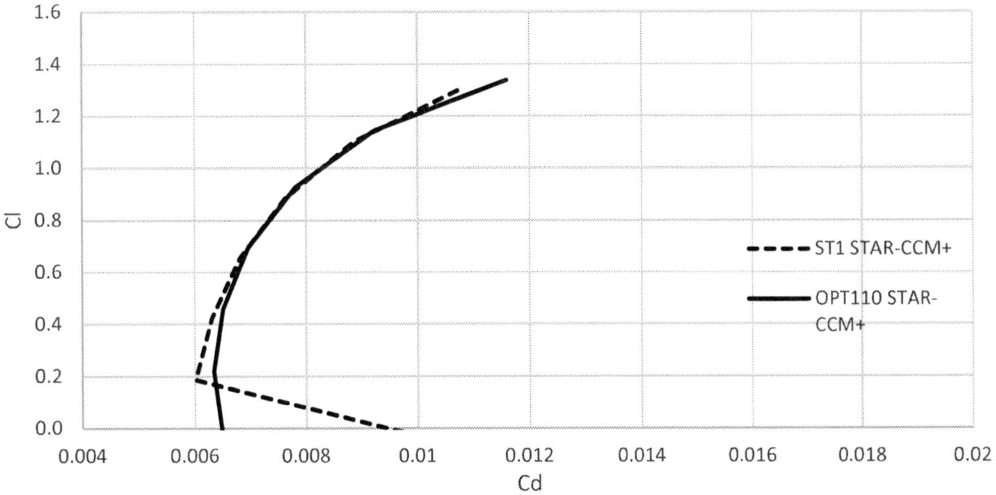

Figure 3: OPT110 and ST1 STAR-CCM+ Cl, Cd results

The problem can thus be summarized as follows:

1. Is the assumption correct that the relative performance calculations for different geometries are accurate when calculated with XFOIL.
2. Does XFOIL or CFD give the more accurate results for these unflapped airfoil section.

3. Wind tunnel experimental setup

To determine which software yielded the most reliable results the two airfoil sections were tested using an open-circuit wind tunnel at the Council for Scientific and Industrial Research (CSIR) with a maximum flow velocity of 33m/s. The airfoils were tested at different angles of attack while the drag was measured with an integrated wake rake [6]. The lift was calculated from the measured pressure distribution over the airfoil surface.

3.1 **Wing Test Sections**

The wind tunnel dimensions and required Re range resulted in an airfoil test section with chord length of 0.54m. The airfoils were manufactured from accurate CNC cut negative moulds. This allowed access to the inside of the sections to install pressure taps from which the surface pressure could be measured. Figure 4 shows the inside of a test section and the installation of the pressure taps.

Figure 4: Interior Airfoil Section View of the Pressure Taps

3.2 **Integrating Wake Rake**

A wake rake that was provided by CSIR was fixed in place a distance of 90cm behind the trailing edge of the airfoil test section in the direction of the air flow.

The wake rake has 40 pressure tubes in the center spaced at 1.65mm. Away from the center the tubes are spaced on 3.3mm as shown in Figure 5. The wake rake was used to measure the pressure distribution of the dynamic pressure that would be integrated to determine the drag coefficient.

Figure 5: Integrating Wake Rake Provided by CSIR.

3.3 **Test rig and Data acquisition**

A custom rig was developed and built to house the wing sections as well as all the equipment and instruments needed to measure the surface pressure distribution of the wing as well as the wake profile. The rig was designed with large endplates to minimize any vortices forming on the edge of the airfoil section to help create more stable airflow for the static pressure taps on the wing surface as well as help stabilize the flow for the wake rake.

The data acquisition was done using a 64-channel Scanivalve MPS4264 pressure scanner that has a range of 1990.7 Pa and has an accuracy of 0.15% of the full scale. The pressure scanner was connected to the pressure taps as well as the wake rake using specialized tubing while also being connected to a laptop, using an ethernet cable, that has the required software installed to control the pressure scanner as well as data recording.

Figure 6: Test rig setup at CSIR 2m Wind tunnel

3.4 **Determining the angle of attack and turbulence intensity**

Due to the rig being custom built and needing to be bolted in place, in this instance it would have been impossible to determine the true level of the wing using conventional methods. Therefore, a different method of determining the angle of attack for each test iteration was needed. An easy method of determining the true angle of attack with regards to the air flow is proposed by Soto-Valle et al. [7] which utilizes comparing the measured experimental pressure distribution and comparing it with the simulated pressure distribution.

XFOIL was set up with a similar calibration procedure to the conditions that the tests took place in, with the Reynolds number needing to be set correctly to ensure match between the pressure distributions. Once the pressure distributions of the experimental data and the XFOIL results match the angle of attack can be reliably assumed to be the same for the airfoil section as it is for XFOIL.

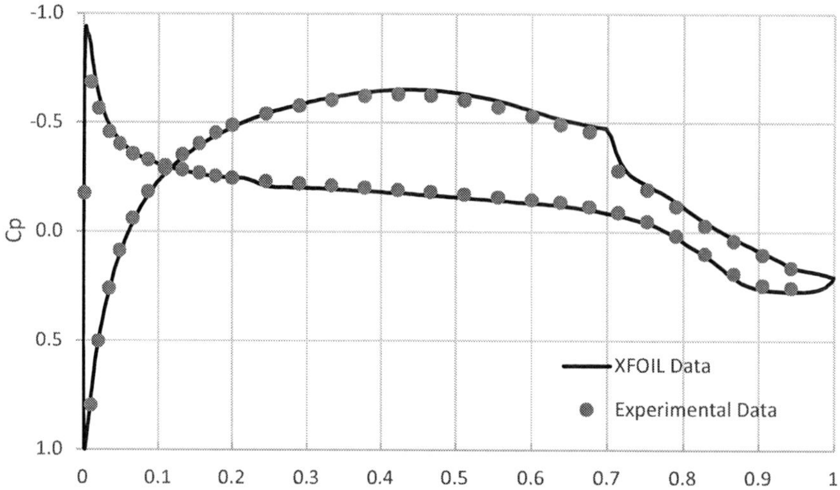

Figure 7: XFOIL and Experimental Data ST1 at α = -2°

The N_{crit} value (turbulence level adjustment) in XFOIL was also determined using the same methodology. After firstly determining each angle of attack by matching the pressure distributions, the transition zone on the experimental data can be estimated by using thermal cameras and heating the wing before starting the wind tunnel to allow the thermal camera to record the transition point where the temperature change will be clearly visible. The N_{crit} can be adjusted until the transition point matches between the experimental data and the XFOIL results.

5 Wind Tunnel Test Results

The wind tunnel tests were conducted over the course of a few days when the weather was stable in terms of outside wind speed and temperature. This helped keep the environmental factors of each test relatively close to each other and allows for all the results to be plotted on the same graph to be compared.

The first results were recorded with small iterations between each angle of attack and had a lower range that was tested. With these tests the data from the pressure taps and the rake were recorded on separate tests to ensure as many ports as possible from the pressure scanner could be dedicated to recording data for either determining the lift or the drag of the airfoil sections.

The second set of experimental results were recorded by splitting the available ports on the pressure scanner between the pressure taps and the rake. This was done to allow the data for both lift and drag to be recorded at the same time. To eliminate the chance that environmental conditions favored one airfoil over the other. These tests were also conducted with larger iterations between each test due to the range of the minimum and maximum angles of attack being greatly increased.

With the pressure scanner ports being split between the pressure taps and the wake, cubic spline interpolation was utilized to help fill in the missing data points, this resulted in an improvement of the results from the lift and drag calculations and was validated by using the STAR-CCM+ simulated pressure distribution at each pressure tap location. Which resulted in more accurate data using cubic spline interpolation with the recorded experimental results.

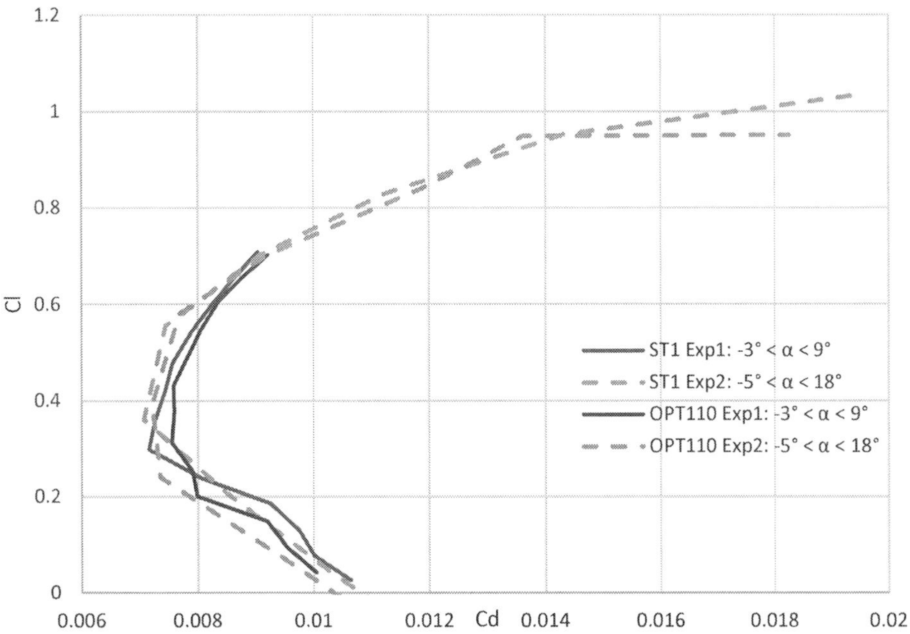

Figure 8: Processed Experimental Wind Tunnel Results

5.4 XFOIL, STAR-CCM+, Experimental Data Comparison

When comparing XFOIL, STAR-CCM+ and the wind tunnel test results, it can clearly be observed that the wind tunnel test lift coefficient of both airfoils was lower than either XFOIL or STAR-CCM+ suggested it would be. This is, in part, due to the data collected from the pressure taps being more spread out over the same airfoil section to compensate for the rake and airfoil needing to collect experimental data at the same time. When comparing both the wind tunnel results of the more focused tests to the wider angle of attack tests it is clear to see that the more focused tests showed the lift trending higher than the wider angle of attack suggests it is. Through validation tests it was also proven to be the case that in these instances the less datapoints are used for determining the lift the calculated lift coefficient will result in lower lift coefficients.

The results from the wind tunnel tests showed in both cases a clear indication that at higher angles of attack the ST1 airfoil consistently produces a better lift to drag ratio or at the very least matches the OPT110 airfoil, as can be seen in figure 10. Therefore, the experimental data supports the simulated results obtained from STAR-CCM+ and it can be stated that XFOIL is under predicting the drag of the OPT110 airfoil as soon as the angle of attack nears or exceeds 0°.

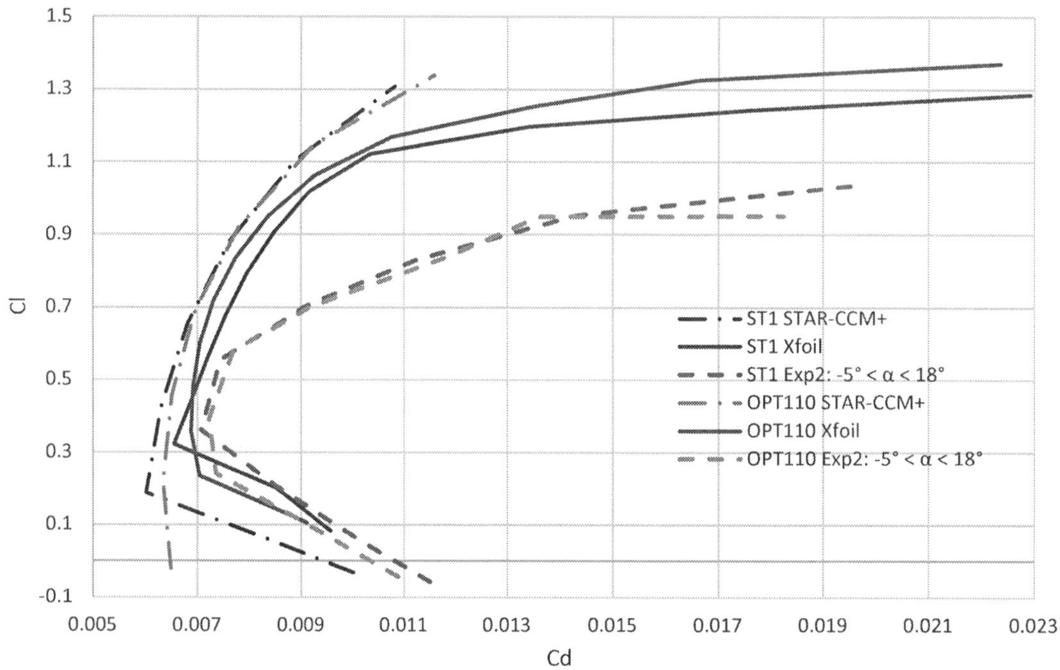

Figure 9: ST1 & OPT110 XFOIL, STAR-CCM+ and Wind Tunnel Test Results

6. Conclusion

After comparing and scrutinizing all the data collected from XFOIL, STAR-CCM+ and the wind tunnel test results, there was a clear correlation between the predicted performance characteristics of the airfoils by STAR-CCM+ and the experimental results obtained from the wind tunnel testing. Due to the ST1 airfoil being an airfoil that has competed in numerous competitions involving sailplanes and has always shown the expected performance characteristics in these competitions, testing and real-world use, it is evident that XFOIL exhibits a tendency to under predict the drag of the OPT110 airfoil.

As XFOIL was utilized in the optimization process of the OPT110 airfoil, it is imperative that a more in-depth investigation needs to be conducted to identify any underlying factors that led to this outcome. Such drastic deviations between the performance characteristics of two airfoils being compared using XFOIL and a computational fluid dynamics (CFD) program like STAR-CCM+ are unusual.

It is plausible that the optimization code utilized in the development of the OPT110 was able to find and exploit a calculation discrepancy in the methodology XFOIL uses to calculate the drag coefficients. Consequently, the optimization process used to develop the OPT110 airfoil may have been adjusting the geometry of the airfoil to influence XFOIL into predicting better performance characteristics than would be reasonably expected from this newly developed airfoil in real world conditions.

This investigation shows the importance of validating results obtained from software simulations to predict the performance characteristics of airfoils. The possibility of the performance predictions being inaccurately skewed in favor of one airfoil over another necessitates in-depth scrutiny. While XFOIL is a very powerful tool that can be used with great efficiency in the development and optimization of airfoils, additional efforts are needed to determine the origins of the exploit that resulted in the inaccurate performance predictions of the OPT110 airfoil.

As mentioned, XFOIL was used for airfoil design work with great success in the past, where improvements over baseline airfoils were pursued. It was, however, assumed that these improvement increments from baseline airfoils shown by the calculations would directly relate to real life performance improvements. This study shows that for certain airfoil geometries, this is not the case and further investigation is necessary.

Nomenclature

α	angle of attack
c	airfoil chord
c_d	section drag coefficient
c_l	section lift coefficient
e^N	boundary layer edge condition
P	pressure
P_s	reference static pressure
ρ	density
C_p	pressure coefficient
U_∞	free stream velocity
N_{crit}	critical amplification factor

References:

[1] Drela, M. & Youngren, H. (2001). XFOIL 6.94 User Guide. Massachusetts: MIT

[2] Liebeck, R. H. (1993). Book Reviews: Airfoil Design and Data- Richard Eppler. AIAA Journal

[3] 37th FAI World Gliding Championships 2022. (2022). Retrieved from URL: https://www.soaringspot.com/en_gb/37th-fai-world-gliding-championships-2022-szeged-2022/results/18m/task-11-on-2022-08-05/total

[4] Melber-Wilkending, S., Heidebrecht, A. & Wichmann, G. (2007) CFD Validation by Wind Tunnel Measurements: Uncertainty Assessment by Numerical Simulation of Complete Wind Tunnel Flows. In Computational Uncertainty in Military Vehicle Design

[5] Menter, F.R., Langtry, R. & Völker, S. (2006). Transition Modelling for General Purpose CFD Codes. Flow Turbulence Combust

[6] Pofer, E.A. & Bramesfeld, G. (2012). Measuring Wing Profile Drag using an Integrating Wake Rake. Saint Louis University

[7] Soto-Valle, et al. (2021) Development and assessment of a method to determine the angle of attack on an operating wind turbine by matching onboard pressure measurements with panel method simulations. University of Florence, Technische Universität Berlin

SPLIT HOPKINSON PRESSURE BAR COMPRESSION TEST SIMULATION AND EXPERIMENTAL VALIDATION OF COMMERCIALLY PURE COPPER SPECIMENS

M.E. Miyambo[1,*], S. Kotole[1], M. Mokalane[1], T. Pandelani[1], A. Clarke[1], and D. Reinecke[1]
[1]Council for Scientific and Industrial Research, Defence and Security – Landward Sciences[*]
mmiyambo@csir.co.za

Abstract: To address the mitigation of blast and ballistic threats, the characterisation of materials under high strain rates is essential for developing precise material failure and impact event simulations. Traditional iterative experimental evaluations are time-consuming and expensive, especially with local constraints on the cost and availability of armour steel. High-fidelity simulation models expedite development, necessitating accurate material models for impact event simulations. In this study, commercially pure copper test specimens, with well-documented properties, served as a pilot for validating experimental materials testing methods. The widely used Split Hopkinson Pressure Bar (SHPB) generated test data to parameterise constitutive material strength models; namely, the Johnson-Cook (J-C) model. Material model parameterization for pure copper, based on curve fitting, required validation before application. Quasi-static compression tests established J-C model parameters. To ensure the robustness of the material model, the Council for Scientific and Industrial Research (CSIR) adopted an approach involving testing at various rates, verified by using a finite element model (FEM) of the SHPB. An LS-DYNA FEM model of the SHPB incorporated Gruneisen Equations of State (EOS) and the J-C material model for the copper specimen material. Experimental SHPB results for commercially pure copper at an average strain rate of 1 533 s^{-1} were compared to the SHPB FEM and published literature results, providing valuable insights into high-strain-rate material behaviour.

Keywords: 3 – 5 keywords

1. Introduction

There is a need in the scientific community dealing with the mitigation of ballistic threats to characterize materials for high strain rates, as modelling of this material behaviour is vital for the development of more accurate material fracture and impact event numerical simulations. Iterative experimental evaluation of developed solutions is both time consuming and expensive, especially due to the cost and availability of armour steel locally. Access to high fidelity simulation models can expedite the development process, and the development of accurate material models is needed for application in the simulation of impact events.

For this work, commercially pure copper test specimens were manufactured as its material properties and various model parameters are widely published [1-3], thus serving as pilot for verification of the experimental and characterization processed implemented.

The Split Hopkinson Pressure Bar (SHPB) is a popular method used for the generation of test data to parameterize material strength and failure models, such as the Johnson-Cook (J-C) material model [4]. The J-C material model has proven to be applicable to a wide range of strain rates, considering the effects of work hardening and material temperatures [5, 6]. The material model parameterization is based on curve fitting techniques thus should be verified before being applied to specific material models [7]. Quasi-static compression tests were also performed to establish J-C model parameters related to the elastic limit and strain hardening of the material.

To provide robustness to the extracted constitutive material model parameters, the CSIR (Council for Scientific and Industrial Research) uses an approach of testing materials in compression at high strain rates by using the SHPB and quasistatic tests are conducted on the material by using a universal compression testing machine. The quasi-static and dynamic results are used as inputs in constitutive material models. A two-dimensional axial symmetry finite element model of the SHPB was developed in LS-DYNA, which consists of a cylindrical striker, incident, and transmitter bar. The finite element model incorporates both the Gruneisen Equations of State (EOS) model [8] and J-C material model, with shell elements and the Lagrange Solver. The J-C model for copper obtained from the LS-DYNA and Ansys Autodyn material library is incorporated in the LS-DYNA model, while the SHPB striker, incident and transmitter bars are manufactured from stainless steel. The test bars are modelled from C350 maraging steel material properties obtained experimentally from the SHPB test bars.

The experimental SHPB results showing the relationship between true stress and true strain of commercially pure copper were compared to other published laboratories for tests executed at a strain rate region of 1 500 s^{-1} [2]. However, the J-C parameters obtained from curve fitting differ from published values, even though good correlation between experimental true stress-strain data and simulation results.

The LS-DYNA simulation results for the velocity of the striker bar at 10 m/s are compared to the experimental SHPB results for the same velocity.

This work is a segment of a long-term strategy to establish projectile-armour steel impact simulation capabilities within the CSIR with the end goal of supporting the SANDF with their protection enhancement needs.

2. Quasi-Static Tests

A set of quasi-static compression tests were executed on pure copper specimens at the CSIR Materials Testing Laboratory. Each specimen material was machined to a length and diameter of 12 mm and 6.25 mm respectively to achieve an approximate length-to-diameter ratio of two (2) for short specimens according to [9]. The tests were conducted using a universal servo-hydraulic testing machine at an estimated speed of 1 mm/min. The servo-hydraulic machine consists of a load cell which is connected to a computer where the load-time curve is displayed, and an extensometer was used to capture the deformation of the specimen.

The force-displacement responses are plotted in terms of engineering stress and strain to calculate ε and σ. The yield stress, σ, is calculated by determining the 0.2% offset yield strain, forming a line with a slope equal to Young's modulus with the yield strain as the x-intercept, and locating the point where this line intersected the stress-strain curve.

The captured raw data displayed negative values during the compression tests, signifying compressive loads. For the two initial compression tests that were conducted on the specimens, the specimens were loaded until they reached 50% of their initial length after the yield strength was exceeded. The loading sequence was increased to 20% of the initial length of the specimen after the initial tests due to the ductility of the material, where no signs of cracking nor fracture were observed.

The final length of the specimen was measured after each test. When the baseline test sequence was defined after the two initial tests which were conducted on the specimens, the remaining specimens were loaded until their yield strengths were exceeded. In total, five copper specimens were tested, but only three of the five tests were recorded.

The compressive load data was captured from the load cell and this load was divided by the specimen's original cross-sectional area to compute the axial engineering stress, σ, exerted on the specimen during the compressive tests, refer to (1). The nominal strain, ε, was determined by dividing the displacement or deformation of the specimen by the specimen's original length, refer to (2).

$$\sigma = \frac{Load}{Area} \tag{1}$$

$$\varepsilon = \frac{Displacement}{Length} \tag{2}$$

The slope within the linear elastic regime is Young's modulus, E, or the ratio of the engineering stress, σ, to engineering strain in the axis, ε, as shown in (3):

$$E = \frac{\sigma}{\varepsilon} \tag{3}$$

3. Split Hopkinson Pressure Bar Tests

The SHPB apparatus consists of a gas gun, striker bar, incident bar, test specimen, transmitter bar, test bars v-block supports, momentum trap, strain gauges and a data acquisition unit. The incident and transmitter bars are supported by the v-block supports while the test specimen is sandwiched by the test bars. In a SHPB test, the striker bar is propelled through a steel barrel to impact the frontal face of the incident bar. A stress wave is generated in the incident bar and propagates through the strain gauge to the front end of the specimen. Incident strain-time signals are produced by the stress wave, measured from the first strain gauge, and displayed on a data acquisition unit. The stress wave continues to propagate through the specimen, the front end of the transmitter bar, the second strain gauge and to the end of the transmitter bar. As the stress wave propagates through the transmitter bar, the transmitted strain-time signal is measured from the second strain gauge and displayed on the data acquisition unit. The incident and transmitted strain-time signals are used to determine the mechanical properties (true stress vs true strain) of the specimen by using the one-dimensional (1D) stress wave propagation theory [10, 11].

Calibration tests were executed to verify the mechanical properties and equilibrium of the bars. First, the physical properties of the bars are established and recorded (length, diameter, mass and density), and the position of the strain gauges are recorded. Then the speed of sound in the bars is determined experimentally [11]. The density and speed of sound of the bars are used to calculate the elastic modulus.

To verify the mechanical properties of the incident and transmitter bars, the striker bar is fired at each bar independently. The equilibrium of the incident and transmitter bars is verified by propelling the striker bar to impact both bars without the test specimen sandwiched in between the bar interfaces as shown in Figure 1.

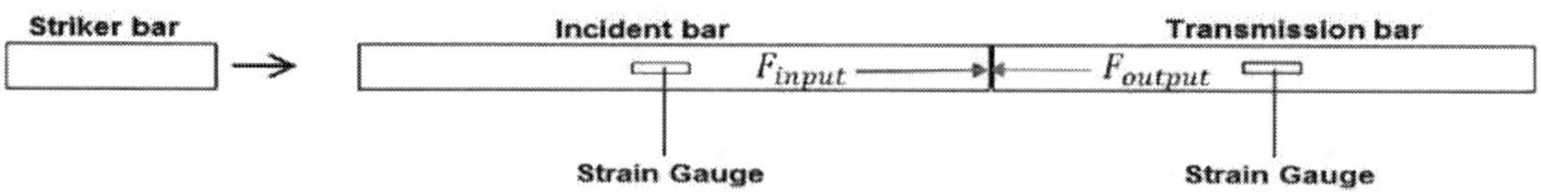

Figure 1: Force equilibrium

The strain signals are used to calculate the input and output forces at the incident/transmitter bar interface. The following formulas are used to verify the equilibrium of the bar system [10]:

$$F_{input} = A_b E(\varepsilon_i(t) + \varepsilon_r(t)) \tag{4}$$

$$F_{output} = A_b E \varepsilon_t(t) \tag{5}$$

In (4) and (5) F_{input} and F_{output} are forces at the incident/transmitter bar interfaces, and Ab and E are the cross-sectional area and Young's modulus of the test bars respectively. The incident, reflected, and transmitted strain signals are denoted by $\varepsilon_i(t)$, $\varepsilon_r(t)$, and $\varepsilon_t(t)$ respectively.

Equation (6) is used to calculate the wave speed of the bar, where C_b denotes the wave speed of the bars and l is the distance measured from the incident/transmitter bar interface to the strain gauge location [11].

$$C_b = \frac{2l}{\Delta t} \tag{6}$$

The Young's Modulus is of the test bars is calculated from equation (7).

$$E = \rho \times C_b^2 \tag{7}$$

A Photron SA4 high-speed camera was used to record the deformation of the copper specimen at a frame rate of 20,000 fps and a shutter speed of 1/35 000 sec during the SHPB tests. A Photron Fastcam Viewer (PFV) software was used to measure the final deformation of the specimen after impact. The specimen dimensions were measured from the high-speed video image a single before the test bars disengaged with the test specimen immediately after the final deformation. The final deformation dimensions are captured in this manner because it was observed from the high-speed video images that the specimen is in equilibrium with the test bars while it is in contact with the bars and before it disengages with the incident/transmitter bar interface. Figure 2, shows the initial and final plastic deformation of the test specimen, with the incident bar on the right, the specimen at the centre and the transmitter bar on the left of the image.

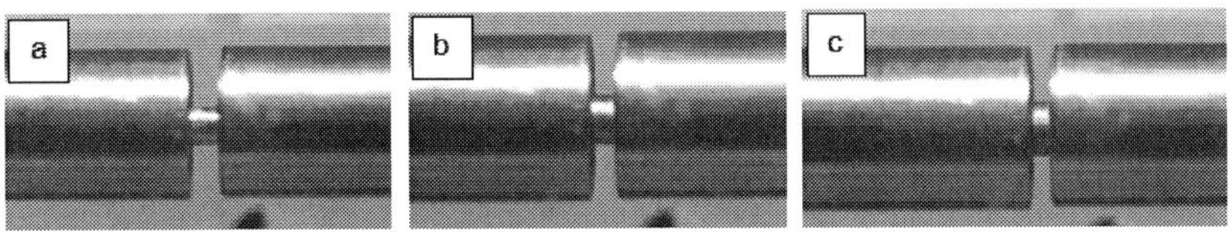

Figure 2: Experimental SHPB; a) before deformation, b) initial deformation, c) final deformation

The initial and final dimensions of the test specimens before and after each test are tabulated in Table 1, where D_0 and l_0 are the initial diameter and length of the specimen respectively, while D_f and l_f are the final diameter and length of the specimen respectively. The dimensions of the specimens before each test were measured using a digital vernier.

Table 1: Initial and final specimen deformation from the SHPB tests

Test Number	Impact Velocity (m/s)	Nominal Strain Rate (s-1)	D_0 (mm)	D_f (mm)	l_0 (mm)	l_f (mm)

1	10.9	2394.394016	6.2	9.5	4.93	3
2	10.7	3051.767152	6.47	8.19	4.81	2.41
3	10.4	2888.952128	5.94	8.42	4.7	2.48
4	10	2799.411605	5.53	7.8	4.61	2.5
Mean	10.5	2780.405512	6.035	8.478	4.7625	2.5975

The nominal strain rate is determined from equation (8):

$$\dot{\varepsilon}_n = \frac{\varepsilon_n C_b}{2L_0} \tag{8}$$

where, ε_n is the nominal strain and L_0 is the length of the striker bar. The nominal strain is given by equation (9):

$$\varepsilon_n = \left|\frac{l_f - l_0}{l_0}\right| \tag{9}$$

4. Results

The true stress-strain results for the quasi-static and SHPB tests that were executed on the copper specimen is shown in Figure 3. The tests executed at an average strain rate of 4 482 s-1 show that the copper material produces higher stresses near the elastic limit when compared to the quasi-static true stress-strain results. The quasi-static true stress-strain results illustrate an unusual behaviour with stresses being higher than the stresses recorded at strain rates in the average region of 1 533 s^{-1} near the elastic limits. The true stress-strain results also show that the copper material is strain rate sensitive when the stresses near the elastic limit are compared to between an average strain rate of 1 533 s^{-1} and 4 482 s^{-1}, respectively in the plastic region.

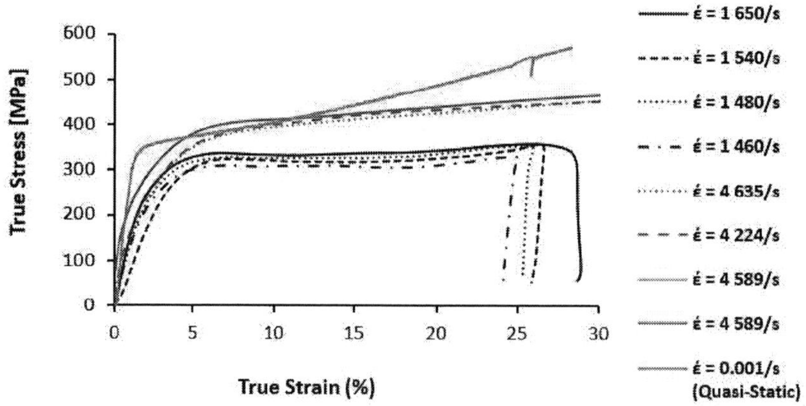

Figure 3: Compression true stress vs true strain for pure copper

5. Constitutive Material Strength Model

Suitable constitutive material models are selected for the test bars and specimen before creating a finite element model of the SHPB in LS-DYNA as part of the verification and validation process of a material model. The MAT_015 Johnson-Cook constitutive material model parameters [12], were extracted from the experimental quasi-static and dynamic mechanical properties. The MAT_001 elastic constitutive material model [12] was used to model the test bars. The J-C constitutive material model is represented by equation (10).

$$\sigma_f = \left[A + B\varepsilon_p^n\right]\left[1 + Cln\left(\frac{\dot{\varepsilon}_p}{\dot{\varepsilon}_0}\right)\right][1 - (T^*)^m] \tag{10}$$

Where, σ_f, A, B, ε_p^n, C, $\dot{\varepsilon}_p$, $\dot{\varepsilon}_0$, T^* and m is the flow stress, the elastic limit, modulus of strain hardening, equivalent plastic strain, exponent of strain hardening (n), strain rate hardening effect, plastic strain rate, normalised reference strain rate, normalised temperature and exponent of thermal softening of the material. The value of $\dot{\varepsilon}_0$ is the strain rate obtained from the quasi-static tests. The normalised temperature is described by equation (11).

$$T^* = \begin{cases} 0, & T < T_{transition} \\ (T - T_{transition})/(T_{melt} - T_{transition}), & T_{transition} \leq T \leq T_{melt} \\ 1, & T_{melt} < T \end{cases} \tag{11}$$

Where, T, $T_{transition}$ and T_{melt} is the deformation temperature, reference temperature of the specimen at room temperature and the melting temperature of the material. For simplified, the room temperature T is inferred to $T_{transition}$, which results in T^*= 0, thus the J-C equation can be simplified as:

$$\sigma = \left[A + B\varepsilon_p^n\right]\left[1 + Cln\left(\frac{\dot{\varepsilon}_p}{\dot{\varepsilon}_0}\right)\right] \tag{12}$$

The extracted experimental versus literature Johnson-Cook constitutive material model parameters are tabulated in Table 2.

Table 2: J-C experimental and literature material model parameters for copper

Parameter	Literature [13]	Experimental
A	90	320 MPa
B	292	920 MPa
n	0.31	1.4
C	0.025	0.008
m	1.09	-
$\dot{\varepsilon}_0$	-	0.001

6. Constitutive Material Strength Model

A computational model of the SHPB setup was developed using the LS-DYNA software. The model incorporates 2D axial symmetry to reduce the simulation run time. Figure 4 shows a partial view of the SHPB test set-up simulation model which comprises the incident bar (green) that is initiated by a 400 mm long striker bar (not shown in the figure), the specimen (blue) and the transmitter bar (red). All the bars have a diameter of 25 mm. The incident and transmitter bars were modelled with lengths of 2 200 mm and 3600 mm respectively. The specimen diameter and length were set to 6.25 mm and 5 mm respectively. These dimensions are similar to the experimental test bars' dimensions.

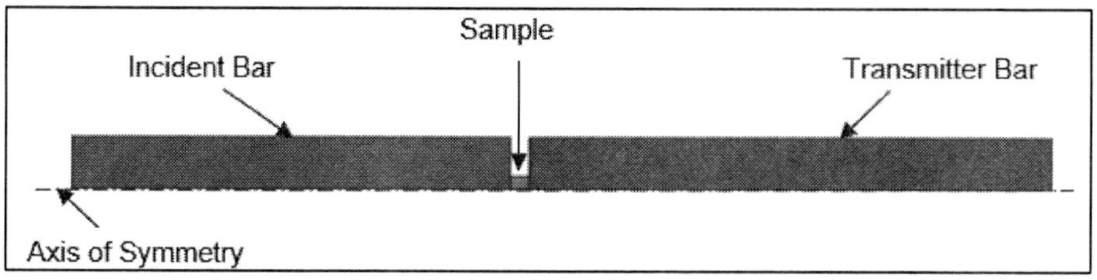

Figure 4: FEM of the SHPB

The striker bar was set to impact the incident bar at an initial velocity of 10.5 m/s. C350 maraging steel material was used for modelling all the bars, i.e. striker, incident, and transmitter bars. The specimen was modelled using oxygen-free copper (OFHC) material properties. The density, Poison's ratio and Young's modulus of the bars were 8020 kg/m^3, 0.33 and 189 GPa, respectively for the MAT_001 material model of the test bars.

The experimental nominal strain-time results obtained from the transmitter bar gauge signal, which represents the nominal strain in the specimen material were compared to the strain-time curve of the SHPB FEM as part of the verification process (see Figure 5). The element corresponding to the transmitter bar's strain gauge location was selected near the surface of the transmitter bar. The transmitted strain signal is used to determine the mechanical properties of the specimen according to the 1D wave theory; thus, the transmitted strain signal is used for the verification process.

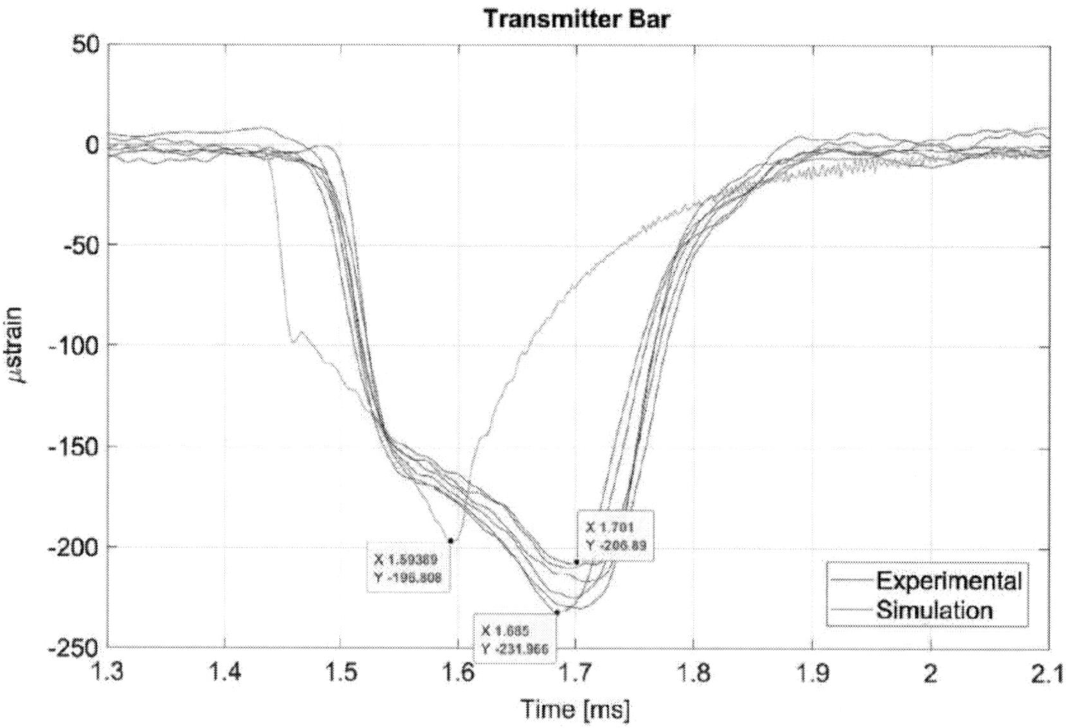

Figure 5: Transmitter bar strain-time SHPB experimental vs simulation

The average SHPB experimental maximum strain measured at the transmitter bar's location was compared to the simulation's maximum strain value at the transmitter bar's strain gauge location (see Table 3). The comparison of the transmitted maximum strain from the simulation illustrates a percentage error of 13 %, which is in good agreement with the experimental maximum transmitted strain.

Table 3: Experimental vs simulation maximum strain at transmitter bar strain gauge location

Test No	Maximum Transmitted Strain (μ Strain)
1	232
2	230
3	225
4	217
Experimental Mean	226
Simulation	197
Percentage Error (%)	13

7. Validation of the Johnson-Cook Constitutive Material Model

The J-C constitutive material model was validated by comparing the experimental final diameter and length of the specimen to the SHPB finite element model's final specimen dimensions. The experimental mean velocity of 10.5 m/s (1 533 s^{-1}) was defined as the striker bar impact velocity in LS-DYNA. The average final dimensions of the specimen are used to validate the constitutive J-C material model for the pure copper specimen are tabulated in Table 4.

Table 4: Copper specimen final deformation dimensions

	Impact Velocity (m/s)	Nominal Strain Rate (s-1)	D_0 (mm)	D_f (mm)	l_0 (mm)	l_f (mm)
Experimental Mean	10.5	2780	6.035	8.478	4.76	2.60
Simulation	10.5	2483	6.25	9.6	5	2.97
Percentage Error (%)	N/A	N/A	N/A	13.24	N/A	14.34

Similarly, the final specimen dimensions were measured from the SHPB finite element model immediately after the specimen disengaged with the test bars' incident/transmitter bar interface. Figure 6, shows the deformation of the specimen as the striker bar is propelled to impact the test bars. The striker bar is omitted from the image to provide a visible view of the specimen deformation.

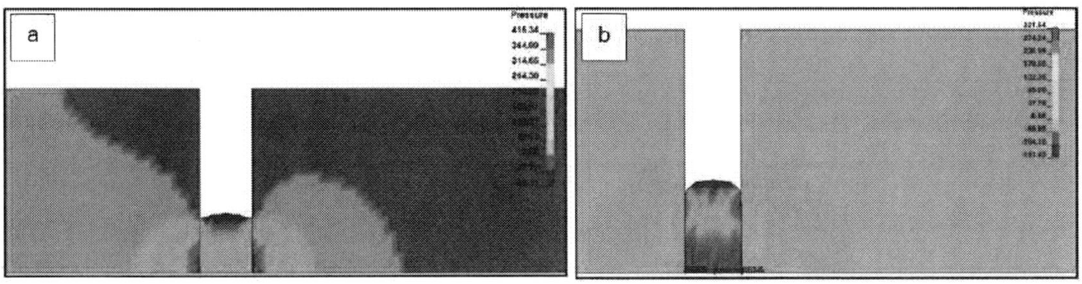

Figure 6: SHPB FEM; a) initial deformation, b) final deformation

During the initial and final deformation of the specimen, the pressure contours from the fringe plots illustrate that the bars are in equilibrium when loaded by the striker bar while the bars are in contact with the specimen, due to similar pressure contours that can be observed on the incident bar/specimen and transmitter bar/specimen interfaces. The pressure contours displayed on the test specimen also appear to be symmetric, which indicates stress equilibrium of the test bars.

8. Discussion and Conclusion

Quasi-static and SHPB tests were performed on pure copper to determine the mechanical properties of the material to develop, verify and validate the material parameters for the Johnson-Cook constitutive material model. The quasi-static and SHPB experimental results were used to extract the constitutive and the parameter for the copper material was compared to literature parameters. The parameters extracted from the experiments is not in good agreement with the published literature parameters [13], thus the discrepancy of these parameters needs to be further investigated.

Although the parameters of the constitutive material model are not in good agreement with the published parameters for pure copper, the SHPB finite element model was verified and validated.

Acknowledgements

The authors would like to thank the DRDB (Defence Research Development Board) and Armscor for funding the research.

Gerrit Visser (SIMTEQ Engineering) is also acknowledged for providing technical support on the LS DYNA computational modelling and simulation of the compression split Hopkinson pressure bar model.

Nomenclature

Greek Symbols

ρ Density
σ Stress
ε Strain
$\dot{\varepsilon}$ Strain Rate

Subscripts

b Bar
f Final
o Initial
i Incident
n Nominal
p Plastic
r Reflected
t Transmitted

Superscripts

n Exponent of strain hardening

Abbreviations

CSIR	Council for Scientific and Industrial Research
DRDB	Defence Research Development Board
EOS	Equation of State
FEM	Finite Element Model
MAT	Material

OFHC	Oxygen Free Copper
SANDF	South African National Defence Force
SHPB	Split Hopkinson Pressure Bar

References

[1] D.A. Gorham, '"The effect of specimen dimensions on high strain rate compression measurements of copper," *J.Phys.D*, vol. 24, no. 8, pp. 1489.

[2] R. Govender, M. Kariem, D. Ruan, R. Santiago, D.W. Shu, M. Alves, G. Lu, G. Nurick and G. Langdon, '"Towards Standardising SHPB Testing-A Round Robin Exercise,", vol. 183, pp. 02027.

[3] W.K. Rule and S.E. Jones, '"A revised form for the Johnson–Cook strength model," *Int.J.Impact Eng.*, vol. 21, no. 8, pp. 609-624.

[4] G.R. Johnson and W.H. Cook, '"Fracture characteristics of three metals subjected to various strains, strain rates, temperatures and pressures," *Eng.Fract.Mech.*, vol. 21, no. 1, pp. 31-48.

[5] A. Banerjee, S. Dhar, S. Acharyya, D. Datta and N. Nayak, '"Determination of Johnson cook material and failure model constants and numerical modelling of Charpy impact test of armour steel," *Materials Science and Engineering: A*, vol. 640, pp. 200-209.

[6] Y.C. Lin, X. Chen and G. Liu, '"A modified Johnson–Cook model for tensile behaviors of typical high-strength alloy steel," *Materials Science and Engineering: A*, vol. 527, no. 26, pp. 6980-6986.

[7] L. Gambirasio and E. Rizzi, '"On the calibration strategies of the Johnson–Cook strength model: Discussion and applications to experimental data," *Materials Science and Engineering: A*, vol. 610, pp. 370-413.

[8] T. Barron, '"Grüneisen parameters for the equation of state of solids," *Annals of Physics*, vol. 1, no. 1, pp. 77-90.

[9] ASTM International, '"Standard Test Methods of Compression Testing of Metallic Materials at Room Temperature," *https://www.astm.org/e0009-09.html*, vol. 2024, no. Jan 10, January 25,.

[10] K.T. Ramesh, '"High rates and impact experiments," *Handbook of experimental solid mechanics*, vol. 933.

[11] W.W. Chen and B. Song, '"Split Hopkinson (Kolsky) bar: design, testing and applications,", 2010.

[12] LIVERMORE SOFTWARE TECHNOLOGY CORPORATION (LSTC), '"LS-DYNA: KEYWORD USER'S MANUAL, VOLUME II, Material Models,".

[13] G.R. Johnson, '"A Constitutive Model and Date for Metals Subject to Large Strains, High Strain Rate and High Temperatures,".

DEEP REINFORCEMENT LEARNING IN FRONTIER-BASED EXPLORATION OF UNKNOWN ENVIRONMENTS

Makhosazana E. Moyo[1,*], Turgay Celik[2]

[1]Mechanical, Industrial and Aeronautical Engineering, University of the Witwatersrand,[*] Makhosazana.moyo@wits.ac.za

[2] Electrical and Information Engineering, University of the Witwatersrand, Turgay.Celik@wits.ac.za

Abstract: This paper explores the potential of combining Deep Reinforcement Learning (DRL), with frontier-based exploration techniques to tackle the challenges of navigating unknown, cluttered environments. Our approach, which involved a multi-robot system, aimed to enhance the adaptability and efficiency of autonomous exploration. This study incorporates two techniques: a frontier-based exploration algorithm, implemented in MATLAB within a grid cell environment; and a DRL-based algorithm that integrates traditional exploration methods with the decision-making of DRL, facilitated by Proximal Policy Optimisation (PPO) agents. We compared the performance of traditional frontier-based exploration and the DRL frontier-based exploration approach before and after training. Our findings indicated that the DRL frontier-based exploration method showed promising results in terms of adaptability, and efficiency, particularly within more complex environments. In conclusion, this study contributes to the field of autonomous exploration by demonstrating the feasibility and potential advantages of these combined techniques. Looking ahead, we anticipate further refining of these methods, particularly by integrating more complex sensory inputs, to better align with real-world applications.

Keywords: Deep Reinforcement Learning (DRL), frontier-based exploration, unknown environments

1. Introduction

Exploring unknown and cluttered environments is a significant challenge in fields such as search and rescue, environmental monitoring, and reconnaissance. Traditional exploration methods often struggle with the unpredictability and complexity of these terrains. One of the primary challenges is the propensity to get stuck in local minima, particularly in environments with intricate obstacles [1]. Local minima refers to a situation where an exploration robot gets trapped in a limited area of the environment and fails to explore further due to the layout of obstacles or terrain. Additionally, the efficiency of exploration is crucial, especially considering the battery and sensor limitations of autonomous vehicles [2].

Deep Reinforcement Learning (DRL) is a type of machine learning where an algorithm, known as an agent, learns optimal decision-making by interacting with its environment through trial and error. The agent improves its actions over time through a system of rewards and penalties. Deep Reinforcement Learning has emerged as a promising solution to these challenges. Its ability to learn optimal policies through interaction with the environment makes it well-suited for complex and unpredictable terrains [3]. DRL's adaptability is particularly beneficial in navigating and mapping unknown environments, addressing the limitations of traditional exploration methods [4]. The integration of DRL with frontier-based exploration methods presents an emerging approach to autonomous exploration. Frontier-based exploration, which focuses on expanding the boundaries of known areas, benefits from the adaptive decision-making capabilities of DRL [5]. This hybrid method allows for more intelligent and dynamic exploration strategies, effectively navigating and mapping unknown environments [6].

The work presented in this paper serves as the preliminary phase of a broader investigation into the exploration of unknown subterranean environments, using micro-aerial vehicles (MAVs). Given the complex and unpredictable nature of these settings, Deep Reinforcement Learning (DRL) was selected over more traditional Reinforcement Learning (RL) techniques or other simpler algorithms. DRL's capability to handle high-dimensional data and generalise from complex sensory inputs makes it exceptionally suited for navigating and mapping these challenging environments. This choice ensures that our exploration strategies are not only adaptive but also scalable, accommodating the vast and varied terrains encountered underground, where traditional methods might falter.

In our work, we employed a multi-robot system to explore an unknown cluttered environment, integrating DRL with frontier-base exploration techniques. The robots were equipped with DRL algorithms, enabling them to adapt their exploration strategies in real time. This approach allowed the robots to effectively navigate and cover substantial portions of the environment, addressing challenges such as local minima and efficiency in exploration [7, 8]. Preliminary results demonstrate that this integrated approach significantly outperforms traditional methods in terms of exploration speed and adaptability to environmental changes. The multi-robot system was able to efficiently explore the environment, showcasing the potential of DRL in enhancing autonomous exploration capabilities [9, 10].

This research not only contributes to the field of autonomous exploration but also lays the groundwork for future applications. The methodologies and findings from this study have the potential to be transferred to other platforms, including aerial and ground robots, opening new avenues in the exploration of unknown environments [11, 12]. Our work contributes to the advancement in the exploration of unknown environments. By leveraging the strengths of both DRL and frontier-based exploration, we have developed a method that is not only effective in navigating complex terrains but also efficient and adaptable to various operational constraints [13, 14].

2. Related Work

The exploration of unknown environments using autonomous systems, particularly aerial vehicles, has gained significant attention in recent research [3, 4, 7]. Current methods face challenges such as being stuck in local minima [1] and extensive time requirements for exploration. Additionally, these methods lack flexibility in adjusting to new, complex environments.

Deep Reinforcement Learning (DRL), especially techniques like Proximal Policy Optimisation (PPO) with actor-critic neural networks, has proven to be promising solutions [15]. DRL adapts to complex environments more effectively than traditional algorithms, overcoming challenges such as local minima. It optimises exploration strategies, leading to faster and more efficient area coverage as shown by Peake et al. [4], Doukhi and Lee [16], and Cai et al. [17].

A hybrid method combining DRL with frontier-based exploration has shown great potential [17]. Frontier-based exploration effectively identifies unexplored boundaries, while DRL optimises decision-making and navigation in these areas [18]. This combination enhances the exploration efficiency and reduces operational time, which is crucial for battery-powered robots [16].

This study introduces a novel exploration approach by combining Deep Reinforcement Learning (DRL) with frontier-based exploration techniques to tackle significant challenges of autonomous navigation in unknown and complex environments. This integration leverages the strengths of DRL for adaptive decision-making, enhancing the system's ability to respond to dynamic changes, and employs frontier-based exploration for systematic and thorough area coverage. Together, these methods significantly improve the operational efficiency and adaptability of autonomous systems, demonstrating marked improvements in navigating and mapping areas where traditional methods fall short.

These advancements suggest promising applications across various robotic platforms. Future work could extend these methods to more complex, real-world environments, leveraging the strengths of DRL and frontier-based exploration to tackle the challenges of autonomous navigation.

3. Methodology

3.1 Frontier-based Exploration Overview

In this study, we developed a frontier-based exploration algorithm and implemented it in MATLAB. It was designed for autonomous robotic navigation in a grid cell environment. During frontier-based exploration, the environment was represented by an occupancy grid where the status of each cell was determined based on its occupancy probability. The grid world consisted of a 2D matrix with a size of 12×12 grid cells.

Robot positions and obstacles were initialised within a matrix. The state of each cell in the grid was represented as follows: S_{ij}= 0 for unexplored, S_{ij} = 1 for explored, and S_{ij} = -1 for obstacles.

3.2 Robot Movement and Frontier Identification

For navigation, the robots located frontiers, defined as the boundaries between explored and unexplored areas. A cell *(i, j)* was classified as a frontier, *F(i,j),* if it met two conditions: firstly, it was unexplored (S_{ij}= 0), and secondly, it had an adjacent cell *(i',j')* that was already explored ($S_{i'j'}$ = 1).

The robots were then assigned to the nearest frontiers based on Euclidean distance. Each robot moved towards its assigned frontier by calculating the shortest path while avoiding collisions. If R_k was the position of the *k*th robot and *Tk* its target, the robot moved in the direction D_k. The movement was constrained by collision avoidance rules. The algorithm terminated when there were no more frontiers to explore, indicating complete coverage of the accessible grid. An overview of the frontier-based exploration algorithm is shown in Algorithm 1.

Algorithm 1: Frontier-based Exploration Algorithm

1.	**Class GenericFrontierExploration** Properties: GridMap, AgentList, ObstacleList, ExploredMap, AgentTargets
2.	**Method initialiseExplorationEnvironment()** Set GridMap dimensions, initialise AgentList with starting positions, place ObstacleList items on GridMap, and set ExploredMap as a matrix of explored status.
3.	**Method performExploration()** *While there are unexplored frontiers:* Identify frontiers as boundary cells between explored/unexplored areas. Assign agents to frontiers based on criteria like proximity or efficiency. *For each agent:* Move towards target, avoiding obstacles and collisions. Update explored status based on agent's new position.
4.	**Method IdentifyFrontiers()** Return a list of frontier cells identified from GridMap and ExploredMap.
5.	**Method AssignTargetsToAgents()** Assign nearest or most strategic frontier cell to each agent in AgentTargets.
6.	**Method MoveAgent(agent)** Move agent towards target if path is obstacle-free and collision-free.

7.	**Method UpdateExploredStatus(agent)** Mark agent's current cell as explored in ExploredMap. **End Class**

3.3 DRL Frontier-based Exploration Overview

The Deep Reinforcement Learning (DRL) frontier-based exploration algorithm we developed combines traditional frontier-based exploration with the decision-making capabilities of DRL. This combination facilitated efficient exploration of unknown environments by autonomous agents. The algorithm identified unexplored frontiers and assigned robots to explore them. This method aimed to efficiently cover the grid, using a heuristic based on proximity to assign frontiers to robots. Control of the robots was managed by three Proximal Policy Optimisation (PPO) agents, each operating in discrete action spaces. These agents were responsible for guiding the robots' movements and making decisions about which frontiers to explore within the grid environment to maximise reward.

Rewards were allocated based on exploration actions. Positive rewards were given for discovering new cells, while penalties were imposed for collisions, attempting to move out of the grid, or inaction. The algorithm updated robot states based on their actions, adjusted the rewards, and checked if the exploration was complete. Observations for each agent were generated, including locations of obstacles and the exploration status.

The exploration concluded when all accessible cells had been explored or after a maximum number of steps, ensuring finite operation time. An overview of the DRL frontier-based exploration algorithm is shown in Algorithm 2.

Algorithm 2: DRL Frontier-based Exploration Algorithm adapted from [19]	
1.	**Class DRLBasedExploration** Properties: AgentStates, ObstacleLocations, MaxIterations, EnvironmentMap, IterationCount, AgentPositions, EnvironmentSize, DisplayHandle, ExploredStatus
2.	**Method initialiseEnvironment()** Define environment size, place obstacles, and mark initial exploration status.
3.	**Method performActionSequence(ActionList)** *For each agent:* Execute action from ActionList, compute rewards, and update environment with new location. Check if exploration objectives are met, formulate observations, and adjust reward scale as needed.
4.	**Method resetEnvironment()** Reinitialise environment for new session, reset iteration count and explored status, and reposition agents.
5.	**Method updateVisualisation()** Display exploration environment state, agent locations, and progress indicators.
6.	**Private Method detectCollision(PotentialPosition, OtherPositions)** Check if a move causes collisions with obstacles or other agents. **End Class**

3.3.1 DRL Agents

The agents employed Proximal Policy Optimisation (PPO) for operating in a discrete action space. They were designed to navigate a *12×12* grid with strategically placed obstacles, making decisions at each step to efficiently explore the environment.

An actor-critic algorithm was employed. The actor network, responsible for policy generation, mapped the observed state of the environment to specific actions. The architecture included convolutional layers for spatial feature extraction from the observation input, followed by fully connected layers. A softmax output layer then provided the probabilities of taking specific actions.

The critic network estimated the value function, assessing the potential long-term rewards from the current state. Its architecture was similar to that of the actor network, featuring convolutional and fully connected layers, but it diverged in its output, providing a single value estimate instead of a probability distribution. This value guided the actor in optimising its policy by evaluating the expected returns of actions.

These networks were integrated with the DRL model, with the actor responsible for selecting actions and the critic for evaluating the potential rewards of those actions.

3.3.2 Action Space

The action space dictates the movement of the agent within the grid, allowing for strategic exploration and obstacle avoidance. The agent had five discrete actions, namely:

- *Wait* represented by 0,
- *Up* and *down* represented by 1 and 2 respectively,
- *Left* and *right* represented by 3 and 4 respectively.

3.3.3 Reward Shaping

The following rewards were applied to guide the agent toward desirable behaviour:

- *Exploration Reward* - The agent received a reward of +1 for moving into an unexplored cell.
- *Discipline Reward* - A penalty of -0.5 was applied for colliding with obstacles or other agents.
- *Performance Reward* - A reward of -0.01 was applied for any action resulting in movement.
- *Efficiency Reward* - A penalty of -0.1 was applied for actions that do not result in motion.

3.3.4 Observation Space

Each agent's observation space was a *12×12* grid that included data on obstacles, the agent's position, the positions of other agents, and areas yet to be explored; an example is shown in Figure 1.

Figure 1: A schematic showing the observation space

3.4 DRL Frontier-Based Model Training Analysis

The three robots were trained, and the episode and average reward for all robots were plotted for each run. The training stopped when the average reward of all three robots equalled or exceeded 100 for 50 consecutive episodes. Figure 2 shows an example of the convergence process for the reward function.

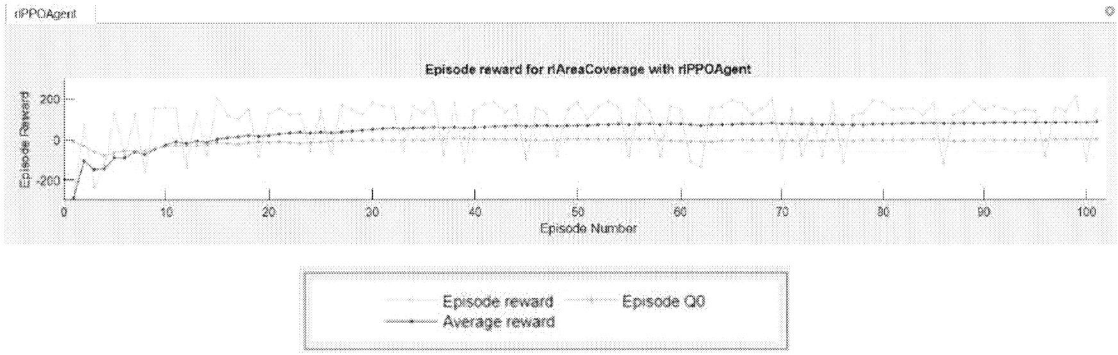

Figure 2: An example of the convergence process of the reward function.

As seen from the figure, episode rewards increased over time as the robots' performance improved. The training concluded after 101 episodes, once the agents had met the stopping criteria. The highest episode reward achieved by any of the robots during this training was 148.5. The average reward smooths out, showing a stable learning process after an initial adaptation period.

4. Results and Discussion

We compared the traditional frontier-based exploration with DRL frontier-based exploration. The performance metrics we used included the total path length, measured as the number of steps taken to achieve coverage, the percentage of the area covered, and the time taken for exploration. Figure 3 shows the performance of the DRL frontier-based exploration model before and after training.

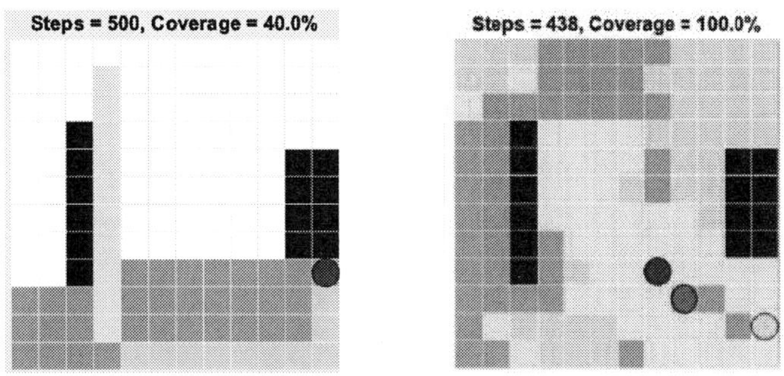

(a) (b)

Figure 3: (a) DRL frontier exploration model without any training and (b) with training.

The DRL frontier-based exploration model without any training achieved only 40% area coverage, indicating it was not as effective in navigating and covering the environment efficiently. The multi-robot system also became stuck in a local minimum. After training, the DRL frontier-based exploration model achieved 100.0% coverage in 438 steps, showing a significant improvement from the untrained model and demonstrating the capability of DRL to learn and adapt. It took 14.3 seconds to explore the area.

A comparison of the performance of the traditional frontier-based exploration and the trained DRL frontier-based exploration model is shown in Figure 4. The frontier-based exploration algorithm achieved nearly complete area coverage of 98.5% in just 195 steps. The time taken to complete the task was 7.89 seconds. This high coverage suggests that the robots effectively explored the grid without the assistance of a learning algorithm.

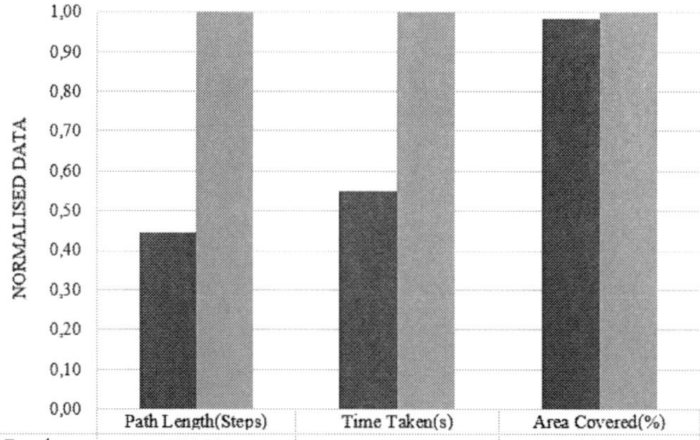

	Path Length(Steps)	Time Taken(s)	Area Covered(%)
Frontier-Based	195,00	7,89	98,50
DRL Frontier-Based	438,00	14,37	100,00

Figure 4: A comparison chart of path lengths, time taken and area covered by the frontier-based exploration and DRL frontier-based exploration.

The difference in speed between the DRL frontier-based exploration algorithm and the frontier-based exploration algorithm can be attributed to the high computational resources required by the neural networks. Additionally, DRL algorithms often spend a significant amount of time exploring the environment, hence the higher number of steps and longer time taken to complete exploration. This exploration is necessary for the algorithm to employ effective strategies; however, it can be time-consuming compared to more direct methods like frontier-based exploration.

In contrast, traditional algorithms like frontier-based exploration are often more straightforward in their approach. They typically involve less computational complexity, less data processing, and do not require the extensive training and continuous updating that DRL algorithms need.

A more complex obstacle pattern, featuring zigzag lines and clusters, shown in Figure 5, was employed. No additional training for the DRL frontier-based exploration algorithm was conducted. A more detailed comparison is presented in Figure 6. It should be noted that since exploration was not completed by either algorithm, the time taken to complete the exploration is not reported.

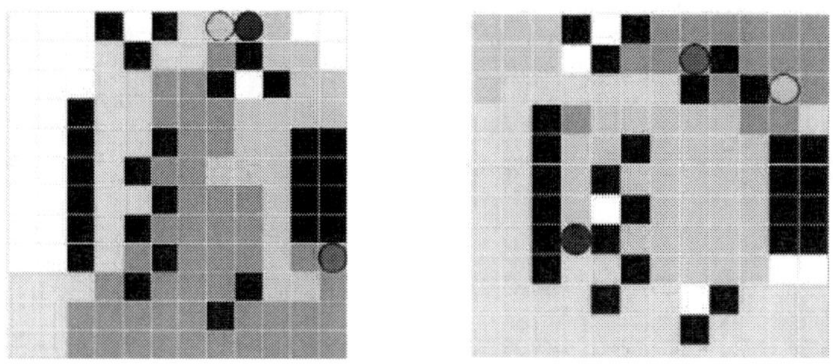

(a) (b)

Figure 5: (a) Area coverage for frontier-based exploration (b) DRL frontier exploration model.

The DRL frontier-based exploration algorithm achieved 94.8% coverage in this complex environment, a notable success compared to the 77.6% coverage achieved by traditional frontier-based exploration. This higher performance of the DRL algorithm can be attributed to its ability to learn optimal policies that generalise across various states within the environment. The traditional frontier-based exploration model encountered difficulties, becoming stuck in a local minimum when the closest frontier was blocked by an obstacle. This incident highlights a key weakness of traditional exploration techniques in complex environments. As the complexity increases, the DRL agent can leverage its past experiences to more successfully navigate through new, unseen configurations of obstacles compared to simpler, heuristic-based methods.

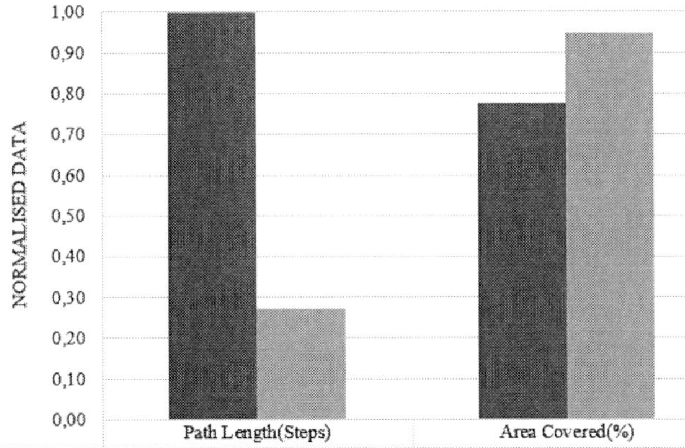

	Path Length(Steps)	Area Covered(%)
Frontier-Based	3672,00	77,60
DRL Frontier-Based	1000,00	94,80

Figure 2: A comparison chart of the frontier-based exploration and DRL frontier-based exploration.

5. Conclusion and Future Work

The simulation results for the grid-based exploration indicate a significant evolution in the performance of the agents, transitioning from initial random movements to a sophisticated exploration strategy. Initially, the agents' lack of training led to inefficient exploration characterised by a high likelihood of collisions and redundant occupancy of cells. However, after undergoing a training regimen, the agents demonstrated an ability to fully explore the environment while avoiding collisions and optimising the number of steps taken.

This progression underscores the potential of DRL to develop complex behaviours such as autonomous exploration and navigation. The final scenario, where the agents achieved 94.8% coverage, is particularly noteworthy as it represents the culmination of learning and adaptation.

The transferability of these results to real-world applications, such as the exploration of unknown environments with Micro-Aerial Vehicles (MAVs), is promising. The simulation models key aspects of exploration tasks that MAVs might encounter, including obstacle avoidance, efficient area coverage, and autonomous decision-making in the absence of human input. Future work should include incorporating more complex sensory inputs into the training process to better mimic the kind of data MAVs would receive, such as LIDAR or visual SLAM (Simultaneous Localisation and Mapping).

6. References

[1] Juliá M, Gil A, Reinoso O. A comparison of path planning strategies for autonomous exploration and mapping of unknown environments. Autonomous Robots. 2012 Nov;33:427-44.

[2] Mokhtar M, El-Badawy A. Autonomous Navigation and Control of a Quadrotor Using Deep Reinforcement Learning. In 2023 International Conference on Unmanned Aircraft Systems (ICUAS) 2023 Jun 6 (pp. 1045-1052). IEEE.

[3] Seel A, Kreutzjans F, Küster B, Stonis M, Overmeyer L. Deep Reinforcement Learning Based UAV for Indoor Navigation and Exploration in Unknown Environments. In2022 8th International Conference on Control, Automation and Robotics (ICCAR) 2022 Apr 8 (pp. 388-393). IEEE.

[4] Peake A, McCalmon J, Zhang Y, Myers D, Alqahtani S, Pauca P. Deep Reinforcement Learning for Adaptive Exploration of Unknown Environments. In2021 International Conference on Unmanned Aircraft Systems (ICUAS) 2021 Jun 15 (pp. 265-274). IEEE.

[5] Hou Y, Ruan X, Zhu X, Li C. Frontier-based exploration on continuous radial basis function neural network map. In 2018 37th Chinese Control Conference (CCC) 2018 Jul 25 (pp. 5534-5538). IEEE.

[6] Perkasa DA, Santoso J. Improved frontier exploration strategy for active mapping with mobile robot. In 2020 7th International Conference on Advance Informatics: Concepts, Theory and Applications (ICAICTA) 2020 Sep 8 (pp. 1-6). IEEE.

[7] Niroui F, Zhang K, Kashino Z, Nejat G. Deep reinforcement learning robot for search and rescue applications: Exploration in unknown cluttered environments. IEEE Robotics and Automation Letters. 2019 Jan 10;4(2):610-7.

[8] Zhu D, Li T, Ho D, Wang C, Meng MQ. Deep reinforcement learning supervised autonomous exploration in office environments. In2018 IEEE international conference on robotics and automation (ICRA) 2018 May 21 (pp. 7548-7555). IEEE.

[9] Ali A, Gul S, Mahmood T, Ullah A. Exploration of Unknown Environment using Deep Reinforcement Learning. In2023 International Conference on Robotics and Automation in Industry (ICRAI) 2023 Mar 3 (pp. 1-6). IEEE.

[10] Deng L, Gong W, Li L. Multi-Robot Exploration in Unknown Environments via Multi-Agent Deep Reinforcement Learning. In 2022 China Automation Congress (CAC) 2022 Nov 25 (pp. 6898-6902). IEEE.

[11] Xue Y, Chen W. Multi-Agent Deep Reinforcement Learning for UAVs Navigation in Unknown Complex Environment. IEEE Transactions on Intelligent Vehicles. 2023 Jul 24.

[12] Chen F, Martin JD, Huang Y, Wang J, Englot B. Autonomous exploration under uncertainty via deep reinforcement learning on graphs. In2020 IEEE/RSJ International Conference on Intelligent Robots and Systems (IROS) 2020 Oct 24 (pp. 6140-6147). IEEE.

[13] Selin M, Tiger M, Duberg D, Heintz F, Jensfelt P. Efficient autonomous exploration planning of large-scale 3-d environments. IEEE Robotics and Automation Letters. 2019 Feb 3;4(2):1699-706.

[14] Qiao W, Fang Z, Si B. Sample-based frontier detection for autonomous robot exploration. In2018 IEEE International Conference on Robotics and Biomimetics (ROBIO) 2018 Dec 12 (pp. 1165-1170). IEEE.

[15] Shin SY, Kang YW, Kim YG. Reward-driven U-Net training for obstacle avoidance drone. Expert Systems with Applications. 2020 Apr 1;143:113064.

[16] Doukhi O, Lee DJ. Deep reinforcement learning for autonomous map-less navigation of a flying robot. IEEE Access. 2022 Mar 28;10:82964-76.

[17] Cai M, Aasi E, Belta C, Vasile CI. Overcoming exploration: Deep reinforcement learning for continuous control in cluttered environments from temporal logic specifications. IEEE Robotics and Automation Letters. 2023 Feb 20;8(4):2158-65.

[18] Hou Y, Ruan X, Zhu X, Li C. Frontier-based exploration on continuous radial basis function neural network map. In2018 37th Chinese Control Conference (CCC) 2018 Jul 25 (pp. 5534-5538). IEEE.

[19] MathWorks. Train 3 Agents for Area Coverage [Internet]. MathWorks; 2023 [cited 2023 Oct 25]. Available from: https://www.mathworks.com/help/reinforcement-learning/ug/train-3-agents-for-area-coverage.html

Thirteenth South African Conference on Computational and Applied Mechanics
SACAM24
Stellenbosch, 22–23 January 2024

ANGLED EXPANDING JET IMPINGEMENT IN SUPERSONIC CROSSFLOW

Sabrina G. Myburgh[*,1], **Craig Law**[*,2]

[*]School of Mechanical, Industrial and Aeronautical Engineering
University of the Witwatersrand, PO Wits, Johannesburg, 2050, South Africa
[1]1466793@students.wits.ac.za,
[2]craig.law@wits.ac.za.

Keywords: Jet injection, supersonic crossflow, asymmetric

Abstract

Significant work has been done on the effect of both perpendicular and spanwise inclined uniform jet injection in supersonic cross flow for boundary layer control. Implicit transitional SST RANS turbulent simulations were performed to investigate the flow field around a streamwise inclined, asymmetrically overexpanded jet in supersonic cross flow at Mach number 3.5. The interaction between the jet impinging on the free stream was observed. The region between the jet exit plane and free stream featured a strong mismatch in pressure and momentum, causing separation at the nozzle wall. The dynamic pressure (measured as momentum) of the system was found to increase by 4.95% on account of the overexpanded jet behaviour.

1 Introduction

Uniform jet injection in supersonic cross flow is widely used in engineering applications. In aviation, jet injection into the high-speed free stream is used as an effective boundary layer control (BLC) mechanism on the wings and stabilisers to reduce stall speeds. Other applications include lift augmentation and circulation control. Figure 1 demonstrates the complex flow structure which emerges when fluid is injected at some angle into a supersonic cross flow. The geometry of the injection nozzle determines the jet exit conditions. The jet flow field acts as an obstruction to the incoming freestream, generating a bow shock. Boundary layer interaction with the jet flow produces a separation shock upstream of the jet and a recirculation region downstream.

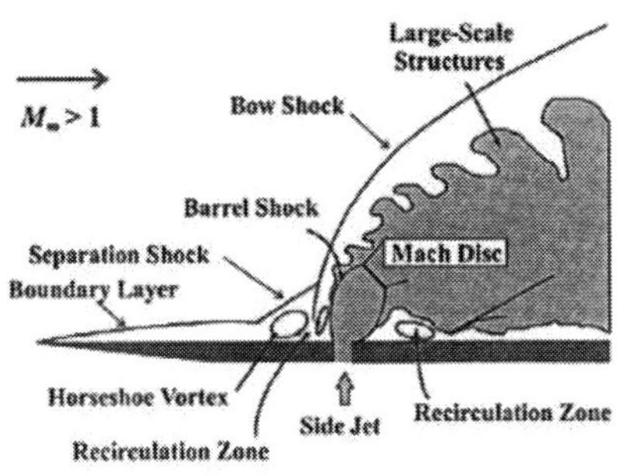

Figure 1: Schematic of flow structure of jet interaction with supersonic cross-flow [1].

Much work has been done on perpendicular and spanwise inclined uniform jets in supersonic crossflow. Papamoschou [3] conducted experimentation on penetration of round sonic and supersonic jets injected normal to the supersonic cross flow. Bhagat [1] carried out a numerical investigation on the interaction between a perpendicular jet and the supersonic cross-flow at Mach 3. Sebastain [2] performed implicit large eddy simulations to understand the flow field around a spanwise inclined jet in supersonic flow. The results were more complex and asymmetrical than those of the perpendicular jet injection depicted in Figure 1. There is still much to be understood regarding inclined, asymmetrically expanding jets in supersonic freestreams.

The primary objective of this work is to numerically investigate the complex interaction between an asymmetrically expanding jet and a crossflow. The overexpanded jet is injected into the crossflow at a spanwise inclination. The regions of interest in the resultant flow field include the region between the jet exit plane and the freestream, the upstream oblique shock wave, fluid wall behaviour and the overall effect of momentum on the flow field.

2. PROBLEM DEFINITION

Figure 2 (a) demonstrates the nondimensionalised computational domain consisting of a solid cylinder, fitted with an angled, asymmetrically expanding nozzle (referred to hereafter as the cylinder-nozzle). Figure 2 (b) shows the cylinder-nozzle alone. The outer radius (R) of the cylinder is 16 mm. The cylinder is located 2.8 R from the domain leading edge. The total length of the quarter-cylindrical domain is 9.4 R, with a height of 4.7 R. The nozzle has an inlet diameter (d) of 1.4 mm, and an exit diameter of 3.78 d. The nozzle inlet is located 0.4 R above the horizontal axis, and the nozzle exit plane is located 5.75 R from the leading edge. The nozzle is inclined 17 ° to the horizontal. The nozzle is of the overexpanded type, with expansion ratio (A/A^*) = 14.31, M_{inlet} = 1, M_{exit} = 4.33. The geometry allows for the precise control of the energy inputs to the flow field in that the total energy of the free stream and that of the jet are essentially the same.

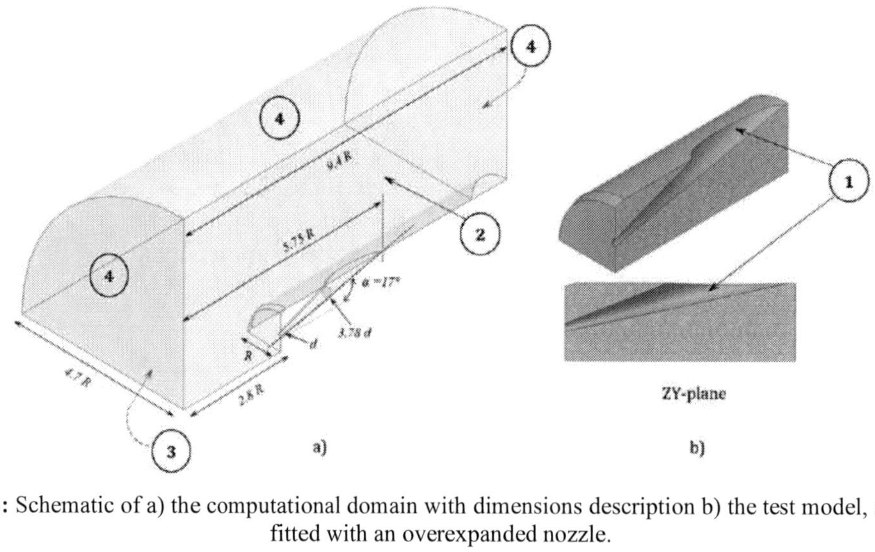

Figure 2: Schematic of a) the computational domain with dimensions description b) the test model, a cylinder fitted with an overexpanded nozzle.

Steady-state numerical simulations were performed to study the impingement of the jet with the supersonic crossflow. The crossflow was generated by placing the blunt cylinder in supersonic flow to create a bow shock. Several other nozzle geometry configurations were also tested. Only the overexpanded case is referred to in this paper. Both the external (crossflow) and internal (jet flow) were modelled.

3. COMPUTATIONAL METHODOLOGY

Numerical simulations were formulated according to the framework outlined by [4], [5]. Boundary conditions, mesh treatment and problem formulation similar to the literature was utilised to aid in ensuring the reliability of the study. The simulation model and conditions closely matched those present in concurrent studies which utilised supersonic wind tunnel testing. Simulations are conducted using a gauge pressure (P_{gauge}) = 9000 Pa for an operating pressure (P_0) = 0 Pa. ANSYS Fluent 2023 R1 computational fluid dynamics (CFD) software was used for the simulations. Steady-state Reynolds-averaged Navier Stokes modelling of the domain was undertaken. The $k - \omega$ SST turbulence model was chosen as it is known for its reliable handling of adverse pressure gradients and separating flow. Both phenomena are present in the chosen jetting configuration.

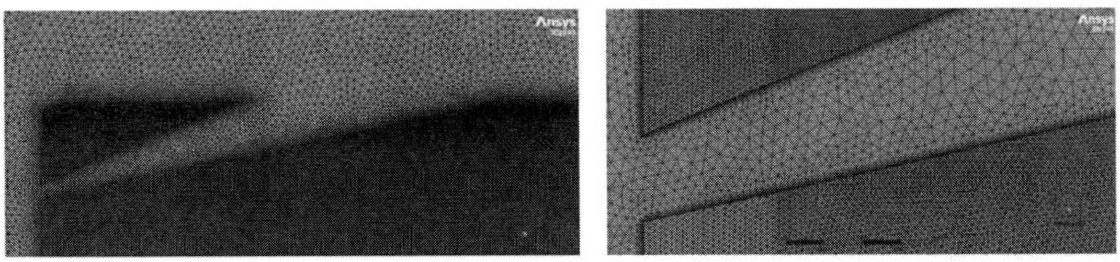

a) b)

Figure 3: Schematic of a) the global mesh b) the mesh at the near-wall.

An unstructured 3D mesh is generated which consists of around 12×10^6 elements. Figure 3 demonstrates (a) the global mesh, and (b) the enlarged view of the nozzle wall mesh. A target y+ value of 5 was set. An inflation layer over the wall boundary was used in conjunction with appropriate wall sizing to capture the near-wall behaviour and viscous effects.

A mesh independence study was carried out by solving four different grids of varying element size, where the solution of the final grid was found to be independent of dimension. The boundary conditions are defined as shown in Figure 2: 1. Cylinder-nozzle – wall (no-slip velocity, zero shear-stress condition). 2. ZY-face – vertical symmetry plane. 3. ZX-face – horizontal symmetry. 4. Cylindrical outer domain surfaces – pressure far-field. Symmetry planes were utilised to reduce the computational nodes and solution time required. The ideal-gas model was used. The freestream conditions were Mach number (M_∞) = 3.5, Reynolds number $Re_D = 1.96 \times 10^4$. The simulation was run using parallel computing on 15 000 cells per node on the Centre for High Performance Computing (CHPC) facility (South Africa).

4. RESULTS AND DISCUSSION

Figure 4 illustrates the flow field around the cylinder-nozzle for the overexpanded nozzle configuration in the ZY plane at Mach 3.5. Figure 4 (a) demonstrates a Mach number flood plot used to identify flow features that are associated with changes in density. Figure 4 (b) is a velocity flood plot used to identify changes in velocity gradients within the flow field.

Downstream of the bow shock, a sonic line separates the subsonic and supersonic regions, as in Figure 4.

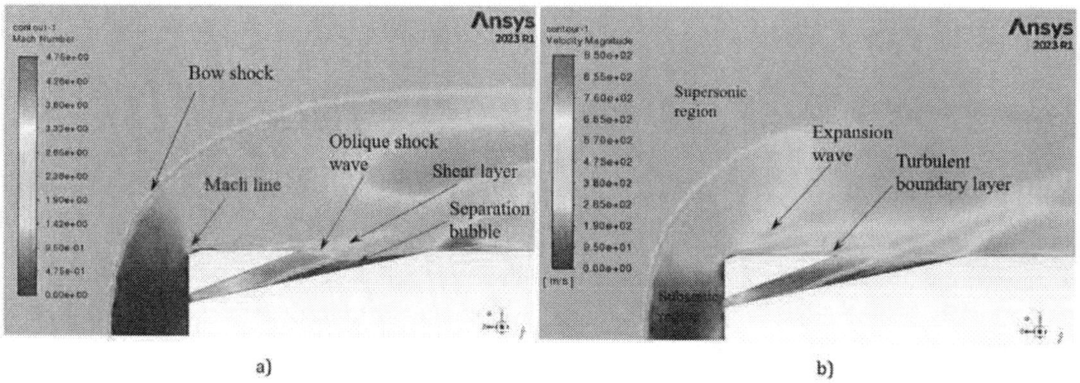

Figure 4: Flooded contour plots of (a) Mach number and (b) velocity for the cylinder nozzle at Mach 3.5, nozzle expansion ration (A/A*) = 14.31

Due to the expansion ratio of the nozzle, an overexpanded supersonic jet forms, that impinges on the supersonic cross flow. At the nozzle exit plane, the static pressure inside the jet is lower than that of the free stream outside of it. This mismatch in pressures causes some of the flow in the interaction region to be pulled back into the nozzle, causing the overexpanded jet to further narrow and separate from the lower nozzle wall. A large separation bubble forms on the wall which reduces the amount of available nozzle area for the flow. The bubble turns the jet further into the oncoming flow, dissipating a greater amount of energy by reducing the jet and free stream momentums.

Neither the flow in the nozzle, nor the resultant jet which impinges into the free stream, are axisymmetric. The shape of the separation bubble induces the asymmetric geometry of the nozzle flow, acting as an asymmetric wall-like boundary for the jet. Due to the shape of the separation bubble, the jet is evenly shaped. Figure 5 (b) shows the velocity streamlines of the separation bubble on the lower wall near the nozzle exit. The streamlines in Figure 5 (b) are more highly concentrated at the top corner of the nozzle, where a shear layer is expected to occur. The streamlines indicate a discontinuous change in direction at the wall, indicating the presence of a shock ahead of the separation bubble.

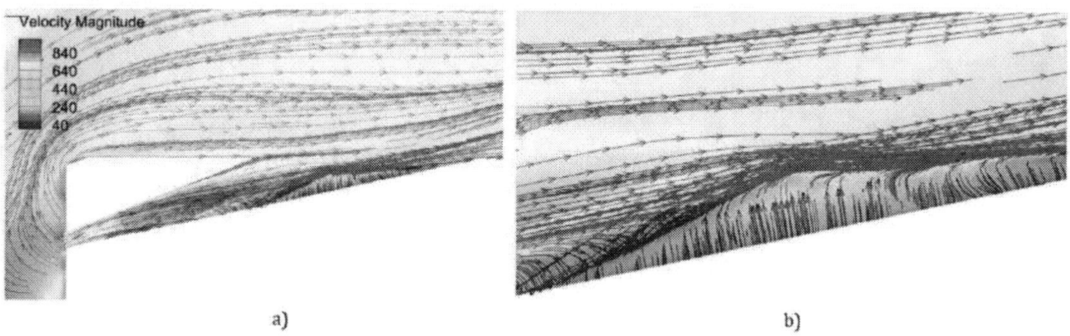

Figure 5: Velocity streamlines of the overexpanded jet (a) global view (b) near-wall view.

When the high-speed jet encounters the upstream nozzle corner, the flow turns in on itself and compresses, forming an oblique shock wave, as demonstrated in Figure 6. Figure 6 illustrates the static pressure flood plot of the jet impingement region for the Mach 3.5, (A/A*) = 14.31 nozzle. The shock wave increases the pressure and decelerates the flow above the nozzle exit, where three distinct pressure regions occur (medium in the free stream, high downstream of the intersection, and low in the nozzle). When the oblique shock wave hits the lower nozzle wall, an adverse pressure gradient is induced by the pressure mismatch. Shock-induced separation occurs on the wall near the exit due to the sharp increase in pressure at the wall. As the streamlines in Figure 5 (b) indicate, the separation bubble acts as a physical boundary that further narrows the shape of the jet.

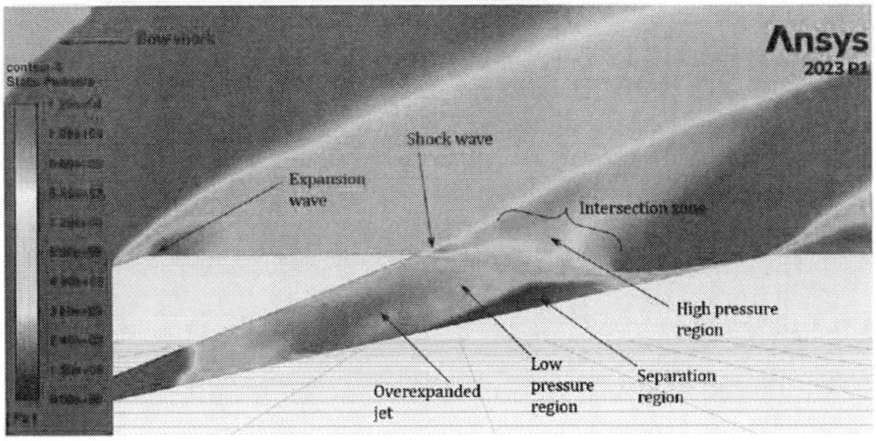

Figure 6: Flooded static pressure contour plot of the overexpanded nozzle jet impingement region for the cylinder nozzle at Mach 3.5, nozzle expansion ration (A/A*) = 14.31.

The global flow field is highly turbulent and complex. The expansion wave generated by the cylinder leading edge accelerates the freestream flow near the wedge surface. As Figure 4 (b) and Figure 6 show, the cylinder boundary layer progressively thickens and increases in turbulent velocity of the jet. At the upstream exit corner of the nozzle, the cylinder boundary layer, crossflow and jet all intersect in the region identified in Figure 6.

As indicated in Figure 4, each of the streams has a different velocity magnitude. A shear layer is formed when each stream separates from the wall it is attached to (i.e. the cylinder, upper nozzle and the lower nozzle wall) and later merges downstream. The pressures of the intersecting flows match across the shear layer, however the velocity, temperature and energy of the flows typically do not match.

A change in momentum across the system is observed due to dissipation in momentum by the jet impinging on the crossflow. The flow inside the nozzle experiences momentum changes due to geometric and direction changes, as well as shock waves and expansions occurring due to changes in pressure conditions. The shock wave at the nozzle exit plane represents an increase in entropy of the flow, and a subsequent loss of momentum. When the jet impinges on the crossflow at 17°, the flow is forced to change direction relative to the governing streamwise flow direction. This change in motion and mixing of the flows causes a momentum loss. The momentum inside the jet is high as the nozzle velocity increases due to the expansion while mass flow remains constant. The freestream flow undergoes limited area change or interference, resulting in the momentum being lower than the jet. Momentum is an additive property, indicating that the jetting positively contributed to the overall increase in system momentum.

When the jet intercepts the crossflow, momentum transfer between the two streams occurs. The greater static pressure of the crossflow causes the disintegration and diffusion of the jet, where it mixes into the free stream. There is also a high vorticity in the interface between the jet and flow (i.e. the shear layer), where the large eddies decay into turbulence. Downstream of the impingement, the different pressures in the freestream equilibrate, mixing with the crossflow near the edge of the bow shock and the expansion wave at the nozzle exit edge.

The area-weighted average of the dynamic pressure ($\rho U_\infty^2/2$) at the inlet and exit planes were calculated. The percentage increase across the system was calculated to be 4.95%. Dynamic pressure is used as an indicator of momentum change as it represents a measure of the average linear momentum of moving gas molecules while the mass flow rate in the system remains constant. The overall increase in momentum (and thus efficiency of the system) is attributed to the jet contributing additional momentum to the system. The flow properties of the jet were a result of the the spanwise inclined, asymmetric nozzle geometry.

5. CONCLUSIONS

The resultant flow field and interaction between the supersonic crossflow and overexpanded jet is highly complex. Key features of interest included the region between the exit plane of the jet and the crossflow, and the behaviour of the separation bubble on the lower wall. This is attributed to the pressure mismatch and the oblique shock wave behaviour at the nozzle exit. The dynamic pressure (and implied momentum) of the system was found to increase by 4.95% between the inlet and outlet due to the configuration of the overexpanded nozzle, affecting the overall flow field.

6. REFERENCES

1. A. Bhagat, H. Gijare, N. Dongari, Computational Analysis of Side Jet Interaction with a Super-sonic Cross-flow, *6th International and 43rd National Conference on Fluid Mechanics and Fluid Power*, 2016.

2. R. Sebastian, T. Lürkens, AM. Schreyer, Flow field around a spanwise inclined jet in supersonic crossflow, *Aerospace Science and Technology*, Volume 106, 2020, 106209, ISSN 1270-9638, https://doi.org/10.1016/j.ast.2020.106209.

3. D. Papamoschou, D.G Hubbard, Visual observations of supersonic transverse jets. *Experiments in Fluids* 14, 468–476 (1993). https://doi.org/10.1007/BF00190201

4. S. G. Mironov, S. V. Kirilovskiy, T. S. Militsina, T. V. Poplavskaya, I. S. Tsyryulnikov and I.R. Valiullin, Effect of the structure of skeleton models of cellular materials on the drag of a cylinder with a frontal gas-permeable insert in a supersonic flow, J. Phys.: Conf. Ser. 1404, 2019, https://doi.org/10.1088/1742-6596/1404/1/012032

5. T. V. Poplavskaya, S. V. Kirilovskiy, S. G. Mironov, I. S. Tsyryulnikov; Numerical simulation of highly porous materials in application to supersonic aerodynamics. *AIP Conf. Proc.* 2017; 1893 (1): 030155. https://doi.org/10.1063/1.5007613

Acknowledgements

The authors acknowledge the Centre for High Performance Computing (CHPC), South Africa, for providing computational resources to this research project. The 'Aerothermodynamics with vortices, vorticity and shear' group (MECH1532) was used for this research.

NICIS CHPC

We are also thankful to Dr. R. Paton for his early contributions to this work.

THREE DIMENSIONAL COMPUTATIONAL FLUID DYNAMICS MODELLING OF PANCAKE SEA ICE ON WAVES

Anand Nair[1,2*], Edward Boje[1], Robyn A. Verrinder[1,2]

[1]Department of Electrical Engineering, University of Cape Town, Rondebosch, South Africa
[2]Marine and Antarctic Research centre for Innovation and Sustainability, University of Cape Town, Rondebosch, South Africa

*Email: anandnair2025@gmail.com

Abstract: Computational fluid dynamics can be used to simulate the small-scale interactions of pancake ice and waves. This paper presents a preliminary investigation of the effect of pancake ice geometry on its dynamics by comparing the motion of a circular and an elliptical pancake under the same wave conditions. From the results, it was seen that the translational motions of the two pancake geometries were largely similar and that most of the variation was visible in the rotations. It was noted that, while the surge and heave motions were sinusoidal as expected, the pitch motion displayed two modes, which are hypothesized to be the waves frequency and the natural frequency of the geometry used. It is also hypothesized that the geometry of the pancake ice floe can be approximated from its dynamic behaviour. A validation experiment will be conducted in a wave tank at Aalto University in Helsinki, to test the fidelity of the model. In future work we hope to develop an algorithm that can approximate the size and geometry of pancake ice using the inertial measurements captured by buoys deployed on them. This will enable us to study the evolution of the size of individual floes.

Keywords: sea ice dynamics, pancake ice, wave-ice-interaction, computational fluid dynamics, openfoam

1 Introduction

The Antarctic Marginal Ice Zone (MIZ) is a highly dynamic sea ice region at the interface between open ocean and consolidated pack ice, where waves propagate into the ice field and cause it to break apart. The MIZ contains various ice types, including pancake ice, which is characterised by a roughly circular shape with raised ridges. The interaction between pancake ice and waves is highly complex and is the subject of ongoing research efforts.

Waves play a key role in the formation of pancake ice. Periodic wave activity causes the unconsolidated ice crystals (frazil ice) forming in the top layer of the ocean, to accumulate and adhere together. This eventually creates a smooth layer called grease ice, that filters out the high frequency components of the wave spectrum, due to viscosity of the frazil slurry and inelastic collisions. As the frazil agglomeration grows, waves prevent a continuous ice sheet from forming, and cause collisions between the growing ice floes, giving them a roughly circular shape and raised edges[1]. The pancake ice floes grow laterally by the welding of adjacent floes[2], and vertically thicken by frazil adhesion on the underside. This gives the floes a convex bottom surface that is rough and porous, compared to the hard, smooth top surface which is subjected to the colder air temperatures[3]. Pancake ice can also increase in thickness by rafting, a process whereby floes are pushed on top each other by wave motion. The maximum diameter of pancake ice is limited by the wave field, due to bending and tensile stresses[4], therefore the floe size distribution is influenced by the wave conditions present. The floe size distribution in turn has an influence on the rate of dissipation of wave energy at different frequencies, which determines how far waves can propagate into the ice field, where they contribute to the breakup of sea ice in the MIZ[5].

The sea ice concentration, thickness and extent in the MIZ undergoes significant variation throughout the year, marked by the seasonal advance and retreat of the ice edge (MIZ boundary)[6]. The seasonal evolution of the pancake ice size distribution is not well documented due to the high variability of the MIZ. Satellite remote sensing provides data on the extent of the MIZ as well as sea ice thickness and concentration, but the resolution is not sufficient to study the interactions of floes on the small scale[7]. Data collected from buoys deployed on pancake ice during cruises into the Southern Ocean can be used to study the dynamics and drift of pancake ice[8], but the deployment and retrieval of these buoys is an arduous process.

Numerical models are useful to predict the small scale dynamics of pancake ice on waves. The simplest numerical model of pancake ice and waves would be where the ice floes follow the wave orbital motion as proposed by Rottier[9].

Experimental work done by Yiew et al. and Meylan et al. found that simple models are sufficient to predict the floe motion, when the diameter of the floe is small relative to the wavelength[10][11]. Smith et al. studied the dynamics of pancake ice in the Arctic, using images captured with shipboard stereo cameras, and found that the net orbital motion of the floes in response to the waves was significantly greater than the relative motion between the floes[12]. Hopkins and Shen proposed a numerical, discrete element model of circular disk shaped floes subjected to water drag, added mass, gravity and buoyancy forces, which included collisions and rafting of floes[13]. It was found that the largest difference between the ice floe velocity and the water surface velocity, occurs in the wave trough, and as the wave amplitude was reduced floe velocity approaches the water velocity. Herman developed a numerical model to study the wave-induced collisions of pancake ice and their contribution to wave attenuation[14]. The response of a pancake ice floe can be characterised by the response amplitude operator (RAO), which is close to 1 for small floes relative to the wavelength, because the floe motion will follow the wave orbital motion. As the floe size increases, its RAO will tend to zero[11][14]. Computational Fluid Dynamics (CFD) models have recently been applied to the context of pancake ice, to resolve the ice floe and wave interactions in greater detail. Marquart et al. presented a small scale CFD model which resolved pancake ice and wave interactions, as well as with the surrounding grease ice, by applying a different viscosity for each ice type. From the results of the model it was found that the responses of floes tend to become independent of the floe size distribution for larger wave periods[15].

Many of these models simplify pancake ice as a cylindrical disk, such as [13] and [14]. Marquart's model is capable of simulating realistic sea ice layouts, but it does not include the responses of ice floes in the vertical direction (z-direction).

The study of small scale interactions of pancake ice and waves would enhance our understanding of sea ice dynamics and could be coupled with data from buoys to quantify the rates of growth and melting of pancake ice in the MIZ. Pancake ice has an influence on the thermodynamic exchange between the ocean and atmosphere, thus plays a key role in the global climate[3].

This paper presents a three dimensional CFD model capable of resolving the six degree of freedom dynamic response of a single pancake ice floe in a wave field. The model was developed with the open-source CFD software OpenFOAM[16]. A preliminary investigation of the effect of pancake ice geometry on the dynamics will be conducted by comparing the response of a circular and elliptical pancake in the same wave field. Section 2 describes the governing equations and implementation of the model in OpenFOAM and the results are discussed in Section 3. The conclusion is given in Section 4.

2 Pancake Ice Model

The model was created using the open-source CFD package, OpenFOAM (Open-source Field Operation and Manipulation). The pancake ice model was modified from the default floating body tutorial case, which features a simple cuboid floating in a sloshing water column. This case was modified to include wave generation and absorption, and the capacity to handle more complex geometries.

2.1 Solver

The 'interFoam' solver, one of many available in OpenFOAM, is used to handle multiphase simulations. It solves the Reynolds Averaged Navier-Stokes (RANS) equations for two incompressible phases using finite volume discretisation and the volume of fluid (VOF) method [17]. In the volume of fluid method, each phase is described by a nondenominational parameter, α, which is also called the volume fraction. Each grid cell occupied by material in the liquid phase (seawater in this case) has a value of $\alpha = 1$, and cells with the gas phase (air in this case) by $\alpha = 0$. The free surface, which is the interface between the gas and liquid phases, is determined using α. Cells that contain values between 0 and 1 are part of this interface. For the simulations of rigid bodies the enhanced version of interFoam, called 'interDyMFoam' is used. The former is only capable of handling static meshes, while the latter can handle dynamic meshes, making it ideal to simulate floating body movement on the free surface.

2.2 Governing Equations

InterDyMFoam calculates the pressure and velocity fields in the domain, by solving the RANS equations, which are shown below for two incompressible, immiscible fluids:

$$\nabla \cdot \mathbf{U} = 0 \tag{1}$$

$$\frac{\partial(\rho \mathbf{U})}{\partial t} + \nabla \cdot (\rho \mathbf{U}\mathbf{U}) - \nabla \cdot (\mu_{eff} \nabla \mathbf{U}) = -\nabla p^* - \mathbf{g} \cdot \mathbf{X} \nabla \rho \tag{2}$$

Where $\mathbf{U}$ is the velocity field, ρ is the fluid density, $\mathbf{g}$ is gravitational acceleration, μ_{eff} is the efficient fluid viscosity, and p^* is the pressure field including the hydrostatic pressure contribution, which is calculated according to:

$$p^* = p - \rho(\mathbf{g} \cdot \mathbf{X}) \tag{3}$$

Where p is the total pressure field, and $\mathbf{X}$ is the position vector. All bold letters indicate a vector field. For this model the energy equation is not required, as thermodynamic effects are not considered. For a relatively small time duration on the scale of seconds, the thermodynamics effects can be considered negligible[15]. An additional equation is required to track the movement of phase, and the conservation of α. In VOF, the velocity and pressure fields are shared by both air and water, while the density varies according to the value of α, as follows:

$$\rho = \alpha \rho_{water} + (1 - \alpha) \rho_{air} \tag{4}$$

The transport equation below is used to solve for α:

$$\frac{\partial \alpha}{\partial t} + \nabla \cdot \mathbf{U}\alpha + \mathbf{U}_c \alpha (1 - \alpha) = 0 \tag{5}$$

Where $\mathbf{U}_c$ is the compressive velocity field. This term does not have a physical meaning and is just used to bound α between 0 and 1, and achieve a sharp interface at the free surface[17][18].

2.3 Overset Mesh Method

Floating body motion on waves was achieved using an overset mesh. The floating body is placed inside a moving component mesh, called the overset, which is superimposed onto a background mesh, which is stationary. This method can be used to handle complicated geometries, while reducing computational cost. The overset mesh used for the pancake ice model can be seen in Figure 1. Each of the sides have a length of 0.6 m. The mesh is divided into 20 cells in the x and y directions, and 30 cells in the z direction, giving a total of 12 000 cells. The geometry for the pancake ice model was created and exported as an .stl file, then generated in the overset mesh with a built-in pre-processing tool, called 'snappyHexMesh'. Figure 1a shows the placement of the circular pancake inside the overset mesh, with the centre of mass located at the centre of the mesh. The cells surrounding the floating body are refined with an iterative process that runs until a specified refinement level is achieved. Figure 1b shows the results of the cell refinement process.

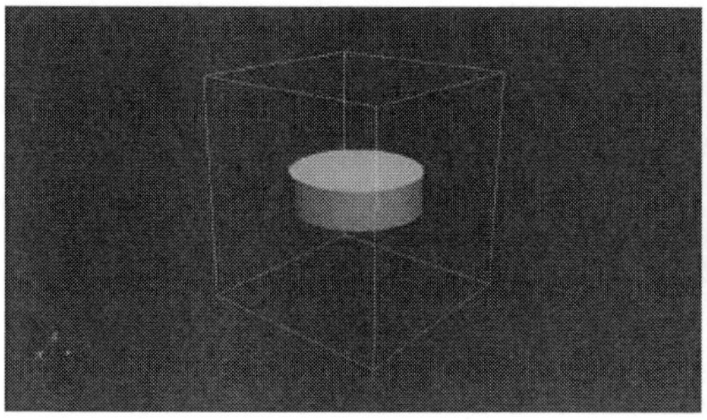

(a)

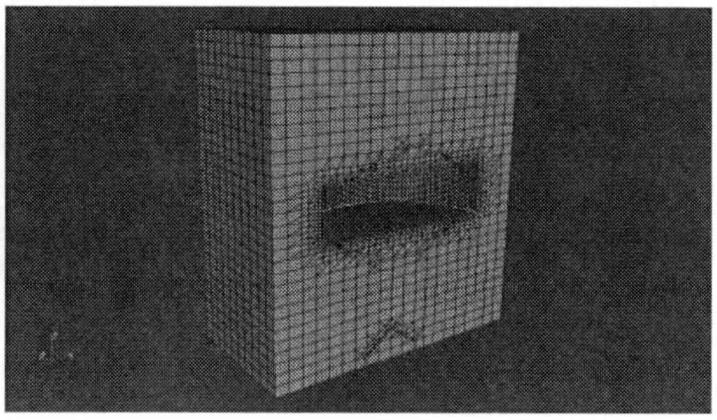

(b)

Figure 1: Overset Mesh with a circular pancake: (a) location of pancake inside mesh, (b) cell refinement around pancake floe geometry

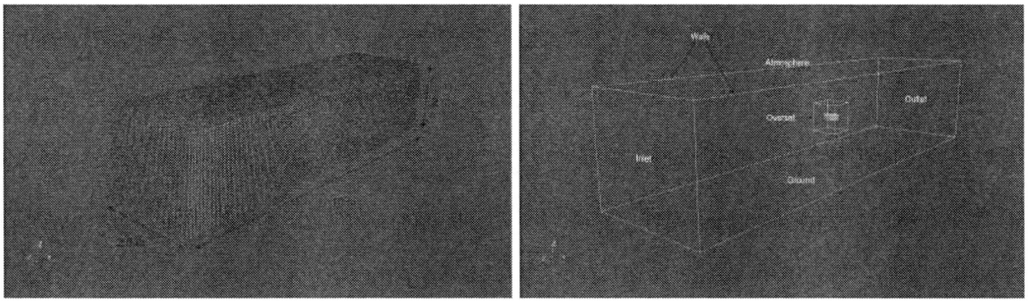

(a) (b)

Figure 2: Domain of pancake ice model: (a) dimensions of the background mesh, (b) location of overset in background mesh.

The background mesh is shown in Figure 2. It has dimensions of 10 m × 2.6 m × 2 m, in the x, y and z directions, respectively. The number of cells in the x, y and z directions is 200, 52 and 40 respectively, making a total of 416 000 cells. The background mesh dimensions were chosen to be comparable to a wave flume, for the purpose of validation with experimental results.

The cells which are occupied by the floating body, called hole cells, are deactivated to exclude them from the flow calculations. The cells surrounding the floating body, called fringe cells, are selected to exchange information between the background and overset meshes. The flow variables are interpolated from so-called donor cells in the background mesh, to the fringe cells located in the overset mesh[18].

2.4 Initial and Boundary Conditions

After the mesh is created, the initial and boundary conditions are applied. The mesh is divided into regions called patches, and the boundary conditions are specified for each patch. Initially the velocity and pressure fields in the domain are uniform. As seen in Figure 2b, the front face of the domain is assigned to the inlet patch, and the far face to the outlet patch. The top face is the atmosphere and the bottom face, the ground. The side faces are walls. The floating object is placed with its centre of mass at the free surface. The pressure and velocity gradients on all patches are set to zero, except the atmosphere and walls. A no slip condition is applied at the walls. The wave generating and absorbing boundary layers are applied to the inlet and outlet patches respectively.

2.5 Simulation Parameters

Two pancake ice geometries were investigated: a circular pancake, and an elliptical pancake as seen in Figure 3a and 3b. Both geometries have the same thickness and were placed in the same wave field. The elliptical geometry was chosen because it is a good approximation of the shape of real pancake ice. The elliptical pancake was placed with its major axis along the x axis. Table 1 displays the dimensions of each pancake ice geometry.

Table 1: Pancake ice geometry parameters.

Geometry	Major Diameter (m)	Minor Diameter (m)	Thickness (m)
Circle	0.4	-	0.1
Ellipse	0.5	0.34	0.1

The Stokes II wave model[19] was used to generate the waves. Stokes II is a non-linear, second order model, therefore the crest and trough portions of the wave are not equal in size. This gives the wave profile a more realistic shape, compared to sinusoidal models. It is also a good approximation of swell waves, which is the dominant form of wave activity in the MIZ. For the comparison with the wave tank experiment, a simple sinusoidal wave model will be used. Regular, unidirectional waves propagate through the domain in the positive x direction. The parameters used in the simulation are summarised in Table 2.

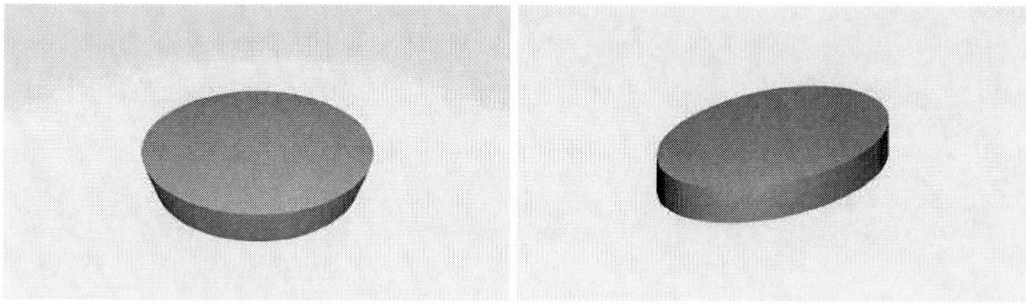

(a) (b)

Figure 3: Pancake ice geometries: (a) circular pancake, and (b) elliptical pancake

Table 2: Summary of parameters used in the simulation.

Parameter	Description	Value	Unit
ρ_p	pancake ice density	900[1]	kg/m^3
h	wave height	0.3	m
T	wave period	2	s
Δt	time step	0.001	s

3 Results and Discussion

The simulation of the circular and elliptical pancake geometries were run for a duration of 18 seconds. The velocity and orientation data were extracted from the simulation log file and processed using Python. The linear and angular velocities for the circular and elliptical pancake floes are shown in Figure 4 below.

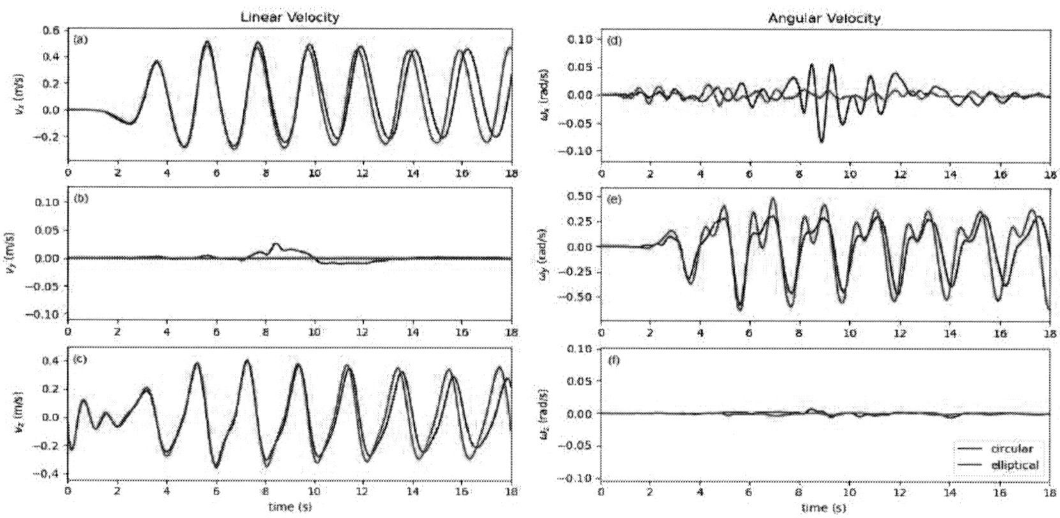

Figure 4: Linear and angular velocity

Since the wave field was unidirectional, only the linear x and z component, and the rotation about the y axis had a significant contribution to the motion. As seen in Figure 4a and 4c, the linear velocities for both geometries are largely similar, but they start going out of phase from about 10 s. In Figure 4e, we see the velocity of the pitch motion of the pancake floes. Both responses display two modes, which is more pronounced in the elliptical pancake. It is hypothesised that there are two main components in this response. One is the wave frequency, and the other is the natural frequency of the geometry. A Fourier analysis will need to be conducted to verify this hypothesis.

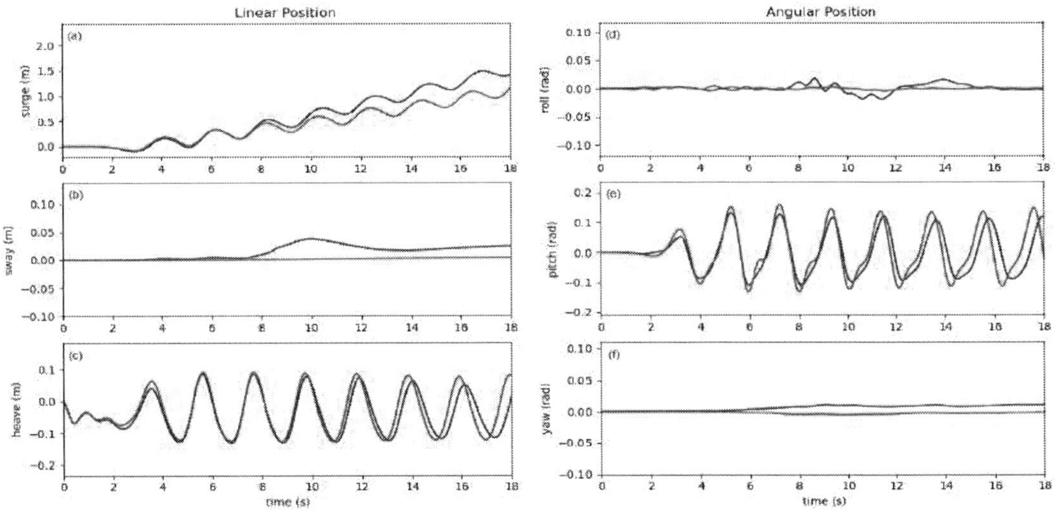

Figure 5: Linear and angular position

The linear and angular positions are shown in Figure 5. The heave (z axis) and pitch (y axis) motions were sinusoidal and largely similar for both geometries, which also starts shifting out of phase at 10 s. However the surge motions gradually tended away from the starting position, due to the drift of the pancake floe. The circular pancake drifted a greater distance. The gradual desynchronisation of the responses between the circular and elliptical floe can be attributed to the difference in drift distance, which starts at about 7 s and becomes more prominent from 10 s onwards.

4 Conclusion and Future Work

A computational fluid dynamics model of pancake ice and wave interaction was implemented in OpenFOAM, and the dynamic response of two pancake ice geometries was investigated. From the analysis of the simulation data, it was seen that the translational motions of two pancake geometries were largely similar and the variation was most significant in the rotations. It was noted that, while the surge and heave motions were sinusoidal as expected, the pitch motion displayed two modes. It is hypothesized that there are two main components in these responses: the waves frequency and the natural frequency of the geometry used. The geometry of the pancake ice floe could be approximated from its dynamic behaviour, but more simulations will need to be run with different sizes of these geometries to provide a more extensive set of data.

The model will be validated using experimental results collected in a wave tank at Aalto University in Helsinki. An artificial pancake floe was tested in multiple wave conditions. Inertial and visual data was collected to compare to the simulation results.

In future work we hope to develop an algorithm that can approximate the size and geometry of pancake ice using the inertial measurements captured by buoys deployed on them. This will enable us to study the evolution of pancake ice size on the scale of individual floes.

Nomenclature

Roman

g Gravitational acceleration [m/s^2]

h Wave height [m]

p Total pressure field [Pa]

p^*	Pressure field, including hydrostatic pressure [Pa]
T	Wave period [s]
t	Time [s]
U	Velocity field [m/s]
v	Pancake ice velocity [m/s]
X	Position vector [m]

Greek

α	Volume fraction
μ	Viscosity [kg/ms]
ω	Angular velocity [rad/s]
ρ	Density [kg/m^3]

Subscripts

x, y, z	Cartesian coordinate directions
air	Gas phase
c	Compressive
eff	Efficient
p	Pancake ice
water	Liquid phase

Abbreviations

CFD	Computational fluid dynamics
MIZ	Marginal ice zone
RANS	Reynold averaged Navier-Stokes
RAO	Response amplitude operator
VOF	Volume of fluid

Acknowledgement

Computations were performed using facilities provided by the University of Cape Town's ICTS High Performance Computing team: hpc.uct.ac.za.

References

[1] H. H. Shen, S. F. Ackley, *et al.*, "A conceptual model for pancake-ice formation in a wave field," *Annals of Glaciology*, vol. 33, pp. 361–367, 2001. DOI: 10.3189/172756401781818239.

[2] L. A. Roach, M. M. Smith, *et al.*, "Quantifying Growth of Pancake Sea Ice Floes Using Images From Drifting Buoys," *Journal of Geophysical Research: Oceans*, vol. 123, no. 4, pp. 2851–2866, 2018. DOI: 10.1002/2017JC013693.

[3] M. J. Doble, M. D. Coon, *et al.*, "Pancake ice formation in the Weddell Sea," *Journal of Geophysical Research: Oceans*, vol. 108, no. 7, p. 3209, 2003. DOI: 10.1029/2002jc001373.

[4] H. H. Shen, S. F. Ackley, *et al.*, "Limiting diameter of pancake ice," *Journal of Geophysical Research: Oceans*, vol. 109, no. 12, pp. 1–13, 2004. DOI: 10.1029/2003JC002123.

[5] D. Dumont, L. Bertino, *et al.*, "A model for waves-in-ice and sea ice dynamics in the marginal ice zone," in *Society of Petroleum Engineers - Arctic Technology Conference 2011*, vol. 1, 2011, pp. 118–123. DOI: 10.4043/22049-ms.

[6] C. Horvat and E. Tziperman, "A prognostic model of the sea-ice floe size and thickness distribution," *Cryosphere*, vol. 9, no. 6, pp. 2119–2134, 2015. DOI: 10.5194/tc-9-2119-2015.

[7] R. A. Massom and S. E. Stammerjohn, "Antarctic sea ice change and variability - Physical and ecological implications," *Polar Science*, vol. 4, no. 2, pp. 149–186, 2010. DOI: 10.1016/j.polar.2010.05.001.

[8] A. Alberello, L. Bennetts, *et al.*, "Drift of Pancake Ice Floes in the Winter Antarctic Marginal Ice Zone During Polar Cyclones," *Journal of Geophysical Research: Oceans*, vol. 125, no. 3, 2020. DOI: 10.1029/2019JC015418. arXiv: 1906.10839.

[9] P. J. Rottier, "Floe pair interaction event rates in the marginal ice zone," *Journal of Geophysical Research: Oceans*, vol. 97, no. C6, pp. 9391–9400, 1992. DOI: https://doi.org/10.1029/92JC00152.

[10] L. J. Yiew, L. G. Bennetts, *et al.*, "Hydrodynamic responses of a thin floating disk to regular waves," *Ocean Modelling*, vol. 97, pp. 52–64, 2016. DOI: 10.1016/j.ocemod.2015.11.008.

[11] M. H. Meylan, L. J. Yiew, *et al.*, "Surge motion of an ice floe in waves: Comparison of a theoretical and an experimental model," *Annals of Glaciology*, vol. 56, no. 69, pp. 155–159, 2015. DOI: 10.3189/2015AoG69A646. arXiv: 1403.1933.

[12] M. Smith and J. Thomson, "Pancake sea ice kinematics and dynamics using shipboard stereo video," *Annals of Glaciology*, vol. 61, no. 82, pp. 1–11, 2020. DOI: 10.1017/aog.2019.35.

[13] M. A. Hopkins and H. H. Shen, "Simulation of pancake-ice dynamics in a wave field," *Annals of Glaciology*, vol. 33, pp. 355–360, 2001. DOI: 10.3189/172756401781818527.

[14] A. Herman, "Wave-Induced Surge Motion and Collisions of Sea Ice Floes: Finite-Floe-Size Effects," *Journal of Geophysical Research: Oceans*, vol. 123, no. 10, pp. 7472–7494, 2018. DOI: 10.1029/2018JC014500.

[15] R. Marquart, A. Bogaers, *et al.*, "A computational fluid dynamics model for the small-scale dynamics of wave, ice floe and interstitial grease ice interaction," *Fluids*, vol. 6, no. 5, 2021. DOI: 10.3390/fluids6050176.

[16] OpenCFD, *OpenFOAM (v2112)*. [Online]. Available: https://www.openfoam.com/, Accessed: 9 October 2023.

[17] P. Higuera, J. L. Lara, *et al.*, "Realistic wave generation and active wave absorption for Navier-Stokes models. Application to OpenFOAM®.," *Coastal Engineering*, vol. 71, pp. 102–118, 2013. DOI: 10.1016/j.coastaleng.2012.07.002.

[18] H. Chen, L. Qian, *et al.*, "Application of an overset mesh based numerical wave tank for modelling realistic free-surface hydrodynamic problems," *Ocean Engineering*, vol. 176, pp. 97–117, 2019. DOI: 10.1016/j.oceaneng.2019.02.001.

[19] R. Dean and R. Dalrymple, *Water Wave Mechanics for Engineers and Scientists*. World Scientific, 1991. DOI: 10.1142/1232.

Estimating Mechanical Properties of AM Lattices Using CT-Scans and FE Models

Minsol Park[1*], Martin Phillip Venter[2], Anton Du Plessis[3,4]

[1] Department of Mechanical and Mechatronics Engineering, University of Stellenbosch, South Africa, minsol.park@mail.mcgill.ca

[2] Department of Mechanical and Mechatronics Engineering, University of Stellenbosch, South Africa, mpventer@sun.ac.za

[3] Department of Physics, University of Stellenbosch, South Africa, anton2@sun.ac.za

[4] *Comet Technologies Canada Inc,* Montreal, Canada, *anton.duplessis@comet.tech*

Abstract. The prevalent reliance on expensive, time-consuming CT (Computed Tomography) scans and traditional mechanical testing presents significant challenges in determining the mechanical properties of additively manufactured (AM) metallic lattices fabricated through laser powder bed fusion (L-PBF). This work integrates CT-scan data with finite element models to correlate process parameters (layer thickness, laser power) with mechanical performance. This correlation allows us to use interpolation methods to predict lattice properties without further CT scans. Two such methods, direct and indirect interpolation, are proposed. Direct interpolation correlates process parameters with mechanical performance, while indirect interpolation uses parameters and manufacturing defects to inform the finite element model. Numerical interpolation results obtained from both direct and indirect methods are compared with experimental results, with the targeted accuracy criterion set at 83 % based on the maximum numerical error of the Finite Element (FE) model. The findings demonstrate promising accuracy rates for both methods, with the direct interpolation method achieving higher reliability with an accuracy rate of 87 %, surpassing the target criterion. The indirect interpolation method also achieves a respectable accuracy rate of 79 %. Both interpolation methods offer cost-effective numerical solutions for material characterisation and build quality monitoring in the context of AM metallic lattices. By reducing reliance on CT scans, this approach significantly saves time and costs associated with the analysis of mechanical properties, presenting a valuable alternative for industry applications.

* Corresponding author: minsol.park@mail.mcgill.ca

1 Introduction

Additive Manufactured (AM) metallic lattice structures have garnered significant attention from researchers due to their high design freedom, enabling the tailoring of mechanical properties [1]. Laser Powder Bed Fusion (L-PBF) is a commonly used technique for producing metallic lattice structures, thanks to its high accuracy and availability of various materials [2]. However, L-PBF faces challenges in terms of manufacturing defects, including residual stress [3], geometrical inaccuracies [4], surface roughness [5], and internal porosity [6]. These defects can have detrimental effects on the mechanical properties of L-PBF metals, particularly in lattice structures due to their intricate small-scale geometry [7]

Therefore, it is crucial to characterise the mechanical properties and monitor the build quality of lattice structures prior to producing them in large quantities [8]. Quasi-static compression tests have been commonly used in the literature to evaluate the mechanical performance and build quality of lattice structures [9–11]. However, this experimental approach is associated with significant costs and time-consuming manufacturing processes, which limit the number of tests that can be performed. In this context, Finite Element (FE) modelling has emerged as a promising alternative approach. FE modelling allows for the creation of virtual models that accurately represent the lattice structures. The iterative nature of FE modelling enables investigations and adjustments to be made without the need for further printing.

The Computed Tomography (CT)-incorporated FE model has been identified as the most accurate numerical solution for evaluating the mechanical performance of lattice structures in the literature [12-14]. By utilising CT-scanned images, this approach reconstructs a 3D digital model of the lattice structure, which is then used for numerical simulations. The CT-incorporated FE model effectively accounts for manufacturing defects and accurately predicts the mechanical properties of the lattice structure. However, it requires the production of physical samples and subsequent CT scan analysis, which are costly and time-consuming.

In this study, alternative methods are proposed for predicting the mechanical performance of diamond lattice structures under compression. These methods leverage various interpolation techniques applied to the numerical results from FE models. By utilising interpolation methods, the mechanical properties of lattice structures can be estimated, reducing the need for expensive and time-consuming CT scans and physical testing. These methods offer promising opportunities for accurately predicting mechanical properties, enabling effective management of build quality by researchers and engineers. The predicted mechanical properties obtained through interpolation can be used as bounds for comparison with predefined criteria, facilitating build quality control. Furthermore, these methods contribute to faster evaluation and optimisation of lattice structures by reducing reliance on physical testing and CT scans. Ultimately, this research advances the field of additive manufacturing by providing efficient tools for evaluating and improving the mechanical performance of lattice structures.

2 Method

The method section is structured into three sub-sections (2.1-2.3) to describe the approach utilised in this study. The first sub-section provides an overview of the development of the FE model, explaining the procedure employed to create a numerical representation of the lattice structure. Moving on, the second sub-section introduces the direct interpolation method, which involves predicting mechanical properties by correlating process parameters with mechanical performance. Lastly, the third sub-section outlines the indirect interpolation method, which leverages correlations between process parameters and CT scan analysis data to predict mechanical properties.

2.1 FE model

For evaluating the mechanical performance of lattice structures, a previously developed simplified FE model was employed in this study [8]. The FE model was created using the non-linear FE software *MSC Marc* 2021.4, chosen for its compatibility with other MSC software packages. This FE model takes into account the influence of manufacturing defects, such as geometry inaccuracies, internal porosity, and surface roughness, on the behaviour of AM lattice structures, utilising information obtained from CT scan analysis. In this study, diamond lattice structures were investigated using five different sets of process parameters. The study adopted a one-factor-at-a-time (OFAT) testing approach, investigating the effects of layer thickness and laser power individually on the mechanical properties of the lattice structures, as illustrated in Fig. 1. In the OFAT testing, the layer thickness was varied across levels of 30 μm, 60 μm, and 90 μm, while maintaining a constant laser power of 280 W. Simultaneously, the laser power was altered at levels of 210 W, 280 W, and 360 W, while keeping the layer thickness fixed at 30 μm. Among these variations, a sample produced with a laser power of 280 W and a layer thickness of 30 μm was designated as the reference lattice (RL), as it encapsulates both laser power and layer thickness changes. The other samples were designated as follows: those with increased layer thicknesses of 60 μm and 90 μm were denoted as medium layer thickness (MLT) and high layer thickness (HLT), respectively. Samples produced with laser powers of 210 W and 360 W were named low laser power (LLP) and high laser power (HLP), respectively. Each sample was replicated three times to ensure statistical reliability and account for variability.

In the current study, the CT scan analysis data obtained from our previous research [7] was used as input for the FE model. This data was employed to incorporate various levels of manufacturing defects, which are detailed in Table 1. The focus was to accurately simulate and predict the mechanical properties of the lattice structures under different process parameter settings. The strut thickness represents the diameter of individual struts, excluding the area where they connect to the nodes. Conversely, the node thickness indicates the thickness of the regions where the struts are joined together. Internal porosity refers to the vacant space within the lattice structure's struts. However, the internal porosity of sample HLT could not be obtained due to poor-quality CT scan images. To quantify the severity of surface roughness, R_a values were used.

The modelling process comprises three sub-modelling approaches: the strut-node model, the internal porosity model, and the surface morphology model. For detailed information on the modelling process, readers can refer to our previous research [8].

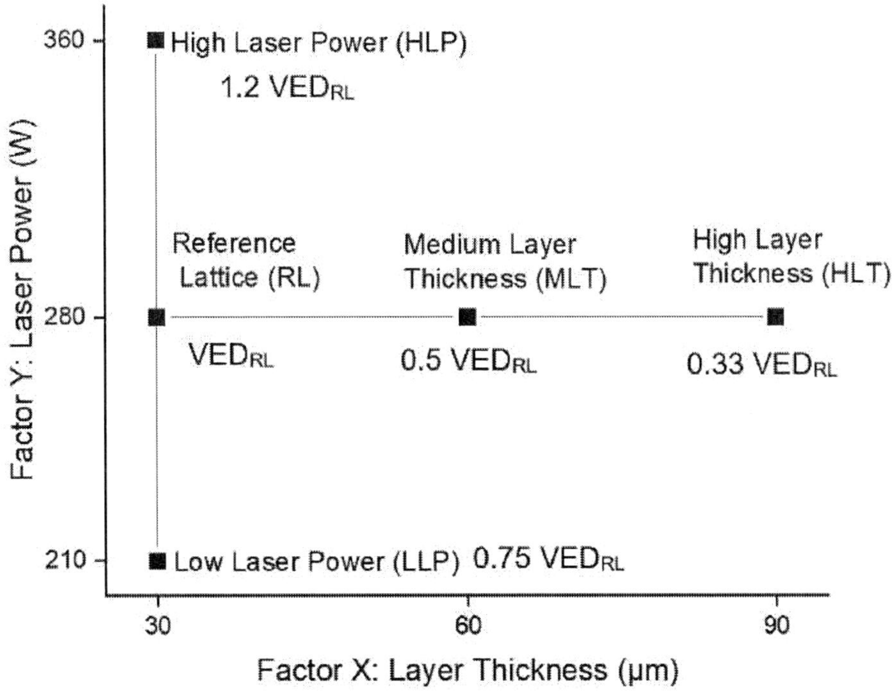

Fig.1. Exploring process design space of OFAT with factors of layer thickness and laser power with corresponding relative volume energy density (VED).

Table 1. Manufacturing defect parameters, including strut thickness, node thickness, internal porosity and surface roughness of lattice structures produced with five different process parameters [7].

Sample Name	RL	MLT	HLT	LLP	HLP
Parameter	Recommend	Thickness 60 µm	Thickness 90 µm	Power 75 %	Power 120 %
Strut thickness (µm)	363	357	344	326	386
Node thickness (µm)	447	463	431	399	504
Internal Porosity (%)	2.17	0.27	-	10.8	0.25
R_a (µm)	17	22	21	21	26

The strut-node model addressed geometric inaccuracies by designing the struts in the lattice structures as lofted cylinders, as depicted in Fig. 2A. The strut thickness determines the diameter of the cross-sectional circle in the middle, while the node thickness determines the diameter of the cross-sectional circle at the outer edge of the cylinder as shown in Fig. 2B.

The internal porosity model considered the impact of internal porosity using the Representative Volume Element (RVE) technique and the *Digimat* 2021.3 software for material homogenisation. *Digimat*, a multiscale material modeling software, enables the homogenisation of materials with voids and is highly compatible with the FE model software *MSC Marc* 2021.4 [15]. The RVE model was created with dimensions of 0.4 x 0.4 x 0.4 mm, and it included randomly distributed voids (Fig. 2C). The volume fraction of the voids in the RVE model was determined based on the internal porosity value obtained from the CT scan analysis (Table 1).

To account for surface roughness, a simplified geometry method proposed by Ghosh *et al.* [13] is implemented on a single strut instead of the entire lattice structure to reduce computational costs. This method incorporated sharp-notch features to represent the irregular surface morphology of AM micro struts (Fig. 2D). The size of the sharp-notch features in the numerical model was selected to align the mean deviation of the profile for the notch feature with the R_a values obtained from the CT scans (Table 1) [13].

The final model integrated all three manufacturing defects and aims to replicate the compression behaviour of the lattice structures. It combined the strut-node geometry with tailored input mechanical properties to incorporate the impact of internal porosity and surface roughness using the internal porosity model and surface morphology model simultaneously.

The final model development process was carried out for five different lattice structures, and the corresponding numerical and experimental data for mechanical properties, including stiffness, yield stress, and peak stress, are presented in Table 2. The proposed model achieved a maximum error of 17 % (excluding the outlier sample LLP with a porosity of 10.8 %) in predicting the mechanical performance of lattice coupons manufactured with varying layer thicknesses and laser powers. However, it should be noted that the proposed model is not effective in predicting the mechanical performance of lattice structures with high internal porosity, as demonstrated in a previous study [8].

While the proposed numerical model has shown effectiveness in predicting mechanical properties, its reliance on CT scanning poses limitations due to cost and time constraints. To overcome this challenge and improve the efficiency of build quality monitoring, this study explores new approaches: direct and indirect interpolation. These interpolation methods, which will be discussed in the following sections, provide approaches for estimating mechanical properties with reduced dependence on CT scans.

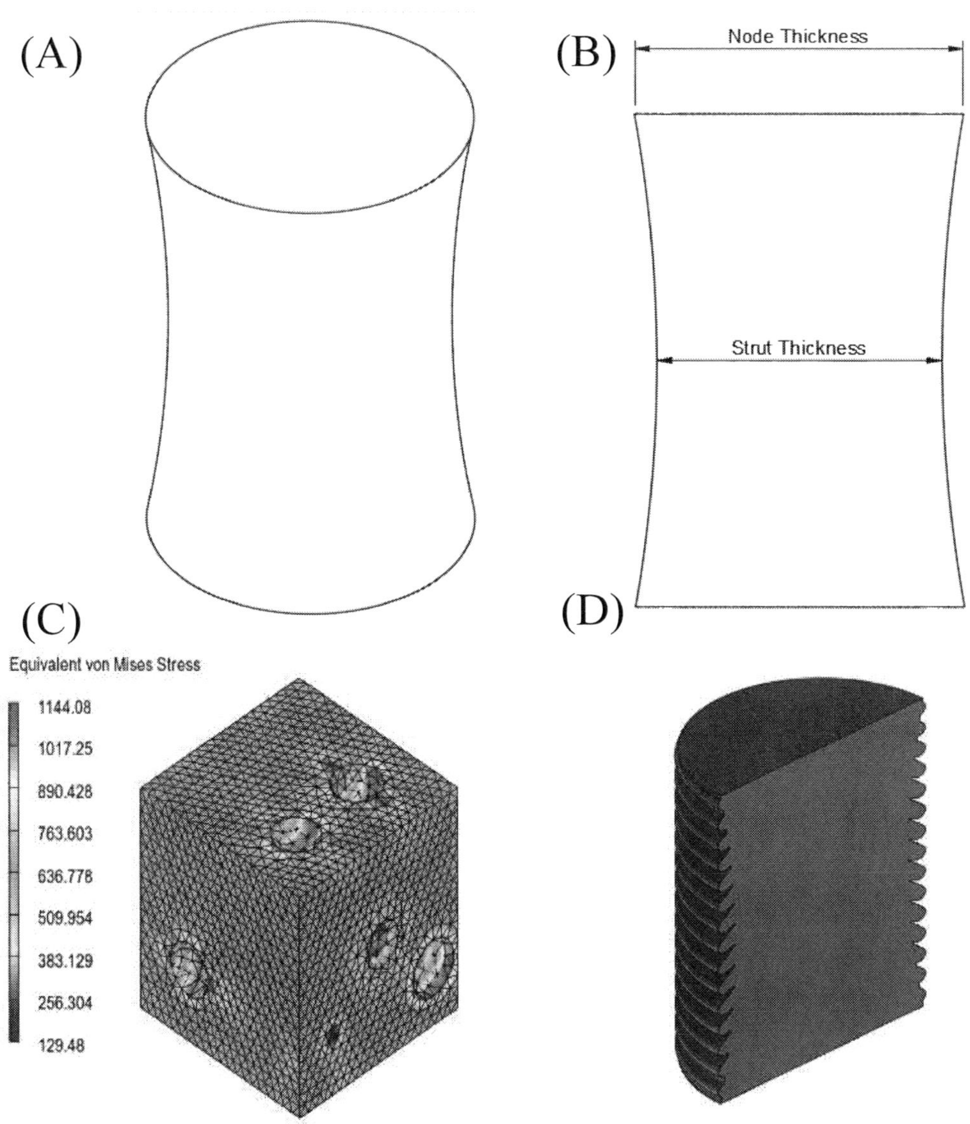

Fig.2. Isometric view of node-strut geometry incorporated strut (A) [8], node-strut geometry of a single strut with its dimension parameters (B) [8], RVE with the ideally spherical voids generated using Digimat (C) [8] , and section view of strut model with notch-feature surface morphology (D).

Table 2 Mechanical properties of lattice structures, including stiffness, yield stress, and peak stress, obtained from both experiments and FE models [8].

Sample	**Stiffness (GPa)**		**Yield Stress (MPa)**		**Peak Stress (MPa)**	
	Experiment	Numerical	Experiment	Numerical	Experiment	Numerical
RL	1.6	1.8	39	35	45	46
MLT	1.8	2.0	43	39	52	52
HLT	1.4	1.6	37	32	43	43
LLP	0.7	0.9	20	19	24	25
HLP	2.6	3.0	59	53	70	69

2.2 Direct Interpolation

Direct interpolation is an interpolation method that involves fitting trend lines to the obtained experimental and numerical data (presented in Table 2) for the five lattice structures. The direct interpolation on experimental data refers to experimental interpolation, to distinguish it from the direct interpolation on numerical data which is simply called the direct interpolation. Through this process, correlations between mechanical properties and process parameters were established. The trend lines, along with their corresponding 95 % confidence intervals, were obtained using the built-in function provided by the graph plotting software *Origin 2023b*. The equations for the trend lines were carefully selected to minimise the R^2 values. The analysis of lattice structures demonstrated a second-order trend in mechanical properties with respect to layer thickness (Fig. 3), with R^2 values ranging from 0.8548 to 0.8675. Conversely, a strong positive linear trend was observed in relation to laser power (Fig. 4), with R^2 values ranging from 0.9785 to 0.9997. Using equations derived from these trend lines, direct interpolation was performed to estimate the mechanical properties of lattice structure models with four specific parameter settings: medium low layer thickness (MLLT: 0.67 VED_{RL}) with a laser power of 280 W and a layer thickness of 45 μm, medium high layer thickness (MHLT: 0.4 VED_{RL}) with a laser power of 280 W and a layer thickness of 75 μm, medium low laser power (MLLP: 0.88 VED_{RL}) with a laser power of 245 W and a layer thickness of 30 μm, and medium high laser power (MHLP: 1.14 VED_{RL}) with a laser power of 320 W and a layer thickness of 30 μm. These lattice structures were not physically printed and CT-scanned; instead, they were treated as virtual samples for the interpolation analysis.

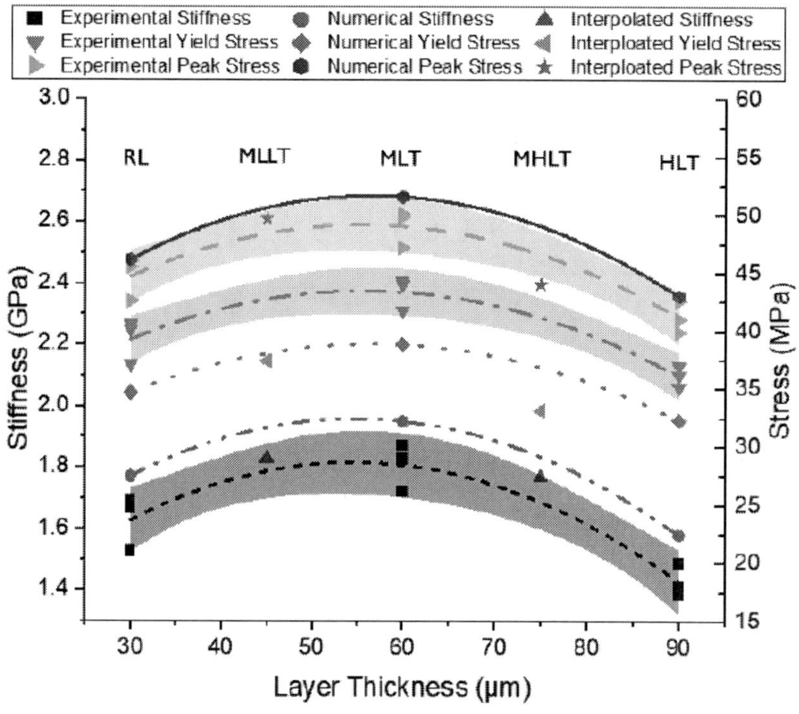

Fig.3. Mechanical properties (stiffness, yield stress, initial peak stress) of lattice produced with different layer thicknesses with second order best-fit lines for both experimental and numerical data and the shaded band for 95% confidence intervals, where the interpolated mechanical properties refers to the predicted mechanical properties from indirect interpolation.

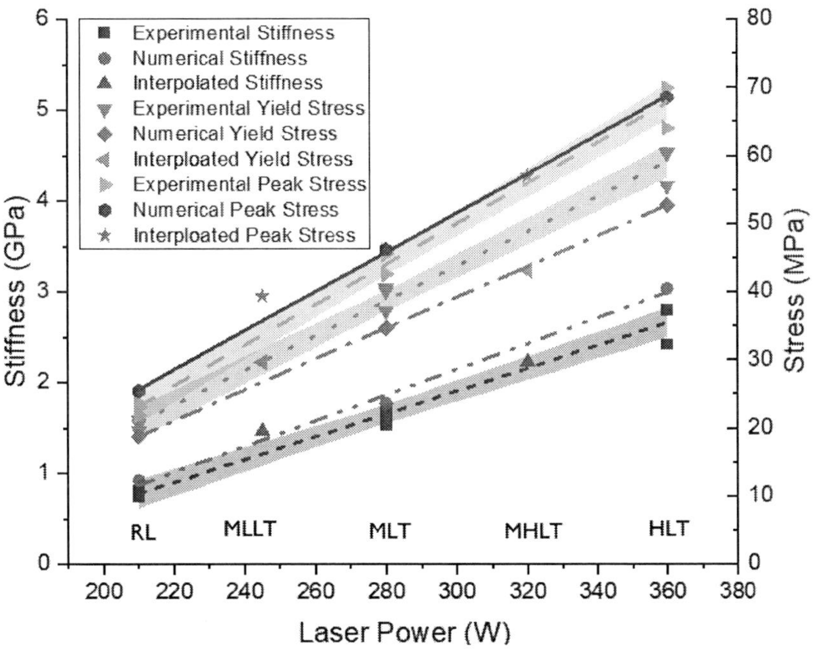

Fig.4. Mechanical properties (stiffness, yield stress, initial peak stress) of lattice produced with different laser power with linear best-fit lines for both experimental and numerical data and the shaded band for 95% confidence, where the interpolated mechanical properties refers to the predicted mechanical properties from indirect interpolation.

2.3 Indirect method

The direct interpolation method focuses solely on the correlation between process parameters and mechanical properties, without considering the influence of manufacturing defects, which play an intermediate role in the relationship. Manufacturing defects, influenced by process parameter variations, can significantly affect the mechanical properties of lattice structures. Therefore, it is crucial to incorporate the impact of manufacturing defects into the correlations. In addition to the direct interpolation method, this study introduces an alternative approach called the indirect method. The indirect method explores the correlations between process parameters and various CT scan analysis metrics, including node thickness, strut thickness, internal porosity, and R_a values. By considering these correlations, the indirect method provides a more comprehensive understanding of the relationship between process parameters, manufacturing defects, and mechanical properties. The analysis revealed that both strut thickness and node thickness exhibited second-order correlations with layer thicknesses (Fig. 5A), while linear correlations were observed for laser power (Fig. 5C). Furthermore, internal porosity demonstrated an exponential correlation with both laser power and layer thickness (Fig. 5B and Fig. 5D). Additionally, second-order correlations were observed between R_a values and both laser power and layer thickness. Based on the established correlations, the CT scan analysis data of virtual samples were estimated. Subsequently, FE models were created for these virtual samples using the interpolated CT scan analysis. Quasi-static compression tests were numerically simulated on these virtual samples to predict their mechanical properties. The obtained results were plotted in Fig. 3 and Fig. 4 for comparative analysis with the direct methods.

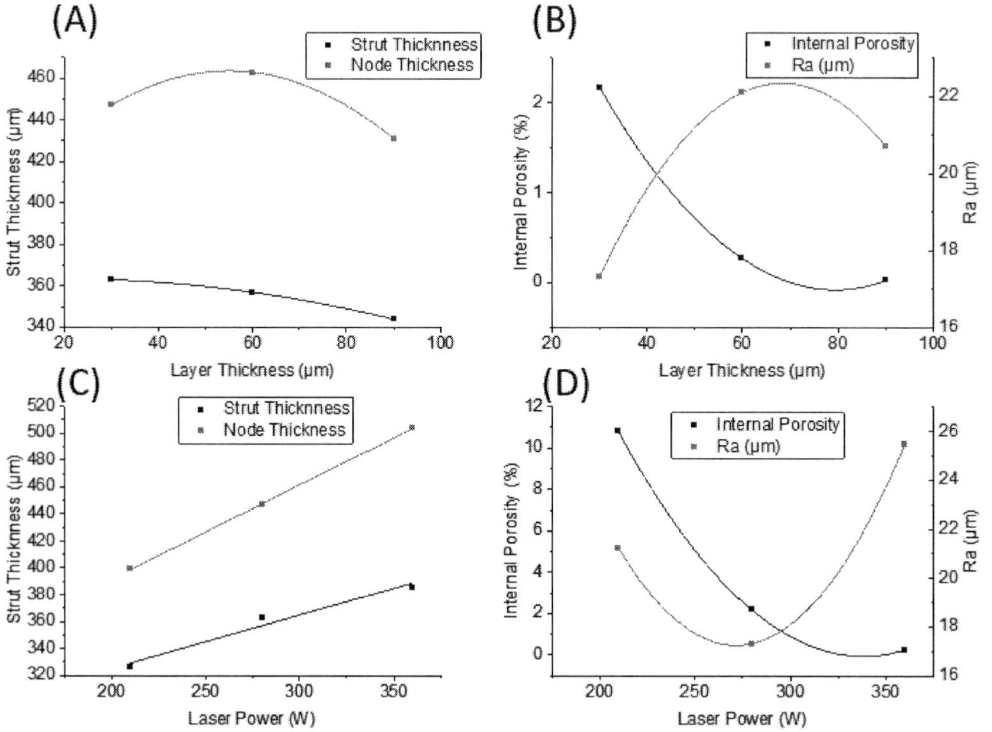

Fig.5. Strut and node thickness interpolations with variation in layer thickness (A), internal porosity and Ra interpolations with variation in layer thickness (B), strut and node thickness interpolations with variation in laser power (C), and internal porosity and Ra interpolations with variation in laser power.

3 Result and discussion

The mechanical properties of the virtual samples were estimated using three different approaches: direct, indirect, and experimental interpolations (direct interpolation using experimental results). The results were compared and depicted in Fig. 6, while the interpolation errors were shown in Fig. 7. It should be noted that there was an overestimation of up to 21 % observed in the estimated mechanical properties. This overestimation can be attributed to the numerical error associated with the FE models, which was approximately 17 % as reported from numerical findings. The direct method outperformed the indirect method, as its errors of 13 % remained below the maximum numerical error of 17 %. The indirect method demonstrated its limitation in prediction particularly in accurately predicting yield stress and for cases like sample MLLP.

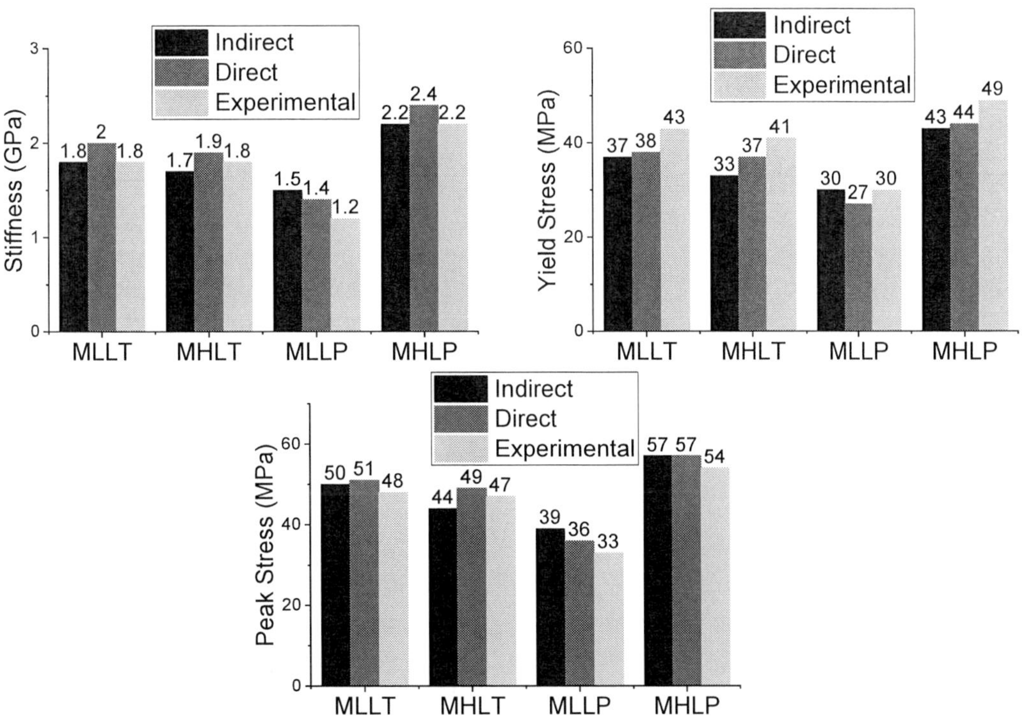

Fig.6. Column graphs for stiffness, yield stress, and peak stress of virtual samples obtained through different interpolation methods including indirect, direct, and experimental interpolations.

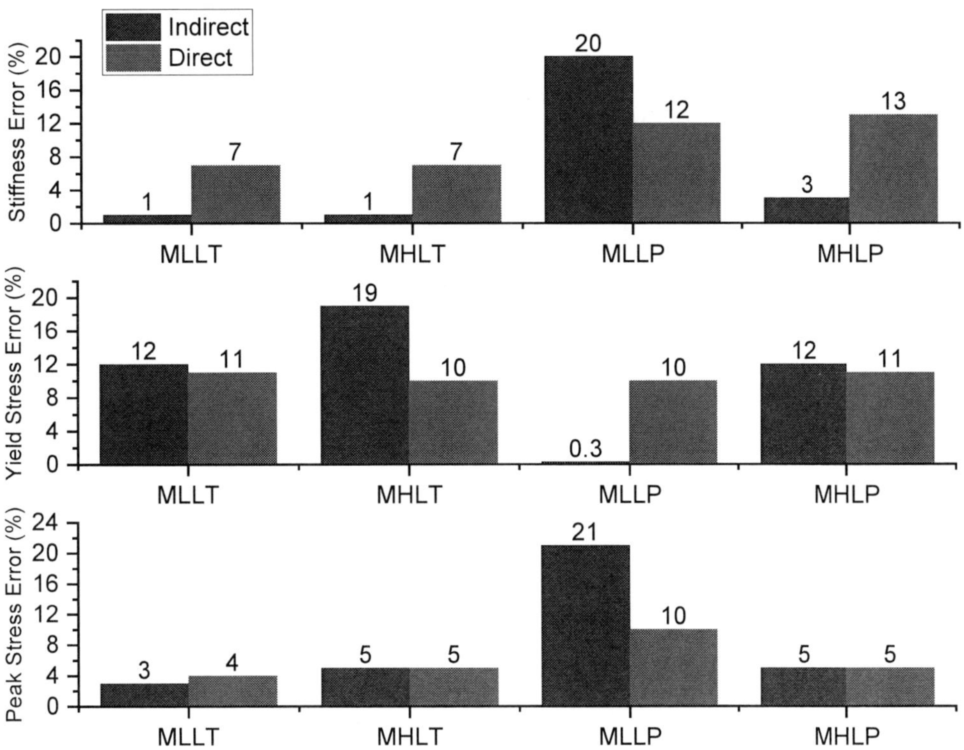

Fig.7. Column graphs for errors in mechanical properties (stiffness, yield stress, and peak stress) of virtual samples obtained by comparing the predicted values from different interpolation methods including indirect and direct to that of the experimental interpolation.

The results for sample MLLP exhibited a relatively high error ranging from 10 % to 21 %, which exceeded the maximum numerical error of 17 %. The numerical findings highlighted the high sensitivity of internal porosity analysis to various factors such as pore size, pore sphericity, pore shape, and pore types. These factors can significantly impact the accuracy of the analysis, especially when the internal porosity percentage is high. In the case of virtual sample MLLP, the internal porosity analysis was conducted without considering data on these factors, as no CT scan analysis was performed for this virtual sample. The absence of such data becomes particularly critical when using the indirect method, which relies on internal porosity analysis. This can result in higher error rates for sample MLLP, as shown in Fig. 7, due to the limited information available for accurate estimation.

The limitation of not having access to these factors underscores the challenge of accurately estimating the mechanical properties for samples with high internal porosity solely using interpolation methods. It is crucial to recognise that interpolation methods alone may have limitations in accurately capturing the intricate characteristics and behaviour of lattice structures with high internal porosity. The indirect method demonstrates its benefits over the direct method, exhibiting a lower interpolation error in terms of stiffness (Fig. 7). Specifically, the indirect method showed a low interpolation error in stiffness at a maximum of 3 %, excluding sample MLLP with high porosity, significantly lower than the maximum interpolation error of direct method, 13 %. This indicates that the indirect method provides a more accurate estimation of the stiffness. It is worth mentioning that the indirect method results in a higher interpolation error, particularly when predicting the yield stress, as depicted in Fig. 7. The specific reason for the discrepancies in the indirect method's ability to predict yield stress remains unclear and could be a subject for future investigation. Nevertheless, it is evident from the results that the indirect method is not a promising approach for predicting yield stress, as it may exceed the required maximum numerical error. Further research and development are necessary to enhance the accuracy and reliability of the indirect method in predicting mechanical properties for lattice structures, especially in cases where yield stress is a critical factor.

The proposed interpolation methods offer promising potential as tools for monitoring the build quality of AM metallic lattice structures, especially when there is variation in the process parameters of laser power and layer thickness within the desired range. To effectively monitor the build quality control while considering variations in laser power and layer thickness, the following steps are recommended. Firstly, the user should define the range of parameter variations for the investigation. This involves specifying the desired minimum and maximum values for laser power and layer thickness. Next, lattice samples and tensile samples should be printed with variations in laser power or layer thickness according to the defined range. A uniform increment in laser power and layer thickness is preferable between each sample to capture the desired range comprehensively. Once printed, the lattice samples should undergo CT scanning for subsequent analysis. The CT scan provides valuable data on the internal structure and properties of the samples, enabling the evaluation of build quality.

When determining the number of increments, increment values, and replicates, careful consideration should be given to the variation range and the desired level of accuracy. In this study, a minimum of 3 increments and replicates were considered for demonstration purposes and cost reduction. However, it is recommended to use a value of 5 for a more conservative analysis, as it provides a greater number of data points and enhances the reliability of the results. When selecting an interpolation method, it is important to consider the desired accuracy and the available budget. Experimental and direct interpolation methods are simpler and less labour-intensive, making them cost-effective options. On the other hand, indirect interpolation takes into account the correlations between manufacturing and process parameters, which can result in higher accuracy, particularly in stiffness. This method provides a more comprehensive understanding of the relationship between the parameters and the predicted properties. A suggested approach is to combine all three interpolation methods. By utilising experimental, direct, and indirect interpolation simultaneously, researchers and engineers can achieve a more reliable estimation of mechanical properties. To address the inherent errors associated with interpolations, it is practical to use the 95 % prediction band (Fig. 8).

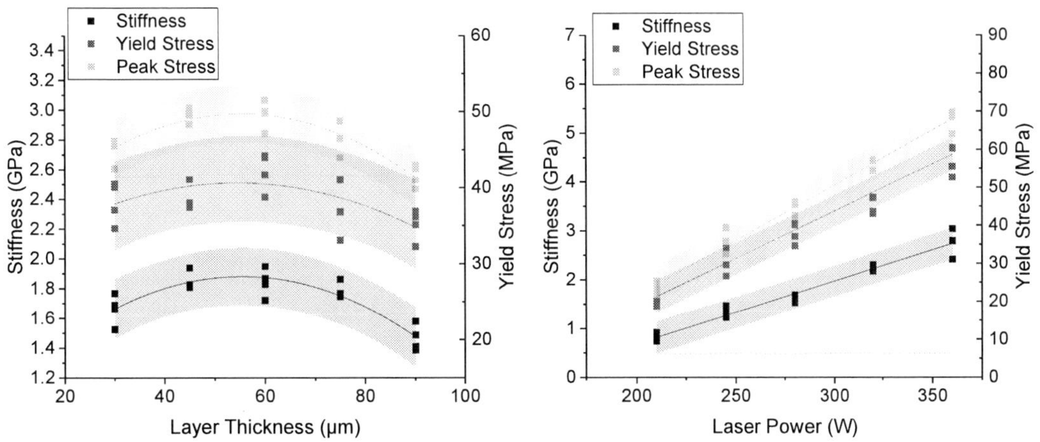

Fig.8. Combined interpolation method for the build quality monitoring with best-fits and 95 % prediction bands.

The generated prediction band, which encompasses the predicted properties from the three interpolation methods, provides valuable information for achieving the desired build quality of lattice structures. As shown in Fig. 9, for example, the lattice structure manufactured with a laser power of 300 W and a layer thickness of 30 μm is expected to exhibit a stiffness ranging from 1.7 to 2.2 GPa, yield stress ranging from 38 to 48 MPa, and peak stress ranging from 46 to 55 MPa. This information can be utilised by the user to establish the maximum and minimum bound of the mechanical properties, thereby enabling effective build quality control prior to the manufacturing process.

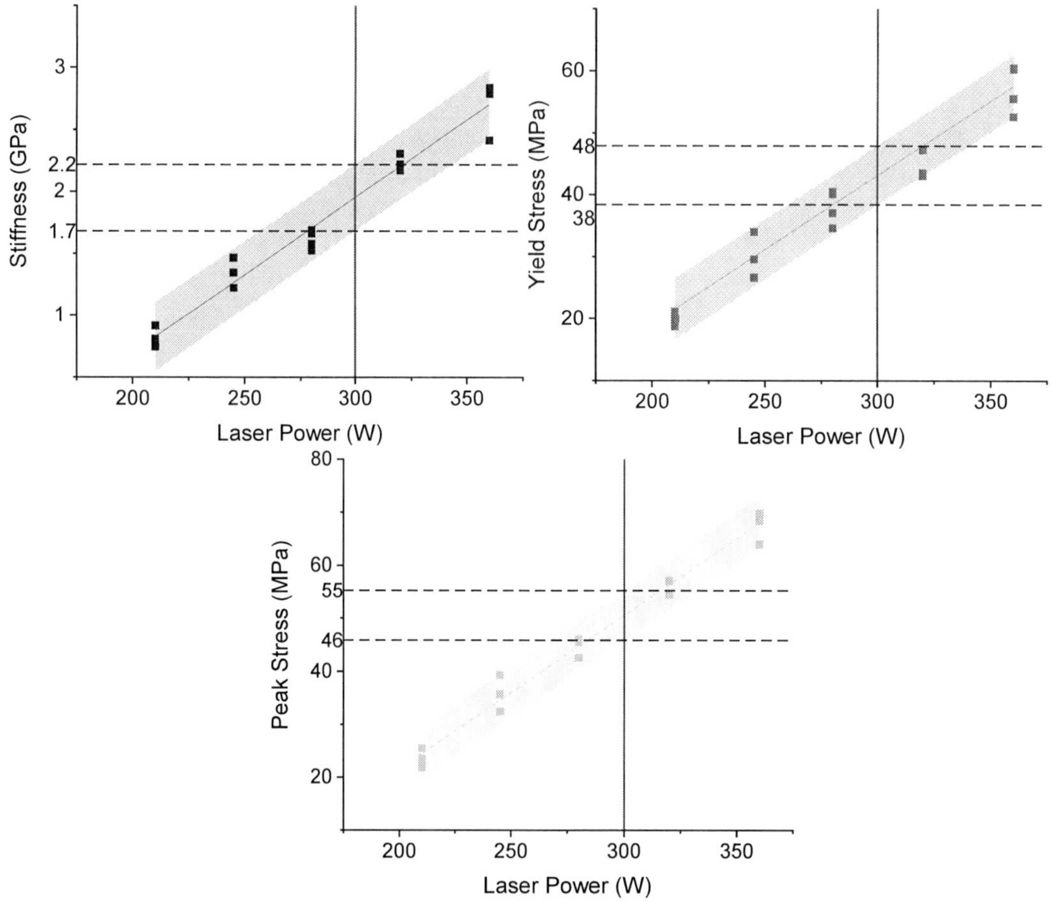

Fig. 9 The plots used for the process parameter selection process. The black, red, and green bands represent the 95% prediction bands for stiffness, yield stress, and peak stress, respectively. The blue horizontal bands represent the mechanical property criteria. The vertical red dotted lines indicate the upper and lower bounds for laser power selection.

4 Conclusion

The present study has proposed numerical interpolation methods, namely direct and indirect methods, for evaluating the mechanical properties of additively manufactured lattice structures with different parameter settings. The results have shown promising accuracy rates for both methods, with the direct interpolation method achieving higher reliability at an accuracy rate of 87 %, surpassing the target criterion. The direct interpolation method also demonstrates a respectable accuracy rate of 79 %. The study highlights the potential of interpolation methods as tools for build quality monitoring in the context of additively manufactured metallic lattices. By reducing reliance on CT scans, this approach significantly saves time and costs associated with the analysis of mechanical properties, making it a valuable alternative for industry applications.

5 References

[1] A. du Plessis *et al.*, "Properties and applications of additively manufactured metallic cellular materials: A review," *Progress in Materials Science*, vol. 125. Elsevier Ltd, Apr. 01, 2022. doi: 10.1016/j.pmatsci.2021.100918.

[2] Yadroitsev, I., Yadroitsev, I., Du Plessis, A., Macdonald, E., 2021Fundamentals of Laser Powder Bed Fusion of Metals, 1st, Elsevier, ISBN 9780128240908, doi: B978- 0-12-824090-8.12001-3.

[3] J. L. Bartlett and X. Li, "An overview of residual stresses in metal powder bed fusion," *Additive Manufacturing*, vol. 27. Elsevier B.V., pp. 131–149, May 01, 2019. doi: 10.1016/j.addma.2019.02.020.

[4] S. Murchio *et al.*, "Additively manufactured Ti–6Al–4V thin struts via laser powder bed fusion: Effect of building orientation on geometrical accuracy and mechanical properties," *J Mech Behav Biomed Mater*, vol. 119, Jul. 2021, doi: 10.1016/j.jmbbm.2021.104495.

[5] G. Strano, L. Hao, R. M. Everson, and K. E. Evans, "Surface roughness analysis, modelling and prediction in selective laser melting," *J Mater Process Technol*, vol. 213, no. 4, pp. 589–597, 2013, doi: 10.1016/j.jmatprotec.2012.11.011.

[6] G. Kasperovich, J. Haubrich, J. Gussone, and G. Requena, "Correlation between porosity and processing parameters in TiAl6V4 produced by selective laser melting," *Mater Des*, vol. 105, pp. 160–170, Sep. 2016, doi: 10.1016/j.matdes.2016.05.070.

[7] M. Park, M. P. Venter, and A. Du Plessis, "A lattice structure coupon sample for build quality control in metal additive manufacturing," *Mater Des*, vol. 235, Nov. 2023, doi: 10.1016/j.matdes.2023.112436.

[8] M. Park, M. P. Venter, and A. Du Plessis, "Simulation of the Compression Testing of Additively Manufactured Lattice Structures Using Inputs from Microcomputed Tomography," *Material Design & Processing Communications*, vol. 2023, pp. 1–15, Nov. 2023, doi: 10.1155/2023/8000727.

[9] J. Parthasarathy, B. Starly, S. Raman, and A. Christensen, "Mechanical evaluation of porous titanium (Ti6Al4V) structures with electron beam melting (EBM)," *J Mech Behav Biomed Mater*, vol. 3, no. 3, pp. 249–259, Feb. 2010, doi: 10.1016/j.jmbbm.2009.10.006.

[10] S. Arabnejad, R. B. Johnston, J. A. Pura, B. Singh, M. Tanzer, and D. Pasini, "High-strength porous biomaterials for bone replacement: A strategy to assess the interplay between cell morphology, mechanical properties, bone ingrowth and manufacturing constraints," *Acta Biomater*, vol. 30, pp. 345–356, Feb. 2016, doi: 10.1016/j.actbio.2015.10.048.

[11] A. du Plessis, S. M. J. Razavi, and F. Berto, "The effects of microporosity in struts of gyroid lattice structures produced by laser powder bed fusion," *Mater Des*, vol. 194, Feb. 2020, doi: 10.1016/j.matdes.2020.108899.

[12] H. H. Cho, Y. Cho, and H. N. Han, "Finite element analysis for mechanical response of Ti foams with regular structure obtained by selective laser melting," *Acta Mater*, vol. 97, pp. 199–206, Jul. 2015, doi: 10.1016/j.actamat.2015.07.003.

[13] A. Ghosh, A. Kumar, X. Wang, A. M. Kietzig, and M. Brochu, "Analysis of the effect of surface morphology on tensile behavior of LPBF SS316L microstruts," *Materials Science and Engineering A*, vol. 831, Jan. 2022, doi: 10.1016/j.msea.2021.142226.

[14] X. Cao *et al.*, "Compression experiment and numerical evaluation on mechanical responses of the lattice structures with stochastic geometric defects originated from

additive-manufacturing," *Compos B Eng*, vol. 194, Aug. 2020, doi: 10.1016/j.compositesb.2020.108030.

[15] T. Trzepieciński, G. Ryzińska, M. Biglar, and M. Gromada, "Modelling of multilayer actuator layers by homogenisation technique using Digimat software," *Ceram Int*, vol. 43, no. 3, pp. 3259–3266, Feb. 2017, doi: 10.1016/j.ceramint.2016.11.157.

OPTICAL RAIL WHEEL WEAR MEASUREMENT UNIT

A.K. Phillips[1], P.D.F Conradie[2]

[1]Department of Mechanical and Mechatronic Engineering, Faculty of Engineering, Stellenbosch University;

21650055@sun.ac.za

[2]Department of Industrial Engineering, Faculty of Engineering, Stellenbosch University;

PieterC@sun.ac.za

Abstract: In the rail industry, the importance of the wheel-rail interface is widely acknowledged. Due to the high stresses in both the wheel and the rail material, surface wear, subsurface defects, and wheel profile wear occur in normal operation.

As part of the condition-based maintenance tactic, regular inspections ensure that railway wheels are within specified wear limits. Currently, inspections are done at set intervals and rely heavily on human involvement and manual equipment, which can lead to the late detection of defects with consequential effects. Therefore, more frequent contactless inspection on railway wheels is needed.

This paper uses a high-performance class 3R laser scanner to detect defects and wear on the railway wheel. The laser uses the principle of triangulation, and the Scheimpflug principle to replicate the incident profile on a sensor matrix. The focus is on the wheel rim, where most defects and wear occur. Single and multiple two-dimensional images are generated to identify and quantify the wear. The optical rail wheel wear measuring and detection unit is tested on a half-scale 3D printed wheel model. Defects from the BS EN 15313:2016 standard are replicated on the scale model, and the unit is used to identify the defects.

The research results could potentially be used in a trackside monitoring system that can inspect the profile of a railway wheel while the train is moving at low speeds.

Keywords: rail wheel, condition-based maintenance, laser scanner, wheel profile measurement

1. Introduction

Due to the high stresses in the material near the rail-wheel interface, surface wear is inevitable during normal operation, (Tunna, Sinclair, and Perez, 2007). To ensure that the wear on rail wheels remains within acceptable parameters, regular condition assessments are performed at designated intervals. Presently, the majority of railway operators in South Africa conduct inspections primarily through visual examination of wheel profiles, supplemented by tools like the Field Gauge and MiniProf, which necessitate the railcar to be stationary. However, the accuracy of these assessments relies heavily on the interpretation of measurements by the operator and, in the case of MiniProf, the effectiveness of its roller's contact with the rail wheel. This dependence raises the possibility of misidentifying defects, potentially leading to unplanned maintenance, downtime, increased labor, or even catastrophic failures such as derailments.

Therefore, there is a pressing need for a more reliable non-contact monitoring system that minimizes reliance on human intervention and ideally allows for wheel inspection while the train is in motion. The proposed system aims to reduce human involvement during operation by automatically capturing predefined wheel profile measurements. Besides assessing the wheel profile without physical contact, the system should evaluate whether the wear meets prescribed standards and regulations for rail wheels and subsequently relay the measurements to maintenance personnel. This approach would allow more frequent inspections with improved accuracy and consistency. Consequently, implementing this enhanced system would lead to reduced maintenance expenses, mitigate the risk of failures, streamline the measurement process for operators, and enhance overall safety by minimizing human error and ensuring a wider and more consistent range of measurements.

2. Wheel profile parameters

Defined parameters for wheelsets enable precise evaluation of the condition of rail wheels. These parameters and associated terminology are largely consistent on an international scale. However, for this study, the terminology is drawn from the BS EN-15313:2016 (BSI Standards Limited, 2016). Figure 1, along with Table 1, provides a summary of the definitions pertinent to the rail industry. Presently, in South Africa, the datum point, $D0$, is 82.5 millimeters from the internal face of the wheel. For this study, a value of 70 millimeters will be utilized to align with the BS EN-15313:2016 standard. The numerical designations one through three in Figure 1 correspond to the three regions of the rail wheel profile: one representing the wheel rim, two denoting the wheel flange, and three indicating the tread surface of the wheel.

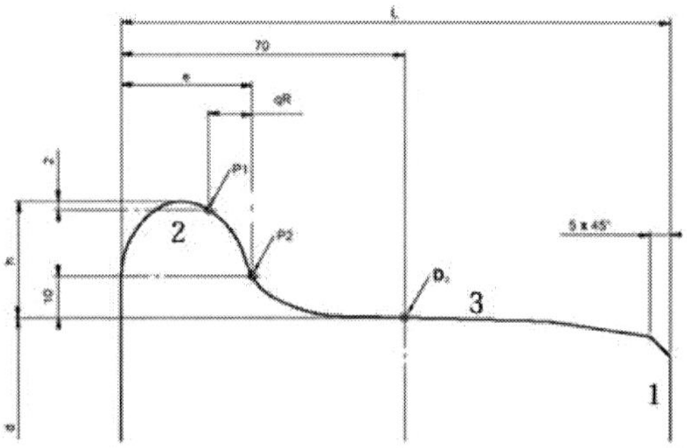

Figure 1-Wheel profile parameters Adapted from (BSI Standards Limited, 2016)

Table 1- Summary of wheel profile parameters

Parameter	Definition
D_0	The location of the wheel tread, 70 mm from its internal face
d	The diameter of the wheel
h (Sh)	The height of the flange
qR	The angle of the flange measured in mm
e (Sd)	The thickness of the flange
L	The width of the rim

3. Technology selection

Laser Light Technology (LLT) and laser scanners were evaluated for accomplishing the measurement task. LLT utilizes laser light for assessing the profile of an object by projecting a band of laser infrared light onto it, which is then observed by a camera to triangulate the angle and distance, i.e., location, of numerous discrete points on two-dimensional surfaces. The laser profile scanner conducts all necessary triangulation calculations within its software, resulting in plotted points of the incident surface with respect to a fixed two-dimensional coordinate system corresponding to the laser's position.

Utilizing high-precision equipment to develop the scanner ensures accurate determination of an object's distance and position information. Additionally, a sensitive sensor matrix allows for high-resolution two-dimensional scans, and when combined with laser light software, facilitates achieving high sampling rates due to the technology's inherent nature.

The LLT profile scanner is integrated with scanCONTROL software, which is programmable, enabling users to scan for various measurement scenarios, configure sensor settings via the developer tool, and trigger measurements on events. Determining wear on a wheel profile becomes highly likely when this technology is integrated with analysis software like MATLAB, which is essential for achieving enhanced accuracy and consistency in profile measurements.

The software facilitates the setup of measurement programs for different scenarios and enables exporting measurements to analysis tools such as MATLAB in a formatted CSV file. Within the analysis software, wear limits can be defined, and scanCONTROL also allows triggering measurements when a measurement exceeds its limit, which can be set up by the user for each measurement program.

4. Methodology

To facilitate testing of the system, an experimental design is being developed to replicate the railway track environment closely. This involves simulating typical rail wheel defects, wheel movement, and the positioning of the profile scanner. Initial testing was conducted in a controlled laboratory setting to minimize costs and delays associated with field visits. The insights gained from these preliminary tests will inform the final field-testing process.

The experimental setup consisted of constructing a rotating rail wheel to be scanned by a laser profile scanner. A frame made of 20mm x 20mm aluminum extrusions provided a solid frame for mounting the equipment. A motorized axle, driven by a timing belt connected to the motor's gear, facilitates rotation. A gear ratio of 2 was selected based on torque requirements, with the tooth counts for the driver and driven gears calculated accordingly. Flange pillow block bearings were utilized to secure the bearings to the aluminum extrusion.

For motorization, a stepper motor was chosen due to its cost-effectiveness and compatibility with a Raspberry Pi interface. The Raspberry Pi 3 model B+ was selected as the microcomputer for its availability. To connect with the stepper motor, a TMC 2160-OC stepper motor driver was employed.

A 3D-printed wheel was created (Figure 2) to replicate various defects outlined in the EN-15313 standard, focusing on wheel profile and surface imperfections. Using PRASAs Profile 22 as a reference, a wheel with an 840-millimetre diameter was designed in CAD software. Each defect was incorporated within the limits set by the EN-15313 standard to ensure detectability by the profile scanner. Additionally, defects related to tread surface, flange, and rim damage were added to the standard wheel profile. To streamline the printing process, the wheel model was scaled down by a factor of 2, reducing the diameter to 420 millimetres, along with scaling down all defect limits accordingly.

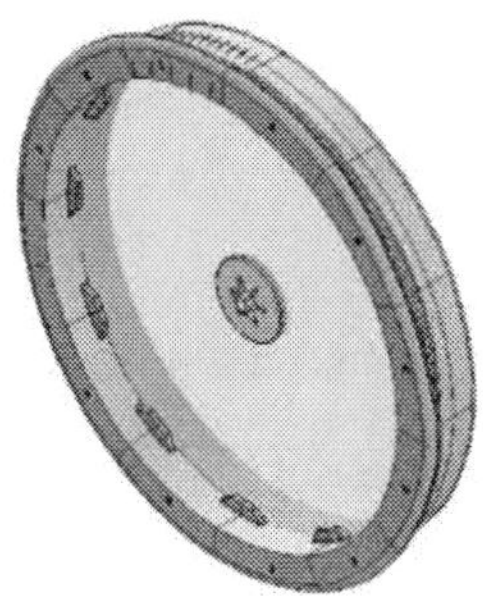

Figure 2-Wheel design

Printing the wheel required segmenting it due to its large volume. The Creality Halot Sky resin 3D printer was chosen for its superior surface finish and detail accuracy. Careful attention was paid to resin viscosity and layer resolution to accurately replicate the designed defects. The wheel was divided into 12 sections, each containing unique defects, totalling 16 defects overall. To secure the sections, a design featuring lip and groove connectors was adopted, reducing material usage and weight by approximately 40%. Additionally, protruding surfaces were added to ensure concentric alignment between sections.

Each section was marked with unique identification for assembly, while a machined Plexiglass sheet served as the wheel's web. Connection to the rotating mechanism was facilitated by a 3D-printed hub connector designed for load distribution and structural integrity. The hub's design focused on maintaining constant thickness and utilizing fatigue-resistant resin to enhance load-bearing capacity while minimizing net volume. A grey resin colour was chosen to improve defect visibility and reduce surface reflectivity for accurate profile scanning, with sensor exposure calibrated based on ambient lighting conditions and wheel reflectivity. Figure 3 shows the final experimental design.

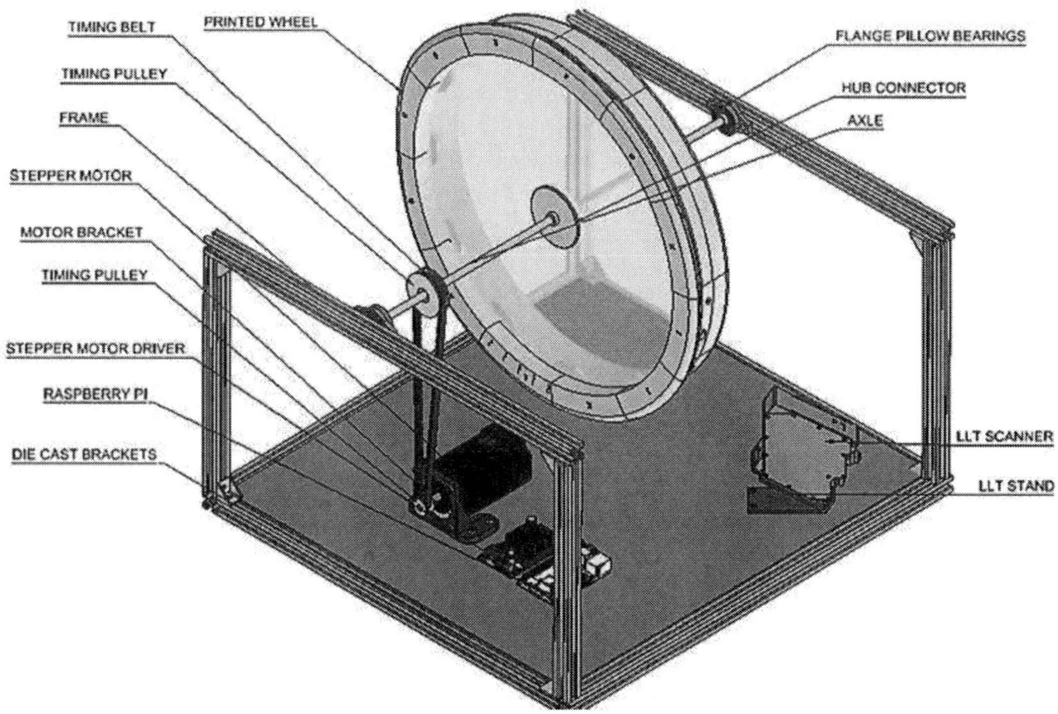

Figure 3- Final experimental design

Using the designed and implemented model, the profile scanner underwent calibration, utilizing user-defined regions of interest and reference lines relative to the laser's angle and offset. Once calibration was completed, measurement programs were established for each wheel profile defect.

To enable defect detection by the profile scanner, specific measurement programs within scanCONTROL were selected and configured. The software utilized regions of interest and references to set up these programs, with a focus on measuring wheel profile defects. Subsequently, programs for detecting surface damage were configured.

After configuring the defect measurement programs, a data acquisition system (DAQ) was implemented for communication. The QuantumX MX840 served as the DAQ, with a gateway facilitating communication between the LLT profile scanner and the DAQ. This gateway incorporated an Ethernet Fieldbus Coupler, an OU Filter module, a 4-20 mA Analogue I/O Module, and a Bus Termination.

The laser profile scanner transmitted results via Ethernet to the Ethernet fieldbus coupler, which then sent them to the analogue output module. The analogue output current was subsequently conveyed to the DAQ through 15-pin D-sub connectors. Within the DAQ, the analogue signal underwent conversion to digital using the integrated 24-bit analogue-to-digital converter. This process conditioned, filtered, and digitized the analogue data. The digital measurements were then accessible to the user via CatmanAP software, where visualizations, measurement procedures, and storage settings were configured. The software was set up for real-time data monitoring, with the capability to trigger data storage and alert the user upon defect detection. The DAQ continuously measured data and saved it upon triggering. The final experimental setup, depicted in Figure 4, comprised the motorised printed wheel, the laser line sensor, a laptop, and the DAQ.

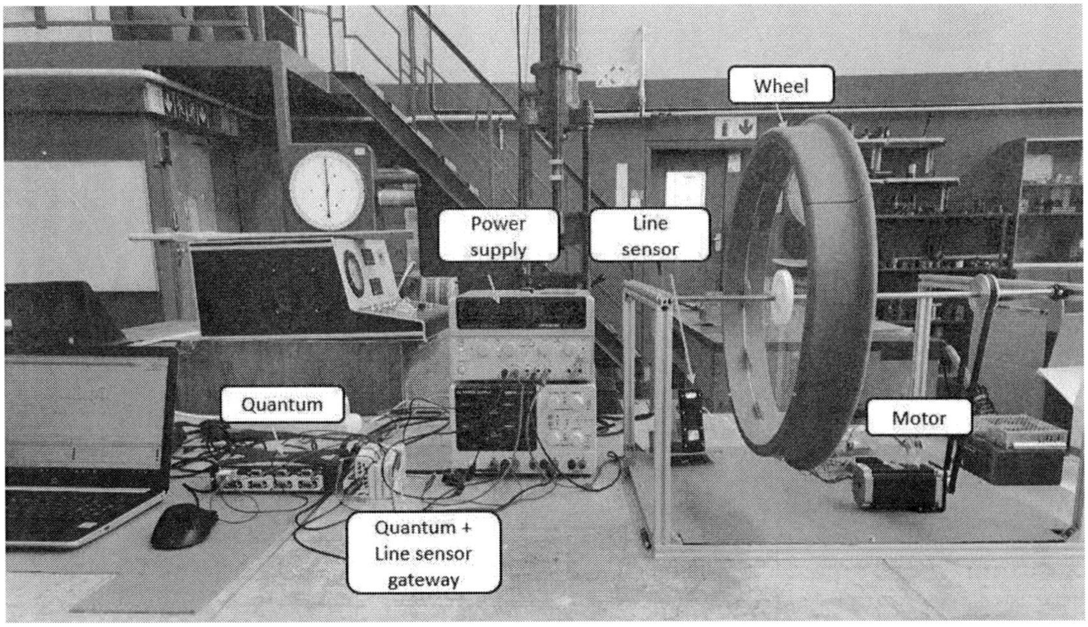

Figure 4-Final wheel design

5. Lab Results

Table 2 presents a comparison between the defect limits designed in CAD software and their measured values by the laser profile scanner while the wheel rotated at 5 km/hr. Of the 16 defects, 13 were detected. The table demonstrates a strong correlation between the set and measured values, with a maximum variation of approximately 0.3 millimeters. This indicates the profile scanner's ability to operate dynamically using predefined measurement programs in a controlled environment. However, further development and consideration are required for detecting flat spots, as the current configuration heavily relies on the wheel's rotational speed.

Table 2- Designed defects and their measured values

Section	Defect	Set value [mm]	Measured value [mm]	Accuracy [%]
A	Flange angle	4.15	4.046	97.49
B	Flange height min	12.591	12.521	99.44
C	Flange height max	18.667	18.354	98.32
D	Flange thickness min	13.226	13.027	98.50
E	Flange thickness max	18.25	18.014	98.71
F	Rim width	63	62.841	99.75
	Hollow wear	1.128	0.91	80.67

G	Standard profile	NA	NA	NA
H	Flange damage	NA	Not detected	Not detected
I	Flat spot	16	Detected for one speed	Detected for one speed
J	Groove defect	2.4	2.403	100.13
K	Tread damage (metal loss)	1.01	Detected	Detected
	Tread damage (metal build-up)	0.89	Detected	Detected
	Rim damage	1.5	Not detected	Not detected
L	FIJ/FEJ cracks	NA	Detected	Detected
	Radial cracks	NA	Detected	Detected

Profile scans of the wheel were captured as the wheel completed one full revolution, with the DAQ recording these measurements. The resulting output is displayed in Figure 5. The maximum values observed in the graphs correspond with both the maximum values set in CAD and those measured in scanCONTROL. The DAQ operates as expected, effectively communicating, storing, and visualizing the data for the user.

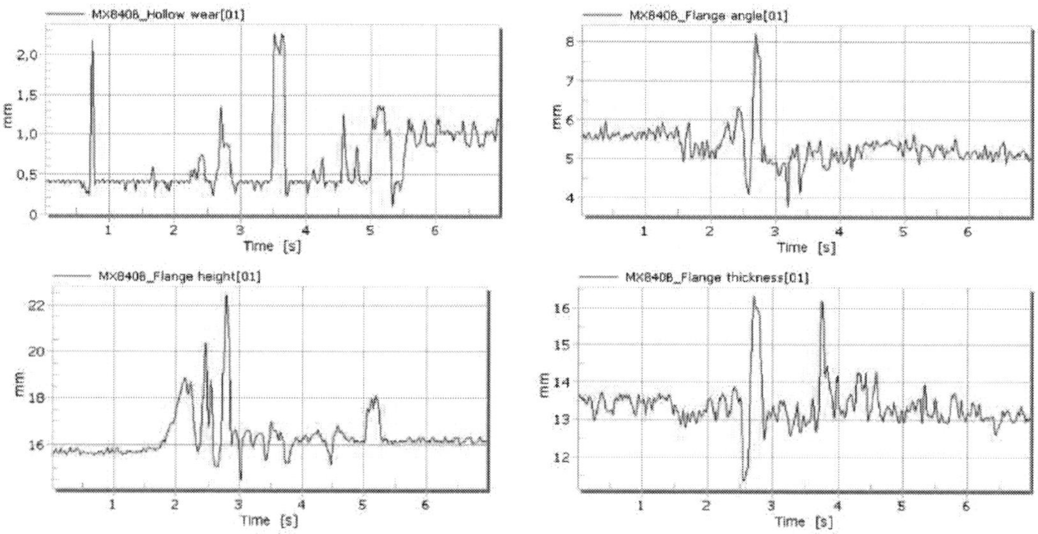

Figure 5- DAQ results for measures A, B, C, and F in Table 2.

The comparison between the two profiles is illustrated in Figure 6. While the profiles align along the tread surface and the rim, there is a discrepancy in the alignment of the flange tip concerning the CAD file, particularly regarding height and thickness. Possible reasons for this misalignment could include the offset and angle at which the laser is positioned relative to the wheel.

An attempt to create a 3D wheel map is depicted in Figure 6. This 3D wheel map was constructed as a visualization tool, intended to aid in understanding spatial data and improve communication. Its purpose is to identify trends and wear anomalies in the data.

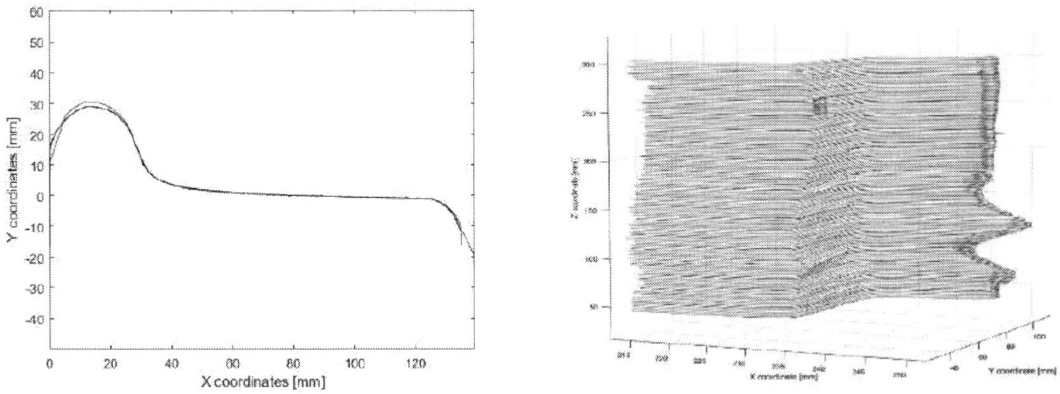

Figure 6-Matlab analysis

6. Field Testing

Figure 7-Laser position

To validate the profile scanner, a defective wheelset was measured, and the results were compared against the results obtained from a MiniProf. The physical system used consisted of a laptop, the LLT profile scanner as well and two defective wheelsets. Four-wheel profile parameters were measured. These parameters are most measured within the rail industry and are measured by the MiniProf. The position of the laser can be seen in Figure 7.

To adhere to the testing standards laid out by the EN Standard, a minimum of three different wheel profiles were taken at separate locations on one wheel to get a comprehensive report of the wheel profile. The same parameters are measured by the MiniProf.

In addition to the two sets of laser scans, a 3D point cloud of one of the wheels was created. The 3D scan was sectioned and measured in Autodesk Inventor to determine the flange profile parameters as several data points were missing for hollow wear detection.

The comparison of the MiniProf, laser tests, and 3D scanned data was graphed. The results of the flange height can be seen in Figure 8. The data is largely consistent for all scans for the laser and MiniProf. A 99.43% accuracy was found between Scan 2 for the second laser scan and the MiniProf. The flange thickness results of the different scans taken by the laser profile scanner differ by a narrow margin, as seen in Figure 9.

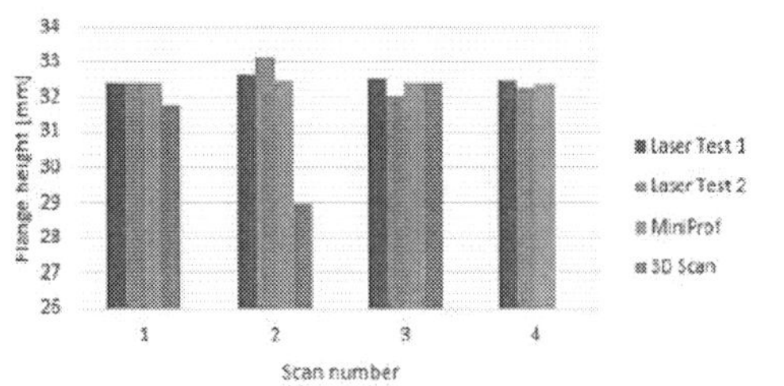

Figure 8-Comparison of flange height (Sh) measures

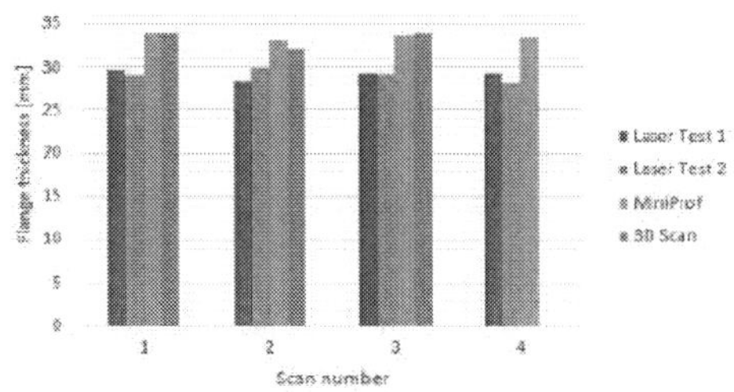

Figure 9- Comparison of flange thickness (Sd) measurements

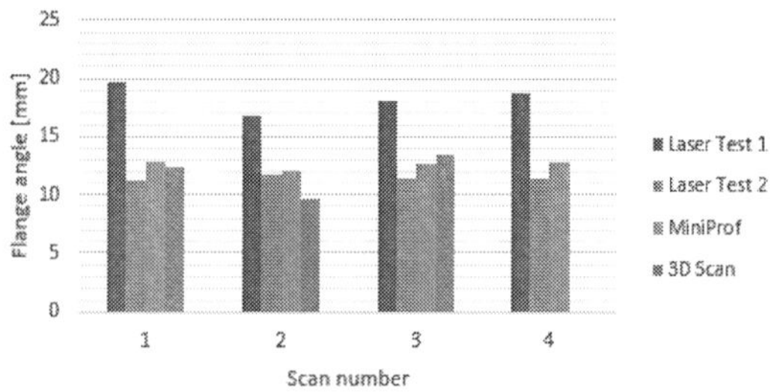

Figure 10- Comparison of flange angle (qR) measurements

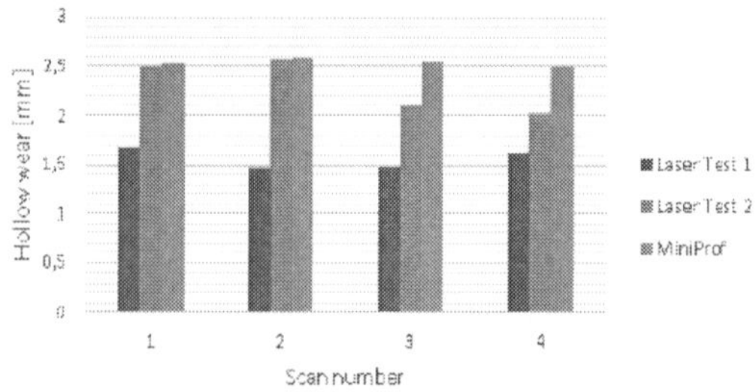

Figure 11- Comparison of hollow wear measurements

The flange angle measurement comparisons can be seen in Figure 10. The flange angle measurements are much more consistent with MiniProf in the second test with the least accurate measurement being 87.29% accurate. The 3D scans approximately illustrate the same results. The comparison of the hollow wear results for test 2 can be seen in Figure 11. The measurement program was updated to span to the end of the tread surface. The results obtained are consistent with the measurements from the MiniProf.

In the first set of laser scans, data collection errors occurred. The error that occurred can be related to misconfiguration of the measurement program, change in geometry of the wheel section and, lastly, the positioning of the laser in the case of flange thickness. However, there were strong correlations between the laser and MiniProf data for flange height with the highest accuracy being 99.91% accurate. Included in the second set of scans was the rim thickness and rim width. It was determined that further study is needed to accurately measure the wheel diameter. Overall, the results of the second set of scans are more consistent with the readings taken by the MiniProf and the 3D scans verified the results. An overall average calculation of all accuracies was taken. The average accuracy was found to be 92% when using the MiniProf measurements as the reference.

7. Conclusions

An experimental setup was devised, featuring a rotating 3D-printed wheel housing 16 defects for scanning by the laser profile scanner. The aim was to assess the scanner's ability to dynamically scan the wheel in a controlled setting, mimicking trackside conditions. Laboratory tests revealed a strong correlation between measured and designed values, validating the scanner's functionality in an idealized non-contact, dynamic manner. Integration of a data acquisition system facilitated communication and storage of measurements.

To validate the prototype, a removed-from-service wheelset with defects was scanned by the LLT and compared to MiniProf results. Some inconsistencies were noted, attributed to LLT configuration errors or positional inaccuracies. A second test on the same wheel confirmed the need to optimize scanner positioning for comprehensive flange scanning. Results from the second test aligned closely with MiniProf measurements, with minimal variation, particularly in flange height. Additional measurements provided insight into rim thickness, width, and wheel diameter, showing slight disparities between MiniProf and LLT measurements, mainly in wheel diameter.

A 3D point cloud of the wheel was captured and compared to MiniProf and LLT measurements, exhibiting alignment with MiniProf data and approximate correspondence with LLT results. Acceptable differences between LLT and point cloud scans validated the latter as a verification tool. The overall accuracy percentage was found to be 92%.

Recommendations for prototype improvement included refining scanner positioning and considering MATLAB as a primary processing unit from exported wheel profile coordinates, potentially enhancing defect detection capabilities. Furthermore, research into using a dedicated server or other cloud-based solutions should be conducted. With the addition of remote monitoring, a stand-alone device can be achieved.

Funding

This research was funded by the Gibela Rail and Transport Consortium.

Credit authorship contribution statement

A.K. Phillips: Conceptualization, Methodology, Investigation, Writing - original draft.

P.D.F. Conradie: Supervision, Writing - review & editing.

References

BSI Standards Publication. (2016). Railway applications — In-service wheelset operation requirements — In-service and off-vehicle wheelset maintenance, BS EN 15313:2016, pp 18-104.

Tunna, J., Sinclair, J. and Perez, J. (2007). A Review of wheel wear and rolling contact fatigue. Proceedings of the Institution of Mechanical Engineers, Part F: Journal of Rail and Rapid Transit, 221(2), pp.271–289. doi: https://doi.org/10.1243/0954409jrrt72

Enhanced Data-Driven Hybrid Approach for Unavailable Damage Scenarios in Welded Rail Tracks with Multiple Reflections and Reflectors for Ultrasonic-Guided Wave Inspection

Dineo A. Ramatlo[1*], Daniel N. Wilke[1] and Philip W. Loveday[2]

[1]Department of Mechanical and Aeronautical Engineering, University of Pretoria, Pretoria, 0002, South Africa
[2]School of Mechanical, Industrial and Aeronautical Engineering, University of the Witwatersrand, Johannesburg, 2000, South Africa

* dineo.ramatlo@up.ac.za

Abstract

This paper addresses the challenge of lacking experimental data for damage in ultrasonic-guided wave monitoring systems, particularly for railway tracks. Laboratory experiments are often impractical for rail tracks, leading to the necessity of synthetic data generation. This study builds upon a deep learning approach using a variational autoencoder (VAE) model to create synthetic ultrasonic inspection data. This data mimics real-world scenarios of welded railway tracks with multiple reflections and reflectors, including damage signatures at specific positions. The model is trained to transform a damage-free baseline signal into an inspection signal indicating damage by incorporating a crack signature in the rail. Training data was derived from a physics-based model, employing Semi-Analytical Finite Element (SAFE) models and traditional finite element methods to simulate virtual experimental response signals. The results demonstrate the feasibility of generating realistic synthetic inspection data for scenarios where actual damage data is unavailable. This advancement is pivotal in developing more reliable ultrasonic-guided wave monitoring systems for rail infrastructure.

Keywords

Ultrasonic guided waves, Deep learning, Synthetic data, Damage, Welded rail track

1. Introduction

Guided Wave Ultrasound (GWU) has become a prevalent method for structural health monitoring, particularly in pipelines and railway lines. Despite its potential, challenges exist in detecting early-stage damages, especially under moving trains, which can lead to derailments. To address this, the Ultrasonic Broken Rail Detection system, utilizing piezoelectric transducers, has been developed to detect complete rail breaks by concentrating energy in the rail head [1,2]. However, early detection of initial damage, crucial for preventing derailments, remains challenging.

Liu et al. demonstrated the use of ultrasonic-guided waves and Receiver Operator Characteristic curves (ROCs) for detecting small damages [3-4]. However, ROCs require extensive monitoring data under various Environmental and Operational Conditions (EOCs), including different stages of damage evolution. This data is challenging to acquire for rail tracks, as damaged sections are replaced immediately, and end reflections from short rail sections limit laboratory experiments.

An alternative approach involves generating synthetic data through physics-based models, using methods like the Finite Element Method (FEM) to model damage reflections. This synthetic data, fused with experimental data collected from undamaged structures over multiple environmental cycles, can overcome data scarcity [4,5]. Digital twin frameworks, such as the Semi-Analytical Finite Element (SAFE) method combined with traditional FEM, as proposed by Ramatlo et al. [6], have been instrumental in simulating realistic virtual guided wave inspections in multi-discontinuous waveguides.

The traditional superposition method for generating synthetic data is insufficient in welded rails, where complexities like dispersion behaviour, modal interactions, and overlapping reflections from various sources are prevalent [5,6,8]. This method falls short due to the intricate nature of rails, which often feature multiple discontinuities, causing double reflections and complicating signal interpretation. Superposition struggles particularly with small or interacting damage reflections and when the transmitted signal is affected. To overcome these limitations, advanced machine learning techniques, such as Variational Auto-Encoders (VAE), are utilised [8]. VAEs are more adept at handling the complexity of welded rails, as they can accurately identify and replicate the nuanced pattern changes caused by damage, making them ideal for fusing synthetic and experimental data in such intricate scenarios.

Ramatlo et al. introduced a hybrid approach, merging a physics-based model [6] with an enhanced Variational Auto-Encoder (VAE) [8] to fuse real and simulated damage signatures in field data. This advanced VAE maps input to low-dimensional outputs and intricately adds new features to the input signal. It is trained using physics-based models to process baseline signals, including reflections from welds, and augment them with simulated damage reflections at specific locations. This method enriches the input data with detailed defect information, enhancing the synthetic data's realism for structural health monitoring.

This paper explores the application of this deep learning strategy [8] to generate synthetic ultrasonic inspection data for rail tracks where traditional data is lacking. It focuses on operational rails with multiple welds. It employs a VAE trained on physics-based model data [6] to encode a damage-free baseline signal and decode it with a crack signature at a specific location. The results are compared to those obtained from traditional superposition methods, demonstrating the effectiveness of the data-driven hybrid approach in simulating realistic damage scenarios for advanced structural health monitoring.

The virtual experimental data used to train the VAE is described in section 2. Section 3 explains the data-driven hybrid approach based on a VAE model. Results are presented in section 4, and the final section gives the conclusions.

2. Simulation of a guided wave virtual inspection in the head of a rail

2.1. Baseline signal for a damage-free rail

A field experiment was conducted on a UIC60 rail, consisting of 240m sections connected by four aluminothermic welds [1]. Guided waves were generated in the rail using a piezoelectric transducer, placed 78m from the closest weld, which emitted a 17.5-cycle Hanning windowed tone burst signal with 35kHz centre frequency. The waves travelled in both directions along the rail, with responses captured by the same transducer. The setup of this experiment is detailed in Figure 1.

The training data for the VAE was derived from a digital model simulating the field scenario shown in Figure 1, based on a physics-validated framework detailed in [6]. Figure 2 illustrates this specific digital model setup. For more comprehensive information, readers are directed to references [6,9].

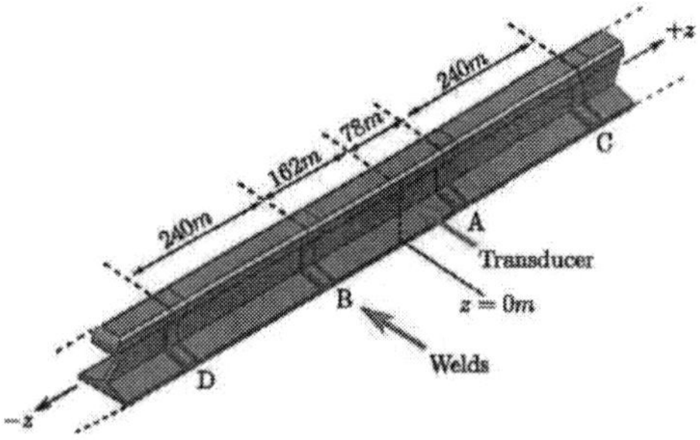

Figure 1. Schematic representation of a section of the field operational rail containing four welds.

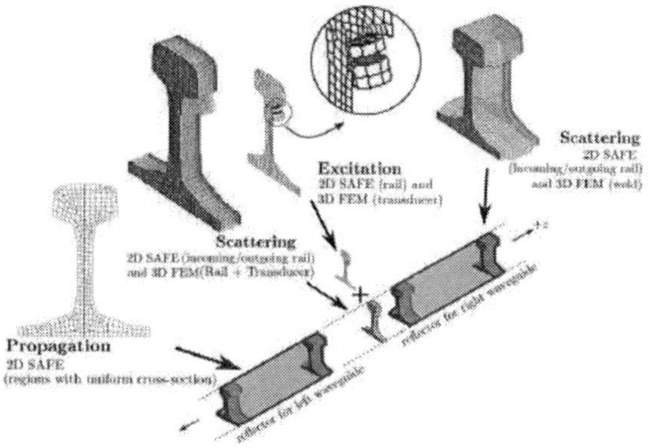

Figure 2. Digital model of the considered section of an operational rail.

The modelling framework initially focuses on the components of the inspection setup. It simulates guided wave excitation using a 3D Finite Element Model (FEM) of the piezoelectric transducer, integrated with a 2D Semi-Analytical Finite Element (SAFE) model for constant cross-section rail regions. For guided wave scattering at welds, a hybrid approach combines a 3D FEM of the weld with two SAFE models, representing incoming and outgoing waveguides. The framework then integrates these elements using Baronian et al.'s method to capture complex reverberations from multiple reflections [8]. This involves linking local discontinuities (welds) to global reflectors on each side of the transducer and connecting these reflectors with the hybrid models to simulate and compute the guided wave reflections received.

Figure 3 presents a spectrogram and corresponding distance domain signals of the simulated virtual experimental response, validated against measurements in [7]. The spectrogram reveals multiple propagating modes reflected from each weld, with a notable double reflection occurring when the waves reverberate between welds A and B. The VAE model training and testing will utilise these distance domain signals with logarithmically scaled amplitudes (as shown in Figure 3(c)), enhancing the visibility of the response signal's features.

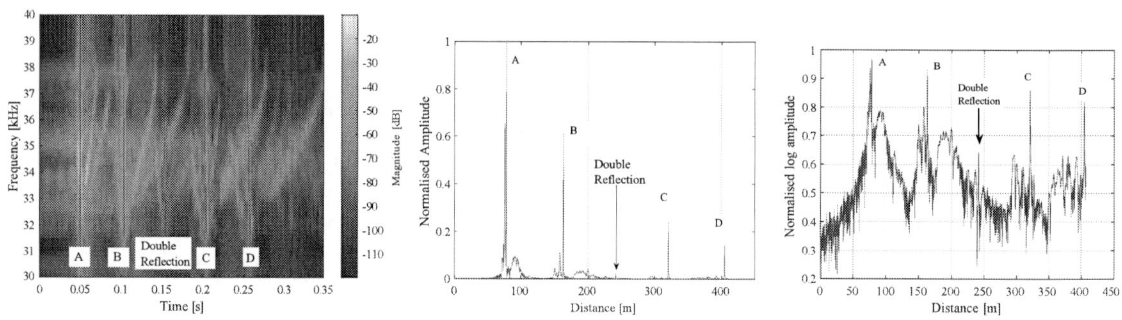

(a) (b) (c)

Figure 3. (a) Spectrogram of the simulated baseline signal for the experimental response and (b)&(c) the equivalent distance domain signals.

2.2. Transverse damage in the head of the rail

Guided wave scattering from rail damage was modelled similarly to weld scattering. Two SAFE cross-sectional meshes flanked a 3D mesh representing a transverse crack in the rail head. Standard displacement-based elements modelled the crack, as shown in the hybrid model of an 8mm radius circular crack in Figure 4. This model simulated inspections for damage at various rail positions.

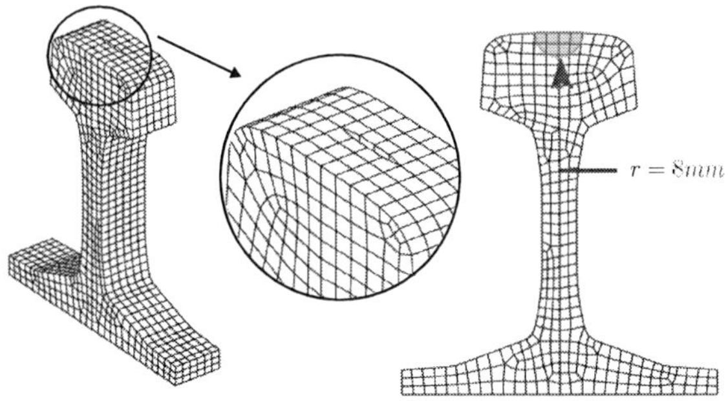

Figure 4. A hybrid model of the defect geometry with an 8mm radius crack in the head of the rail.

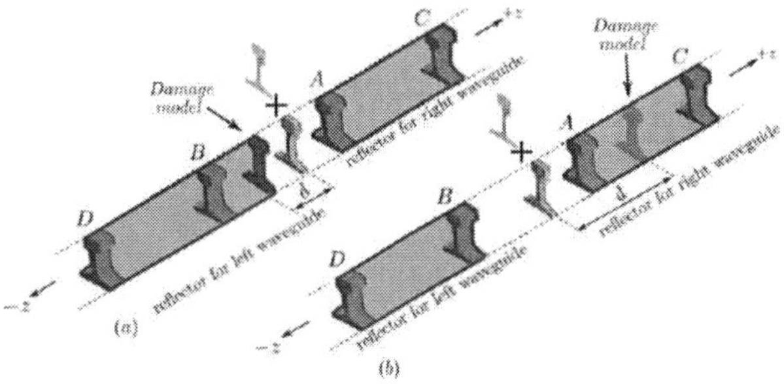

Figure 5. Digital models of the considered section of an operational rail with fake damage at specified distances *d*.

2.3. Virtual experimental response with damage in the head of the rail

Incorporating the crack model from Figure 4 into the operational rail's digital model (Figure 2), virtual responses with damage were simulated as per section 2.1. This involved models of four welds (as in Figure 2) and a variable-position crack, demonstrated in Figure 5 with two damage scenarios. Figure 6 shows the varied simulated response signals resulting from different crack locations.

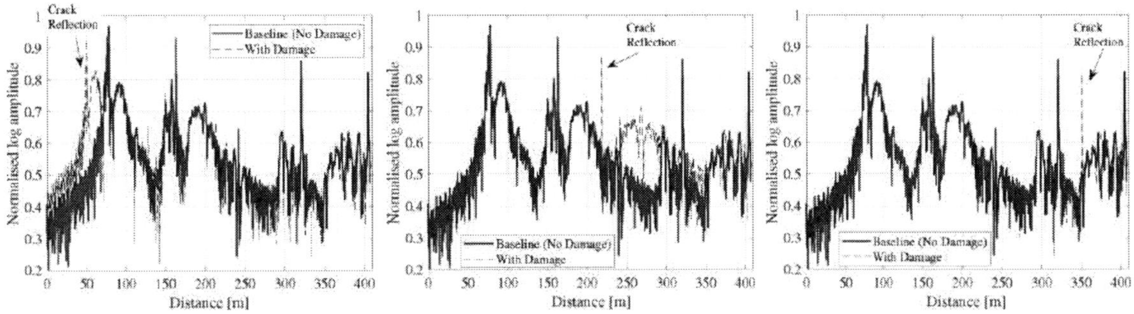

Figure 6. Virtual experimental response signals (**y***i*) containing damage reflections at user-specified distances of (a) 50m, (b) 220m, and (c) 350m.

3. Enhanced data-driven hybrid approach for fusing damage signatures with virtual experimental data

Synthetic damage data was created using a hybrid approach, combining superposition with a physics-based VAE model integration, as shown in Figure 7. This process involved preparing two training data sets: one with baseline, damage-free signals (Section 2.1, Figure 3(c)) and another with signals containing damage reflections at various distances (Section 2.3). The VAE's structure, detailed in [8], includes designed encoder and decoder layers (Table 1), facilitating the fusion of field and simulated data.

In the second phase of the hybrid approach, the VAE is trained to create synthetic data with damage signatures at specified distances from the transducer. This involves learning to map baseline signals to target signals with damage. A dataset of 3173 paired baseline and corresponding damaged signals was simulated for this.

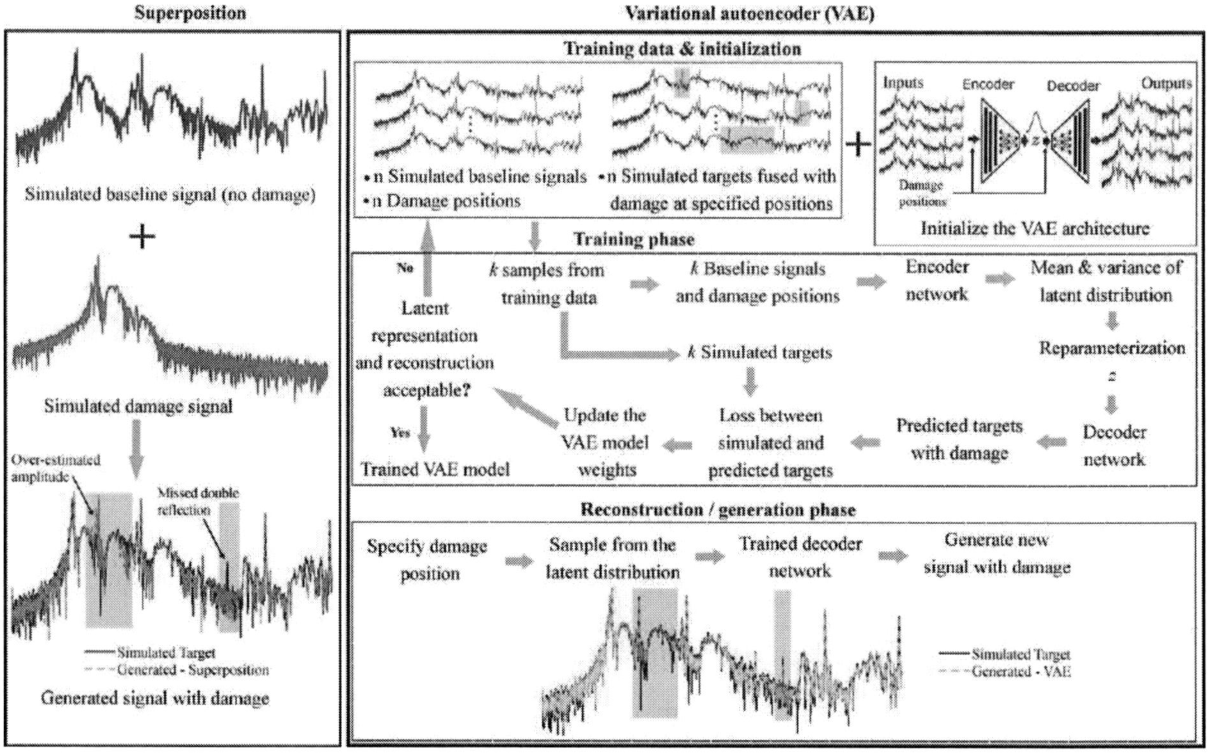

Figure 7. The generation of synthetic data containing damage signatures using the VAE and the superposition principle.

Table 1. VAE architecture

Layer	Layer description	Activation
	Encoder	
Embedding ×2	Embeddings for the distance and side indicator	-
Concatenate	To concatenate the baseline signals and the two embeddings	-
Time Distributed Dense	Fully connected; nodes = 1 for each time step	Sigmoid
Conv1D	Filters = 8; kernel size = 3; stride length = 2	eLu
Conv1D	Filters = 16; kernel size = 3; stride length = 2	eLu
Conv1D	Filters = 32; kernel size = 3; stride length = 2	eLu
Dense	Fully connected; nodes = 150	Sigmoid
Dense ×2	Latent dimension = 2	-
	Decoder	
Dense	Fully connected; nodes = 150	Sigmoid
Dense	Fully connected; nodes = 150×2×32	Sigmoid
Conv1DTranspose	Filters = 16; kernel size = 3; stride length = 2	eLu
Conv1DTranspose	Filters = 8; kernel size = 3; stride length = 2	eLu
Conv1DTranspose	Filters = 1; kernel size = 3; stride length = 2	eLu
Concatenate	To concatenate the output from the previous layer and the two embeddings	
Time Distributed Dense	Fully connected; nodes = 1 for each time step	Sigmoid

The VAE training used 70% of this data, with the remaining 30% for testing. In each training iteration, 128 pairs of baseline and target signals are selected. The encoder of the VAE compresses these baseline signals into a lower-dimensional space, represented by a distribution's mean and variance. This compressed data is then reparametrised and inputted into the decoder, which is trained to introduce damage into the baseline signals at the specified positions, aiming to replicate the target signals. The

VAE is trained by minimising the mean absolute error (MAE) between the original target $\boldsymbol{y}_i$ from Figure 6 and the approximation $\widetilde{\boldsymbol{y}}_i$:

$$VAE\ Reconstruction\ loss = \frac{\sum_{i=1}^{k} |\boldsymbol{y}_i - \widetilde{\boldsymbol{y}}_i|}{k} \quad (1)$$

Suppose the reconstruction ability of the VAE is unsatisfactory. In that case, the training phase is repeated with a different subset of training data to minimize the reconstruction loss measured by equation 1. Low reconstruction loss indicates better fidelity in reconstructing the input data.

Finally, in the last phase, the trained VAE model generates response signals with damage fused at user-specified positions.

4. Results

In Figure 8, response signals reconstructed using a VAE model are compared against those from superposition and the original target responses across various crack locations in the rail. While superposition accurately captures complex patterns, it overestimates amplitudes near damage reflections. Conversely, the VAE model replicates these patterns without such overestimation, closely matching the reflections from welds in the original simulated targets. This indicates the VAE's proficiency in generating synthetic data by blending weld and crack reflections.

Notably, the VAE predictions include double reflections resulting from damage interaction. For instance, in Figure 8(a), alongside the crack reflection at 92.6m, a double reflection at approximately 256m is observed, caused by reverberation between the crack and weld B, which the VAE accurately predicts. In contrast, limited to direct reflections, superposition misses this double reflection.

Furthermore, the VAE effectively captures overlaps in reflections when a crack is near a weld or on opposite sides but closely positioned. Figure 8(b) shows that a crack at 154.75m and weld A at 170.1m on the right of the transducer create an overlapping reflection well approximated by the VAE. The VAE model's success, indicated by a low reconstruction loss of 0.0093, contrasts with superposition's inability to replicate overlapping reflections in synthetic signals. accurately.

5. Conclusions

This study demonstrates the efficacy of a data-driven hybrid approach, integrating a physics-based FEM model with a VAE, in generating synthetic ultrasonic inspection data. The VAE model successfully reconstructed response signals with damage signatures that were closely aligned with the original simulations from the physics-based model. It excelled in capturing complex signal interactions within a multi-discontinuous waveguide, a significant improvement over the traditional method. Comparative analysis revealed the limitations of the superposition method, particularly its inability to capture double reflections from cracks and its tendency to overestimate reflection amplitudes near damages. In contrast, the VAE accurately replicated these complex signal features, precisely approximating overlapping reflections between closely positioned cracks and welds. The findings highlight the VAE's potential to generate realistic synthetic data in scenarios lacking actual damage data. Its proficiency in modelling intricate signal interactions in rail infrastructure signifies a notable advancement in ultrasonic inspection technology. The results affirm the VAE's capacity for producing highly accurate inspection data for challenging damage scenarios, contributing significantly to enhanced structural health monitoring in railway systems.

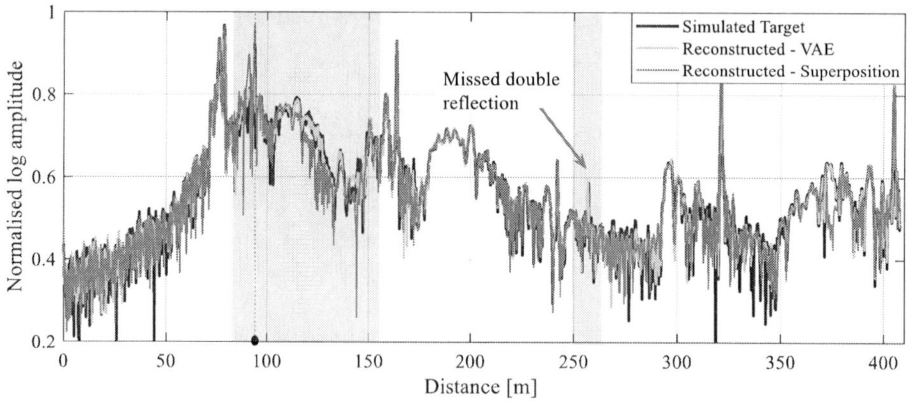

(a)

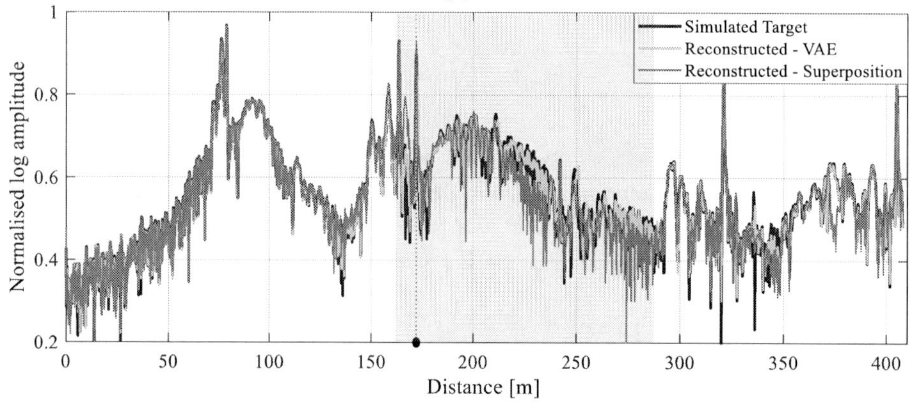

(b)

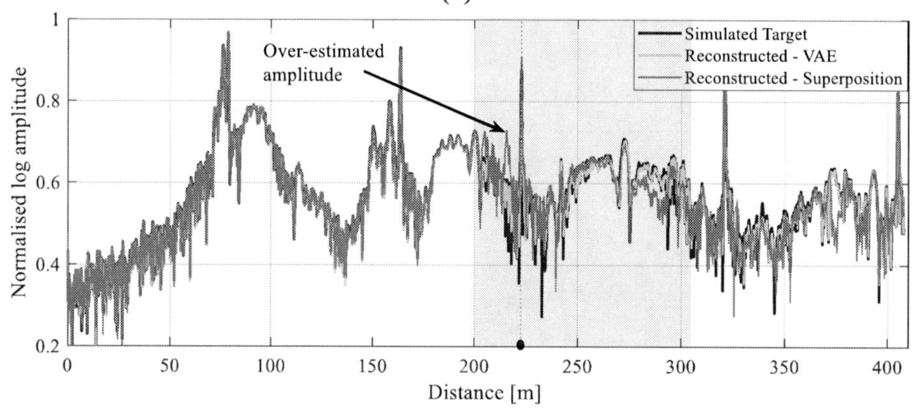

(c)

Figure 8. Response signals reconstructed from the VAE compared to superposition and the original target response for a crack located at (• · · ·) (a) 92.6m from the transducer on the left side, (b) 170.1m from the transducer on the right side, and (c)220m from the transducer on the left side. The area shaded in green indicates the region over which the energy reflected from the damage is significant.

References

[1] Loveday PW, Long CS and Ramatlo DA. Ultrasonic guided wave monitoring of an operational rail track. Struct Heal Monit 2020; 19(6): 1666–1684. DOI:10.1177/1475921719893887.

[2] Burger FA, Loveday PW and Long CS. Large scale implementation of guided wave based broken rail monitoring. In AIP Conf. Proc., volume 1650. ISBN 9780735412927, pp. 771–776. DOI:10.1063/1.4914679.

[3] Dobson J and Cawley P. Independent component analysis for improved defect detection in guided wave monitoring. Proc IEEE 2016; 104(8): 1620–1631. DOI:10.1109/JPROC.2015.2451218.

[4] Liu C, Dobson J and Cawley P. Efficient generation of receiver operating characteristics for the evaluation of damage detection in practical structural health monitoring applications. Proc R Soc A Math Phys Eng Sci 2017; 473(2199): 1–26. DOI: 10.1098/rspa.2016.0736.

[5] Mariani S, Heinlein S and Cawley P. Compensation for temperature-dependent phase and velocity of guided wave signals in baseline subtraction for structural health monitoring. Struct Heal Monit 2020; 19(1): 26–47. DOI:10.1177/1475921719835155.

[6] Ramatlo DA, Long CS, Loveday PW et al. Physics-based modelling and simulation of reverberating reflections in ultrasonic guided wave inspections applied to welded rail tracks. J Sound Vib 2022; 530(116914): 1–20. DOI:10.1016/j.jsv.2022.116914.

[7] Heinlein S, Cawley P and Vogt T. Validation of a procedure for the evaluation of the performance of an installed structural health monitoring system. Struct Heal Monit 2019; 18(5-6): 1557–1568. DOI:10.1177/1475921718798567

[8] Ramatlo DA, Wilke DN and Loveday PW. A data-driven hybrid approach to generate synthetic data for unavailable damage scenarios in welded rails for ultrasonic guided wave monitoring. Struct Heal Monit 2023;0(0). doi:10.1177/14759217231197265.

[9] Ramatlo DA, Long CS, Loveday PW et al. A modelling framework for simulation of ultrasonic guided wave-based inspection of welded rail tracks. Ultrasonics 2020; 108(106215): 1–18. DOI:10.1016/j.ultras.2020.106215.

[10] Baronian V, Lhémery A and Jezzine K. Hybrid SAFE/FE simulation of inspections of elastic waveguides containing several local discontinuities or defects. AIP Conf Proc 2011; 1335(June): 183–190. DOI:10.1063/1.3591855.

Computational modelling of alternating current conduction effects in packed beds

Quinn G. Reynolds[1,2,*] and Alfred E.J. Bogaers[3,4]
[1]Pyrometallurgy Division, Mintek, Randburg, South Africa, *quinnr@mintek.co.za
[2]Department of Chemical Engineering, Stellenbosch University, Stellenbosch, South Africa
[3]Ex Mente Technologies, Pretoria, South Africa
[4]Department of Mechanical and Aeronautical Engineering, University of Pretoria, Pretoria, South Africa

Abstract: Alternating current (AC) is used as a source of electrical power in a wide variety of household and industrial applications. In particular it is used to provide resistive heating in certain types of pyrometallurgical smelting furnaces, where AC is passed between large graphite electrodes mounted into refractory-lined vessels in order to melt raw materials and perform thermochemical reactions to manufacture metallic products such as ferro-alloys and other products. In order to heat the furnace current must pass through the charge burden, an electrically conductive bed of particulate matter. In the present work, the authors discuss the merits of different numerical approaches for solving AC electromagnetic field problems in both the time and phasor domains. One such method, the finite-volume time domain algorithm, is implemented in the OpenFOAM® computational mechanics framework and used to examine current distribution effects and overall impedance of packed beds of conductive particles. Regular particle packings are generated in pre-processing and used as a template for the creation of computational meshes and domains for the electromagnetic solver. The effect of parameters such as domain size, particle contact area, and electrical conductivity are investigated. The results are also compared to those obtained from continuum models, a commonly-used simplification in which the packed bed is approximated as a material with an average electrical conductivity.

Keywords: computational electromagnetics, electrical conduction, pyrometallurgy

1. Introduction

The invention of alternating current (AC) in the early 20th century revolutionised the generation, transmission, and application of electrical power in modern society and industry. Today we take for granted the availability of AC grid electricity in a variety of forms, and it is used to power everything from household appliances to large-scale industrial equipment. In particular it is used as a power source for metallurgical submerged arc furnaces (SAF), which are ubiquitous worldwide in the production of economically-important materials such as ferrochromium, ferromanganese, silicon, platinum group metals, and many others [1].

SAFs are available in two main designs, circular or rectangular. In both cases, each phase of a three-phase AC power supply is connected to one or more vertical graphite electrodes which enter through the roof of the furnace. Electric current passes between each pair of phases via the contents of the furnace, heating it to temperatures of 1500°C or greater in the process. This causes the raw materials fed to the unit to melt and react thermochemically to generate the desired products. It is not uncommon to have multiple different current paths between electrodes due to the heterogeneous structure of the furnace interior – in a typical SAF, the bath of molten products is covered by a deep particulate layer of solid raw materials (the "burden") which is continuously replenished from feed ports in the roof and consumed in the hot zones around the electrode tips. A schematic of the interior of a circular SAF used in ferro-alloy production is shown in Figure 1. Inter-electrode conduction can occur via the molten bath of alloy and slag at the base of the furnace, or through the burden via electrically conductive components of the solid raw materials. Depending on the metallurgical process electric arcs may also be present in small spaces between the electrode tips and the surrounding material, and act in parallel or series with the main resistive paths [2].

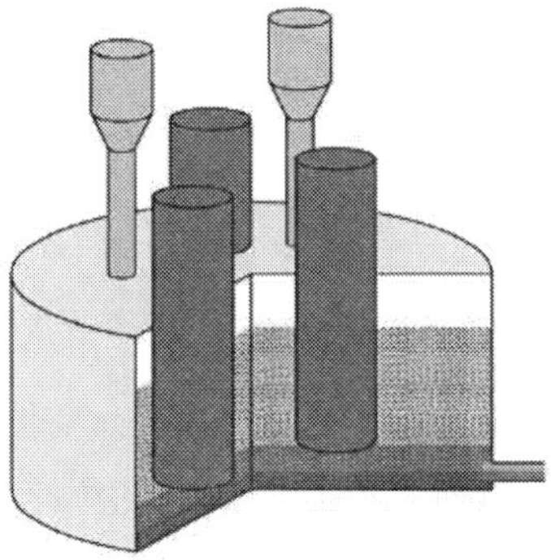

Figure 1: Schematic diagram of a submerged arc furnace unit

Distributed-parameter numerical and computational modelling of electric smelting furnaces requires a multiphysics approach due to the phenomenological complexity of the real world process. In particular, a sub-model for the electromagnetic fields inside the vessel is needed in order to provide accurate source terms for ohmic heating, magnetohydrodynamic fluid flow, and other effects. Electromagnetic fields are governed by Maxwell's equations [3] – in the low- to medium-frequency domain these can be simplified and expressed succinctly in terms of electric and magnetic potentials ϕ and $\mathbf{A}$ as shown in (1) and (2). Here, the gauge invariance has been fixed using the standard Coulomb gauge $\nabla \cdot \mathbf{A} = 0$. Electric current density is then obtained from the potential fields using the definition in (3).

$$\nabla \cdot \sigma \left(\nabla\phi + \frac{\partial \mathbf{A}}{\partial t} \right) = 0 \tag{1}$$

$$\nabla^2 \mathbf{A} = \mu\sigma \left(\nabla\phi + \frac{\partial \mathbf{A}}{\partial t} \right) \tag{2}$$

$$\mathbf{j} = \sigma \left(-\nabla\phi - \frac{\partial \mathbf{A}}{\partial t} \right) \tag{3}$$

Key to the behaviour of the electromagnetic fields is the distribution of the electrical conductivity field σ. In continuum media it is reasonable to assume that σ is constant, or at least varies smoothly. However, in particulate beds such as those found in SAF burdens current must pass between adjacent particles in contact, and the value of σ can vary by many orders of magnitude between the interior and exterior of the particles. In addition to this, AC systems exhibit a peculiarity called the skin effect in which time-varying electric and magnetic fields can couple together and bias the current density field toward the boundaries of the conductor. In engineering applications this is often discussed in terms of the skin thickness δ_s, which is dependent on the sinusoidal frequency of the AC supply as well as physical properties of the conductor [4]:

$$\delta_s = \sqrt{\frac{2}{\omega\sigma\mu}} \tag{4}$$

A plot of (4) indicating the typical range of sizes in SAF burdens is given in Figure 2. It can be seen that between approximately 10^3 and 10^7 S/m, skin effects start to become relevant at either particle or furnace scales. Metallurgical reductants typically have conductivity values within this range, and it is therefore reasonable to ask whether AC effects are important in this case.

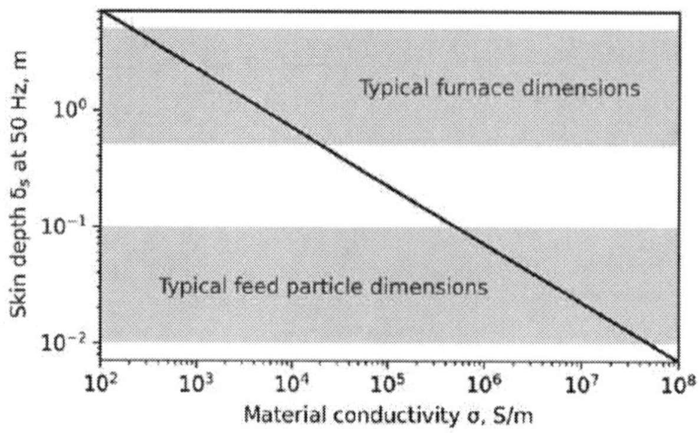

Figure 2: Skin depth at 50 Hz showing relevant dimensions in SAF processes

Conduction mechanisms in beds of metallurgical coke particles were studied extensively by Eidem and co-workers [5,6], with the conclusion that particle-to-particle contact was a significant contributor to overall bed resistance and bulk conductivity. It should be noted however that their work did not consider AC effects. Moving to the engineering scale, a number of other studies have performed detailed computational mechanics simulation of SAF interiors of varying complexity. These generally account for both particle conduction effects and AC effects by treating various parts of the furnace (including the particulate burden) as unresolved continuum materials with an average conductivity [7–11]. Tesfahunegn et al [12] also compared results obtained from direct current and AC simulations of silicon smelting furnaces, but only at full furnace scale and not at the level of individual particles in the burden.

As mentioned above it remains an open question whether particle-level effects are important in AC conduction through particle beds, and hence whether the continuum approximation is in fact suitable for particulate media. This is the focus of the present study.

2. Model description

The solution of Maxwell's equations can be handled in a number of different ways depending on the nature of the problem at hand. If the excitation voltage or current are pure sine waves, a sinusoidal ansatz can also be assumed for the time dependence of all fields at each point in space. After some manipulation the governing equations may then be presented in terms of time-independent complex-valued (phasor) fields for ϕ and $\mathbf{A}$. This can save considerably on computational expense as the time dependence in (1-2) does not need to be explicitly evaluated [3]. The real and imaginary components of the fields also have practical interpretations in electrical engineering applications (resistance and inductance respectively).

The alternative to the phasor method is direct solution in the time domain, which is more expensive but allows more general time dependence behaviour. This option was selected for the present work as it was desirable to verify that sinusoidal responses are in fact obtained from sinusoidal inputs for packed bed problems. A segregated time-domain solution approach was taken in which the fields for each time step are calculated by solving (1) for ϕ, followed by (2) for each of the three spatial components of $\mathbf{A}$. The calculations are iterated several times at each time step to resolve the coupling terms.

Although boundary element or finite-difference time domain formulations are more common in traditional electromagnetic field solvers, in multiphysics problems it is often preferred to use finite volume methods to facilitate easy coupling with additional physics. With this in mind, the single-region time domain solver used in the present work was developed and implemented in the OpenFOAM® computational mechanics framework [13].

Geometries and computational meshes for the solver were constructed using OpenFOAM®'s `blockMesh` and `snappyHexMesh` tools. For particle packing cases the location of each spherical particle is first calculated using a hexagonal close packing (HCP) layout. In addition to the cubic region size L and particle diameter D_p, an overlap fraction between adjacent particles is specified by $1 - D_o/D_p$. During mesh generation, refinement is performed in the overlap regions to capture the sharp variations in fields and geometry. After the mesh is generated, the conductivity field is set by assigning the specified material conductivity to the cells inside the particles, and a very low value to the cells in the empty spaces between them. Explicit contact resistances were not considered in this work. An example mesh and σ field is shown Figure 3.

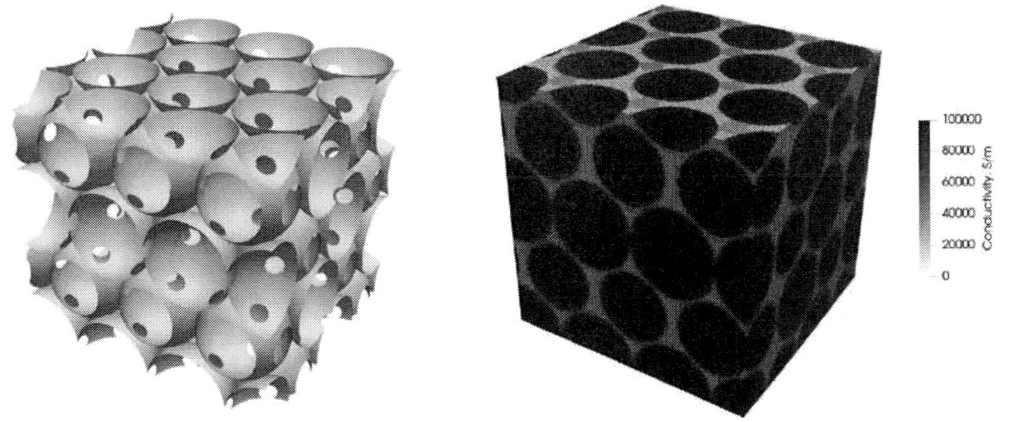

Figure 3: Example model geometry for packed bed cases, (l) particle boundaries showing contact overlap, (r) computational mesh and conductivity field

Boundary surfaces in the model can be either anodes (electrically grounded), cathodes (electrically excited), or insulating walls. Boundary conditions are applied to the various fields accordingly as shown in Table 1.

Table 1: Boundary conditions for electromagnetic field solver

Field	**Anode**	**Cathode**	**Insulating walls**
ϕ	$\phi = 0$	$\phi = V_p \sin \omega t$	$\frac{\partial \phi}{\partial \mathbf{n}} = 0$
A	$\frac{\partial(\mathbf{A} \cdot \mathbf{n})}{\partial \mathbf{n}} = 0, \quad \mathbf{A} \cdot \boldsymbol{\tau} = 0$	$\frac{\partial(\mathbf{A} \cdot \mathbf{n})}{\partial \mathbf{n}} = 0, \quad \mathbf{A} \cdot \boldsymbol{\tau} = 0$	$\frac{\partial(\mathbf{A} \cdot \mathbf{n})}{\partial \mathbf{n}} = 0, \quad \mathbf{A} \cdot \boldsymbol{\tau} = 0$

After a simulation is complete, the time-dependent total current I passing through the anode and cathode boundaries is extracted using (3). These values are compared to ensure that current conservation is satisfied across the domain, and then fitted to a sinusoidal curve with a peak current and phase delay. The quality of the fit indicates how close the output current is to a sine wave. The fit parameters are then converted into a complex-valued current phasor and used to obtain the impedance of the system from $Z = V_p/I$. The real component of Z is interpreted as a resistance, and the imaginary component as an inductance.

3. Results and discussion

In order to validate the time-domain solver, a simple test case was first considered. This case consists of a portion of a cylindrical conductor, across the opposite ends of which an AC voltage is applied. The internal impedance of such a conductor has a known analytical solution [4]:

$$Z = \frac{kL_c}{2\pi\sigma R_c} \frac{J_0(kR_c)}{J_1(kR_c)}, \quad k = \frac{1-i}{\delta_s} \tag{5}$$

Here the J_n are Bessel functions of the first kind, and $i = \sqrt{-1}$. To analyse this case, a 3D model of a conductor of 0.25 m radius and 1 m length was constructed and solved for a range of material conductivities at an AC mains frequency of 50 Hz. A relatively coarse mesh resolution with elements approximately 0.01 m in size was used with a fixed time step of 10^{-4} s. 0.2 s of total time was simulated in each case. Numerical calculations of the impedance were obtained from the time-varying current and applied voltage in each simulation as described above. Very good agreement was obtained for both the resistance and inductance components of Z, as shown in Figure 4.

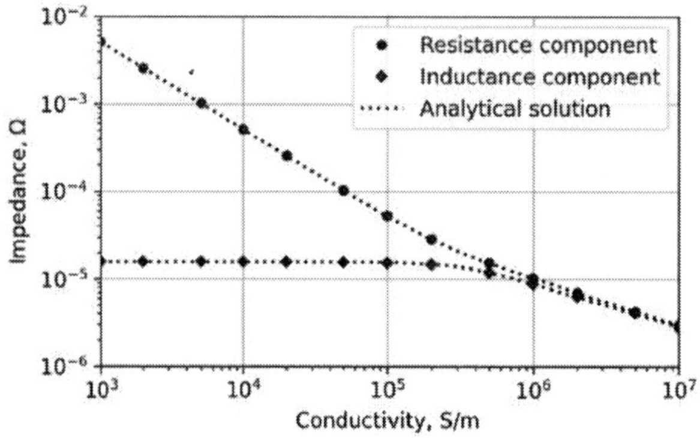

Figure 4: Numerical (points) and analytical (lines) solution for cylindrical conductor validation case

Mesh dependence of the time-domain solver was then examined in a simplified problem chosen to be representative of particle contact geometry in particle beds. Two spherical particles are placed in oblique contact in a rectangular domain, and an AC voltage is applied across the upper and lower boundaries of the region. Different levels of mesh refinement in the contact region were then applied, and the convergence behaviour studied using the impedance results. For this case, a particle size of 0.1 m was chosen together with contact overlaps of 2, 4, and 6%, and a range of particle conductivities between 10^3 and 10^7 S/m. An AC frequency of 50 Hz, a total simulation time of 0.2 s, and a time step of 10^{-4} s were used as before. Errors were then calculated relative to the highest resolution case. Selected results are shown in Figure 5, with the observed first-order convergence being similar for other conductivity values.

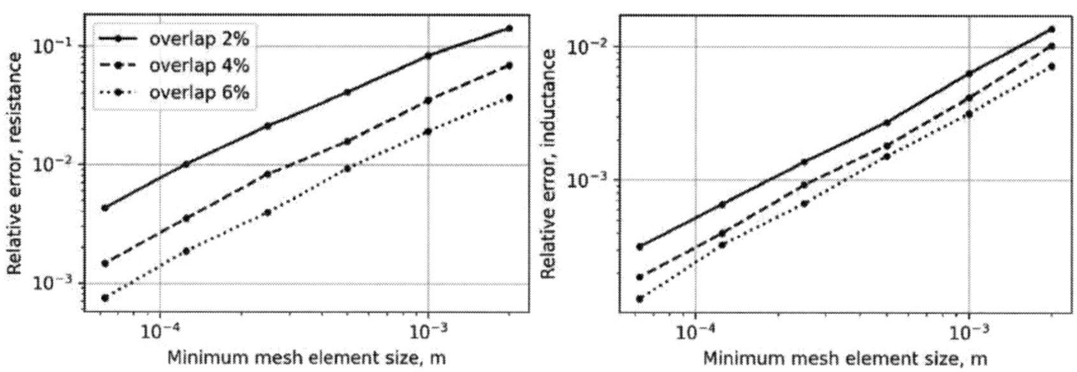

Figure 5: Mesh dependence for two-particle contact case, $\sigma = 10^7$ S/m

A mesh refinement of four cell subdivision levels, equating to a minimum mesh element size of 0.00025 m, produced an error of less than 2.5 % and was chosen for the remainder of the study.

In order to study AC conduction behaviour in packed beds, geometries of HCP ensembles of spheres 0.1 m in diameter were generated for a range of cube-shaped domains (L = 0.3, 0.4, and 0.5 m) and particle contact overlaps (2, 4, and 6%). This resulted in computational meshes of between 8 and 40 million elements in size. In each case the electromagnetic solution was performed at 50 Hz and particle conductivities between 10^3 and 10^7 S/m, for a total time of 0.2 s at a time resolution of 10^{-4} s.

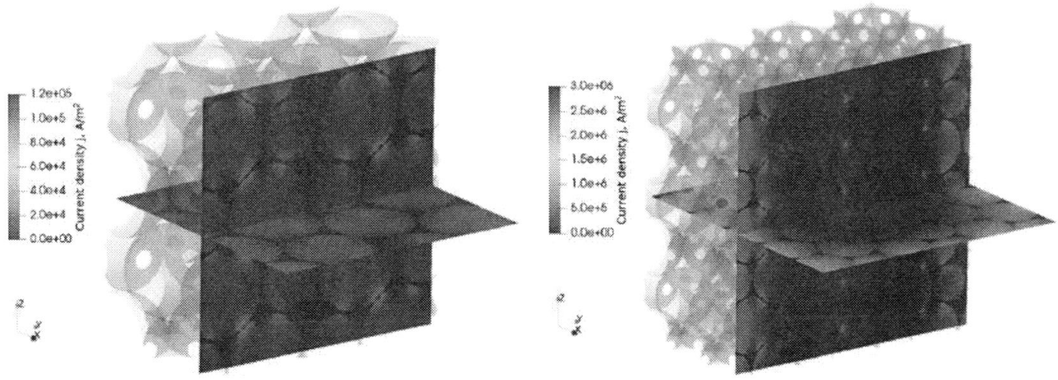

Figure 6: Current density fields at 0.195 s for resolved models at 2% particle overlap (l) $L = 0.3$ m, $\sigma = 10^4$ S/m, , (r) $L = 0.5$ m, $\sigma = 10^6$ S/m

Examples of the current density field near the end of the simulations are shown in Figure 6. At the individual particle scale it can be seen that the current concentrates at contact points where the conduction paths are constricted. At the scale of the region the current density field is distributed evenly through the volume for smaller regions and lower conductivities, while for larger regions and higher conductivities it is biased toward the outer edges of the domain as the AC skin effect becomes significant.

As before, system impedance was calculated from the transient current response to the applied voltage, and separated into resistance and inductance components. As shown in Table 2 the quality of the sine-wave fits to the current response were very good with R^2 values above 0.997 in all cases, indicating that the use of the phasor approximation would be reasonable for packed-bed simulations.

Table 2: R^2 metrics for sine-wave fits to current in all resolved simulation cases

σ	Contact overlap 2%			Contact overlap 4%			Contact overlap 6%		
	L=0.3 m	L=0.4 m	L=0.5 m	L=0.3 m	L=0.4 m	L=0.5 m	L=0.3 m	L=0.4 m	L=0.5 m
10^3 S/m	1.000	1.000	1.000	1.000	1.000	1.000	1.000	1.000	1.000
10^4 S/m	1.000	1.000	1.000	1.000	1.000	1.000	1.000	1.000	1.000
10^5 S/m	1.000	1.000	1.000	1.000	1.000	1.000	1.000	1.000	1.000
10^6 S/m	1.000	1.000	1.000	1.000	1.000	1.000	1.000	1.000	1.000
10^7 S/m	1.000	1.000	1.000	1.000	0.999	0.999	0.999	0.998	0.997

Together with the fully-resolved cases, a separate set of simulations using a constant uniform conductivity across the entire domain were also run – this represents the continuum approximation, where an average conductivity is applied over the entire region rather than explicitly resolving the individual particles. Figure 7 shows the impedance relationships generated by assuming uniform material properties, which were then used as reference curves for comparisons to the fully-resolved cases.

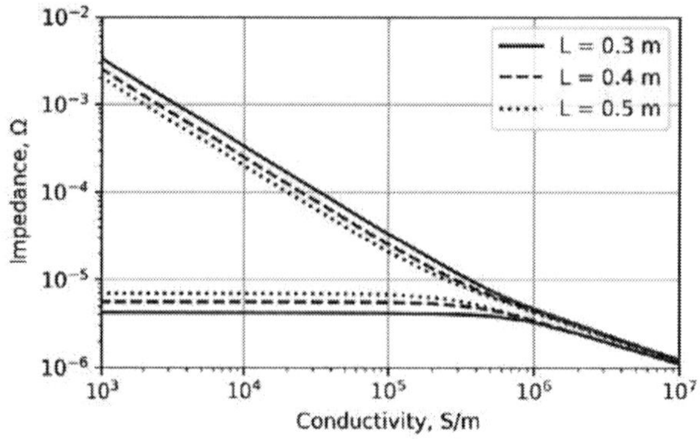

Figure 7: Impedance curves for AC conduction through a cubic region of uniform conductivity (upper set shows resistance component, lower set shows inductance component)

These curves were then fitted to the results from the fully-resolved simulations by applying a scaling ratio for the average conductivity (this amounts to shifting the curves in Figure 7 left or right by a fixed amount on logarithmic axes), to assess whether the continuum approximation is reasonable. This is shown in Figure 8. It can be seen that the match between the fitted curves and the resolved model data is very good across the entire AC regime, and is consistent across domain sizes for a given particle contact overlap – this demonstrates that the continuum approximation is a viable alternative to resolved simulations for conduction through packed beds.

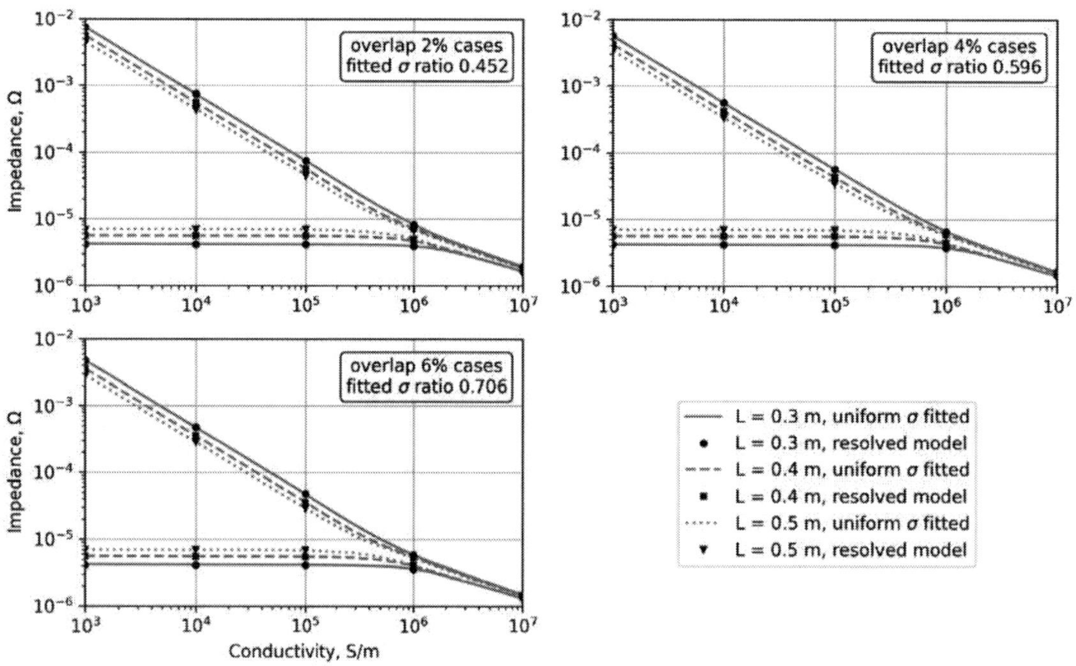

Figure 8: Impedance-conductivity results showing resolved models (points) and fitted uniform-σ curves (lines)

It is, however, important to note that the assumed contact overlap has a significant influence on the resulting equivalent uniform conductivity, with smaller overlaps between particles leading to lower equivalent conductivities. In other words, as the assumed overlap reduces, the effective electrical impedance of a packed bed system increases. The bulk conductivity value should therefore be chosen very carefully in continuum simulations of packed bed materials to ensure that correct results are obtained. For example if a simple volume average were used for bulk conductivity, the σ ratio would be approximately 0.740 for all of the cases presented in Figure 8 – this would grossly under-predict the impedances unless the overlaps between particles were unrealistically large.

4. Conclusions

A time-dependent single-region solver for electromagnetic field problems in low- to medium-frequency range was successfully developed and demonstrated, and applied to the problem of AC conduction through packed beds of spherical particles.

Validation of the solver against an analytical test case showed good performance, and the mesh dependence of the particle contact problem was seen to be approximately first order in minimum element size. Resolved simulations of packed particle beds at a range of length scales, overlap geometries, and particle conductivities demonstrated that sinusoidal excitation of packed bed structures resulted in sinusoidal outputs with little or no distortion of the waveforms – the phasor approximation is therefore valid in these cases. Resolved and unresolved/continuum simulations of the same region size were then compared, and it was found that for a given particle contact geometry the continuum approximation is a good approximation to the behaviour across a range of length and conductivity scales provided that the correct average conductivity is used.

It is important to be aware that the choice of average conductivity is strongly dependent on particle contact geometry. Since this is likely to be affected by numerous factors and may vary significantly in different regions of a SAF furnace burden, it is advisable to study the relationships with particle size and shape, packing types, and contact geometry further in the future to enable reliable a priori estimations to be made in engineering calculations. Particle contacts in packed beds may be estimated using Hertzian contact laws together with methods such as discrete element modelling, but care should be taken to account for complications such as temperature-dependent mechanical properties.

In addition, it is advisable to investigate the development and testing of more performant phasor solvers and multi-region solvers for applications in both resolved and continuum models of conductive packed beds.

Nomenclature

$\mathbf{A}$	Magnetic vector potential, m·T
D_o	Distance between particles in contact, m
D_p	Particle diameter, m
δ_s	AC skin thickness, m
ϕ	Electric potential, V
I	Current, A
$\mathbf{j}$	Current density, A/m^2
L	Domain size, m
L_c	Length of cylindrical conductor, m
$\mathbf{n}$	Unit normal vector
$\boldsymbol{\tau}$	Unit tangent vector
μ	Magnetic permeability, H/m
R_c	Radius of cylindrical conductor, m
σ	Electrical conductivity, S/m
V_p	AC peak voltage, V
ω	AC frequency, radians/s
Z	Impedance, Ω

Acknowledgement

This paper is published by permission of Mintek. The authors acknowledge the Centre for High Performance Computing (CHPC), South Africa, for providing computational resources to this research project.

References

1. Gasik M. Handbook of Ferroalloys. Elsevier; 2013.
2. Barker IJ, Rennie MS, Hockaday CJ, Brereton-Stiles PJ. Measurement and control of arcing in a submerged-arc furnace. In: INFACON XI: Innovations in Ferroalloys Industry. New Delhi, India: IFAPA; 2007. p. 685–94.
3. Griffiths DJ. Introduction to electrodynamics. 3rd ed. Upper Saddle River, N.J: Prentice Hall; 1999. 576 p.
4. Weeks WL. Transmission and distribution of electrical energy. New York, USA: Harper & Row; 1981.
5. Eidem PA, Tangstad M, Bakken JA. Determination of Electrical Resistivity of Dry Coke Beds. Metall Mater Trans B. 2008 Feb;39(1):7–15.
6. Eidem PA, Runde M, Tangstad M, Bakken JA, Zhou ZY, Yu AB. Effect of Contact Resistance on Bulk Resistivity of Dry Coke Beds. Metall Mater Trans B. 2009 Jun;40(3):388–96.
7. Tesfahunegn YA, Magnusson T, Tangstad M, Saevarsdottir G. Effect of electrode shape on the current distribution in submerged arc furnaces for silicon production - A modelling approach. J South Afr Inst Min Metall. 2018;118(6):595–600.
8. Herland EV, Sparta M, Halvorsen SA. Skin and Proximity Effects in Electrodes and Furnace Shells. Metall Mater Trans B. 2019 Dec;50(6):2884–97.
9. Tesfahunegn YA, Magnusson T, Tangstad M, Saevarsdottir G. The Effect of Side Arcs on Current Distributions in a Submerged Arc Furnace for Silicon Production. In: Lee J, Wagstaff S, Lambotte G, Allanore A, Tesfaye F, editors. Materials Processing Fundamentals 2020 [Internet]. Cham: Springer International Publishing; 2020 [cited 2023 Dec 27]. p. 177–88. (The Minerals, Metals & Materials Series). Available from: http://link.springer.com/10.1007/978-3-030-36556-1_16
10. Sparta M, Varagnolo D, Stråbø K, Halvorsen SA, Herland EV, Martens H. Metamodeling of the Electrical Conditions in Submerged Arc Furnaces. Metall Mater Trans B. 2021 Jun;52(3):1267–78.
11. Halvorsen SA, Herland EV, Sparta M, Risinggård VK. Understanding Scale-Up of High-Current Electrodes. Metall Mater Trans B. 2023 Aug;54(4):1985–97.
12. Tesfahunegn YA, Magnusson T, Tangstad M, Saevarsdottir G. Comparative Study of AC and DC Solvers Based on Current and Power Distributions in a Submerged Arc Furnace. Metall Mater Trans B. 2020 Apr;51(2):510–8.

13. OpenCFD Ltd. OpenFOAM v2306 [Internet]. 2024 [cited 2023 Jan 12]. Available from: https://www.openfoam.com/

13th South African Conference on Computational and Applied Mechanics (SACAM2024)
22 – 23 January, 2024, Stellenbosch, South Africa

AEROACOUSTIC SIMULATION AND IMPROVEMENT OF A PROPELLER FOR AN ELECTRIC SELF-LAUNCH SAILPLANE

Karin Schutte[1*], Johan Bosman[2†], Jan-Hendrik Kruger[3]

[*] Presenting author
[†]Corresponding author: johan.bosman@nwu.ac.za
[1,2,3]North-West University, School of Mechanical Engineering
Private Bag X6001, Potchefstroom, 2520

Abstract: Propeller design for various aviation type applications has always been a challenging task where airframe, engine and propeller properties must match in order to obtain adequate thrust at the required flight speeds. This challenge becomes particularly intricate in applications like electric sailplanes, where propeller geometries face restrictions due to fuselage constraints, limiting the propeller's radius. Consequently, compromises must be made such as high rotational speeds when adequate thrust is needed. These compromises usually lead to secondary adverse effects such as inefficiency and excessive noise. Regulations govern the noise levels of electric-powered sailplanes, and certification of these propulsion systems requires strict adherence to specified standards. Propeller designers must, therefore, incorporate noise levels as a crucial parameter to minimise. Aeroacoustic CFD models have the potential to predict noise levels of propellers, where the pressure fluctuations from the fluid flow simulation are used as sources and coupled with equations to describe the propagation of small perturbations in pressure and velocity that produces sound waves. The study investigates the complexities of propeller design, exploring strategies such as incorporating humpy chords, sweeps, and reducing the tip airfoil thickness of the blades to achieve noise reduction without significantly compromising efficiency. An electric propulsion system of a sailplane was used as a baseline validation case, where the propeller produced significant noise levels when measured. The research integrates principles of aerodynamics and fluid mechanics to propose systematic geometric modifications, addressing airflow characteristics and vortex formation. The study also considers the unique constraints imposed by sailplanes, such as diameter limitations based on fuselage dimensions and takeoff requirements. The article concludes by highlighting the promising results of a modified propeller, showcasing a 6 dB reduction in noise levels through a combination of reduced RPM and geometric alterations. Leveraging advanced simulation techniques and embracing geometric modifications hold the promise of achieving a balance between thrust, efficiency, and noise reduction for electric self-launch sailplanes.

Key words: Aeroacoustics, CFD simulations, Propeller design, Electric propulsion

1 Introduction

Noise pollution poses significant threats to both human and animal well-being, with aircraft noise emerging as a prominent factor. The primary source of aircraft noise typically emanates from the main propulsion system, namely jet engines or piston engines driven by propellers. In the pursuit of mitigating the environmental impact, a transition from piston engines to electric propulsion systems is being explored to curtail noise emissions. However, this shift brings attention to a persistent challenge - propeller noise.

In addressing the impact of aircraft noise, both the European Aviation Safety Agency (EASA) and the Federal Aviation Administration (FAA) have established airworthiness standards pertaining to noise. These standards align with the criteria outlined in the International Civil Aviation Organisation (ICAO) Standards and Recommended Practices for aircraft noise certification, as specified in Annex 16, Volume I [17].

In the context of an electric self-launch sailplane, stringent space and weight limitations impose constraints on the propeller size (Figure 1). This limitation, in turn, results in higher Revolutions Per Minute (RPM) compared to conventional propellers, leading to significant noise challenges. The heightened noise levels result from elevated tip speeds, causing air particles travelling over the tip surfaces to reach near transonic velocities. The certification standards set by EASA necessitate adherence to specific noise emission limits, outlined in directives aimed at mitigating environmental noise [14].

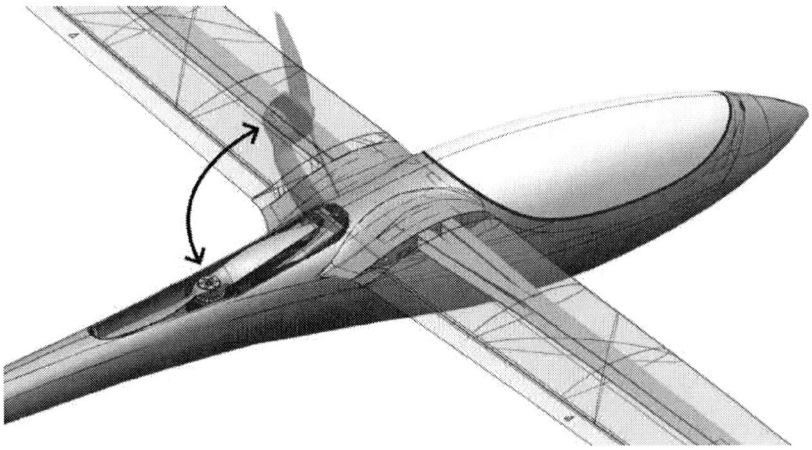

Figure 1: An electric self-launch system configuration for sailplanes

Reducing propeller noise is essential for both mitigating its impact on noise pollution and meeting regulatory noise certification s tandards. In this research article the existing literature to identify optimal design strategies for propellers that minimise noise production is examined and a propeller design case study is presented.

2 Literature review

Propellers serve the function of converting rotational power derived from the engine into forward thrust. The fundamental concept governing thrust, the propelling force propelling an aircraft in the horizontal direction of flight, adheres to Newton's Third Law, stipulating that for every action force, there exists an equal-magnitude reaction force in the opposite direction. In this context, the propeller exerts an action force on the air, effectively drawing it behind the propeller. This induced movement of air initiates a pressure differential, resulting in a forward pull on the aircraft. Consequently, the magnitude of thrust generated by the propeller is directly proportional to the quantity of air drawn through it [12]. Various theoretical frameworks prove instrumental in comprehending propeller efficiency and facilitating design considerations.

2.1 Propeller design methodology

A combination of Blade Element theory and Momentum theory can be used in the preliminary design phases of propellers. Blade Element theory serves to predict the performance of a propeller through the computation of forces exerted on the flow field, by using geometrical properties such as airfoil shape and twist distribution. This involves utilising annular stream-tube control volumes to divide the full length of the blade into infinitesimally narrow e lements. Momentum theory, on the other hand, centers around an actuator disk model such as a propeller, replacing the rotor with a round surface or disk of zero thickness. This disk can thus support a low-pressure differential in front of the propeller and accelerate air moving through it. Blade Element theory and Momentum theory can be combined to derive a model that can determine the performance of a propeller with specified size, twist distribution and airfoil properties [7].

An additional propeller design theory pioneered by A. Betz, focuses on optimising the geometry of the propeller under specific operating c onditions. The primary objective is to minimise the power needed to attain a predetermined propulsive force within those particular operational parameters [1]. In the modern context of propulsion, particularly with electric propellers, Betz's insights serve as valuable guidance for optimising performance and energy efficiency across diverse operational scenarios, although the theory was developed over a century ago.

2.2 Maximising efficiency while mitigating noise

While various theories and models play a crucial role in propeller design, many of them overlook the aspect of propeller noise. Succi [2] and Patrick *et al.* [3], drawing on Betz's theory [1], employed an approach to define an aerodynamically optimal propeller and subsequently modified it to mitigate noise. However, Peixun *et al.* [13] highlighted a flaw i n this i terative p rocess, e mphasising i ts sub-optimal nature; attempting to reduce noise compromised efficiency, leading to diminished overall performance.

Wu *et al.* [8] introduced a methodology aimed at minimising noise while enhancing blade geometry in the spanwise direction. The Ffowcs Williams and Hawkings (FW-H) model, unsteady slip mesh, and Large Eddy Simulation (LES) methods were employed to predict aerodynamic noise. Their investigation revealed a significant influence of blade load noise on aerodynamic noise, with the potential for reduction by modifying the blade shape along the spanwise direction.

Benoit [6] conducted a study demonstrating that incorporating a humpy chord and sweep in high-speed propellers effectively decreases the sound pressure level generated by the propeller. A humpy chord refers to a region with an increased chord width at the tip of a blade. The primary role of a hump is to alter the span-wise distribution of the load on the blade.

Ultimately, the key challenge lies in finding a balance between optimising performance and minimising the noise produced.

2.3 Propellers for electric aircraft

Considering propellers for electric aircraft, extensive research has focused on propeller design for this application. D'Angelo *et al.* [4] proposed two numerical procedures. The first procedure is as follows. Given an operative speed and profile distribution along the blade, the chord distribution, angle of twist, efficiency and torque and thrust coefficients of the optimally efficient propeller can be determined. The second algorithm assesses the efficiency, thrust, and torque coefficients of the optimally designed propeller at various operative speeds and blade pitches [4].

In a study on noise generated by small electric aircraft, Albarran *et al.* [9] observed that a single electric motor aircraft produces more noise compared to a single-engine piston aircraft, pinpointing the propeller as the primary noise source. They recommended increasing the number of electric engines to four and reducing the maximum propeller tip Mach number.

During a study by Keueter *et al.* [15] the powertrain combination of an electric motor and propeller for an electric aircraft was investigated. The study yielded two noteworthy conclusions: firstly, the optimisation of propeller pitch angle and speed emerged as a viable strategy to enhance overall efficiency. Secondly, the optimisation of both the hub diameter and outer diameter of the electric motor results in an optimal hub-to-tip ratio ranging between 0.16 and 0.20. This specific range indicates an advantageous configuration for the powertrain, providing valuable guidance for achieving an effective balance in design parameters.

2.4 History of aeroacoustic models

In the era preceding the 1950s, studies on flow-generated noise primarily focused on understanding the connection between the frequency of fluid fluctuations and the resulting emitted sound. However, a pivotal shift occurred with the contributions of Sir James Lighthill. His groundbreaking work involved developing a theory to estimate the intensity of sound emanating from turbulent flow, marking the initiation of modern aeroacoustic science. Lighthill's insights reshaped the landscape of research in this field, redirecting the trajectory of flow-generated noise studies. Lighthill first introduced the concept of taking an equivalent system of noise sources as a replacement of the actual flow field responsible for generating noise. The governing principles of standard acoustic propagation equations apply to these noise sources, acting within a uniform and motionless fluid. Consequently, the focal challenge in predicting noise shifts towards effectively characterising the aerodynamics of these sources [11].

Curle expanded upon Lighthill's concept in 1955 by incorporating the influence of flow-body interaction on sound generation. The Curle noise source model demonstrates effectiveness in predicting the noise generated by rotating components. This model assesses the noise originating from the turbulent boundary layer flow around a solid body, capturing dipole sources of noise. These dipole sources arise from fluctuating surface pressure induced by the interaction of solid boundaries with the surrounding fluid. Lighthill's theory was further extended in 1969 by Ffowcs Williams and Hawkings to incorporate arbitrary surface motion. This extension was specifically applied to predict the noise generated by the aerodynamics of rotor blades [11].

2.5 Noise prediction using Computational Fluid Dynamics (CFD)

The verification and validation of CFD simulations are crucial steps to ensure the accuracy of results. Following this, CFD becomes a powerful tool for simulating various designs and configurations. Greschner *et al.* [5] employed CFD to compute flow field information, utilising a Discrete Fourier Transform (DFT) to determine the frequency domain distribution of the sound source. This approach enabled the prediction of sound generated by a rod-airfoil configuration.

Schulz A. [16] conducted an investigation into a drone propeller, employing CFD to anticipate noise levels. The trailing edge of the propeller underwent modifications through the addition of various forms of serrations. The simulation, executed in OpenFOAM, incorporated LES turbulence models in conjunction with an aeroacoustic analogy. The results demonstrated a discernible reduction in propeller noise, as evidenced by the CFD analysis.

Schenk [10] used CFD as the primary research method, leveraging its ability to provide detailed flow field information and adaptability once a simulation is e stablished. He utilised an Unsteady Reynolds-averaged Navier-Stokes (URANS) simulation in STAR-CCM+ with an aeroacoustic solver to forecast noise generation from rotors. For external aerodynamics simulations in STAR-CCM+, it is recommended to use a bullet-shaped domain. The spherical end of the domain should have a radius equal to ten times the length of the simulation scale. The downstream domain should extend twenty times the length of the simulation scale in the downstream direction. It is advisable to position the propeller at the center of the spherical end for optimal simulation setup. As per the STAR-CCM+ documentation [11], it is advisable to employ a polyhedral mesh with a surface remesher for curved surfaces, particularly in the region surrounding the propeller. Additionally, a prism layer mesh is recommended for this volume. The remainder of the volume can be effectively meshed using a trimmed cell mesh combined with a surface remesher, contributing to improved processing time for the simulation. The selection of physics models is depended on the specific simulation s cenario. For propeller simulations, an implicit unsteady model is preferred, enabling the utilisation of turbulence models. The Reynolds-Averaged Navier-Stokes turbulence model is particularly recommended, given its established reputation in the aerospace industry. Aeroacoustic challenges necessitate the use of transient calculations due to the intrinsic transient physics involved. Nevertheless, it is crucial to acknowledge that transient simulations come with elevated computational run times. The STAR-CCM+ User Guide [11] advises executing aeroacoustic simulations as three-dimensional cases with compressible flow. To conduct an aeroacoustic simulation on a propeller, the recommended physics models to incorporate include aeroacoustics, broadband noise sources, Curle, and Proudman [10].

Consequently, CFD proves valuable in simulating propellers for predicting both thrust and noise production. BEM theory will be initially employed as a design tool, serving as a preliminary stage for propeller design. Following the BEM analysis, adjustments will be implemented to mitigate noise. The primary assessment of the effectiveness of these modifications will then be carried out using CFD simulations. The study underscores the importance of choosing appropriate simulation approaches, such as coupled flow models and broadband noise source models like Curle, to comprehensively capture the dynamic and aerodynamic characteristics influencing noise generation in propeller systems.

3 CFD aeroacoustic simulations

The original propeller generates a significant degree of noise pollution. To pinpoint the specific areas of concern and determine necessary modifications, a CFD simulation specifically employing aeroacoustic models was conducted on the propeller.

Simple two-cylinder mesh regions were setup in STAR-CCM+ where the large flow domain consisted out of polyhedral cells and the inner rotating mesh had tetrahedral cells with prism layers formed around the propeller surface (Figure 2). Wake refinement was implemented behind the propeller rotating disk to improve moment (power) and thrust predictions.

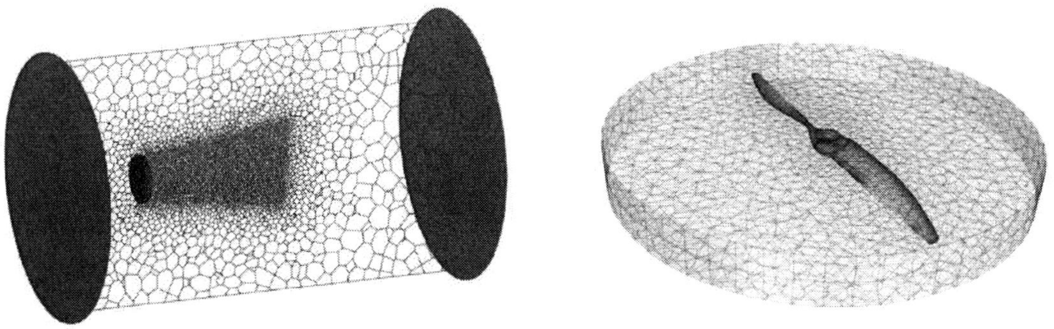

(a) A cylindrical flow domain with inner rotating volume

(b) Rotating mesh disk with prism layers around propeller

Figure 2: Mesh and flow domain setup in STAR-CCM+

The motion of the propeller was simulated using a rigid body moving reference frame for the propeller disk to establish rotational movement between the two regions. Interface boundaries between the polyhedral and tetrahedral meshes were created and an implicit unsteady solver with time steps of 0.001s was used. URANS simulations are well-suited for simulating propellers and forecasting the associated noise generation, due to their time-dependent characteristics. This aspect is crucial for capturing the dynamic nature of propeller noise. These simulations incorporate turbulence models to account for turbulence fluctuations in the flow. Additionally, URANS is more computationally efficient compared to LES models [11].

In this study, a coupled flow model is preferred over a segregated flow model when choosing an aeroacoustic simulation approach. The nature of the problem at hand makes coupled flow more applicable. While coupled flow is computationally more demanding compared to segregated flow, it stands out by simultaneously addressing multiple physics aspects. This is particularly crucial for scenarios where interactions between fluid flow and blade vibrations play a significant role in influencing the produced sound, a consideration effectively captured by the coupled flow model.

In order to compute the position and magnitude of the primary noise sources, broadband noise source models were chosen. Broadband noise exhibits a continuous spectrum across the frequency domain, wherein acoustic energy is uniformly distributed across all frequencies within a specified range [11]. This choice enables a comprehensive analysis of noise sources, offering insights into their locations and intensities across the entire frequency spectrum.

In adopting the framework of the broadband noise source model, the selected model of choice is the Curle noise source model. This particular model characterises noise emanating from the fluctuating surface pressure induced by the interaction of solid boundaries with the fluid [11]. The Curle noise source model finds its applicability in unsteady analyses and integrates with a spectrum of Reynolds-Averaged Navier-Stokes (RANS) Turbulence models. Notably, it is made available specifically for simulations employing segregated flow approaches.

Figure 3 displays the outcomes of the Curle noise source model. Evidently, the primary noise source is localised at the propeller tips, distinguished by the brightest red colour in comparison to the remaining sections of the propeller. This analysis reveals that the tips generate a peak acoustic power of 142 dB.

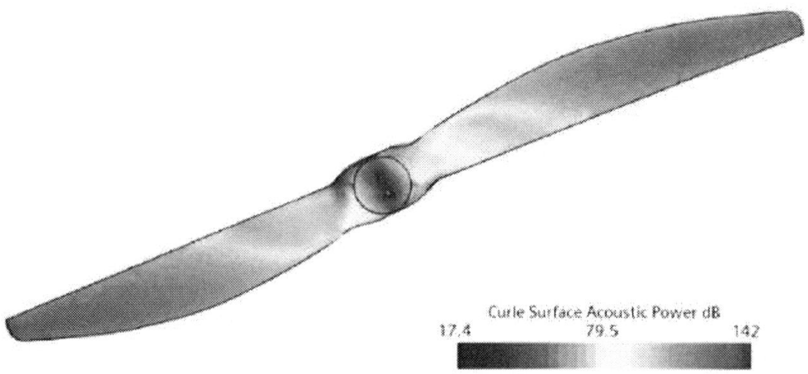

Figure 3: Aeroacoustic noise results for the original propeller (4000 RPM)

As depicted in Figure 4, the tip region exhibits elevated Mach numbers characterised by substantial velocity gradients. This phenomenon is attributed to the high RPM and accelerated motion of air particles over the upper surface of the airfoil, making a notable contribution to the overall noise production. The calculated maximum Mach numbers approach transonic speeds, indicating the presence of high-speed airflow. Additionally, the simulations revealed heightened noise levels emanating from the trailing edge of the airfoil.

4 CFD validation with experimental data

An experimental evaluation of the noise generated by the original JS3 RES propeller was conducted. The experimental setup provided a foundation for comparing the noise characteristics of the new propeller and validating the CFD program developed in STAR-CCM+.

During the experiment, the following procedure was systematically executed: First, the glider was securely positioned and tethered to the ground, with the decibel meter and GoPro strategically placed to capture both the decibel meter's screen and the propeller in a single frame. Figure 5 depicts the experimental arrangement placement of the decibel meter at the front of the glider.

After activating the decibel meter, recordings were initiated on both the GoPro and a smartphone inside the cockpit; the operating pilot was seated in the cockpit. The propeller's RPM was systematically

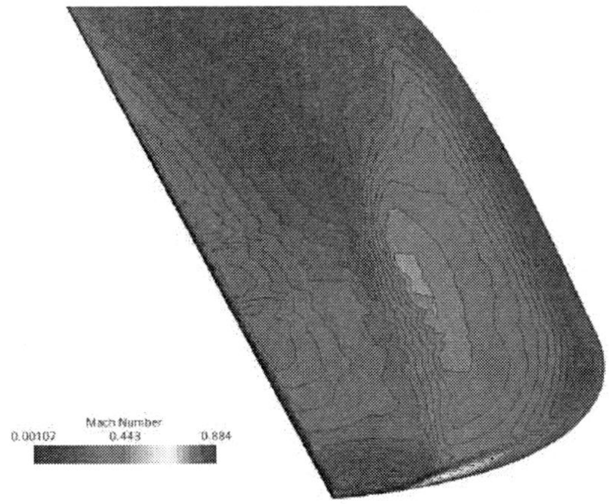

Figure 4: Mach number distribution in tip region for the original propeller (4000 RPM,72 kg Thrust)

Figure 5: Decibel meter positioned at the front of the glider

varied from 0 RPM to 4000 RPM, with recording equipment paused once the motor was safely switched off. Subsequently, the decibel meter and GoPro setup were relocated to a new position, and the entire sequence was repeated, ensuring a comprehensive dataset for analysis. This process was iterated through subsequent repetitions for a thorough examination of noise levels at different propeller speeds.

Applying the principles of the Inverse Square Law, the sound pressure level derived from the CFD results

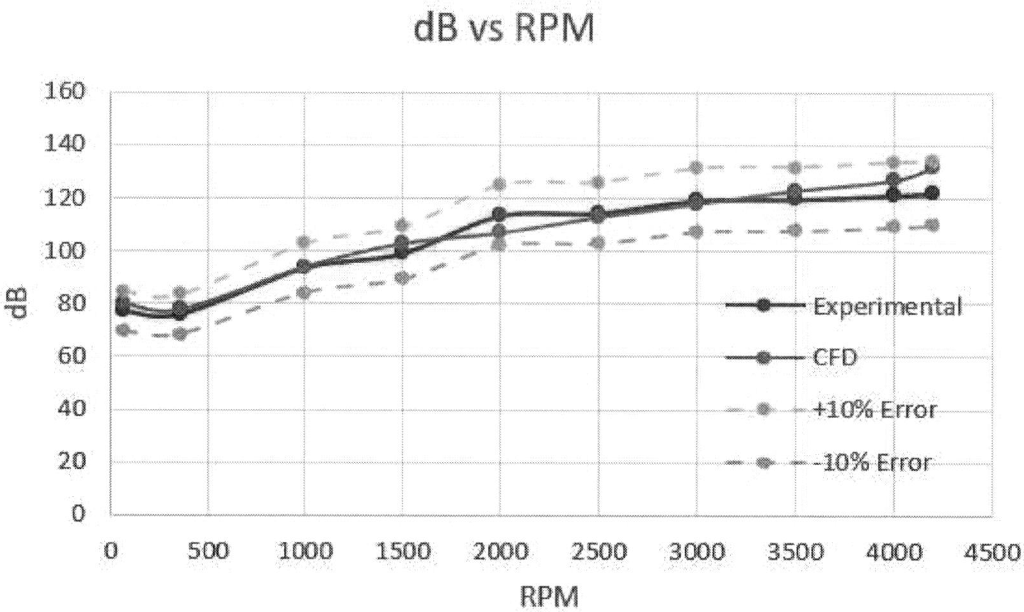

Figure 6: Experimental vs CFD results

underwent conversion to facilitate a meaningful comparison with experimental measurements taken at a distance from the source. The visual representation of this comparison is illustrated in the graph shown in Figure 6. The $\pm 10\%$ error lines are indicated by the dashed lines. The experimental results exhibit a significantly higher degree of deviation compared to the trendline displayed by the CFD r esults. This variance can be attributed to the conditions under which the experimental measurements were taken, including factors such as wind intensity on the day and the presence of surrounding buildings. The structures may have caused sound to bounce off, potentially leading to an increase in the measured sound levels. Nevertheless, the outcomes remain within the acceptable margin of error, falling within the $\pm 10\%$ error range, thus affirming the adequacy of the results.

5 Proposed modifications to reduce noise

When implementing modifications to the propeller design, it is imperative to account for the constraints dictated by the particular application, in this instance, sailplanes. The diameter of the propeller is limited by the fixed dimensions of the opening through which it extends and retracts in the fuselage (Figure 1). These dimensions are determined based on stiffness calculations and predetermined layup schedules for the fuselage. Additionally, there are constraints on the breadth of the propeller, as it cannot be excessively thick to ensure compatibility with the designated opening. Furthermore, the sailplane must adhere to takeoff requirements within a specified runway length stipulated by the certification authority, necessitating the propeller to produce sufficient thrust to meet these specifications.

From the initial simulations conducted on the original propeller, it is evident that the noise originates mainly from the tip regions, with the possible contribution of small transonic shock waves. To address this issue, the design emphasis for the new propeller focused on the reduction of RPM whilst maintaining the same thrust, ensuring operation within the allowed power delivery of the electric motor. This section further delves into the secondary modifications made to the planform and tip airfoil to refine the propeller design.

The production of thrust in propellers relies on the solidity ratio, defined as the ratio between the projected area and the outline netting. A higher solidity ratio correlates with increased propeller thrust. Additionally, augmenting the frontal surface area of the propeller results in enhanced thrust production, as it necessitates the displacement of a greater volume of air particles. These factors are crucial considerations, as an increase in propeller thrust could facilitate a reduction in RPM, subsequently lowering the velocity of air particles over the propeller tip. With the increased blade area, a new chord and twist distribution were used for optimum efficiency at the lower operating RPM.

A backward swept planform was also incorporated and is beneficial for noise reduction as it alters the load distribution along the blades, delays flow separation, and influences vortex formation—key factors contributing to the generation of noise.

The generation of noise from propellers is most pronounced at the tip surfaces, attributable to fast-traveling air particles over these regions. Recognising that velocity is dependent on the displacement of air particles along the upper surface of the airfoil, an approach to mitigate noise involves reducing the thickness of the blade tip. A reduction in thickness results in a decrease in the acceleration of air particles over the upper surface of the blade. Additionally, diminishing thickness corresponds to a reduction in negative pressure in accordance with Bernoulli's principle. Small changes to the standard Clark-Y airfoil was made to reduce surface velocities in the tip region, by reducing the thickness of the airfoil. Although reducing the thickness of blade tip can decrease the upper surface velocity, the focus shifted to a more effective solution: lowering the RPM while maintaining sufficient thrust.

Incorporating serrations, particularly in the tip regions, offers a potential avenue for noise reduction. Serrations play a multifaceted role in this context: they alter the formation of significant tip vortices, influence the noise spectrum by redistributing frequencies to less bothersome ranges for the human ear, and serve as miniature vortex generators. These miniature vortex generators interact with larger vortices, effectively absorbing and dissipating their energy.

Based on the research, several of the modifications detailed above were incorporated into the design of the modified propeller. Figure 7 shows a side-by-side comparison of the original propeller and the modified propeller. This figure clearly illustrates the augmentation in the solidity ratio and the integration of a swept blade.

6 Aeroacoustic simulation results of modified propeller

The modified propeller underwent aeroacoustic simulations using the same parameters as the prior assessments in STAR-CCM+. The results, depicted in Figure 8, show that a reduction in the noise level of 6 dB in comparison with the original propeller was obtained.

The pressure gradient on the top surface, as illustrated in Figure 9, displays a significant reduction through the geometric modifications and lower operating RPM in contrast to the original propeller's Mach distribution (Figure 4).

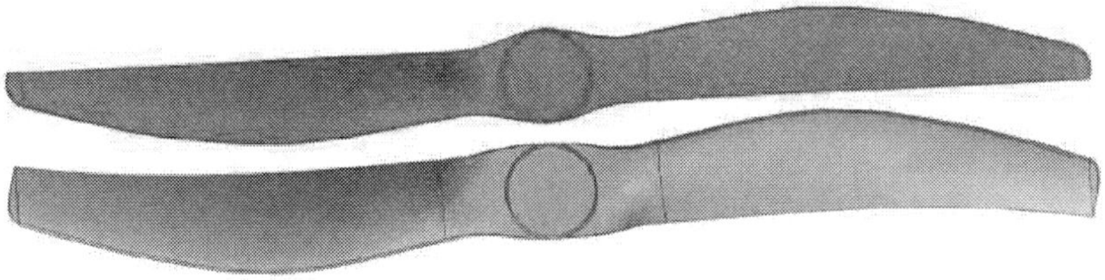

Figure 7: Modification comparison (Old propeller at the top)

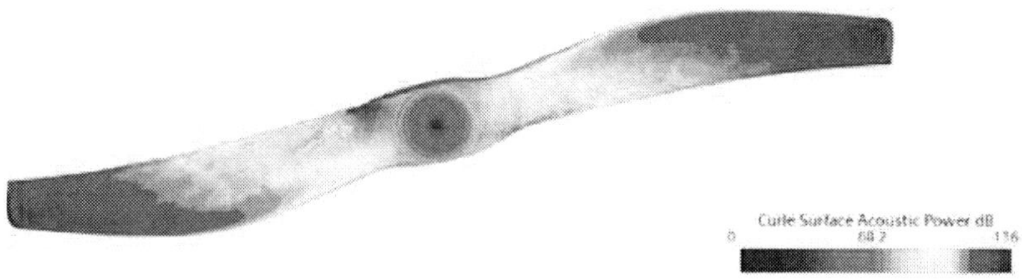

Figure 8: Aeroacoustic noise prediction results for the new propeller (3750 RPM)

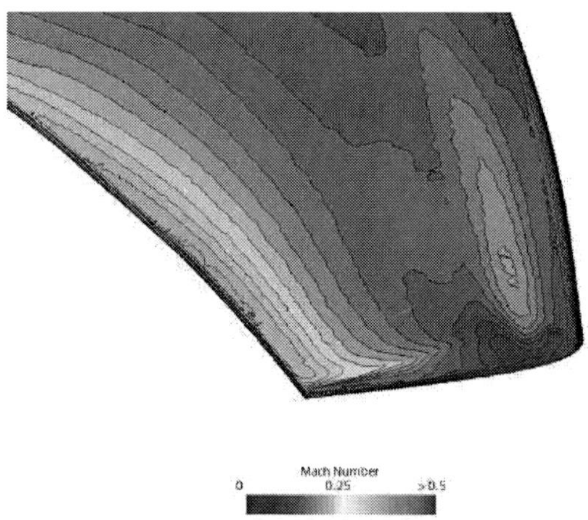

Figure 9: Mach number on tip region of new propeller (3750 RPM, 68kg Thrust)

7 Conclusion

The exploration of propeller design for electric self-launch sailplanes presents a multifaceted challenge, particularly in the context of stringent space and weight limitations. The necessity for optimal thrust, high rotational speeds, and compliance with noise emission standards poses a complex design problem. This article delves into the aeroacoustic improvement for such propellers, emphasising the pivotal role of CFD simulations in predicting both thrust and noise production.

Propeller modifications, including reducing the tip thickness, introducing swept blades, and increasing the solidity ratio, emerge as promising strategies to address noise concerns without drastically compromising efficiency. It is crucial to note that these design considerations must navigate the constraints imposed by the unique characteristics of sailplanes. The limitations on the propeller's diameter, determined by fuselage dimensions and takeoff requirements, underscore the need for design strategies that balance performance and compliance.

The original propeller was analysed using CFD aeroacoustic simulations. Results indicated substantial noise at the tips, associated with transonic speeds and significant velocity gradients. Validation with experimental data was conducted, considering factors such as wind and surrounding structures, with results falling within an acceptable margin of error. After the proposed modifications were implemented, the modified propeller underwent further aeroacoustic simulations, resulting in a 6 dB reduction in noise and adequate thrust compared to the original propeller.

This article underscores the intricate interplay between propeller design, aeroacoustics, and the challenges posed by electric self-launch sailplanes. Leveraging advanced simulation techniques and embracing geometric modifications hold the promise of achieving a balance between thrust, efficiency, and noise reduction. Future research endeavours should continue to refine and validate these design strategies through simulations and empirical data.

Acknowledgement

The authors would like to express their gratitude for the support and resources provided by the North-West University. This research was conducted in collaboration with Jonker Sailplanes, and their expertise greatly enriched the project.

References

[1] A. Betz. "Schraubenpropeller mit geringstem Energieverlust. Mit einem Zusatz von L. Prandtl". In: *Nachrichten von der Gesellschaft der Wissenschaften zu Göttingen, Mathematisch-Physikalische Klasse* 1919 (1919), pp. 193–217. URL: http://dml.mathdoc.fr/item/GDZPPN002505371.

[2] G. P. Succi. "Design of Quiet Efficient Propellers". In: *SAE Business Aircraft Meeting*. 1979.

[3] H. Patrick, W. R. Finn, and K. C. Stich. *Two and Three-bladed Propeller Design for the Reduction of Radiated Noise*. Report AIAA-1997-1710. Reston: AIAA, 1997.

[4] Salvo D'Angelo, F Berardi, and E Minisci. "Aerodynamic performances of propellers with parametric considerations on the optimal design". In: *The Aeronautical Journal* 106.1060 (2002). Department of Aeronautical and Space Engineering, Politecnico di Torino, Turin, Italy, pp. 313–320.

[5] Björn Greschner et al. "Prediction of sound generated by a rod–airfoil configuration using EASM DES and the generalised Lighthill/FW-H analogy". In: *Computers & fluids* 37.4 (2008), pp. 402–413.

[6] Benoıt G Marinus. "Comparative study of effects of sweep and humps on high-speed propeller blades". In: *AIAA journal* 52.4 (2014), pp. 739–746.

[7] M. K. Rwigema. "Propeller Blade Element Momentum Theory with Vortex Wake Deflection". In: *27th International Congress of the Aeronautical Sciences* (2019).

[8] Yuhang $_2$ Wu et al. "A novel aerodynamic noise reduction method based on improving spanwise blade shape for electric propeller aircraft". In: *International Journal of Aerospace Engineering* 2019 (2019). $_2$ School of Aerospace Engineering, Shenyang Aerospace University, Shenyang 110136, China $_3$ Liaoning Key Laboratory of Advanced Measurement and Test Technology for Aviation Propulsion System, Shenyang Aerospace University, Shenyang 110136, China, pp. 1–10.

[9] MY Pereda Albarrán et al. "Noise evaluation of battery powered small aircraft". In: *CEAS Aeronautical Journal* 11 (2020), pp. 125–135.

[10] Austin R Schenk. *Computational investigation of the effects of rotor-on-rotor interactions on thrust and noise*. Brigham Young University, 2020.

[11] Siemens. *Simcenter STAR-CCM+ Help Guidelines*. Simcenter STAR-CCM+ Software is a Computational Aided Engineering (CAE) solution for solving multidisciplinary problems in both fluid and solid continuum mechanics, within a single integrated user interface. © 2020 Siemens. This software and related documentation are proprietary and confidential to Siemens. Other trademarks belong to their respective owners. 2020.

[12] Southern Wings. *Aircraft Propeller Basics*. Published on July 9, 2020. 2020. URL: https://www.southernwings.co.nz/aircraft-propeller-basics/#:~:text=The%20propeller%20works%20by%20displacing.or%20forward%20propulsion%20is%20generated.

[13] Peixun Yu et al. "Aeroacoustic and Aerodynamic Optimization of Propeller Blades". In: *Chinese Journal of Aeronautics* 33.3 (2020). Received 8 January 2019; revised 4 November 2019; accepted 4 November 2019, pp. 826–839. DOI: 10.1016/j.cja.2019.11.005.

[14] European Commission. *Section 9 – Noise Legislation*. Date of access: 13 Oct. 2022. 2022. URL: https://ec.europa.eu/environment/archives/enlarg/handbook/noise.pdf.

[15] Ralf Johannes Keuter et al. "Design Decisions for a Powertrain Combination of Electric Motor and Propeller for an Electric Aircraft". In: *IEEE Access* 11 (July 2023). Received 22 June 2023, accepted 10 July 2023, date of publication 28 July 2023, date of current version 2 August 2023, pp. 79144–79155. DOI: 10.1109/ACCESS.2023.3299816.

[16] Axel Schulz. *Development of a Low Noise Drone Propeller Using CFD Simulations*. Degree project for the degree of Master of Science in Engineering
Supervisor: Robert-Zoltán Szász
Examiner: Prof. Johan Revstedt
Opponent: Mitra Malekpour. Division of Fluid Mechanics, Department of Energy Sciences, Faculty of Engineering, Lund University, June 2023.

[17] *International Civil Aviation Organization (ICAO) Annex 16 - Environmental Protection - Volume I - Aircraft Noise*. Standards and Recommended Practices for aircraft noise certification. Accessed on January 8, 2024. 2017. URL: https://store.icao.int/en/annex-16-environmental-protection-volume-i-aircraft-noise.

13th South African Conference on Computational and Applied Mechanics (SACAM2024)
22 – 23 January, 2024, Stellenbosch, South Africa

CFD ANALYSIS OF CONVECTION COOLING OF THE NWU SOLAR CAR PV WING

Düter Struwig[1*], Jan-Hendrik Kruger[2†], Johan Bosman[3]

* Presenting author
†Corresponding author: jandhendrik.kruger@nwu.ac.za
[1,2,3]North-West University, School of Mechanical Engineering
Private Bag X6001, Potchefstroom, 2520

Abstract:

Optimising the efficiency of solar vehicles is highly dependent on the performance of their photovoltaic (PV) cells used to harvest solar energy. The operating temperature, which increases to 60-80°C, significantly impacts efficiency, dropping with each degree above 25 °C, which is the standard operating temperature. The early designs of NWU solar cars relied on air passing over the top of the wing to cool the PV cells, while the latest model, Naledi 2.0, features internal airflow through the wing to cool cells and increase efficiency. However, the precise impact on the drag and temperature distribution across the PV panel remains unknown.

Ansys Fluent software was used to accurately model drag and temperature. The simulations focused first on aerodynamic drag only, employing momentum and boundary layer equations, and precisely modelled the turbulent flow around the car and the wing. Heat-transfer simulations then integrated momentum and thermal equations, a solar ray-tracing algorithm, and the discrete-ordinate radiation model to accurately represent energy and fluid flow domains simultaneously.

For the particular test case, it was found that the modification yielded an increase of 8.15% in the drag coefficient, a 54.13% increase in the heat transfer from the wing, a 3.35% decrease in the panel temperature, and a 3.11% increase in the efficiency of the photovoltaic panel, which yielded a higher power output of 124.4W from the PV panel.

Key words: Solar car, PV panel efficiency, convection cooling, solar energy modelling

1 Introduction

1.1 Background

The automotive industry is currently undergoing a significant transformation, driven by the urgent need for sustainable transportation solutions. Electric vehicles (EVs) are widely recognised as a critical component of this shift, offering a zero-emission alternative to conventional gasoline-powered vehicles. The drawback of most current electric vehicles is their range limitations.

Solar energy presents a promising solution to eliminate this dependency, allowing electric vehicles to directly use or store the energy generated by photovoltaic (PV) panels for extended operation. The Naledi 2.0 solar car, developed by North-West University (NWU), demonstrates this approach, as seen in Figure 1 [1], also showing the CAD geometry used for simulations and the internal geometric features that allow internal flow and cooling of the PV wing. This impressive vehicle showcases a well-designed integration of PV panels and a lightweight and aerodynamic frame, maximising its efficiency and performance. By harnessing the energy of the Sun, the Naledi 2.0 can travel considerable distances without requiring any external charging.

(a) Naledi 2.0 [1] (b) CAD Geometry (c) Internal Structure

Figure 1: Geometric Representations of the Naledi 2.0 Solar Car

To extract maximum performance from Naledi 2.0, it is crucial to have a design that optimises the collection efficiency of the PV panels. The increase in panel efficiency directly translates into increased power generation for use by the solar car. Photovoltaic cells are very sensitive to a change in temperature, with a decrease in temperature resulting in an increase in their electrical efficiency [2]. Therefore, effective cooling systems can possibly enhance the performance of the solar car.

The Naledi 2.0 design achieves this cooling by an opening that runs across the leading and trailing edge of the wing. This forces air to flow internally, between the top and bottom halves of the wing, essentially allowing air to pass through the wing, which cools the panels by means of convection.

1.2 Problem Description

The effectiveness of the PV panels used for the NWU solar car can be improved if cooled sufficiently. Therefore, the construction technique used to manufacture the wing plays a crucial role in the heat transfer from the PV cells, with the internal flow within the wing providing the heat removal from the panel surface. The problem is that the detailed flow patterns and its effects on heat transfer from the panel are unknown.

1.3 Purpose

The purpose of this study is to use CFD analysis and experiments to get better insights into additional drag, heat transfer, and panel efficiency, due to internal flow patterns through the modified PV wing of the NWU Naledi 2.0 solar car.

2 Simulation Methodology

Two software packages were used for the simulations performed during the course of this study, XFOIL [3] and Ansys Fluent [4]. Each package offers a certain simulation approach for a specific purpose.

Validation: To ensure the accuracy of the Ansys Fluent simulation setups, a validation simulation was performed on a basic aerofoil within XFOIL and Ansys Fluent. Subsequent simulations primarily utilised Ansys Fluent, due to XFOIL's limitations in only being able to analyse two-dimensional aerofoils.

Focus on Ansys Fluent: Ansys Fluent's broader capabilities, such as its thermal simulation capabilities, were crucial for analysing heat transfer in the intricate solar car geometry.

Exploratory Simulations: Prior to commencing simulations for the Naledi 2.0 solar car, a series of exploratory simulations were conducted on a modified version of the basic aerofoil used for validation purposes. These simulations aimed to establish concrete setups and procedures for the Naledi 2.0 simulations, exploring two key areas of interest, that is, the drag caused by the internal flow and the effect of the internal flow on the temperature distribution across the wing surface.

Internal Flow Simulations: Dedicated simulations investigated internal flow patterns within the Naledi 2.0 wing that focused on the drag coefficient and the temperature distribution throughout the wing surface.

Comprehensive Simulations: These comprehensive and integrated simulations of the entire vehicle provided valuable information on the complex fluid dynamics and thermal behaviour of the wing surface and its inside volume.

2.1 Naledi 2.0 Model Development

Two wing configurations were considered, one with and one without internal flow. To facilitate a meaningful comparison between the two configurations, simulations were conducted for both configurations. Consequently, the solar car model incorporated two wing designs: one that impeded internal flow and the other that facilitated it. For temperature simulations, the PV wing was required to be represented as a surface body with zero thickness, as seen in Figure 1b. Siemens NX [5] was used for all geometric modifications of the solar car and Ansys SpaceClaim [4] was used for the creation of the flow domain, as discussed later in this paper.

2.2 The Simulation Problem

From here on, the article will shift focus to the procedures performed and the results obtained from the temperature simulations on the Naledi 2.0, as this was the more complex problem of the project.

Computational Fluid Dynamics (CFD) simulations for bodies without internal flow are relatively straightforward. The process involves creating a flow domain that encompasses the body and then subtracting the body itself, resulting in a void that allows the analysis of the flow characteristics around the body. However, the introduction of heat energy into the equation significantly increases the complexity of the flow domain and the simulation.

2.2.1 Baffle Surfaces

The primary challenge arises from the need to account for temperature and heat transfer across the Naledi 2.0 wing. The conventional method of creating a void in the domain becomes unsuitable to simulate heat transfer through a wall. This is because thermal conductivity, which describes the ability of a material to conduct heat, is a fundamental property of the wall [6]. Therefore, meshing the structure of the PV wing becomes essential to provide both thermal conductivity and a thermal boundary condition.

The meshing of the PV wing thickness as a separate solid-cell zone could be a potential solution. However, the photovoltaic wing itself comprises three distinct material layers: a gel top layer, photovoltaic cells, and finally the aluminium wing structure. These layers are very thin, necessitating an extremely fine mesh to accurately represent each one. This impracticality introduces the requirement of a surface body with zero thickness. In Ansys Fluent, such a surface is referred to as a baffle, which is essentially a shell-type element with no thickness.

2.2.2 Challenges Associated with Baffle Implementation

Flow Continuity: Baffle surfaces are disconnected from the main flow domain, preventing accurate drag simulations. This discontinuity allows air to flow through the baffle, invalidating its intended effect of restricting flow. To overcome this issue, the "share topology" function in Ansys Fluent can be employed to obtain a conformal mesh between the domain and baffle.

Thermal Conductivity: Baffles, with zero thickness, lack the necessary physical properties to accurately represent heat transfer through the walls they model. This is particularly problematic for simulating conductive heat transfer, which requires the thickness of the material to be accurate. The solution lies in the use of one of the conductive heat transfer settings in Ansys Fluent.

2.2.3 Conductive Heat Transfer through Walls

Ansys Fluent offers three distinct approaches for simulating conductive heat transfer through a wall:

Solid Cell Zone: This method involves explicitly meshing the wall as a separate solid cell zone, allowing heat transfer in both the normal and the planar directions. It is considered the most accurate and comprehensive approach [7]. However, as previously mentioned, the presence of three thin layers in the Naledi 2.0 PV wing would necessitate an extremely fine mesh, rendering this option impractical due to computational limitations.

One-Dimensional Thin Walls: This approach involves meshing only the fluid region while assigning a thickness to the wall surface [7]. This option accounts for heat transfer only in the normal direction and is therefore not suitable for the Naledi 2.0 PV wing, as heat transfer is required in all directions.

Three-Dimensional Conduction Shell Walls: This method employs Ansys Fluent's "shell conduction" function as a boundary condition for the baffle surface. Shell conduction enables simulation of heat transfer through a wall without the need for explicit meshing. This approach also allows the assignation of an effective thermal conductivity to the wall surface [7]. When enabled, 1 or more layers of virtual cells (prism or hex) are automatically generated on the baffle surface. The shell conduction setting requires specifying the wall thickness and material type or thermal conductivity, which are then used for simulation purposes. This approach allows an accurate simulation of heat transfer through the photovoltaic wing, in all directions, while avoiding the computational demands associated with meshing thin layers.

2.2.4 Energy Source

The application of shell conduction as a boundary condition to a wall introduces a crucial limitation: it prevents the simultaneous application of a heat flux to the same wall. To overcome this constraint and provide an energy source for the simulation, alternative methods must be considered. Ansys Fluent offers a valuable tool, known as solar load. This setting allows for the application of a realistic representation of solar energy as an energy source within the flow domain [8]. The solar irradiance value for the solar load can be either specified directly by the user or automatically calculated using an integrated solar calculator.

For more accurate simulations, another tool inside Ansys Fluent was used, which is the solar ray-tracing algorithm. The solar ray tracing algorithm in Ansys Fluent predicts the Sun's direct impact on the solar car model. This is achieved by considering the Sun's position, date, time, and model orientation. Two solar radiation models are used: one for direct sunlight, accounting for infrared (IR) and visible bands, and the other for diffuse sunlight. Specification of how boundary conditions interact with solar radiation, including absorption, transmission, and scattering, is allowed [8]. The algorithm also determines the shadow positions and computes the heat flux on all surfaces. Together with the discrete-ordinates (DO) radiation model, which accounts for the radiative heat transfer within the domain itself, the simulation could be performed under conditions very close to what would be experienced in the real-world scenario.

3 Simulation Setup

Flow Domain: The size of the flow domain, as shown in Table 1, was carefully selected to ensure that it is large enough, minimising its effect on the simulation results, but also as small as possible to reduce the computational expense of the simulations. A guideline from [9] was used to select the size of the flow domain as a multiple of the length of Naledi 2.0, which is five metres.

Meshing: Figure 2 shows the named selections used for the temperature simulations. These are added

Table 1: Flow Domain Size for Temperature Simulations

Direction	Body Lengths	Domain Length [m]
Upstream	0.2	1
Downstream	0.4	2
Above	0.4	2
Below	0.2	1
Side	0.4	2

to the geometric features of the solar car to be able to separate the regions from each other during the setup and solution of the simulation and to provide adequate mesh refinements to the regions of interest. Considerations were made between the mesh size and the computational expense of the simulations. This was done to ensure that accurate results were obtained without having unnecessary long simulation times. For adequate mesh sizing near the solar car, a body of influence (BOI) was used for mesh refinement. The final size of the meshed domain and geometry included 10 million cells, 46 million faces, and 27 million nodes.

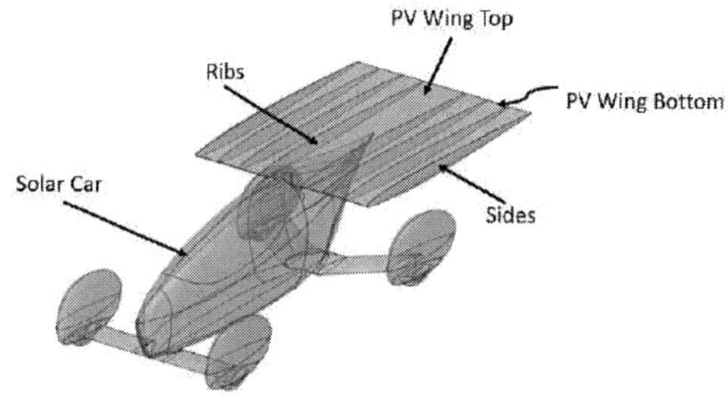

Figure 2: Named Selections used for the Temperature Simulations

Simulation Settings: Finally, the simulation was setup inside Ansys Fluent, using the $k - kl - \omega$ turbulence model, as this was found to be the most accurate in modelling the flow from laminar to turbulent, in and around the solar car [10]. For all simulations, the fluid used was air, with the density calculated under the specific atmospheric conditions present on the day of experimental testing.

4 Temperature Simulation Results

Thermal simulations were conducted under two sets of boundary conditions. First, experimental conditions (measured average speed and irradiance) were applied to validate the model against experimental results. Subsequently, the standardised conditions (65 km/h airspeed, 1000 W/m^2 irradiance) allowed a direct comparison between the closed and open configurations of the wing. Under these conditions, the closed-wing configuration exhibited an average PV panel temperature of 39.98°C, while the open-wing configuration reached 38.64°C.

The impact of internal flow on the thermal distribution within the wing was further analysed using simulated temperature contours. The comparison between Figures 3a and 3b and Figures 4a and 4b clearly shows that internal flow effectively cools the wing surface (including the PV panel).

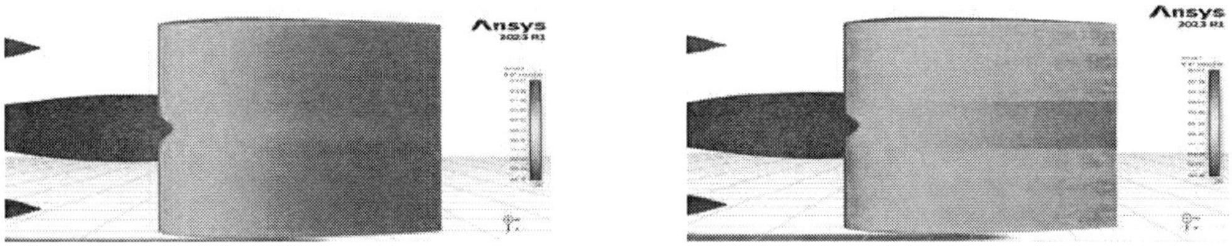

(a) Closed Configuration (b) Open Configuration

Figure 3: Temperature Distribution across the PV Wing for both Configurations

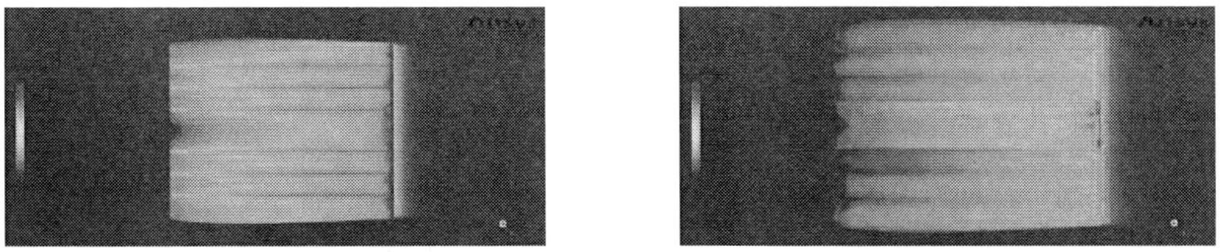

(a) Closed Configuration (b) Open Configuration

Figure 4: Air Temperature Inside the Wing for both Configurations

5 Experimental Results and Theoretical Calculations

5.1 Temperature Measurements

Eight temperature sensors were used to measure the temperature of the air and structure at several different locations during the experiment. Four experimental runs measured the temperatures of the outlet air at the trailing edge (TE), through six internal channels, and two locations beneath the wing for both closed and open-wing configurations. Comparison of measured outlet temperatures with those obtained during simulations produced differences of 2.77% and 3.79% for closed and open configurations, respectively, demonstrating a close correspondence between the experimental data and the simulated results, thus validating the thermal simulations.

6 Theoretical Calculations

6.1 Heat Transfer

Using the validated temperature simulation models, similar boundary conditions were applied to the closed and open configuration. These simulations provided results that were further used to perform the necessary heat transfer calculations. The energy balance in both configurations of the wings is represented in Figure 5. The two main modes of heat transfer from the PV wing are convection and radiation, which can be calculated by Equations 1 [11] and 2 [6], respectively.

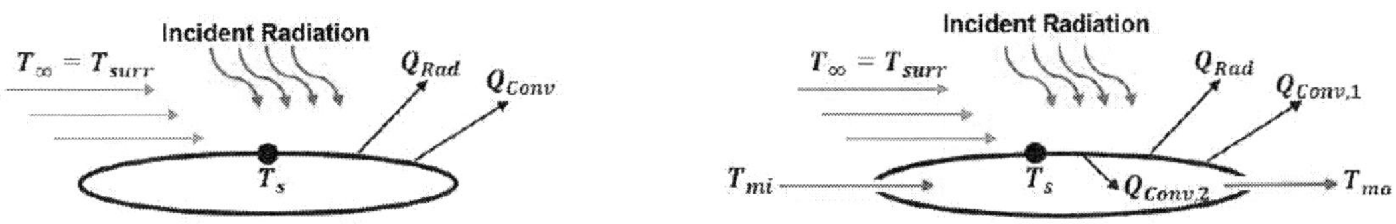

(a) Closed Configuration (b) Open Configuration

Figure 5: Energy Balance on the PV Wing for both Configurations

$$\dot{Q}_{Conv1} = hA(T_s - T_\infty) \tag{1}$$

$$\dot{Q}_{Rad} = \varepsilon\sigma A(T_s^4 - T_{surr}^4) \tag{2}$$

For internal flow, an additional mode of heat transfer is introduced, which is the heat transferred by convection to the air inside the wing, which can be calculated with Equation 3, which is the simplified steady-flow thermal energy equation for internal flow [12].

$$\dot{Q}_{Conv2} = \dot{m}c_p(T_{mo} - T_{mi}) \tag{3}$$

The assumptions that make this equation valid are as follows [11]:

- The fluid moves at a constant mass flow rate $\dot{m}$,
- Viscous dissipation is negligible and
- The fluid can be modelled as either an incompressible liquid or an ideal gas with negligible pressure variation.

Calculating the heat transfer from both configurations of the wing yields the results shown in Table 2. Observing the total heat transfer from the wing, there is a substantial increase in the heat transfer when comparing the closed to the open wing configuration, highlighting the effect of the internal flow.

Table 2: Theoretical Heat Transfer Results

Heat Transfer Mode	Closed Configuration [W]	Open Configuration [W]
$\dot{Q}_{Conv1}$	1921.11	1663.17
$\dot{Q}_{Conv2}$	-	1442.42
$\dot{Q}_{Rad}$	212.03	182.31
$\dot{Q}_{Total}$	**2133.14**	**3287.90**

6.2 PV Panel Efficiency

During the experimental runs, the power output of the PV panels and solar irradiance were also recorded. Solar irradiance measurements were taken at thirty-minute intervals. These measurements were used to estimate solar irradiance for each minute during the temperature experiment, as provided in Figure 6.

Equation 4 [13] provides the means of calculating the efficiency of solar panel electric collection:

$$\eta = \frac{P_{PV}}{(G_n \cdot A)} \tag{4}$$

Calculating the average efficiency over four experimental runs, an efficiency of 15.33% was obtained for the closed configuration, while for the open configuration it was found to be 18.44%. Clearly, the efficiency of the panel is improved, improving the performance of the solar car.

7 Conclusion

From the study, it was found that internal air flow through the photovoltaic wing of the Naledi 2.0 had several effects on drag, temperature and electric collection efficiency. Although not discussed in detail in this paper, additional simulations and experiments found that the drag coefficient of the solar car increased by 8.15% with the open wing. Using validated simulations in Ansys Fluent, together with measured experimental results and theoretical calculations, it was observed that the heat transfer from

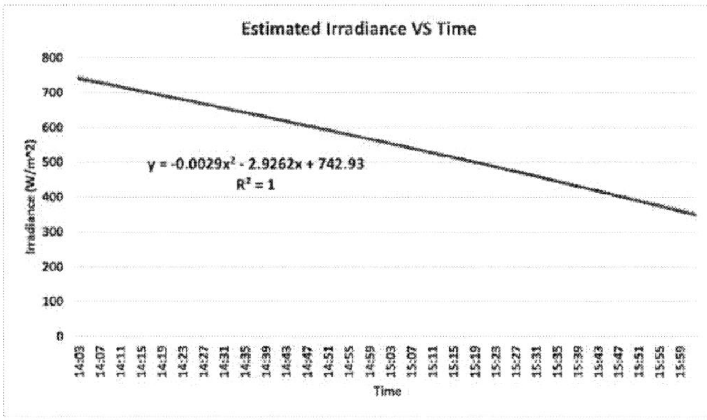

Figure 6: Estimated Solar Irradiance Calculated with Measured Data

the wing increased by 54.13% due to the cooling effects of internal flow. This resulted in a decrease in the average temperature of the PV panel of 3.35%, with an energy source of 1000 W/m^2 and a speed of 65 km/h. Furthermore, the decrease in the temperature of the PV panel produced an operating temperature closer to the standard operating temperature of 25°C. An increase of 3.11% was observed in the efficiency of the PV panel, which, based on theoretical calculations, would result in a 124.4W increase in the power output of the PV panel at 1000 W/m^2.

It can be concluded that the design modification of the Naledi 2.0 to allow internal flow has both positive and negative effects on the overall performance of the car. To fully comprehend these effects, it is recommended that further investigation is performed over longer duration events (such as the cross-country challenge) to determine if the positives outweigh the negatives and to deem the modification successful.

Nomenclature

Acronyms and Abbreviations

BOI	Body of Influence
CAD	Computer Aided Design
CFD	Computational Fluid Dynamics
DO	Discrete Ordinates
EVs	Electric Vehicles
IR	Infrared
NWU	North-West University
PV	Photovoltaic
TE	Trailing Edge

List of Symbols

$\dot{m}$	Mass flow rate	kg/s
$\dot{Q}$	Rate of heat transfer	W
ε	Surface emissivity	-
η	Efficiency	-
σ	Stefan-Boltzmann constant	W/m^2K^4

A	Area	m^2
c_p	Specific heat	kJ/kgK
G	Incident irradiance	W/m^2
h	Convective heat transfer coefficient	W/m^2K
P	Power	W
T	Temperature	°C or K

List of Subscripts

∞	Indicating the value of the property at the far-field
Conv	Convection
i	Inlet conditions
m	Mean
n	Global property
o	Outlet conditions
PV	Photovoltaic
Rad	Radiation
s	Indicating the surface of an object
surr	Indicating the value of the property at the surroundings

Acknowledgement

I wish to extend my sincere thanks to my project leader, Dr. Jan-Hendrik Kruger. Thank you for your unwavering support throughout this project. Your guidance and assistance were invaluable in helping me complete this project successfully. It was an honour to have you as my project leader. I would like to acknowledge my peer student, Etienne Kruger, who also did a project on the Naledi 2.0 solar car. Thank you for the many conversations and support that you have provided throughout the year. Your help is greatly appreciated. Lastly, I also would like to thank my family and friends for their support. It kept me motivated throughout the project.

References

[1] NWU. *Home — engineering.nwu.ac.za*. 2023. URL: https://engineering.nwu.ac.za/solarcar (visited on 04/11/2023).

[2] Ahmed Hussien, Abdelgalil Eltayesh, and Hesham M. El-Batsh. "Experimental and numerical investigation for PV cooling by forced convection". en. In: *Alexandria Engineering Journal* 64 (Feb. 2023), pp. 427–440. ISSN: 11100168. DOI: 10.1016/j.aej.2022.09.006. URL: https://linkinghub.elsevier.com/retrieve/pii/S1110016822005932 (visited on 02/21/2023).

[3] M Drela and H Youngren. *XFOIL Subsonic Airfoil Development System*. 2013. URL: http://web.mit.edu/drela/Public/web/xfoil/ (visited on 03/31/2023).

[4] ANSYS. *Ansys Fluent — Fluid Simulation Software*. en-US. 2023. URL: https://www.ansys.com/products/fluids/ansys-fluent (visited on 03/31/2023).

[5] Siemens. *Siemens NX*. Munich, 2020.

[6] Frank P Incropera et al. "Introduction". English. In: *Incropera's Principles of Heat and Mass Transfer (Global Edition)*. 8th ed. Singapore: John Wiley & Sons, 2017, pp. 2–9. ISBN: 978-1-119-38291-1.

[7] Qfinsoft. "Ansys Fluent Getting Started, Lecture 12: Heat Transfer". PowerPoint. PowerPoint. Ansys FLUENT Student Training – Day 1. Online, June 6, 2023. URL: https://qfinsoft.co.za/ (visited on 06/06/2023).

[8] *ANSYS FLUENT 12.0 User's Guide - 13.3.9 Solar Load Model.* ANSYS FLUENT 12.0/12.1 Documentation. Jan. 29, 2009. URL: https://www.afs.enea.it/project/neptunius/docs/fluent/html/ug/node481.htm (visited on 01/10/2024).

[9] Gestione. *CFD computational domain.* en-GB. Sept. 2020. URL: https://www.idealsimulations.com/resources/cfd-computational-domain/ (visited on 07/04/2023).

[10] ANSYS. *ANSYS FLUENT 12.0 Theory Guide - 4.6.2 Transport Equations for the - Model.* 2009. URL: https://www.afs.enea.it/project/neptunius/docs/fluent/html/th/node71.htm (visited on 03/31/2023).

[11] Frank P Incropera et al. "Internal Flow". English. In: *Incropera's Principles of Heat and Mass Transfer.* 8th ed. Singapore: John Wiley & Sons, 2017, pp. 475–491. ISBN: 978-1-119-38291-1.

[12] Frank P Incropera et al. "Internal Flow". English. In: *Incropera's Fundamentals of Heat Transfer.* 7th ed. USA: Don Fowley, 2011, pp. 517–592. ISBN: 10987654321.

[13] D. Lopez-Pascual et al. "Experimental characterization of a geothermal cooling system for enhancement of the efficiency of solar photovoltaic panels". en. In: *Energy Reports* 8 (Nov. 2022), pp. 756–763. ISSN: 23524847. DOI: 10.1016/j.egyr.2022.10.154. URL: https://linkinghub.elsevier.com/retrieve/pii/S235248472202090X (visited on 03/22/2023).

START-UP AND TRANSIENT RESPONSE CHARACTERISTICS OF NATURAL DRAFT DIRECT DRY COOLING SYSTEMS AT VARIOUS SCALES

Wian Strydom[1,*], Johannes Pretorius[1], Ryno Laubscher[1]

[1]Department of Mechanical and Mechatronic Engineering, University of Stellenbosch,[*]wian.str.98@gmail.com

Abstract

Natural draft direct dry cooling systems (NDDDCSs) combine the benefits of two traditional dry cooling systems, air-cooled condensers (ACCs) and indirect natural draft dry cooling systems, to offer advantages such as reduced system complexity, high thermal efficiency, and low parasitic power losses and operational costs. These scalable systems can potentially operate in both large coal-fired power plants (900 MWt) and medium-scale concentrated solar power (CSP) plants (100 MWt). While prior research focused on NDDDCS steady-state performance, this study delves into the transient behaviour and start-up characteristics of this cooling system, crucial for power plants with rapid start-up requirements or load ramps during electricity demand peaks. The paper employs a transient 1-D mathematical model to characterize the NDDDCSs' response to start-up and load ramp conditions under no-wind, presenting results for both large- and medium scales while identifying limiting factors.

Keywords: Natural draft, transient, start-up, condenser, dry cooling

1. Introduction

Climate change challenges the energy generation industry, through increasing water scarcity [1], energy demand, and global population growth [2]. These factors necessitate a shift towards more resource-efficient energy generation technologies. Dry cooling technologies, with reduced water usage compared to wet cooled systems [3], are natural candidates.

Two prevalent dry cooling systems, forced draft air-cooled condensers (ACCs) and indirect natural draft dry cooling systems, offer distinct advantages. The ACCs operate using a multitude of large fan drives and have a high thermal efficiency. However, the fans run virtually continuously in harsh conditions and require regular maintenance. These systems also have significant parasitic losses. Indirect natural draft dry cooling systems, while avoiding fan drives, face challenges in reduced thermal efficiency and high complexity which demands substantial initial capital investments [4].

The natural draft direct dry cooling system (NDDDCS) combines the strengths of ACCs and indirect systems, offering higher thermal efficiencies and lower operational costs. The absence of fan drives eliminates several failure modes, as a natural draft drives passive airflow. Limited research has been performed on NDDDCSs to date, and have focused on steady-state performance. Studies by Kong et al. using 3-D computational fluid dynamics (CFD) simulations highlight the superiority of vertically arranged heat exchanger bundles, with increased performance associated with larger apex angles [5, 6]. Normal and radiator type windbreaker walls further enhance performance under crosswind conditions [7, 8].

Following from Kröger's work on ACCs and indirect systems [9], 1-D modelling techniques have proven viable for NDDDCS investigations. Strydom et al.'s sensitivity analysis emphasized the impact of geometric variation on NDDDCS performance, while also showcasing its potential as a standalone cooling system for CSP plants [10, 11].

This study addresses a gap in the literature by exploring the start-up characteristics of NDDDCSs, crucial for power generation cycles during rapid start-ups or demand peaks. Transient performance is gaining significance as intermittent renewable technologies become more prevalent on the grid. Existing transient investigations have focussed on small-scale indirect natural draft dry cooling towers [12, 13, 14], leaving a knowledge gap concerning NDDDCS start-up. The NDDDCS relies on the interdependence of the condensation process and system draft, unlike indirect systems which mechanically pump water into the heat exchangers.

This research develops transient 1-D mathematical models to quantify NDDDCS start-up and transient load ramp characteristics at two scales: large-scale coal-fired power plant (900 MWt) and medium-scale CSP plant (100 MWt). Results shed light on NDDDCS response time and potential bottlenecks in start-up and load ramp scenarios.

2. Reference Tower Geometry and Sizing

A schematic of the NDDDCS tower and finned tube geometry is shown in Figure 1, with the reference tower dimensions used in this study listed in Table 1. The tube dimensions, heat transfer- and pressure drop characteristics are identical to those used in previous studies [10, 11]. The total tower height is given by H_5, the inlet height is represented by H_4 and the inlet- and outlet diameters are indicated by d_3 and d_5 respectively.

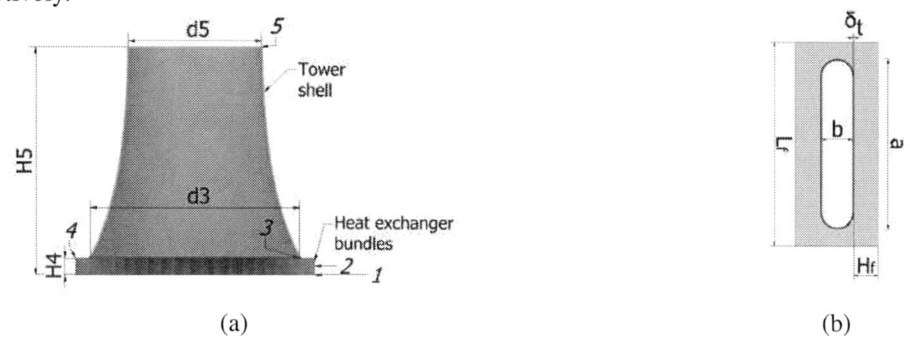

Figure 1: Schematic of NDDDCS (a) and effective single-row finned tube (b)

Table 1: NDDDCS reference dimensions at the two considered scales

Dimension	Symbol	Large-scale	Medium-scale
Total tower height (m)	H_5	165.0	65.0
Inlet height (m)	H_4	16.5	6.5
Inlet diameter (m)	d_3	165.0	65.0
Outlet diameter (m)	d_5	82.5	32.5

3. Mathematical Model

The steam- and air-side of the 1-D model are not spatially discretised. The governing equations are solved simultaneously, where required. Transient continuity- and energy equations are derived for the steam-side, while transient continuity-, momentum- and energy equations are derived for the air-side. It is assumed that there is no steam-side pressure drop within the condenser tubes. The density at the next time step on the steam-side is calculated via:

$$\rho_{si}{}^{t+\Delta t} = \frac{\Delta t}{V_c}\left(\dot{m}_{si} - \dot{m}_{co}^{\;t}\right) + \rho_{si}{}^{t} \quad (1)$$

where $\rho_{si}{}^{t+\Delta t}$ is the steam density at the next time step, Δt is the time step, V_c is the total steam-side volume, $\dot{m}_{si}$ the supplied steam mass flow rate boundary condition from the power cycle, $\dot{m}_{co}^{\;t}$ is the mass flow rate of exiting condensate and $\rho_{si}{}^{t}$ is the current steam density. The internal energy of the steam is calculated by:

$$u_{si}{}^{t+\Delta t} = \frac{\Delta t}{\rho_{si}{}^{t} V_c}\left(\dot{m}_{si} h_{si}{}^{t} - \dot{m}_{co}^{\;t} h_{co}{}^{t} - \dot{Q}_r{}^{t} - u_{si}{}^{t}(\dot{m}_{si} - \dot{m}_{co}^{\;t})\right) + u_{si}{}^{t} \quad (2)$$

where $u_{si}{}^{t+\Delta t}$ is the internal energy of steam at the next time step, $u_{si}{}^{t}$ is the current internal energy, $h_{si}{}^{t}$ and $h_{co}{}^{t}$ are the respective enthalpies of entering steam and exiting condensate and $\dot{Q}_r{}^{t}$ is the steam-side heat transfer rate. The latter is given by:

$$\dot{Q}_r{}^{t} = \dot{m}_a{}^{t} c_{pa25}{}^{t} e^{t} (T_w{}^{t} - T_{a2}) = h_c{}^{t} A_c (T_{si}{}^{t} - T_w{}^{t}) \quad (3)$$

where $T_w{}^{t}$ represents the tube wall temperature, T_{a2} the inlet air temperature, $T_{si}{}^{t}$ represents the saturated inlet steam temperature, $\dot{m}_a{}^{t}$ is the current air mass flow rate, $c_{pa25}{}^{t}$ is the average specific heat capacity of air, e^{t} is the heat exchanger effectiveness, $h_c{}^{t}$ is the condensation heat transfer coefficient and A_c is the total steam-side wetted area. The tube wall temperature is assumed to be uniform between the inner and outer surfaces of the condenser tubes. Rearranging Equation 3 to calculate the tube wall temperature yields:

$$T_w{}^{t} = \frac{h_c{}^{t} A_c T_{si}{}^{t} + \dot{m}_a{}^{t} c_{pa25}{}^{t} e^{t} T_{a2}}{h_c{}^{t} A_c + \dot{m}_a{}^{t} c_{pa25}{}^{t} e^{t}} \quad (4)$$

The heat exchanger effectiveness, e^{t} is given by:

$$e = 1 - \exp\left(\frac{UA^{t}}{\dot{m}_a{}^{t} c_{pa25}{}^{t}}\right) \quad (5)$$

The overall heat conductance, UA, is equivalent to the air-side heat conductance. The mass flow rate of condensate leaving the condenser tubes is calculated as:

$$\dot{m}_{co}^{\;t} = \dot{Q}_r{}^{t} / h''^{\;t}_{fg} \quad (6)$$

where $h''^{\;t}_{fg}$ is the latent heat of vaporization, accounting for superheating effects. The air-side heat transfer, $\dot{Q}_a{}^{t}$, is given by:

$$\dot{Q}_a{}^{t} = \dot{m}_a{}^{t} (h_{a5}{}^{t} - h_{a2}) \quad (7)$$

where ${h_{a5}}^t$ and h_{a2} are the enthalpies of entering and exiting air, respectively. The air mass flow rate at the next time step is calculated via:

$$\dot{m}_a{}^{t+\Delta t} = \frac{A_{35}\Delta t}{L_a}\left(\Delta {P_a}^t - \Delta {P_L}^t\right) + \dot{m}_a{}^t \tag{8}$$

where $\Delta {P_a}^t$ is the system driving force, $\Delta {P_L}^t$ is the total air-side pressure loss, A_{35} is the average tower flow area and L_a is the air travel distance. The pressure balance within the bracketed term of Equation 8 is equivalent to the draft equation taken from the work of [9]. Thus, the driving force is:

$$\Delta {P_a}^t = p_{a1}\left[\left(1 - 0.00975\frac{H_4}{2T_{a1}}\right)^{3.5}\left(1 - 0.00975\frac{H_5 - \frac{H_4}{2}}{{T_{a3}}^t}\right)^{3.5} - \left(1 - 0.00975\frac{H_5}{T_{a1}}\right)^{3.5}\right] \tag{9}$$

where p_{a1} is the ground-level atmospheric pressure, T_{a1} is the corresponding temperature and ${T_{a3}}^t$ is the outlet air temperature after the heat exchangers. The flow resistance is given by:

$$\Delta {P_L}^t = \frac{(K_{il} + {K_{he\theta}}^t + {K_{ct}}^t + {K_{ts}}^t)_{he}}{2{\rho_{a25}}^t}\left(\frac{\dot{m}_a{}^t}{A_{fr}}\right)^2\left(1 - 0.00975\frac{H_5 - \frac{H_4}{2}}{{T_{a3}}^t}\right)^{3.5} + \frac{({K_{to}}^t + {a_{e5}}^t)}{2{\rho_{a5}}^t}\left(\frac{\dot{m}_a{}^t}{A_5}\right)^2 \tag{10}$$

where K_{il} (constant value of 2.5), ${K_{he\theta}}^t$, ${K_{ct}}^t$, ${K_{ts}}^t$, ${K_{to}}^t$ and ${a_{e5}}^t$ represent the non-dimensional inlet louvre-, heat exchanger-, tower-, tower support-, outlet pressure loss and velocity correction factors. The tower outlet air density is represented by ${\rho_{a5}}^t$, the corresponding area is given by A_5 and the frontal area of the heat exchangers is represented by A_{fr}. The enthalpy of air leaving the NDDDCS is calculated via:

$${h_{a5}}^{t+\Delta t} = \frac{\Delta t}{{\rho_{a25}}^t V_a}\left(\dot{m}_a{}^t(h_{a2} - {h_{a5}}^t) + \dot{Q}_r{}^t\right) + {h_{a5}}^t \tag{11}$$

where V_a is half of the internal flow volume. The equations for the loss factors and overall heat conductance are presented in previous papers [10, 11].

4. Validation and Sensitivity Analysis

The 1-D steady state NDDDCS model (SS) was previously validated against the work of Kong et al. [10]. Therefore, this study validates the transient 1-D NDDDCS model by comparing the results of the transient model, once it reaches a steady state, to the results of the steady-state model. The large scale tower is used as reference and the same effective boundary conditions are imposed to allow comparison. It is important to note that two transient models were developed: transient steam-side combined with steady-state air-side (TS) and transient steam- and air-side (TT). The SS model was adapted to make use of Equation 3 and to neglect the steam-side pressure drop for the sake of consistency between the three models. A constant ambient temperature of 293.15 K (20 °C) was selected. The steam mass flow rate (444.98 kg/s) attained by the SS model for a constant saturated steam temperature boundary condition of 323.15 K is assigned as a boundary condition for the TS and TT models. The TT model requires an initial steam temperature and air mass flow rate, specified as the ambient temperature and as effectively zero, respectively. The results are summarized in Table 2. It is clear from Table 2 that the TS and TT models converge to the same steady-state condition as the SS model. A sensitivity analysis revealed that a time-step of 0.2 s is suitable for this simulation.

Table 2: Summary of steady state results for cross validation of TT model

Model	T_{si} (K)	T_w (K)	T_{a5} (K)	$\dot{Q}$ (MW)	$\dot{m}_a$ (kg/s)	ΔP (Pa)
SS	323.15	322.35	315.45	1059.92	47033.71	137.64
TS	323.15	322.35	315.45	1059.92	47033.70	137.64
TT	323.17	322.38	315.48	1059.90	46986.32	137.76

5. Results

All properties are calculated using the CoolProp library within the Python environment. The steam mass flow rate ($\dot{m}_{si}$, step input), ambient temperature (T_{a1}, 293.15 K) and pressure (p_{a1}, 101325 Pa) at ground level are boundary conditions, while the outlet air temperature (T_{a3}, 293.15 K), air mass flow rate ($\dot{m}_a$, $\approx$0 kg/s) and saturated steam temperature (T_{si}, 295.15 K) are initial conditions. Figure 2 serves to illustrate why simulating only the steam-side transiently (TS model) is not adequate to capture the full spectrum of transient start-up effects in NDDDCSs.

In line with operational power plants, the maximum allowable internal pressure is selected as 40 kPa(g) (141.3 kPa(a)), which avoids classification of the system as a pressure vessel. Results indicate that the large-scale NDDDCS can handle 6.25% of its maximum steam admission rate as a step input without exceeding this value. Under this admission rate, the TS and TT models calculate vastly different results. Due to the air-side being calculated steadily for the TS model, the airflow through the NDDDCS can respond instantaneously to changes in the steam-side to provide adequate heat rejection. This is not the case for the TT model, as the response time of the air-side is significantly slower than the steam-side. The initial air mass flow rate is not enough to reject the thermal energy provided by the steam, and an acute build up of steam back pressure is observed until the air mass flow rate is sufficient.

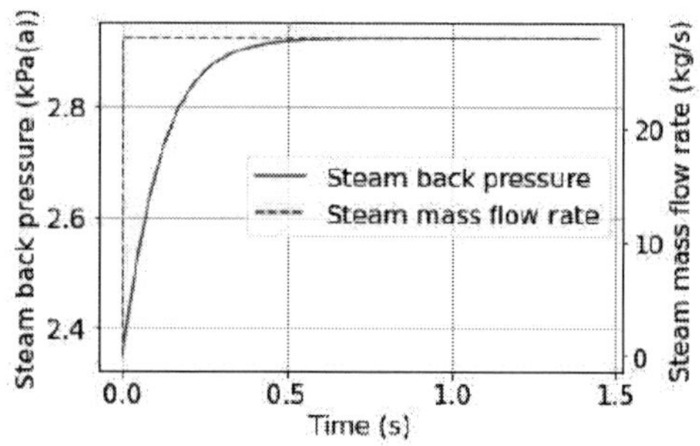

(a)

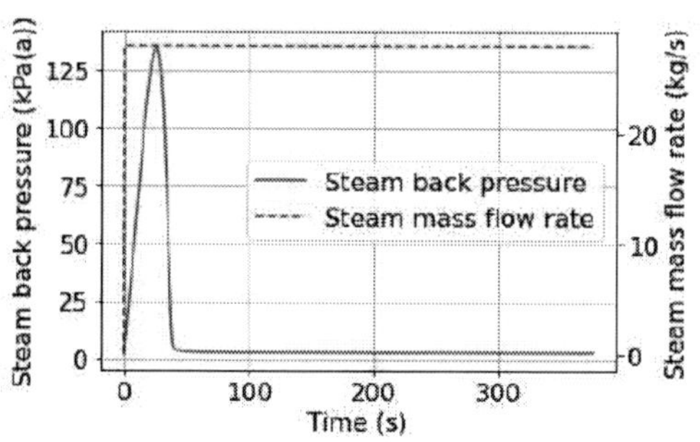

(b)

Figure 2: Large-scale system steam back pressure predicted by TS (a) and TT models (b)

Figure 3 (a, b) shows the transient heat balance and tube wall temperature of the large-scale NDDDCS under the same step input previously mentioned. The heat addition to the tube wall increases acutely due to a rapid increase in steam saturation temperature, accompanied by a build up in tube wall temperature due to a lack of airflow. An inflection point is reached in the heat addition curve, where the heat removal from the tube wall increases and the heat addition to the tube wall settles as the imbalance in heat transfer starts to reduce, leading to a reduction in steam back pressure. The steam-side settles about five times faster (30 s) than the air-side (150 s). NDDDCS performance under two turbine islanding scenarios are also investigated (but not shown), employing common exhaust pressure protection limits of 75 kPa(a) and 55 kPa(a) for prior operation at full and low loads respectively. Our model shows that the NDDDCS can respectively accept a 150% and 125% load increase for these scenarios, indicating dramatically reduced sensitivity to back pressure spikes once operational.

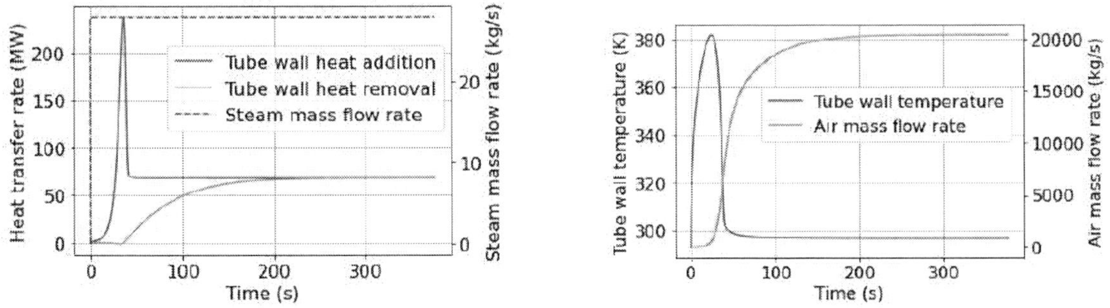

(a) (b)

Figure 3: Large-scale heat balance (a) and tube wall temperature and air mass flow rate (b) graphs using TT model under maximum steam admission rate step input

The large- (coal-fired) and medium scale (CSP) NDDDCS back pressure and air mass flow rate trends are shown in Figure 4 for typical start-up steam admission rates, acquired from proprietary sources.

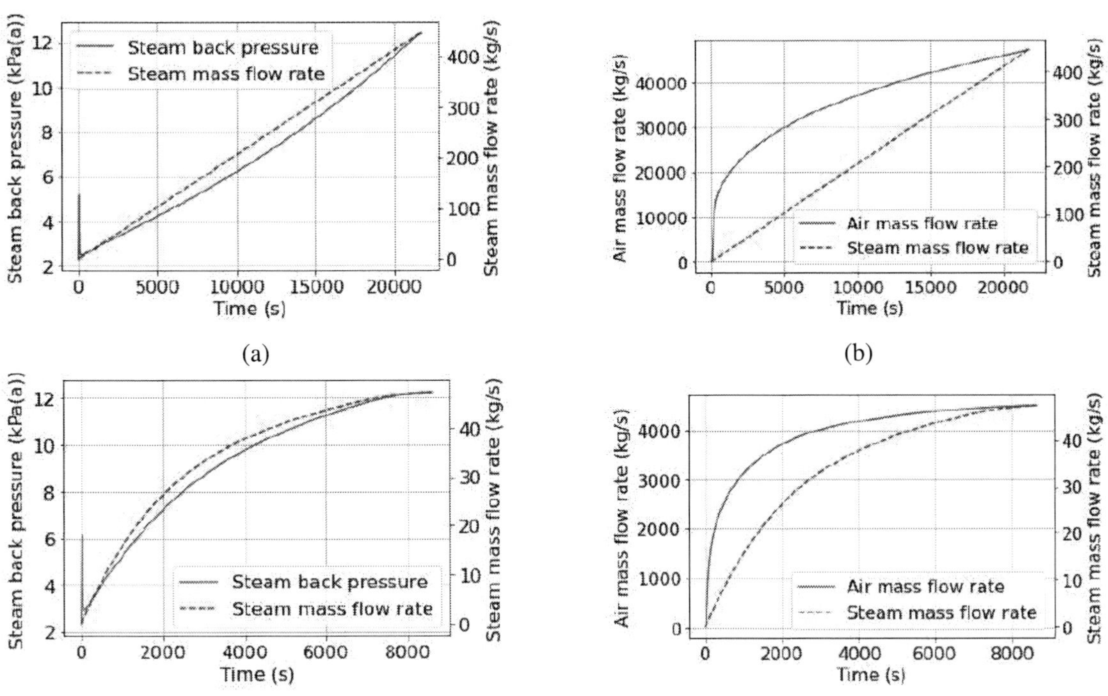

(a) (b) (c) (d)

Figure 4: Steam back pressure (a, c) and air mass flow rate (b, d) curves under typical start-up load ramp conditions for coal-fired power plant (a, b) and CSP plant (c, d) using TT model

The large- and medium-scale NDDDCSs experience controlled spikes in steam back pressure (Figure 4 (a, c)) initially due to airflow in the tower not being established yet, after which the air mass flow rate increases rapidly, as seen in Figure 4 (b, d). As the air mass flow rate continues to increase, the NDDDCSs can easily supply the required heat rejection rate to match the transient cycle requirements, which controls the back pressure. This indicates that for a typical coal-fired or CSP plant, the NDDDCS will not impose transient operation limits, as the fastest cycle load ramp rates are still slower than the response time of the cooling system. This belays past concerns about the transient response time of natural draft steam condensing systems, which rely on the interplay between the condensing process and draft creation.

6. Conclusion

This study investigated the transient start-up characteristics of NDDDCSs at large- and medium-scales under transient start-up and load ramp scenarios, for windless atmospheric conditions. Results indicate that the steam-side responds much faster than the air-side of the system during unsteady start-up conditions. The study finds that for the typical transient scenarios experienced at a coal-fired or CSP plant, the NDDDCS is able to respond rapidly enough to avoid back pressure increases that would limit turbine output during the load ramp. Additionally, once start-up is achieved, the system can manage substantial load increases without exceeding the low pressure turbine protection limits.

Acknowledgements

This work is supported by the National Research Foundation (NRF) of South Africa and the Solar Thermal Energy Research Group (STERG) of the University of Stellenbosch.

Nomenclature

Greek symbols

Δ	Change or thickness (– or m)
δ	Thickness (m)
ε	NTU-effectiveness (–)
μ	Dynamic viscosity (kg/m · s)
ρ	Density (kg/m^3)

Parameters

A	Area (m^2)
a	Major axis or velocity correction factor (m or –)
b	Minor axis (m)
c	Specific heat capacity (J/kg · K)
d	Diameter (m)
g	Gravitational acceleration (m/s^2)
H	Height (m)
h	Heat transfer coefficient or latent heat of vaporization (W/m^2 · K or J/kg)
K	Loss factor (–)
k	Thermal conductivity (W/m · K)
L	Length (m)
$\dot{m}$	Mass flow rate (kg/s)
p	Pressure or pitch (Pa or m)
Pr	Prandtl number (–)
$\dot{Q}$	Heat transfer rate (W)
T	Temperature (K)
UA	Overall heat conductance (W/K)
v	Velocity (m/s)

Subscripts

3	Tower inlet diameter
4	Tower inlet height
5	Tower outlet diameter
a	Air, Atmospheric
c	Condensate, Cooling
f	Fin
fr	Frontal
he	Heat exchanger
i	Inlet, In
L	Loss
l	Louvre
o	Outlet, Out
r	Radiator

s	Steam, Support
t	Tube, Tower
w	Wall
Superscripts	
Δ	Change
t	Time
Abbreviations	
3-D	3-dimensional
ACC	Air-cooled condenser
CFD	Computational fluid dynamics
NDDDCS	Natural draft direct dry cooling system

References

[1] H.J. Schellnhuber and J. Heinke et al. Multimodel assessment of water scarcity under climate change. *Proceedings of the National Academy of Sciences - PNAS*, 111(9):3245–3250, 2014. ISSN 0027-8424.

[2] WEC. World energy scenarios. Technical report, World Energy Council, 2013.

[3] J. Macknick and R. Newmark et al. Review of operational water consumption and withdrawal factors for electricity generating technologies. 2011. doi: 10.2172/1009674.

[4] A.E. Conradie and J.D. Buys et al. Performance optimization of dry-cooling systems for power plants through sqp methods. *Applied Thermal Engineering*, 18(1):25–45, 1998. ISSN 1359-4311.

[5] Y. Kong and W. Wang et al. Thermo-flow performances of natural draft direct dry cooling system at ambient winds. *International Journal of Heat and Mass Transfer*, 116:173–184, 2018. ISSN 0017-9310.

[6] Y. Kong and W. Wang et al. Annularly arranged air-cooled condenser to improve cooling efficiency of natural draft direct dry cooling system. *International Journal of Heat and Mass Transfer*, 118:587–601, 2018. ISSN 0017-9310.

[7] Y. Kong and W. Wang et al. Wind leading to improve cooling performance of natural draft air-cooled condenser. *Applied Thermal Engineering*, 136:63–83, 2018. ISSN 1359-4311.

[8] M. Goodarzi and R. Keimanesh. Heat rejection enhancement in natural draft cooling tower using radiator-type windbreakers. *Energy Conversion and Management*, 71:120–125, 2013.

[9] D.G. Kröger. *Air-cooled heat exchangers and cooling towers*. Penwell Corp., Tulsa, Okla, 2004. ISBN 9780878148967.

[10] W. Strydom and J.P. Pretorius et al. Natural draft direct dry cooling system scaling for diverse applications. *16th International Conference on Heat Transfer, Fluid Mechanics and Thermodynamics*, 2022.

[11] W. Strydom and J.P. Pretorius et al. Sensitivity analysis on the performance of a natural draft direct dry cooling system for a 50mwe csp application. *8th World Congress on Momentum, Heat and Mass Transfer*, 2023. doi: 10.11159/enfht23.138.

[12] P. Dong and X. Li et al. The transient start-up process of natural draft dry cooling towers in dispatchable thermal power plants. *International Journal of Heat and Mass Transfer*, 123:201–212, 2018. ISSN 0017-9310. doi: https://doi.org/10.1016/j.ijheatmasstransfer.2018.02.114.

[13] P. Dong and X. Li et al. Numerical investigation of the influence of local effects on the transient start-up process of natural draft dry cooling towers in dispatchable power plants. *International Journal of Heat and Mass Transfer*, 133:166–178, 2019. ISSN 0017-9310. doi: https://doi.org/10.1016/j.ijheatmasstransfer.2018.12.066.

[14] P. Dong and X. Li et al. The crosswind effects on the start-up process of natural draft dry cooling towers in dispatchable power plants. *International Journal of Heat and Mass Transfer*, 135:950–961, 2019. ISSN 0017-9310. doi: https://doi.org/10.1016/j.ijheatmasstransfer.2019.02.039.

NEARFIELD DISPERSION MODELLING OF DISCHARGED EFFLUENTS INCORPORATING FAR-FIELD HYDRODYNAMICS

T. L. Surajpal[1], G.P.J. Diedericks[2], G.J.F. Smit[3*]

[1]Division of Applied Mathematics, Department of Mathematical Science, Stellenbosch University, Stellenbosch, South Africa

[2]Division of Applied Mathematics, Department of Mathematical Science, Stellenbosch University, Stellenbosch, South Africa

[3]Division of Applied Mathematics, Department of Mathematical Science, Stellenbosch University, Stellenbosch, South Africa, fsmit@sun.ac.za

Abstract: As water scarcity continues to rise, many companies are responding by planning and constructing desalination plants along coastlines. To ensure the environmental impacts of these desalination plants are thoroughly assessed, engineers employ nearfield models for designing the intakes and outflows of these facilities. These models can vary in complexity, ranging from empirical to advanced Computational Fluid Dynamics (CFD) models. Comprehensive environmental impact assessments also require the modelling of the dispersion of the effluents in the far-field, ensuring they do not have adverse effects. Precisely modelling the characteristics of the outfall plume at different mixing stages is thus crucial for accurately determining recirculation, spreading, and the resulting environmental impacts. In this study, we investigated the use of Ansys Fluent for the nearfield simulations, and additionally to evaluate the feasibility to incorporate far-field hydrodynamics. The CFD model proved to be effective in simulating the scenarios with considerable detail. This suggests that Ansys Fluent could be a valuable tool for assessing the environmental impacts of desalination plant effluents on marine ecosystems. The offline one-way coupling between a far-field model and Ansys Fluent was successful, but further advancements in two-way coupling methods may enhance the accuracy of such assessments in the future.

Keywords: Effluent, CFD, near-field, discharge, computational fluid dynamics, modelling

1. Introduction

The 2015 to 2018 drought in South Africa's Western Cape Province highlighted the freshwater scarcity issue, leading to the consideration of desalination plants as a solution. These plants convert seawater into freshwater while discharging brine and heated water back into the sea. The environmental impacts of these discharges must be carefully assessed, especially regarding their effects on the marine ecosystem. To achieve this, near-field models, ranging from empirical to advanced Computational Fluid Dynamics (CFD) models, are used to study the dispersion of the effluent plume near the outfall point.

The dynamics of plume behaviour are categorized into three zones, namely the nearfield, intermediate zone, and far-field. The nearfield of an outfall refers to the vicinity immediately surrounding the outfall, i.e., the region in which the jet characteristics dominate the behaviour of the discharge, and hence the discharge at this point is referred to as a jet. Nearfield behaviour is primarily influenced by outfall characteristics, while far-field behaviour is governed by ambient conditions. The intermediate zone represents the transitional region between these two zones. Figure 1, from Bleninger and Jirka [1], displays the near-field, intermediate zone and far-field for a particular example.

In the near-field region, effluent mixing with seawater is carried out, predominantly, by jet diffusion. As the momentum generated by buoyancy exceeds that of the momentum of the jet, the jet transitions to, and is referred to as, a plume whose behaviour is characterised by plume characteristics. This area of plume-dominant behaviour may be referred to as the intermediate or transition field. Buoyant spreading of the effluent dominates mixing in this region. As the distance from the outfall point increases, the effect of the ambient conditions outweighs that of the plume characteristics and a region described as the far-field is introduced. At this stage ambient diffusion, and advection, control the mixing of the effluent with the seawater [1]. Advection by the ambient currents plays a significant part in mixing from as close as the near-field region (but still a distance away from the outfall point), as described in Figure 2.

The nearfield and far-field processes occur at different spatial and temporal scales. Modelling of these processes is generally conducted by different types of models. Nearfield modelling is conducted in isolation and focusses on the details of the outfall and effluent characteristics. The output from the nearfield models is used as input to the far-field models in an offline one-way manner. Improved modelling techniques require the effective coupling between nearfield and far-field simulations. This coupling process can be offline or online, static or dynamic, and one-way or two-way.

This study's main objective is to evaluate the performance of a CFD model in conducting nearfield simulations and determine its suitability for offline, one-way coupling with a far-field model to incorporate far-field hydrodynamics [2].

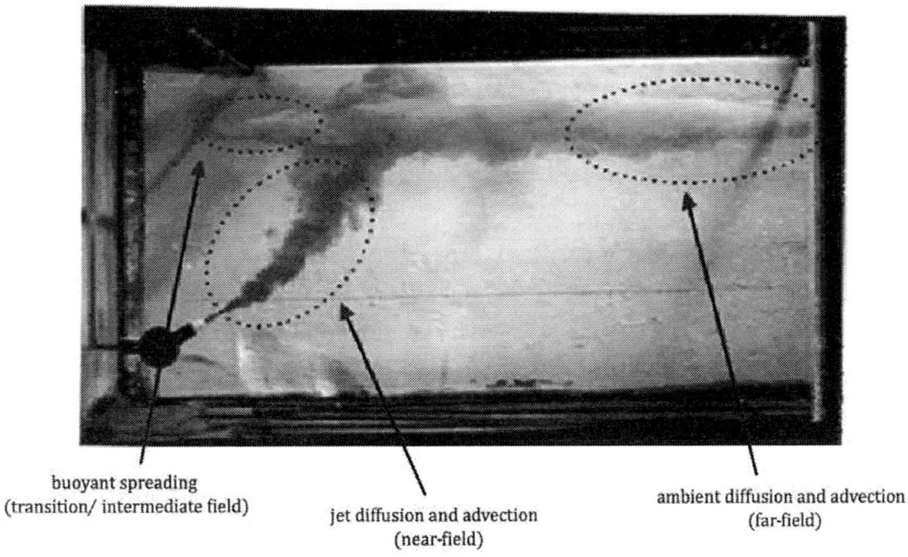

Figure 1: Near-field, intermediate field and far-field regions displayed for a plume developing from a submerged outfall point, along with the physical processes which dominate the plume behaviour during each region [1]. This image is reproduced and edited from Bleninger and Jirka [1].

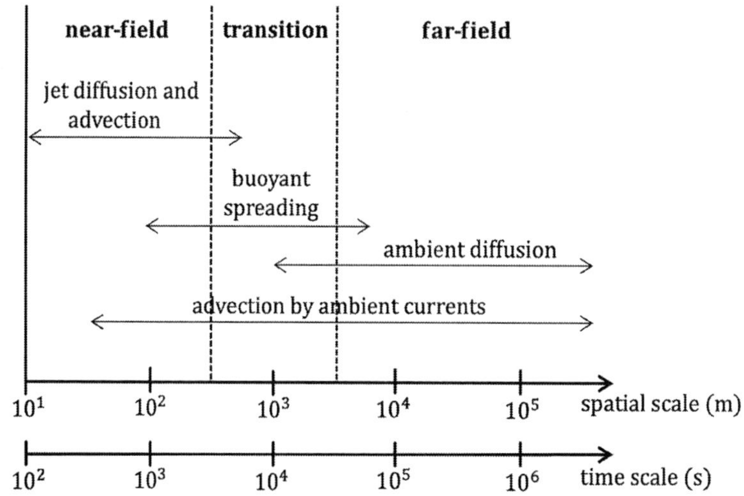

Figure 2: Spatial and time scales of near-field, intermediate field, and far-field regions and mixing processes, reproduced from Bleninger and Jirka [1].

2. Near-field simulation in Ansys Fluent

Mortensen et al. [3] conducted a study on buoyant plume dispersion from offshore oil and gas platforms. Using a three-dimensional CFD model in OpenFOAM, they simulated two scenarios with variations in discharge rate, pipe radius, wastewater type, and domain dimensions. The study observed helical vortices causing the plume to split, influenced by the interaction between ambient velocity ("side current") and the "vertically aligned jet". Two stages of plume development were identified, namely the rising-stage near the outlet with "jet-like" characteristics and the surface-stage with the effluent spreading laterally while in contact with the surface. Transition between stages occurred approximately 30 m and 40 m away from the pipe outlet in the first and second scenarios, respectively. Their CFD results were compared to CORMIX, an industry-standard empirical model.

In this study the numerical set-up of Mortensen et al. [3] was used as a basis for the work carried out in Ansys Fluent. The geometry simulated in Ansys Fluent was a rectangular box with length, width and depth given respectively as 220 m, 100 m, and 14 m, as illustrated in Figure 3. The geometry was constructed in Ansys DesignModeler, and Ansys Meshing was utilised to create the mesh for the domain. Various combinations of meshing control features were experimented with to generate a mesh with good quality, which also allowed for a reasonable computational time. The geometry was built up, firstly using a rectangular prism surrounding the pipe and secondly, using a cylinder surrounding the pipe. This was done with the goal of gaining a structured mesh. It was found that the orthogonal quality of the mesh with the rectangular prism geometry was better than that of the mesh with the cylinder in the corresponding geometry. Therefore, the former was chosen to be used for the simulations. This investigation of the mesh was carried out to determine which method allowed for the maximum control over the mesh so that refinement could occur in specified areas, for example in areas where the plume is expected to be formed. Figure 4 illustrates the grid used in this study, showcasing local refinements.

Ansys Fluent allows the choice of solvers, i.e. pressure-based, and density-based solvers, respectively, and discretisation schemes, i.e. spatial discretisation, pressure interpolation, and temporal discretisation schemes, respectively. A detailed discussion of all the options available in Ansys Fluent is beyond the scope of this study. The solver and scheme choices for the Ansys Fluent simulations for this study are listed in Table 1.

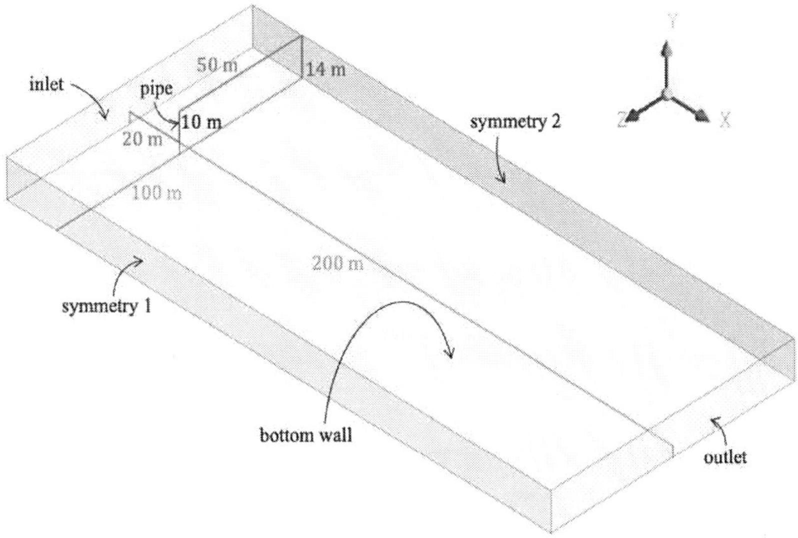

Figure 3: Diagram displaying a schematic representation of the physical model. Note that the pipe inlet is located at the bottom of the pipe, vertically submerged 10m into the domain, but is not labelled on the diagram; additionally, the top wall, representing the ocean surface, is not labelled on the diagram.

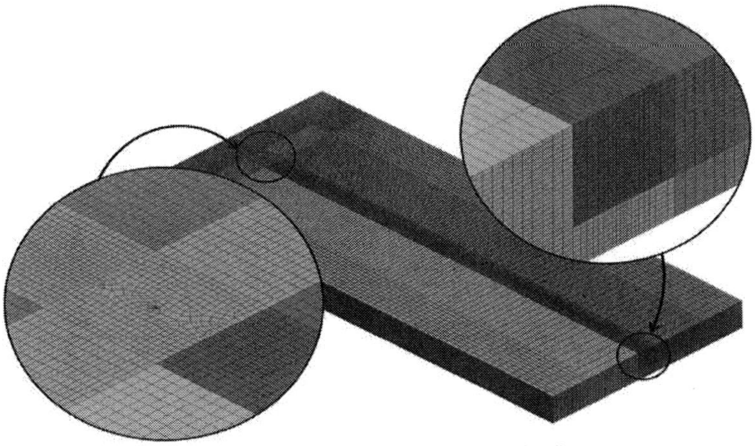

Figure 4: Diagram displaying the grid and local refinements.

Table 1: Solver and scheme choices for the Ansys Fluent simulations.

Steady state / transient simulation	Transient
Solver type	Pressure-based
Pressure-velocity coupling method	SIMPLE
Spatial discretisation schemes	
Gradient	Least squares cell based
Momentum	Second-order upwind
Turbulent kinetic energy	First-order upwind
Turbulent dissipation energy	First-order upwind
Species transport	Second-order upwind
Energy	Second-order upwind
Pressure interpolation scheme	Second-order
Temporal discretisation scheme	First-order implicit

A species transport model was employed in Ansys Fluent to model the mixing of two fluids, namely a heated discharge freshwater and seawater. This requires the specification of a mixture material, composed of various species, in terms of its constituent species, reactions and physical properties. Thereafter it is necessary to specify the material properties of each constituent specie of the mixture. The material properties of the seawater and discharge used in the simulations are listed in Table 2.

Table 2: Table displaying material properties of seawater and the discharge.

Ambient fluid	
Domain length (in direction of flow)	220 m
Domain depth	14 m
Domain width	100 m
Velocity	0.22 m/s
Temperature	28°C
Salinity	34 ppt
Discharge	
Pipe diameter	0.1524 m
Pipe length	10 m
Velocity	1.6 m/s
Flow rate	2500 m^3/day
Temperature	98°C
Salinity	7 ppt

Since the specified discharge is less dense than the ambient seawater, it is expected that the plume will reach the surface at some point. The rising of the plume occurs gradually and over a distance, in the direction of ambient flow, and is influenced by the ambient velocity i.e. the velocity of the seawater. As depicted in Figure 5, the plume rises until it reaches the surface upon which it moves predominantly along the surface. Figure 5 shows the volume rendering of the discharge mass fraction as obtained from the Ansys Fluent simulation.

Figure 5: Side view of volume rendering of discharge mass fraction.

An approach to interpreting simulation results involves quantifying the point where the discharge plume makes contact with the surface. This method may however be open to interpretation. On one hand, the location at which the plume strikes the surface may be seen as the point of maximum discharge mass fraction, minimum seawater mass fraction, and minimum density along the surface midline (considering the pipe's alignment along this line). Although these individual points may be identifiable, a more meaningful analysis may be to consider their interrelation. It is crucial to note that discrepancies between the locations of maxima and minima might stem from numerical computations rather than physical phenomena. Ideally, these extremes should coincide. Alternatively, the location at which the plume strikes the surface may signify the point where the discharge initiates an influence on the surface seawater along the midline in the flow direction. This influence becomes apparent with the first changes in density, discharge mass fraction, and/or seawater mass fraction along the midline and across the surface in the horizontal direction. As stated earlier, the initial changes in these parameters should be considered simultaneously for a robust interpretation. Depending on the chosen interpretation of the plume-surface interaction, the expected results may vary. For the purpose of this study, the location where the discharge begins to influence the surface must undergo a simultaneous change in discharge mass fraction, seawater mass fraction, and density relative to each other. Specifically, as the discharge mass fraction increases, the seawater mass fraction and the density of the fluid mixture decrease. To quantify this relationship, the study introduces the Talee number ($\mathbb{T}$), defined as the measure of the relationship between the density, temperature, and mass fractions of two species in a mixture:

$$\mathbb{T} = \frac{\rho_{\text{bulk}} \gamma_{\text{bit}} \text{T}_{\text{m}}}{\rho_{\text{m}} \gamma_{\text{bulk}} \text{T}_{\text{bulk}}},$$

where ρ_{bulk} is the density of the most abundant specie, ρ_{m} is the density of the mixture at a point in the field, γ_{bit} is the mass fraction of specie present in minority, γ_{bulk} is the mass fraction of the specie present in bulk, i.e. the most abundantly present specie, T_{m} is the temperature of the mixture at a point in the field and T_{bulk} is the temperature of the most abundant specie. Note, ρ_{bulk} and T_{bulk} are initial values and hence are constants. The locations where the plume strikes the surface, along the surface midline, were analysed for maximum $\mathbb{T}$, $\mathbb{T} \geq 0.0005$, $\mathbb{T} \geq 0.0001$, and $\mathbb{T} \geq 0.001$, by adjusting the appropriate initial conditions.

Mortensen et al. [3] distinguish a rising-stage and a surface-stage in the plume development. The rising-stage exhibits jet-like characteristics, while the surface-stage denotes lateral spreading in contact with the surface. Mortensen et al. [3] approximate the transition from rising-stage to surface-stage at approximately 30 m from the pipe, or 50 m from the domain inlet, considering it as the location where the plume contacts the surface. However, the precise point of the plume hitting/reaching the surface is not explicitly provided by Mortensen et al. [3]. Furthermore, the study does not delve into the significance of the plume reaching the surface, whether it is the initial surface impact or the point of maximum plume fluid concentration. This study, for comparison purposes, considers the 50 m mark as the approximate location where Mortensen et al. [3] describe their plume making contact with the surface.

Figure 6 displays a plot of the salinity at various positions along the surface midline in the direction of the flow, as well as the locations corresponding to the various interpretations at which the plume hits the surface, including the location at which Mortensen et al. [3] describe their plume to hit the surface. The first encounter of $\mathbb{T} \geq 0.0001$, $\mathbb{T} \geq 0.0005$ and $\mathbb{T} \geq 0.001$ occurs at 43.8 m, 48.4 m and 55.8 m, respectively. With reference to Figure 6, it can be seen that the location at which the plume hits the surface, as given by Mortensen et al. [3], is closest to the location at which the plume hits the surface, as given by the first encounter of $\mathbb{T} \geq 0.0005$. The maximum discharge fraction, or minimum salinity, has been reached at 67.0 m (which it the location of maximum $\mathbb{T}$ along the surface midline).

Figure 6 displays a steep decrease in salinity up until a certain point, which is then followed by a gentle increase in salinity. The increase in salinity, along the midline, after the minimum salinity has been reached, can be explained by referring to Figure 7. It can be seen that after the maximum discharge mass fraction displayed by the pink dot, the plume splits into two streams which deviate from the midline. As the distance from the maximum discharge mass fraction increases, the salinity increases; this occurs as the plume splits to become concentrated in two steams away from the midline.

The splitting of the plume into two channels or columns may be understood by considering tangential velocity vectors at transverse planes across the domain, normal to the direction of flow, as depicted in Figure 8. The enlargements in Figure 8 display velocity vectors present in a circular layout. The enlargement on the left-hand side of Figure 8 (within the pink box) contains anti-clockwise, rotational arrows of velocity vectors. Similarly, the enlargement on the right-hand side of Figure 8 (within purple box) contains clockwise, rotational arrows of velocity vectors. Ultimately, this suggests the presence of rotational velocities. The rotational velocities are found to originate around the discharge being released from the pipe and may be described as rotating outwards, away from the middle of the top of the discharged fluid, around and then back to the middle, at the bottom of the discharge.

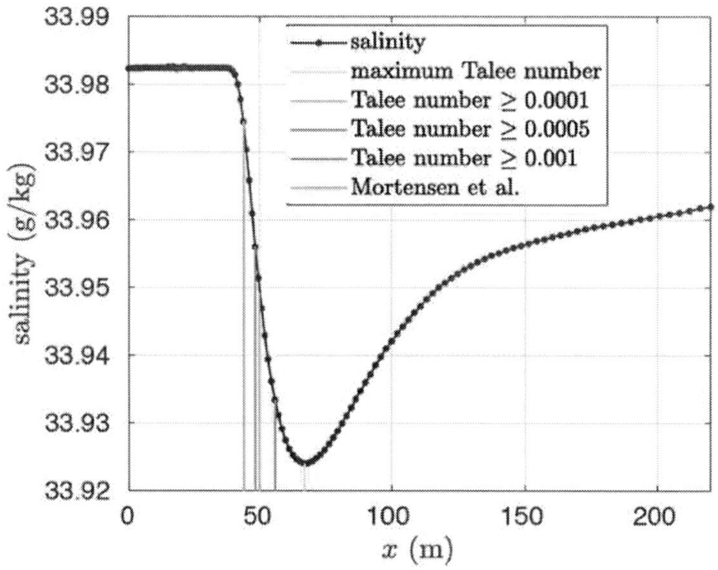

Figure 6: Graph displaying salinity along the surface midline and the various interpreted locations at which the plume makes contact with the surface.

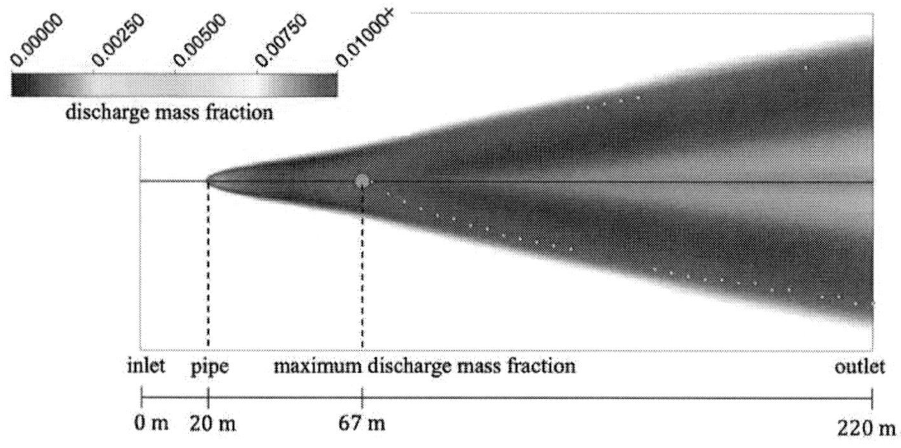

Figure.7: Graphic displaying an aerial (surface) view of the discharge mass fraction volume rendering, for values between [0, 0.01], where the pink dot displays the maximum discharge mass fraction along the midline, which is displayed by the purple line. The white dots are displayed at the point of maximum discharge mass fraction of transverse planes obtained at 5 m intervals across the length of the domain.

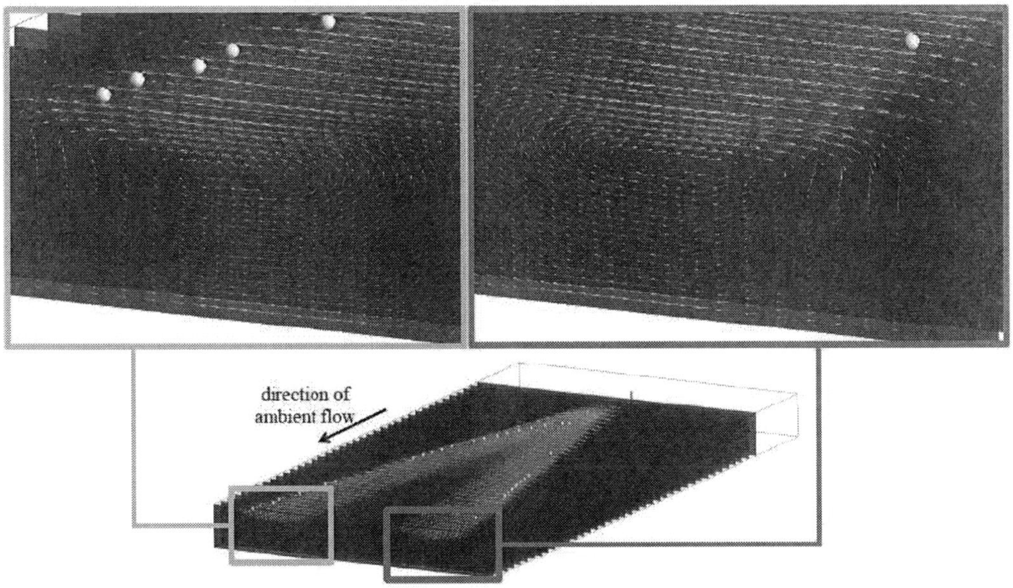

Figure 8: Graphic displaying transverse planes of contours of discharge mass fraction, at 5 m intervals across the domain, where each plane contains tangential velocity vectors (arrows) and the location of maximum discharge mass fraction (white dots).

3. Far-field to near-field coupling

The integration of ambient flow from a far-field model with a nearfield model (Ansys Fluent) was achieved through the incorporation of results obtained from the far-field model into Ansys Fluent [1]. These results were transformed into a format that aligned with the requirements of Ansys Fluent. In essence, the data imported into Ansys Fluent comprised a series of oscillating velocities recorded at various time intervals, effectively emulating the characteristics of an M2 tide.

The velocities employed in our tidal boundary condition generally exhibit magnitudes that surpass what one might typically encounter in a South African tide. South African tidal velocities tend to be relatively modest when compared to tides in other parts of the world. Velocities of such magnitude, as incorporated into our tidal boundary condition, are not commonly encountered in the open ocean but are more commonly associated with enclosed or restricted water bodies. For the purpose of exploring the hydrodynamics, via Ansys Fluent, these velocities with a larger magnitude were utilised.

Subsequently, these velocities were applied as a transient boundary condition at the inlet, while keeping the surface as a rigid lid. This boundary condition is informally referred to as a tidal boundary condition. In Figure 9, a graphical representation of the velocity variations is depicted constituting the tidal boundary condition, as they evolve over the course of the flow time.

In essence, the tidal boundary condition imposed encompasses one 'ebb' and one 'flow' phase. In other words, it involves a series of positive velocities (flow into the domain) followed by a series of negative velocities (flow of the domain).

The tidal impact of our boundary condition is presented in Figure 10. Initially, at $t = 60$ s, the discharge shows a near-vertical ascent due to low ambient velocity and positive buoyancy compared to seawater. As ambient seawater velocity increases in the positive x-direction over time, the discharge spreads laterally as it entrains with the flow. As velocity peaks in the positive x-direction and then decreases, the discharge accumulates closer to the pipe's surface, causing less lateral spreading. When velocity approaches 0 m/s around 22500 s, the discharge starts to spread vertically upward again. As ambient velocity becomes negative and increases toward its maximum magnitude, the discharged jet angles away from the pipe in the negative x-direction. From Figure 10, the highest discharge mass fractions consistently appear in the jet-like section of the plume, confined within a fixed length range around the pipe. A brief analysis suggests this range to be approximately 14 m to 20 m on both sides of the pipe, aligned with the flow direction.

To consider the effect of the tidal boundary condition on the surface spreading of the discharge, in all directions, a simulation was carried out in which the pipe is shifted towards the middle of the domain. This was to prevent the effluent from leaving the computational domain. It can be seen, from Figure 11, that over time, discharge mass fractions of approximately 0.009 and greater, are roughly contained within a 5 m radius of the pipe but almost totally contained within a 10 m radius of the pipe.

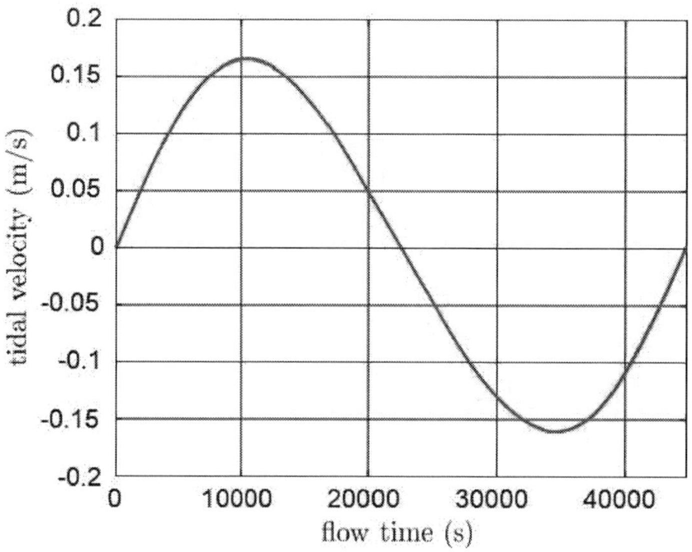

Figure 9: Diagram displaying the tidal velocities with respect to flow time, for the tidal boundary condition.

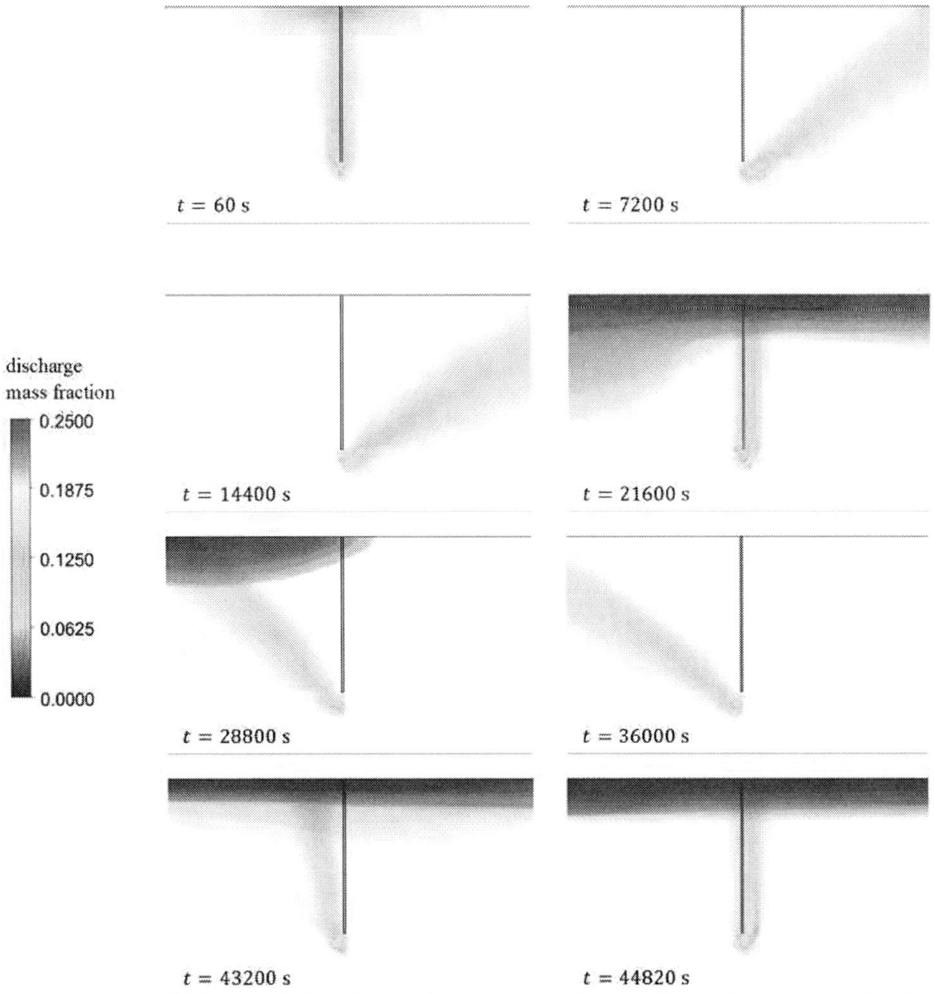

Figure 10: Images displaying volume renderings of the discharge mass fraction at various times and at a fixed location in the domain. Note, the diagram is such that the domain inlet is to the left side of the page.

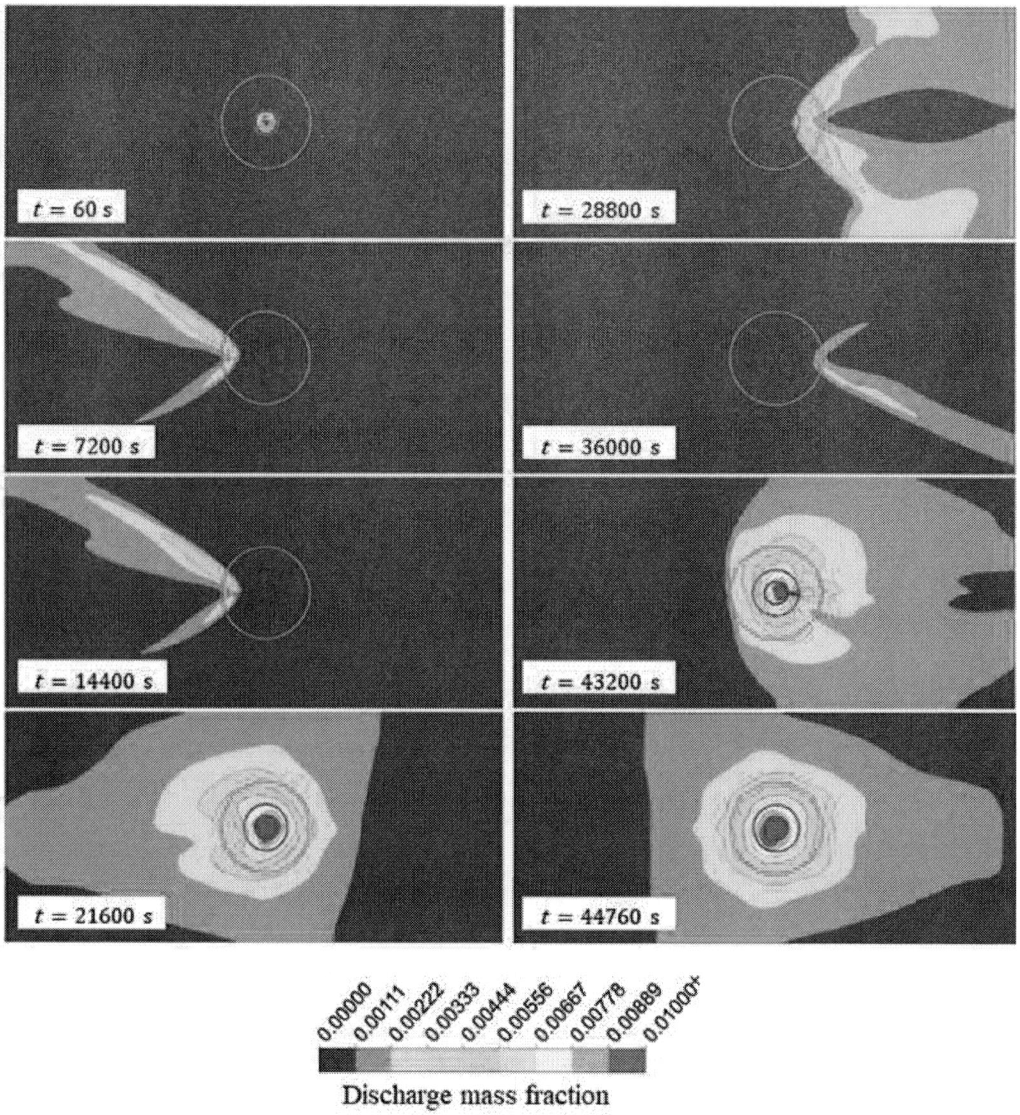

Figure 11: Surface discharge mass fraction contours at various times (approximately 2-hour intervals), with circles of radius 5 m, 10 m and 20 m depicted in purple, dark green and pink, respectively. Note, the diagram is such that the domain inlet is to the right side of the page.

4. Conclusions

Ansys Fluent successfully reproduced the modelling scenario of Mortensen [3] for an effluent of heated fresh water in seawater. There is some uncertainty of how a discharge of brine can be simulated in Ansys Fluent. Even though the discharge of brine was not conducted, for the case of modelling brine being discharged into seawater, two different approaches may be considered in terms of the specification of the species for the species transport model, namely specifying two fluids of either brine and seawater or fresh water and a saturated aqueous solution of sodium chloride. The first approach entails the specification of two fluids, with one fluid having the properties of brine and the other having the properties of seawater. The location of the brine and seawater in the domain is then assigned by controlling the mass fraction of brine imposed at each inlet and initialised throughout the domain. Furthermore, the mass fraction specified cannot be greater than 1 and at each point, the mass fractions of brine and seawater must sum to 1. Note, since brine is present in a smaller quantity than seawater, its mass fraction is specified instead of the seawater mass fraction, as per Ansys Fluent standard procedure. In other words, upon mixing, since seawater is more abundantly present than brine, it is the bulk specie and should be specified as such.

Including the tide in the simulation changed the behaviour of the plume dramatically. This shows that, for certain applications, the inclusion of ambient conditions from a far-model into the nearfield is necessary to account for the proper nearfield behaviour.

References

[1] Bleninger, T. and Jirka, G. (2004). "Near- and far-field model coupling methodology for wastewater discharges". DOI: 10.1201/b16814-73.

[2] Surajpal, T.L. (2022). "Near-field dispersion modelling of discharge effluents under the various hydrodynamic conditions". M.Sc. thesis, University of Stellenbosch.

[3] Mortensen, S., Jackson, C., Petersen, O. and Wake, G. (2013). "Comparing CFD and CORMIX for Buoyant Wastewater Near-Field Dispersion Modeling".

WAKE REDIRECTION TECHNIQUES FOR WIND FARMS

Kayla Surujhlal[1,*], Ayodimeji Biobaku[2], Nick Zang[3], Wei Hua Ho[4,*]
[1]University of the Witwatersrand,* kayla36252@gmail.com, * weihua.ho@uct.ac.za,
[2]University of the Witwatersrand, [3]University of Bristol, [4]University of Witwatersrand.

Abstract: In a wind farm, the upstream turbine has several negative effects on the downstream turbine which leads to uneven loads exerted on the downstream turbine and reduced power outputs of the wind farm as a whole. Wake redirection techniques could reduce these negative effects and improve the overall power output of a wind farm. A Reduced Order Model (ROM) was created and trained using an open-source LES dataset of a two-turbine array using Dynamic Mode Decomposition (DMD). During the DMD process, pairs of modes were generated where each pair is linked to a physical characteristic of the flow field. Analysis was conducted on these modes during the selection process to determine which combination of the modes resulted in the most accurate reconstruction of the flow field. It was determined that using the modes linked to the mean flow and the tip vortex shedding resulted in the most accurate reconstruction with and error of 6.38%. It is hypothesized that this same combination of modes may also be the most suitable data for the training of a machine learning algorithm to predict flow outside of the current dataset.

Keywords: Wind turbines, Dynamic Mode Decomposition, Wake dynamics

1. Introduction

As the world aims to produce more energy from renewable sources, wind energy production continues to grow rapidly [1]. Wind turbines are used to harness the energy of the wind to generate electrical power that can be supplied to power stations. To harness the full potential of the wind energy in a certain location and generate power, wind farms are set up and configured to maximise the power output. Wind farms are an array of wind turbines that are connected to each other via underground cables from where the electricity generated is then channelled into the electricity grid [2]. Wind turbines in a wind farm are usually aligned behind one another in order to maximize the number of turbines that can be placed in the available area. This can result in the wake from an upstream turbine affecting the performance of the downstream turbine. Figure 1 in the paper by Stevens and Gayme [3] gives an excellent illustration of the effect of the upstream wake on the downstream turbine. The wake of an upstream turbine can have the following negative effects [4]:

1. After an upstream turbine extracts energy out of the wind, it leaves less energy remaining in the wake, which impinges on the downstream turbine. If a downstream turbine is completely in the wake of the upstream turbine, less energy can be extracted by the downstream turbine – leading to a lower power output by this turbine.
2. There is a significantly higher turbulence intensity of the air in the wake. This causes uneven loading on the downstream turbine which increases fatigue loads and leads to a reduced lifetime and generally more maintenance requirements.
3. Downstream turbines operating in the wake of upstream turbines also experience increased noise emission, which is a major social factor that often limits the establishment of wind farms.

Wake redirection techniques could reduce these negative effects and possibly improve the efficiency of the wind farm as a whole. There exist two main wake redirection techniques – tilt misalignment and yaw misalignment. Tilt misalignment is the rotation of the wind turbine rotor to face either upwards to downwards. Yaw misalignment is the rotation of the wind turbine rotor to the left or right of the freestream. Tilt misalignment has been found to be effective in redirecting the wake and increasing the power output of the wind turbine, however, studies have shown that tilt misalignment causes significant loading on the downstream turbine and an unacceptably high risk of tower strikes [5]. It was chosen to investigate yaw misalignment because it has also proven to be effective in wake redirection and it exerts less loads on the downstream turbine compared to tilt and most existing turbines already have the ability to yaw. This means that the development of a yaw control strategy could be implemented in existing wind farms.

The scenario that will be investigated is two turbines aligned directly behind one another with both turbines at non-zero yaw angles. When looking at a turbine from the front view, when it is yawed such that it is angled towards the left of the freestream, it causes the wake to be redirected to the left of the centre-line stream. Similarly, when the turbine is yawed towards the right of the freestream, the wake is redirected to the right of the centre-line stream. It is expected that redirecting the wake of the upstream turbine away from the downstream turbine and steering the downstream turbine away from the upstream turbine's wake, as seen in Figure 1, could reduce these negative effects and possibly improve the efficiency of the wind farm as a whole.

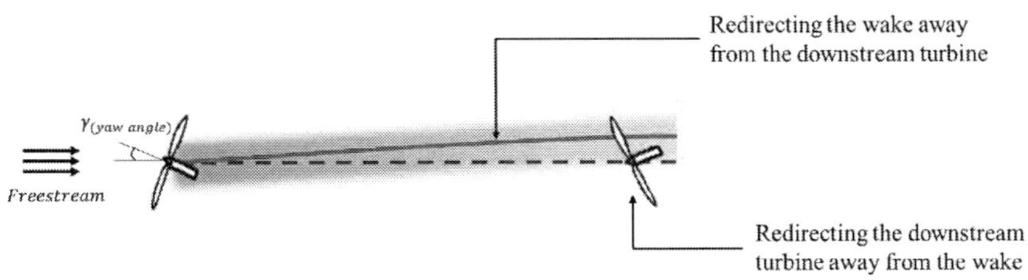

Figure 1. Configuration of two yawed turbines

2. Dynamic Mode Decomposition (DMD) Method

Wake prediction is a complex task because several parameters such as the turbine rotor diameter, rotor speed, blade pitch, tower height and control strategy contribute significantly to the instability of the wake. There is also no simple relationship between the yaw angle and the wake redirection [6]. Correct implementation of the yaw is also imperative otherwise it could lead to incorrect conclusions about the effect of yaw on the wake characteristics and thus compounding effects on the wind farm control strategy. Traditional methods of wake prediction such as full-scale CFD simulations require very extensive computational resources and significant time to complete.

It was chosen to use the method of surrogate modelling because it has proven to be an accurate tool in predicting wind turbine wake dynamics. A surrogate model is created by first reducing the order of a complex model and then applying a machine learning algorithm for prediction. This paper will investigate only the first part of the entire surrogate modelling technique, which is the Reduced Order Model (ROM). It is hypothesized that the modes that result in the best reconstruction of the original data from the identification data set should also be the best modes for the machine learning model and this will be the premise for the remainder of this paper.

A reduced order model is a simplification of a complex model. It is created to capture and analyse the dominant dynamic behaviour of the complex model while using less computational resources [7].

This can lead to shorter design cycles as models can be run in a shorter time without sacrificing significant accuracy [8].

The dataset that was used to train the ROM was an already existing open-source dataset containing Large Eddy Simulations (LES) of a two-turbine array and it was generated using the Simulator for Wind Farm Applications (SOWFA). This dataset was split into two sets – namely an identification dataset and a validation dataset for training and testing of the model.

It was chosen to use the method of Dynamic Mode Decomposition (DMD) because it is able to capture more information of the flow field such as the frequency and growth/decay rate [9] which are important for understanding the factors that influence the wake dynamics and thus the loading on the downstream turbine in the wake.

To create the ROM, the original LES dataset was first resampled to reduce the size of the 4-dimensional dataset (x, y, z and time) into 2-dimensional matrices that are more suitable and less computationally expensive to work with [10]. The 4-dimensional flow field is shown in Figure 2 below, where x, y and z refer to the spatial positions and t refers to the time step. Single snapshots of the dataset (X_k) were then generated and contained the velocity field information at different spatial points at single time steps. A matrix (X) was created by arranging these snapshots in sequential columns so that each column contained velocity field data at a single time step.

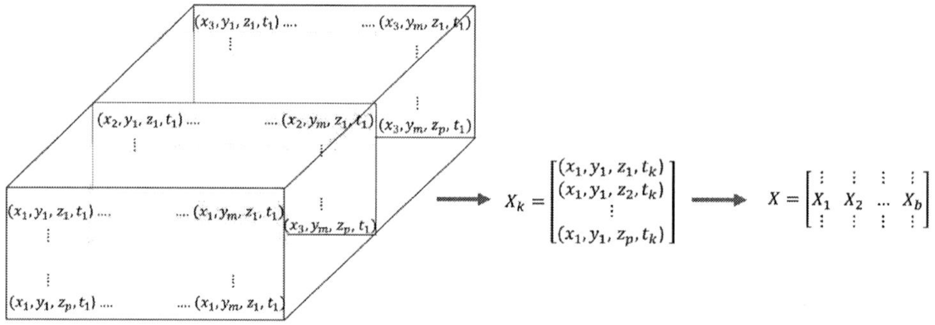

Figure 2. Processing of the Flow Field

In order to be able to work with this snapshot matrix X, its size needed to be reduced and an optimal low-rank approximation of the matrix needed to be calculated that would be able to still capture the dynamics of the dataset. This was done through the method of Singular Value Decomposition (SVD). Performing SVD on matrix X defined above decomposes the matrix into left (U), right (V) and diagonal (Σ) singular value matrices [11], as seen in Equation 1 below. Truncation was then applied to the SVD in order to further reduce the computational resources required, resulting in the truncated matrices $\tilde{U}$, $\tilde{V}$ and $\tilde{\Sigma}$ and matrices containing the remaining singular values $\tilde{U}_{rem}$, $\tilde{V}_{rem}$ and $\tilde{\Sigma}_{rem}$. This process results in a state-space system defined by the number of singular values specified for truncation.

$$X = U\Sigma V = [\tilde{U} \quad \tilde{U}_{rem}] \begin{bmatrix} \tilde{\Sigma} & 0 \\ 0 & \tilde{\Sigma}_{rem} \end{bmatrix} \begin{bmatrix} \tilde{V} \\ \tilde{V}_{rem} \end{bmatrix} = \tilde{U}\tilde{\Sigma}\tilde{V} \qquad (1)$$

Validation is then performed by processing the second dataset (the validation dataset) using the same steps outlined for the identification dataset. The performance of the validation dataset was evaluated in comparison to the identification dataset. The Variance Accounted For (VAF) was plotted for 100 DMD modes and is show in Figure 3 below. The VAF is a quantitative measure of how much the model differs from the original data. A higher percentage of VAF indicates a stronger correlation.

The equation used to calculate the VAF is shown in Equation 2 below.

$$VAF_i\big(y_i(k), \hat{y}_i(k)\big) = \left(1 - \frac{\frac{1}{N}\sum_{K=1}^{N}|y_i(k) - \hat{y}_i(k)|^2}{\frac{1}{N}\sum_{K=1}^{N}|y_i(k)|^2}\right) \tag{2}$$

Where $y_i(k)$ is the ith output of the model at time step k and $\hat{y}_i(k)$ is the ith output of the original simulation at time step k.

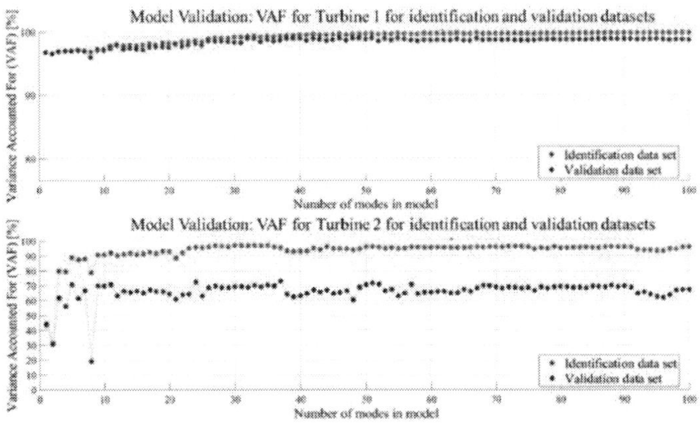

Figure 3. VAF for 100 DMD Modes for the Identification and Validation Datasets for Turbine 1 (top) and Turbine 2 (bottom)

Figure 3 above shows that the model is much better at predicting the power generated for turbine 1 as opposed to turbine 2. This is because clean, undisturbed air impinges on turbine 1 whereas the turbulent air influenced by the dynamics of the upstream turbine wake impinges on turbine 2, which is further away from the linear function used to predict the trajectory of the power.

DMD modes were then generated, where each mode represents a particular physical characteristic of the flow field. The energy, frequency and damping ratio of these modes were analysed to understand how each mode or group of modes contribute to the total flow field and thus which combination of modes would best capture the wake dynamics and result in the most accurate reconstruction. The flow field was then reconstructed using the particular modes decided on and the accuracy of the full reconstruction was then quantified.

These modes will be used to train a machine-learning model that is hypothesized to be able to predict the wake of the turbines when both the upstream and downstream turbines are at non-zero yaw angles. This will allow for better optimization of wind farm wake redirection so that the loads on turbines can be reduced and the total power generation of the wind farm can increase.

3. Results and Discussion

For Dynamic Mode Decomposition, the time-resolved data is decomposed such that the physical characteristics can be evaluated. This yields the generation of DMD modes. Each mode or collection of modes capture a different physical property. DMD modes are characterized by their frequency, energy, wavelength and growth rate and by analysing this information about each mode, it can be understood how each mode contributes to the wake shape and dynamics.

100 DMD modes were extracted to allow for a more detailed analysis. In Table 1 below, information about the energy, frequency and damping ratio is summarized for some pair of modes. The modes

shown in Table 1 were chosen due to their defining values of energy contribution, frequency and damping ratio.

Table 1. Selected DMD mode pairs and their corresponding information

DMD Modes	Energy Contribution (%)	Frequency (Hz)	Damping ratio
3,4	0.05	0.248	0.0561
72,73	0.91	0.020	0.0025
95,96	0.76	0.002	0.847
97	0.97	0	1

It was observed from the information generated about the modes that the DMD modes exist in eigenvalue pairs. Therefore, the modes are shown in pairs in Table 1 above and will be referred to in pairs onwards. From Table 1 it can be observed that the higher numbered modes have more energy than the lower numbered modes. Modes with higher energy are more unstable and thus add more kinetic energy to the system. Given their higher energy, these modes are expected to contribute to tip vortex shedding and tower vortex shedding. The modes with less energy are more stable and more damped and therefore are expected to contribute to the mean flow of the flow field.

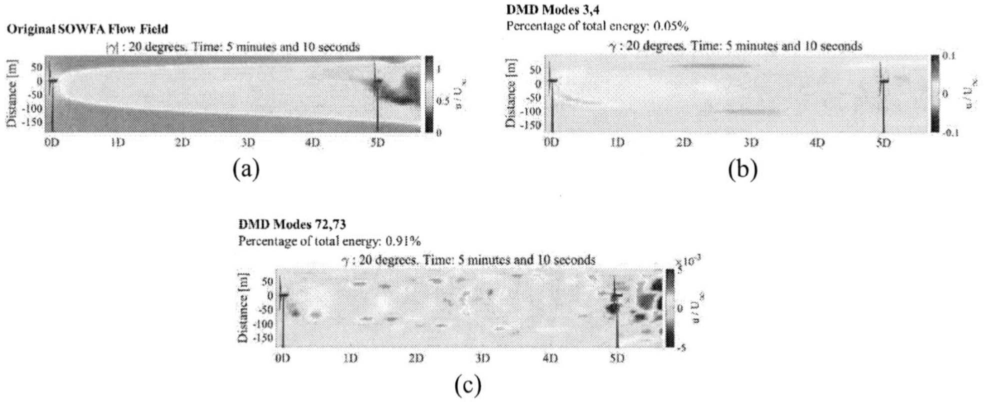

Figure 4. Side View of the Original SOWFA Flow Field (a), the DMD Mode Pair 3,4 (b) and DMD Mode Pair 72,73 (c)

From Table 1 it can be observed that certain modes are completely damped. It was found that these modes were stable and contributed to the mean flow, even though their normalized velocities were very small. Figure 4b shows that DMD mode pair 3,4 captures the steady-state flow of the flow field well. Figure 4c shows that DMD mode pair 72,73 captures the flow moving in alternating directions throughout the flow field with varying velocities. These are the helicoidal tip vortices that form at the tip of the blades and move downstream as the blade is rotated. This shows that a pair with a higher energy is likely to be linked to the dynamics of the flow, whereas a mode pair with a lower energy is linked to the mean flow.

From analysing the side view and top view of the flow field reconstructions made using the different pairs of modes, it was observed that multiple pairs of modes exhibit very similar characteristics and therefore can be linked to the same physical feature of the wake. The modes were grouped visually according to the physical feature they capture. Group 1 modes were linked to the mean flow because they have a smooth steady-state structure throughout, as circled in Figure 5 - Group 1. Group 2 modes were linked to the vortex shedding at the tip of the blade because, as circled in Figure 5 - Group 2, there are alternating sections of positive and negative flow that exist in an oscillating pattern. Group 3 modes were linked to the vortex shedding at the tower because the higher velocity flow occurs only at the tower location, as highlighted in Figure 5 - Group 3. Group 4 modes were linked to the wake deflection due to the yawing of the upstream turbine because it can be observed that from the top view in Figure 5 - Group 4, there is a distinct deflection of the wake. Group 5 modes were linked to the noise of the system because there is a low velocity throughout the entire flow field, as seen in Figure 5 - Group 5. In Figure 5 below, the side view and top view is shown for one pair of modes in each group.

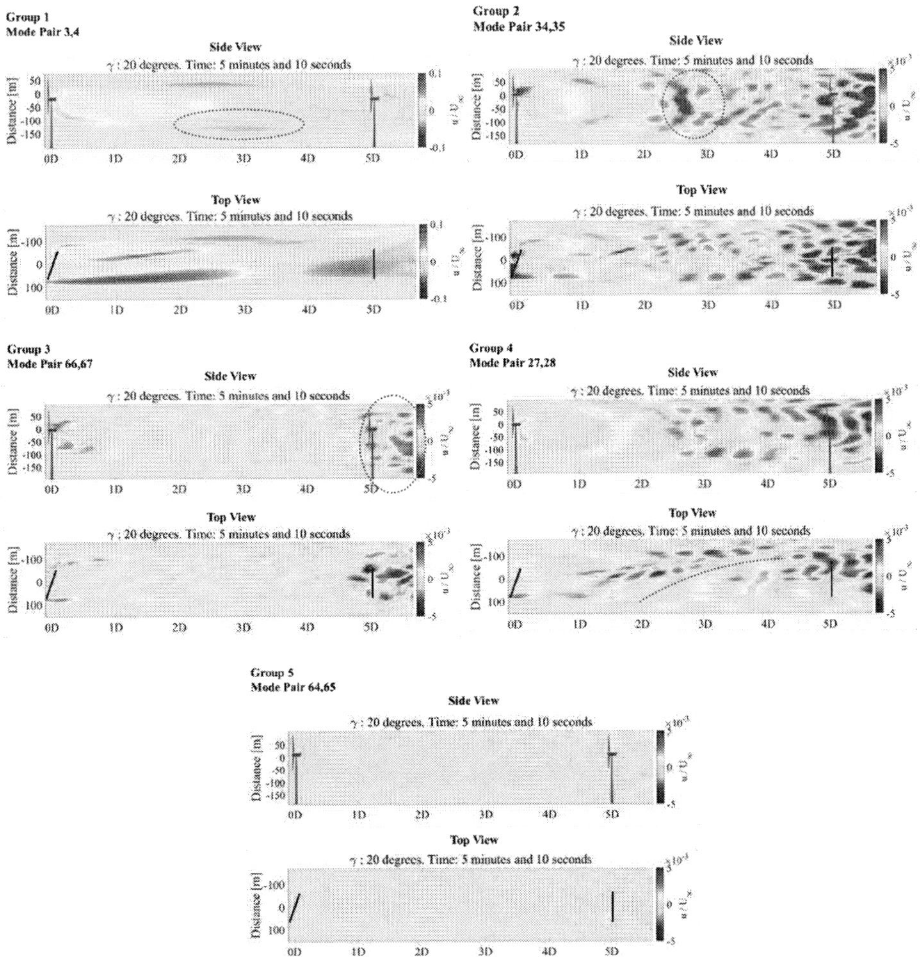

Figure 5. Side View and Top View of One Pair of Modes from Each Group

Below is a collection of some of the pairs of modes from group 4 that were all linked to the redirection of the flow due to the yaw angle of the upstream turbine. In these mode pairs, the wake is seen to be deflected upwards at around 3D and continues to flow around the downstream turbine instead of directly impinging on it.

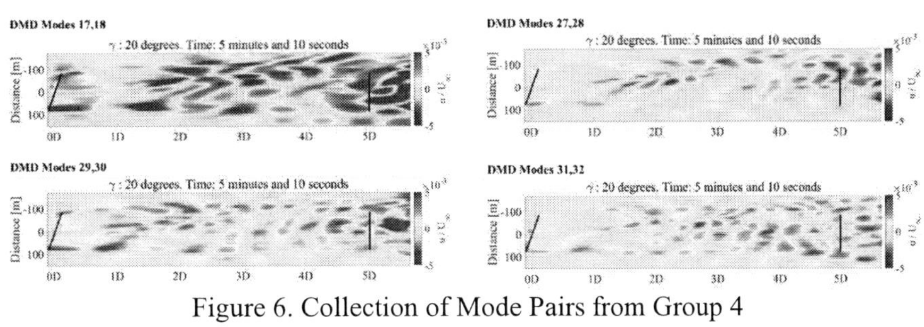

Figure 6. Collection of Mode Pairs from Group 4

The flow field was then reconstructed using each group of modes separately. Figure 7 shows that group 1 reconstructs the flow field very accurately and it has an average error of 6.39%. This also means that the vortex shedding does not contribute a significant amount to the reconstruction of the flow field and accurate reconstructions of the shape of the flow field can be made without it.

Reconstructions of group 2, group 3, group 4 and group 5 do not capture the full flow field well, with all these reconstructions having an average error of 100%. Although the combination of all the modes within the group reconstruct the flow much better than any singular pair, these modes alone are not enough to capture the flow field sufficiently. It is important to have modes that can capture the mean flow since the turbines are in a steady state at this time step and are not yawing continuously.

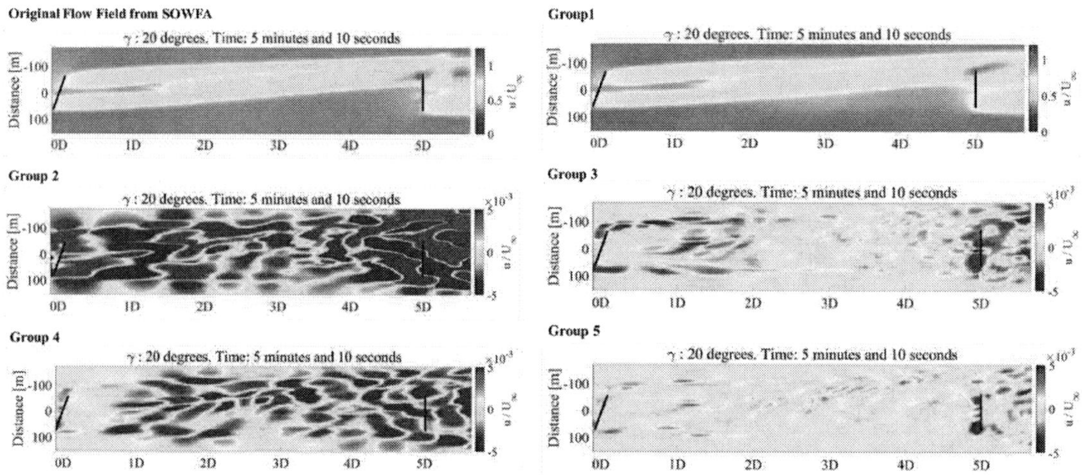

Figure 7. Reconstructions of the Flow Field using Different Groups of Modes

The flow field was then reconstructed using different combinations of groups to see which combination would result in the most accurate reconstruction. The images of the reconstructions are shown in Figure 8 below. Groups 1, 2 and 4 were combined because it is expected that combining the tip vortices, mean flow and flow deflection would result in majority of the flow dynamics being captured. However, this reconstruction had a slightly higher error compared to the reconstruction with just group 1. This could mean that capturing the dynamic properties of the flow field has a negative effect on the reconstruction of a flow field that is in a steady-state. Groups 1, 3 and 5 were combined to compare if the reconstructions are more accurate with less dynamic properties being captured. This reconstruction was more accurate than both group 1 individually and groups 1, 2 and 4 combined.

Table 3. DMD mode groups and the average percentage error of the reconstruction

DMD Mode Group	Average Error (%)
1	6.39
1,2,4	6.43
1,3,5	6.38

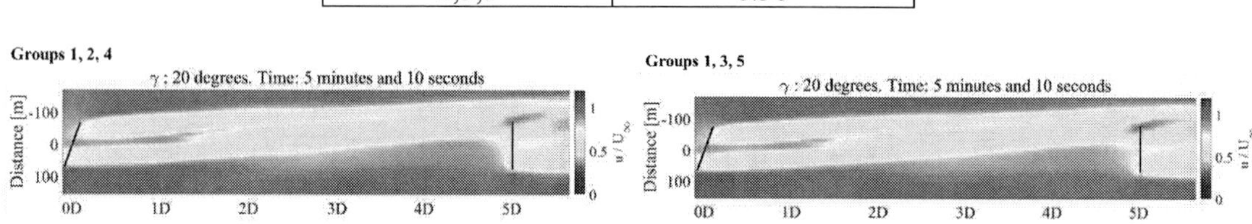

Figure 8. Reconstructions of the Flow using Different Combinations of Mode Groups

4. Conclusion

From the analysis conducted on the individual modes, it was observed that the DMD modes are linked to a specific physical characteristic of the flow field such as the mean flow, vortex shedding from the tower, from the tip of the blades and system noise.

DMD modes exist in pairs and reconstructions are more accurate when the modes are used in their pairs. Some pairs of modes are linked to the same physical characteristics and can be grouped accordingly. Reconstructions made with modes from the same group do not capture the full flow field accurately, except for the group linked to the mean flow. Reconstructions made with a combination of modes from different groups, specifically the group that is linked to the mean flow and the system noise are the most accurate in capturing the flow field, with an average absolute error of 6.38%.

One study chose to use the first 37 modes to reconstruct the flow field [10]. The reconstruction had an average absolute error of 6.52%. However, this paper has shown that individually analysing the modes and specifically choosing certain combinations results in a slight increase in accuracy of 0.14%. Although the increase is small, it is expected that the advantages will increase in larger models. It is therefore important to conduct an analysis on the modes to choose the modes that best capture the flow field, as the first modes are not necessarily the most useful.

Future Work

It is hypothesized that the combination of modes that result in the best reconstruction of the flow field will also result in the best prediction of the flow field when a parameter such as the yaw angle is changed. For future work it is planned to use the modes to train a machine learning algorithm to predict the wake of a two-turbine array.

References

[1] Porté-Agel, F., Bastankhah, M. et al. (2020). Wind-Turbine and Wind-Farm Flows: A Review. Boundary-Layer Meteorology, 174(1), 1–59.

[2] Enel. (N.A). Wind Farm. [Online]. Available at: https://www.enelgreenpower.com/learninghub/renewable-energies/wind-energy/wind-farm. [Accessed 09/01/2023].

[3] Stevens, R. J. A. M., Gayme, D. F. et al. (2016) Effects of turbine spacing on the power output of extended wind-farms. *Wind Energy.*, 19: 359– 370.

[4] van Treuren, K. W., Burdett, T. A. (2014). *Experimental Testing of Wind Turbines Using Wind Tunnels With an Emphasis on Small-Scale Wind Turbines Under Low Reynolds Numbers* (pp. 67–110).

[5] Scott, R. (2000). Characterizing Tilt Effects on Wind Plants. *Journal of Renewable and Sustainable Energy*, 12(4).

[6] Howland, M; Bossuyt, J, et al. (2016). Wake Structure of Wind Turbines in Yaw under Uniform Inflow Conditions. *Journal of Renewable and Sustainable Energy*.

[7] Mathworks. (2023). Reduced Order Modeling.

[8] Kulp, B. (2019). What is a Reduced Order Model and What's Its Product Development Role?

[9] Zhong, Y., Ma, H., et al. (2022). Effects of different suction angles on flow structures and dissipation in the wake of a circular cylinder. *Ocean Engineering*, 266, 112521.

[10] Rodrigues Cassamo, N., van Wingerden João Miguel da Costa Sousa, J.-W.,et al. (2020). *Model Predictive Control for Wake Steering: a Koopman Dynamic Mode Decomposition Approach Mechanical Engineering Examination Committee.*

[11] Taira, K, Dawson, S.T.M, et al. (2017). Modal Analysis of Fluid Flows: An Overview. *AIAA Journal*, 55(2).

MAKING FUNDING PROPOSALS MORE EFFECTIVE: PLAIN LANGUAGE GUIDELINES

Antoinique van Staden[1,*], Idette Noomé[2]
[1]University of Pretoria,[*] ant_vst@yahoo.com
[2]University of Pretoria

Abstract:

Funding is critical to solving complex problems in the face of globalization and rapid technological advances, but opportunities may be lost if scientists fail to communicate across disciplines and across national, cultural and linguistic borders. Communication is mainly in English, the global language of science, but a second language for many scientists in South Africa. Interdisciplinary collaboration, especially in terms of writing and reviewing funding proposals, involves navigating discipline-specific details, terminology, and concepts. However, reviewing panels seldom consist only of experts from the same field of expertise as the proposers. This article proposes a definition of Plain Language for funding proposals and eight Plain Language guidelines. As a comprehensive communication strategy, Plain Language is meant to enhance the clarity and accuracy of documents without losing essential details. These guidelines aim to assist scientists and engineers in crafting more effective funding proposals, ensuring clarity and enabling a multi-disciplinary reviewing panel to assess the merit of a particular research topic. The guidelines are derived from an exploratory, mixed-methods study on the use of Plain Language principles in interdisciplinary peer-to-peer science communication and how they can help to clarify funding proposals. The study included an extensive literature review, a survey and follow-up interviews with scientists from different specialisations.

Keywords:
Funding proposals
Plain Language definition
Plain Language guidelines

1. Introduction

Rapid technological advances and complex global problems often require scientists, including engineers, to work together on projects, [1] and to attract funding for large, expensive projects. To draw effectively on the expertise of colleagues across disciplines, and to convince peers in panels that allocate funding, scientists need to communicate clearly and accurately. However, even if all the scientists concerned communicate in English and appear to use a shared set of technical terms, several factors can hamper communication. One problem arises from the high degree of specialisation required in the sciences and the application of scientific findings and inventions. This specialisation requires very specific terminology, which may not overlap disciplines, leading to confusion. Other challenges include limitations in second-language speakers' ability to express themselves in English, uncertainty around idiomatic usage, or the kinds of "code-mixing and code-switching [that…] establish identity and belonging to a speech community" [2] in different "Englishes" around the world, and differences in how a specific term might be used in each specialisation.

One important area of interdisciplinary science communication is the writing and reviewing of funding proposals. The higher the funding, the stricter the review process. What makes a given funding proposal unique is what it offers and requires, formulated in terms of discipline-specific details, using the terminology and concepts of a particular specialisation. If the readers come from the same field of expertise, these terms and concepts assist in effective communication, but the reviewers on a panel are often not experts from the same field of expertise as the person or team writing the proposal. This implies that the content needs to be communicated to a group of educated peers, who are usually from various fields. A proposal should not lead to misconceptions or omit crucial information which may seem self-evident to those who write the proposal, but are not similarly obvious to reviewers. A further challenge may be making information clear in the rigid templates that are often instituted by funding panels with the intention of creating an equal platform and uniform layout to simplify comparison of applications.

This article argues that carefully applying a selection of appropriate Plain Language principles for peer-to-peer communication as it relates to funding proposals is one possible solution to ensure clear and unambiguous communication across disciplines. The article also offers a Plain Language definition to help guide scientists in drafting funding proposals. When Plain Language principles are applied correctly, they do not undermine the specificity of scientific vocabulary, but help to bridge areas of scientific communication where misunderstanding might arise from unclear expression and organisation. The use of Plain Language principles should aid in making a text accessible in ways that highlight the meaning in the most effective manner possible.

Although communication is regarded as a "soft skill" in terms of the sciences, there is a definite need to develop this skill for science communication [3]. Some effort to promote more effective science communication is being made in the form of training programmes focusing on this skill: several universities already include technical communication in their prescribed coursework for scientists and engineers [4,5,6]. However, in practice, the skills taught may not be enough to enable the writing of funding proposals that get the required results. Nor do popular online resources such as Grammarly offer a fail-safe solution. This resource mainly helps to determine whether information is pitched at the right level for the general public, rather than for specialised peer-to-peer communication. It is a tool that "can help you replace complex words with simpler ones that appeal to a wider audience" [7], so it focuses primarily on vocabulary (but can also help to indicate where sentences might be unclear and give suggestions of how to reword a sentence). A disadvantage of an online tool such as this is that if the specialised vocabulary (or the specialised meaning of a *seemingly* ordinary word) is unknown to the tool's database, it can provide incorrect alternatives, or flag words that are crucial for the interpretation of the particular science content in a funding proposal.

2. Why consider Plain Language?

The Plain Language movement has gained popularity in recent years, but the ideals that the movement promotes are not new [8]. They are connected to the notion of a *lingua franca* or "bridge" or "trade" language, in the sense that Plain Language, like a *lingua franca*, is about wider (and thus global) access to information. English as a *lingua franca* is "a contact language between persons who share neither a common native tongue nor a common (national) culture and for whom English is the chosen foreign language of communication" [9].

Before English became the dominant language of science, and after the decline of Latin, top-level scientists were expected to be well-versed in more than one language and scientific research was conducted in an array of different languages. According to Gordin, "most of humanity for most of its existence has been to a greater or lesser degree multilingual" [10]. The spread of English due to British and later American imperialism altered this situation.

For language to be clear, Orwell [11] insists that a writer should "let the meaning choose the word and not the other way around". In the 1940s, he already suggested a set of rules for clear language which have influenced thinking about Plain Language ever since. Sir Ernest Gower, in the prologue to *The Complete Plain Words* states: "Writing is an instrument for conveying ideas from one mind to another; the writer's job is to make [the] reader apprehend [the] meaning readily and precisely" [12]. Gower's statement seems obvious, but how should a writer convey this message? Exactly how should a writer support the reader and ensure that the reader understands the meaning easily? What is plain to one set of readers could be confusing to another. The different Englishes around the world also affect understanding, and, as English develops, what might be plain in meaning today might not be understandable in future [13]. In the case of science, a specialist term might mean one thing to one specialisation and another thing to other disciplines.

3. Prior Plain Language awareness in the sciences

Although there is some awareness amongst scientists of the benefits of using Plain Language to communicate important scientific information to the general public, there has been no assessment yet on the possible value that Plain Language can have when applied to funding proposals as a form of interdisciplinary peer-to-peer communication.

Scientists from all fields are increasingly made aware of Plain Language to communicate with a wider public. Several journals publishing scientific research from various specialities require researchers to submit their articles along with a Plain Language summary. Gudi et al. [14] regard Plain Language summaries as an essential component to what they call knowledge translation. Kerwer et al. [15] and Locke et al. [16] have also discussed the importance of Plain Language summaries in the scientific community. They believe that Plain Language summaries have the potential to communicate complex (critical) scientific concepts to a wider audience that might otherwise not have had access to this information. One of the benefits of writing Plain Language summaries is that it helps to promote "knowledge transfer" [14] to everyone, from the general public to the expert.

To investigate the use of Plain Language in the sciences, we conducted an extensive literature review on science communication (including Plain Language abstracts) and Plain Language. We also investigated several definitions of Plain Language to construct a specific definition of Plain Language for funding proposals (what does "plain" imply in a funding proposal?). The literature review allowed us to identify possible Plain Language guidelines applicable to finding proposals. These guidelines were then used to analyse two existing funding proposals [17]. Example passages were selected from the two funding proposals. The examples were then rewritten in Plain Language. A survey sent out via SACNASP in 2020 to scientists then tested scientists' preferences regarding the randomly organised examples and their rewritten versions [17]. The majority preferred the rewritten Plain Language versions because these could be understood more easily. The survey and follow-up interviews with some of the scientists also probed participants' writing experiences, their perceptions on Plain Language and their willingness to collaborate with language practitioners in science communication [17].

4. Defining Plain Language for funding proposals

Martin Cutts [8] defines Plain Language as follows:

> The <u>writing</u> and <u>setting out</u> of <u>essential information</u> in a way that gives a cooperative, motivated person a good chance of <u>understanding it at first reading</u>, and in the <u>same sense</u> that the writer meant it to be understood. (our emphases)

The aspect that is most important in writing for the reviewers of funding proposals is that the reviewers should be able to understand the proposal at "first reading" (because of time limitations) also "in the same way the author[s] meant the information to be understood" (because miscommunication could prejudice the outcome of the review). An important aspect to consider regarding scientific communication is that specialist terms may have different meanings in different specialisations. Moreover, words that seem to be ordinary may have specialised meanings. Given that funding proposals aim to show that the proposed research or application is necessary and worth funding, we propose the following definition of Plain Language for funding proposals:

A funding proposal is in Plain Language if it presents essential information in such a way that the sequence of the scientific story, organisation and structuring work together to give a reviewer a good chance of understanding the funding proposal at first reading. A funding proposal's language is clear and concise to ensure that a reviewer also understands the content in the same sense that the writer(s) meant it to be understood.

5. Plain Language guidelines

In this section, selected Plain Language guidelines for writing funding proposals are discussed. The basic guidelines are selected from those that Cutts [13, 18] proposes, adjusted for funding proposals, and considering their relevance based on the wider Plain Language literature. We investigated what other guidelines might apply to funding proposals. To strengthen the guidelines, we drew on Greene [19] for her perspective on Plain Language as it applies to science writing, on Viljoen-Smook [20] for her South African perspective, and on PLAIN's [21] guidelines for an international perspective.

The overarching intention of a funding proposal is to ensure that reviewers grasp what the research is about as early as possible. Organising content in a way that supports this is crucial.

5.1 Guideline 1: Organise your material so that readers can grasp the important information early and navigate through the document easily

Funding applicants need to consider what gets reviewers excited about the research so that they grant funding. This implies presenting the content in such a way that it gives reviewers a chance of understanding the research quickly and being convinced of its importance [22]. Reviewers may have many proposals to read, and a tight deadline. Cutts [18] acknowledges that most readers only skim documents and that "they'll be asking two questions about everything they read: "'So what?' and 'How does this affect me?'" He warns that merely answering these questions in the text might not be enough. Viljoen-Smook [20] states that the purpose of a document should be clear from the start and suggests writing a document in a way that achieves the intended goal of the documentation. Content should be organised in a way that answers the following two questions [23]:

- Is the purpose of the writing clearly stated?
- Has the information been arranged in a logical and reader-friendly manner?

Keeping these two questions in mind helps with the sequencing and logical order of the content, as logical flow is a way for readers to follow from sentence to sentence and from paragraph to paragraph [20]. It is also important to ensure that paragraphs address only one topic at a time and that clear transitions connect sentences as well as paragraphs.

5.2 Guideline 2: Use words your readers are likely to understand

Funding proposals are usually aimed at a panel of reviewers and contain discipline-specific vocabulary that might not be clear to reviewers who are not in the same field. Some specialist terms can have different meanings in different fields, so technical terms should be clearly defined and consistently used throughout the proposal. Cutts [13] suggests using only words that ensure that "busy readers" do not miss the point. This is valid in respect of the general descriptions, but does not fully address the fact that some technical terms or scientific expressions can form part of a "shortcut" in communication between scientists. Where that is the case, and the technical term is the most efficient one, it should be kept. The point is not to cut out technical terms, but to make sure that (if they have more than one meaning), the right meaning is clear from the (known) context. In the scientific context, specific words (jargon) are better than using general words, but only if the audience is guaranteed to know the vocabulary.

PLAIN [21] defines jargon as "unnecessarily complicated, technical language used to impress, rather than to inform". By this PLAIN does *not* mean that all necessary *technical* terms should be removed, but rather that the words around the term should be as clear as possible, for example, opting for "use" instead of "utilise". This does not alter the technical specificity of the text: it focuses readers on the essence. The words chosen in the connecting argument should thus aim at clarity.

Cutts [13] argues that where technical terms are essential, "clear explanations are vital" and states that in

> a technical document, technical words will usually be clear enough to technical people… (S)ometimes an unusual word is perfect for the job, expressing just what you want to say; then you should use it and either give an explanation or trust the context to explain.

What Cutts [18] terms "trust the context" links to the first guideline in the sense that you can only create this necessary context by organising your information properly. Technical vocabulary is not eliminated. The aim in evaluating the vocabulary is to ensure maximum clarity, not to sacrifice precision.

5.3 Guideline 3: Write concisely

For this guideline, Cutts [18] emphasises that busy readers want to get to the main points as quickly as possible. "Wordiness often comes from trying to make a simple procedure sound impressive" [18]. It is important to look at the content in a document from the reader's perspective. Cutts [18] suggests the following:

- remove useless words (overly descriptive words that have a single word equivalent, or straight repetition);
- remove words that do not show the action; and
- shorten prepositional phrases (these are phrases that start with prepositions such as "for", "to", "by", "with", "from", "in", and "of").

5.4 Guideline 4: Use vertical lists to break up complicated text

Vertical lists are bullet or numbered lists, instead of horizontal lists (where elements of the list are simply strung into a sentence in a text chunk or paragraph). The purpose of this structural and formatting strategy is to make information-dense documents easier to follow [20]. Vertical lists help readers to get to the essence at first glance.

According to PLAIN, [21] the advantages of using vertical lists are that they

- highlight levels of importance;
- help users to understand the order in which things happen;
- make it easy for users to identify all the necessary steps in a process; and
- add white space for easier reading.

Cutts [13] confirms that vertical lists make information "easier" to grasp. Some words of warning, however:

- Vertical lists should always include a lead-in (platform) sentence.
- An important point to ensure understanding is that a list must follow parallel structure [13]. A parallel structure refers to using the same grammatical construction or sequence. For example, in the bullet list on the advantages above, all the points start with a verb, and every point completes the sentence "the advantages of using vertical lists are that they…". That sentence functions as a platform statement. Parallel structure is often overlooked by scientists, who tend to focus on the content rather than the grammatical form. For example, they might switch confusingly between starting bullets with nouns and verbs, or they may use different forms of verbs, such as infinitives in some bullets and participles in others. They may make some bullet points a full sentence and others a sentence fragment.
- The punctuation in a vertical list should also be used consistently [13,18].
- Use simple bullet formats [21]. Large bullets with "creative" shapes and sizes distract readers from the content.

5.5 Guideline 5: Over the whole document, make the average sentence length 15 to 20 words

Complex ideas are often written in long, overly complicated sentences. Cutts [13] recommends variation in sentence length. He asks what length of a sentence is too long and suggests that sentences between 15 and 20 words are easiest to read. He emphasises that the focus should be on the average, so it is not necessary to rewrite every sentence in a document to fit between 15 and 20 words.

Although PLAIN [21] does not suggest an explicit number of words to define a "short" sentence, PLAIN does suggest presenting "only one idea in each sentence". The reason for keeping to the 15 to 20 word range is that readers are used to it. One way to keep sentences short [13] is to create a vertical list (see Guideline 4).

5.6 Guideline 6: Prefer the active voice unless there is a good reason for using the passive

Scientific breakthroughs are often communicated using abstract nouns and the passive form. The preference for the passive voice is diminishing, however. Using the active voice in a proposal ensures that the value of the research is clearer.

Although the process is usually the key to any scientific research, and is sometimes conveyed more effectively in the passive, texts should be written in the active unless it changes the context of what the researcher is trying to say. For science, the "doer" is often irrelevant, and for this reason scientists often prefer to write in the passive form, [19] but it can add up to 30% to a document. One advantage of using the active voice is that it helps to write more concisely.

In the following example, taken from Cutts, [18] the first sentence is in the passive, and the second is in the active:

> Concern has been raised about arrangements for gaining immediate access to the chimney. Winch failure or the presence of debris between the platform edge and the chimney internal wall may necessitate access.

Cutts [18] explains that in the first sentence, the focus is on gaining access, although it has been placed towards the end of the first sentence. However, because the second sentence is in active, it seems to take the focus away from gaining access, so in this case rewriting the second sentence in the passive form ensures that the focus stays on gaining access. The example can be rewritten as follows:

> Concern has been raised about arrangements for gaining immediate access to the chimney. Access <u>may be needed</u> if the winch fails or there is debris between the platform edge and the chimney internal wall. (Cutts's emphasis) [18].

An even shorter and clearer version would involve using active in the first sentence:

> Immediate access to the chimney may become necessary if the winch fails, or if there is debris between the platform edge and the chimney internal wall.

The passive can also be used if the intention is "(t)o focus attention on the receiver of the action by putting it first". The passive can also help to place already known information at the start of a sentence or clause.

5.7 Guideline 7: Reduce cross-references to a minimum

Funding proposals often require using fixed templates. The risk is then that authors keep referring the reader to later sections of the document. The challenge is to ensure that all relevant information is kept together and explained where it is first mentioned as far as possible because "(c)ross-references mean cross readers" [18] and that if the point (purpose) of a document is not clear, then it might not be worth making the point at all.

PLAIN [21] concurs that "(m)ost users consider [cross-references] a bother, and just skip over them". This can be a problem if the information being referred to is critical to the argument that the author is trying to convey at a given point. However, repeating content again and again "can be equally annoying to users" [22]. Where it might be impossible to remove all the cross-references, PLAIN [21] suggests:

> Be sure that the reference you insert clearly describes the referenced material. That way, users can decide if they need to read it to know how the rule [or section] affects them. Sometimes just including the title of the referenced section is enough.

If cross-references are necessary, it is best rather to include them at the end of a text so that it does not distract readers. Including the cross-reference at the end of the text also gives readers "a chance to absorb your main message before your references elaborate on it" [21].

5.8 Guideline 8: Sequence process steps for clarity

This guideline comes from Cutts's [18] instruction to "(d)evote special effort to producing lucid and well-organised instructions". Although a process is not, strictly speaking, a set of instructions, it follows the same kind of structuring, because one step leads to the next. The following principles can help to ensure that sequencing is clear:

- remember the readers;
- split the information into chunks;
- favour a basic style of language; and
- use separate, headed sections.

For this last point, we needed to consider an alternative, because funding proposals often use templates with prescribed headings. Although templates may provide different sections, containing questions that the panel specifically wants answers to, PLAIN [21] suggests addressing a process in a chronological order and clearly indicating the steps that the process follows. Sequencing steps clearly help a reviewer to follow what the scientists have done, clearly and without any misunderstandings.

6. Conclusion

Even though English is not the home language of most scientists, English is likely to remain the *lingua franca* of the science community, including that community in multilingual countries such as South Africa, for the foreseeable future. Finding ways for scientists to communicate more effectively across disciplines will become even more of a priority in future. Using Plain Language is one way to bridge countries and disciplines more easily, but not all Plain Language principles are equally useful in writing funding proposals. This article has selected and discussed the principles and strategies likely to be the most useful to scientists preparing a funding proposal.

The language of science contains technical terminology and abstract concepts, which affects the clarity of the content in documents such as funding proposals. For the scientific vocabulary to have meaning, the context within which it is included in funding proposals becomes crucial. To help scientists to communicate well in funding proposals, a focused definition of Plain Language and a set of eight guidelines are provided. The aim in peer-to-peer communication remains clarity, precision and effectiveness, bearing in mind Quintillian's dictum, quoted by Bulaon: "We should not write so that it is possible for the reader to understand us, but so that it is impossible for him [or her] to misunderstand us." [24]

References

[1]Phillip, R. *Five Ways to Tackle the World's Wicked Problems.* World Economic Forum. Available from: https://www.weforum.org/agenda/2012/07/five-ways-to-tackle-the-worlds-wicked-problems/ [Accessed 26 October 2017].

[2]Kirkpatrick, A. & McLellan, J. World Englishes and/or English as a lingua franca. In: Gee, J.P. & Handford, M. (eds.), *The Routledge Handbook of Discourse Analysis.* New York, NY: Routledge Handbooks; 2012. p. 654-669.

[3]Taylor, E. Investigating the perception of stakeholders on soft skills development of students: Evidence from South Africa. *Interdisciplinary Journal of e-Skills and Life Long Learning,* 12; 2016. p. 1-18.

[4]Baldauf, R.B. Jr. Speaking of science: The use by Australian university staff of language skills. In: Ammon, U. ed., *The Dominance of English as a Language of Science: Effects on Other Languages and Language Communities.* Berlin: Mouton de Gruyter; 2011. p. 139-165.

[5]Grady, H.M. & Davis, M.T. *Integrating technical communication into engineering education: A case study.* IEEE Professional Communication Conference, 23-25 September 1998, Quebec City, Canada; 1998. p. 265-270.

[6]Jameson, D.A. Management consulting and teaching: Lessons learned teaching professionals to control tone in writing. *Business Communication Quarterly,* September; 2009. p. 333-338.

[7]Romanyshyn, M. *Plainly speaking, a linguistic approach to simplifying complex words.* Available at https://www.grammarly.com/blog/engineering/plainly-speaking-a-linguistic-approach-to-simplifying-complex-words/ [Last accessed 28 July 2022].

[8]Cutts, M. *Oxford Guide to Plain English.* 3rd ed. Oxford: Oxford University Press; 2004.

[9]Firth, A. The discursive accomplishment of normality: On "lingua franca" English and conversation analysis. *Journal of Pragmatics,* 26; 1996. p. 237-259.

[10]Gordin, M.D. *Scientific Babel: How science was done before and after global English.* London: University of Chicago Press; 2015.

[11]Orwell, G. Politics and the English language. In: Maxwell-Mahon, W.D. & Titlestad, P.J.H, and colleagues (eds) *Ways with Words. A Collection of Essays.* Johannesburg: Lexicon; [1946] 1992. p. 107-112.

[12]Gower, E. *The Complete Plain Words.* London: Penguin; [1954] 1966.

[13]Cutts, M. *Oxford Guide to Plain English.* 4th ed. Oxford: Oxford University Press; 2013.

[14]Gudi SK, Tiwari KK, Panjwani K. Plain-language summaries: An essential component to promote knowledge translation. *International Journal of Clinical Practice.* 2021 Jun;75(6):e14140. doi: 10.1111/ijcp.14140.

[15]Kerwer, M., Chasiotis, A., et al. Straight from the scientist's mouth: Plain Language summaries promote laypeople's comprehension and knowledge acquisition when reading about individual research findings in Psychology. *Collabra: Psychology,* 7(1); 2021. p. 1-24.

[16]Locke, J.N., Whiteman, L. & Mitrany, D. Plain Language in science: Signs of intelligible life in the scientific community? *Science Editor,* 24(6); 2001. p. 194.

[17]Van Staden, A. Using Plain Language for more effective interdisciplinary science communication [thesis]. Pretoria: University of Pretoria.

[18]Cutts, M. *Oxford Guide to Plain English.* 5th ed. Oxford: Oxford University Press; 2020.

[19]Greene, A.E. *Writing Science in Plain English.* Chicago, IL: University of Chicago Press; 2013.

[20]Viljoen-Smook, E.E. *Plain Language: The Practical Guide to Easy-to-Understand Language.* Cape Town: NB Publishers; 2020.

[21]PLAIN. *Federal Plain Language Guidelines.* Available at https://www.plainlanguage.gov/guidelines/ [Last accessed 28 July 2022].

[22]Ness, R. *Writing Science: The Story's the Thing.* Available at https://www.science.org/content/article/writing-science-storys-thing [Last accessed 31 July 2022].

[23]Cornelius, E. Defining "Plain Language" in contemporary South Africa. *Stellenbosch Papers in Linguistics,* 44; 2015. p. 1-18.

[24]Bulaon, M.A.M. Why is the language of mathematics confusing to students? Available at https://www.researchgate.net/publication/325968025_Why_is_the_language_of_mathematics_confusing_to_students [Last accessed 25 June 2022].

A STRAIN GAUGE-BASED METHOD FOR CORRECTING OUT-OF-PLANE MOTION ERRORS IN 2D DIGITAL IMAGE CORRELATION

Carl-Hein Visser[1,*], Gerhard Venter[1], Melody Neaves[1]

[1]Department of Mechanical and Mechatronic Engineering, Stellenbosch University, Private Bag X1 Matieland, 7602, South Africa
*Correspondence: chvisser56@gmail.com

Abstract: A strain gauge-based compensation method for strain errors from out-of-plane motion in 2D digital image correlation (DIC) measurements was previously proposed by the authors. In this paper, further work conducted on the functionality of the strain gauge-based error compensation method is presented. A sensitivity analysis was performed to study the influence of the number and the size of the strain gauges used in the error compensation method. While the number of strain gauges used proved to have minimal effect on the strain error that was compensated for, using larger strain gauges significantly improved the accuracy of the data. Furthermore, possible limitations of the error compensation method were investigated. This investigation determined that when there is a non-linear variation in out-of-plane motion over the region-of-interest, such as at a stress concentration, the error compensation is not as effective. A study was also done on the effect of using virtual strain gauge measurements from stereo-DIC, instead of measurements from physical strain gauges, to apply the error compensation method to 2D-DIC data.

Keywords: digital image correlation; full-field strain measurements; error compensation; strain gauges; out-of-plane motion

1 Introduction

A versatile measurement tool for non-contact full-field surface deformation readings is found in digital image correlation (DIC). DIC can either be performed as single-camera 2D-DIC or as multi-camera stereo-DIC, where stereo-DIC is the more widely used option due to its improved accuracy. Despite its relatively lower accuracy, multiple factors still exist that make 2D-DIC an attractive measurement option. These include 2D-DIC being both financially and computationally less expensive than stereo-DIC, stereo-DIC having stricter testing requirements and a more complex test setup than 2D-DIC, 2D-DIC being the more user-friendly option, and stereo-DIC being confined to very large and very small fields-of-view, for example [1, 2]. Various error sources exist in DIC measurements, with lens distortions [3] and out-of-plane motions [4] being the most significant. Unlike lens distortions which can relatively simply be corrected for [5], a method does not currently exist that readily corrects for errors from out-of-plane motion. This is of particular concern as out-of-plane motion is both a common occurrence and often unavoidable in mechanical testing scenarios which is where 2D-DIC is most often used.

A variety of error compensation methods have therefore been developed to correct for displacements and strain errors from out-of-plane motion in 2D-DIC measurements. Examples include the Region of Compensation method [6] which was originally developed by Pan, Yu, and Wu [7], the Image Rectification method initially proposed by Lava et al. [8] and the method of using bilateral telecentric lenses as first implemented by Sutton et al. [4].

A strain gauge-based compensation method for strain errors from out-of-plane motion in 2D-DIC measurements was previously proposed by the authors [9]. The first component in the compensation method's development was the derivation of an equation that describes the theoretical strain error from out-of-plane motion in a 2D-DIC measurement:

$$\varepsilon_{yy,\ total\ error} = A + BY + CX \tag{1}$$

Where $\varepsilon_{yy,\ total\ error}$ is the total out-of-plane motion strain error, A, B and C are unknown constants, and X and Y are the locations in the region-of-interest. Using this equation, as well as at least three sets of (physical) strain gauge data and 2D-DIC virtual strain gauge measurements to find $\varepsilon_{yy,\ total\ error}$ at the same known X and Y locations, a system of linear equations can be set up to find A, B and C, as follows:

$$\begin{bmatrix} 1 & Y_1 & X_1 \\ 1 & Y_2 & X_2 \\ 1 & Y_3 & X_3 \end{bmatrix} \begin{bmatrix} A \\ B \\ C \end{bmatrix} = \begin{bmatrix} \varepsilon_{yy,\ total\ error\ 1} \\ \varepsilon_{yy,\ total\ error\ 2} \\ \varepsilon_{yy,\ total\ error\ 3} \end{bmatrix} \tag{2}$$

With A, B and C known, and using Equation 1, an estimate of the strain error over the 2D-DIC data field can be obtained. Error compensation can then be performed by subtracting the estimate of the strain error from the 2D-DIC full-field strain data to generate a more accurate 2D-DIC measurement. This is finally repeated at each time step to perform error compensation over the entire 2D-DIC measurement.

Following the derivation of the compensation method, a set of two experiments was performed to practically validate its functionality. In these experiments, tensile tests were performed on two different dog-bone specimen geometries: the one being a smaller, plain design that produced a homogeneous strain field, referred to as the "simple" test specimen, and the other being a larger design with two holes and a slot cut out of it that produced a heterogeneous strain field, referred to as the "complex" test specimen. All of the specimens had DIC speckle patterns applied to one of their surfaces and had strain gauges attached to their other surfaces. With this setup, 2D-DIC, stereo-DIC and strain gauge measurements could be performed simultaneously during the tensile tests. Although it is not needed for applying the error compensation method, stereo-DIC measurements were also captured to use as a full-field strain measurement reference to compare the 2D-DIC measurements against.

The error compensation method provides both a relatively inexpensive and relatively user-friendly way to produce more accurate 2D-DIC measurements. This therefore makes 2D-DIC a more applicable measurement tool, and in situations where stereo-DIC is either unavailable or not practically possible, makes 2D-DIC a possible alternative to stereo-DIC. Additional details regarding the derivation of the error compensation, the specimen geometry, the experimental setup and procedures, as well as the results from the practical experiments can be found in [9].

In this work, the results from the experimental validation were further analysed, and a sensitivity analysis was conducted, to gain further insight into the functionality of the strain gauge-based error compensation method. The sensitivity analysis was performed to study the effect of varying the number of strain gauges and the size of the strain gauges used in the error compensation method. Error compensation was also performed using stereo-DIC virtual strain gauge data, instead of physical strain gauge data, to study its impact. Ultimately, this work provides greater insight into the viability of the strain gauge-based error compensation method for improving the accuracy of 2D-DIC measurements.

2 Sensitivity Analysis

A sensitivity analysis was performed to determine the effect of the number and size of the strain gauges used in the strain gauge-based error compensation method. Before the analysis was conducted, a set of strain gauge locations was determined for each of the various numbers of strain gauges being analysed. From this, error compensation was simulated for different numbers of strain gauges, for different strain gauge sizes, and also for the simultaneous impact of both the number of strain gauges and their sizes. Using this analysis, both a suitable number and size of strain gauges to use in the error compensation method were determined. The analysis was performed on both of the specimen geometries used in the experimental validation of the error compensation method [9].

2.1 Data Generation

The DIC data (both 2D-DIC and stereo-DIC) obtained during the practical implementation of the error compensation method was used here in the sensitivity analysis. Experimental DIC data, instead of simulated DIC data (such as Finite Element (FE) analysis data) was employed as the data includes the effects of the physical phenomena that the specimen experiences during experimentation. Furthermore, the strain gauge measurements used in this analysis needed to be obtained from the stereo-DIC data as a limited amount of physical strain gauges were used for measurements, and the same size strain gauges were used, throughout the practical experiments. Here, the term "virtual strain gauge" defines a strain gauge measurement obtained from stereo-DIC data. The virtual strain gauge measurements were obtained by:

1. Choosing a subset of values from the stereo-DIC full-field strain data, which correspond to the specified size of the strain gauge and its position in the region-of-interest.
2. Determining the average of the selected strain values to find a single strain value.
3. For the full-field strain data from each image in the stereo-DIC measurement, apply Steps 1 and 2 to generate a set of virtual strain gauge measurements.
4. For each strain gauge position and size being analysed, repeat Steps 1 to 3.

2.2 Strain Gauge Locations

For the sensitivity analysis, a D-Optimal design of experiments was conducted to find an optimal set of locations for the strain gauges to be placed, within the region-of-interest. Using a D-Optimal design should improve how the strain errors are fitted to a plane, which in turn should lead to an increase in the amount of error compensation. A version of the Fedorov exchange algorithm [10], implemented according to de Aguiar et al. [11]'s proposed method, was used to perform the design of experiments.

2.3 Effect of the Number of Strain Gauges Used

First, an analysis was done on the effect that the number of strain gauges used has on the error compensation method, where the strain gauge size was constant. Specifically, error compensation was simulated for a range of three to twelve strain gauges, where strain gauge size of $6\,\mathrm{mm} \times 8\,\mathrm{mm}$ was used as it is the same size as the physical strain gauges used in the experimental validation of the compensation method. The strain gauge locations used in the analysis were obtained from the aforementioned D-Optimal design, according to the number of strain gauges being analysed. The results from the analysis, for both test cases, are shown in Figure 1. Here, the mean RMS error, which is the average value of the RMS error from each of the DIC images in the measurement was plotted against the corresponding number of strain gauges being analysed. For reference, the RMS error is found by subtracting the 2D-DIC data from the stereo-DIC data.

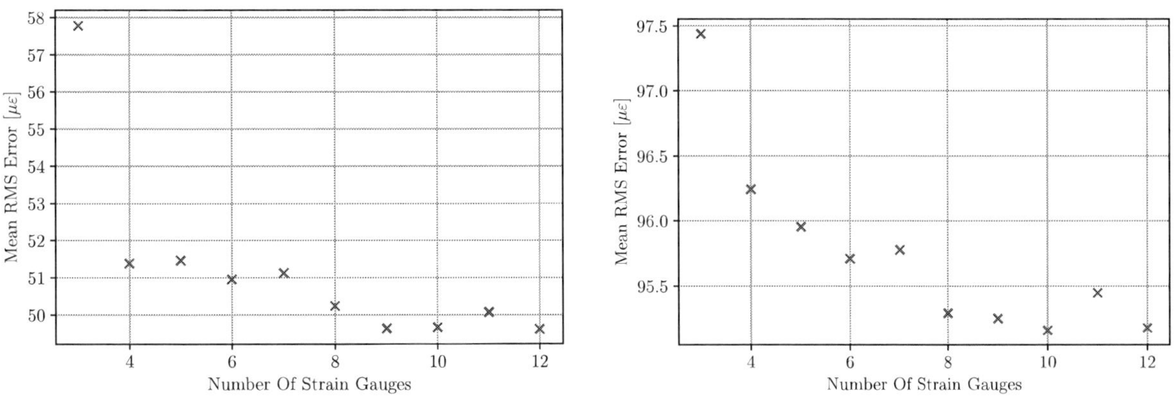

(a) Simple specimen (b) Complex specimen

Figure 1: The influence which the number of strain gauges used has on the error compensation method.

For both test cases, an increase in the number of strain gauges used led to an improvement of less than $7\,\mu\varepsilon$, as shown in Figures 1a and 1b. These results demonstrate for these two test scenarios, the number of strain gauges used for error compensation has an insignificant impact on the error compensation produced. This is probably due to using more strain gauges not leading to a greatly improved plane fit of the strain errors, particularly for four or more strain gauges.

2.4 Effect of the Strain Gauge Size Used

Following this, an analysis was performed on the strain gauge size's influence on the error compensation method, while the number of strain gauges used remained the same. The number of strain gauges used matched that used in the experimental validation, namely three strain gauges and four strain gauges for the analyses of the simple specimen test and complex specimen test, respectively. The strain gauge locations used in this analysis were also found using the D-Optimal design. Four different sizes of strain gauges were studied here, specifically: $0\,\mathrm{mm} \times 0\,\mathrm{mm}$ (strain measurement at a single datapoint), $2.5\,\mathrm{mm} \times 3\,\mathrm{mm}$, $2.8\,\mathrm{mm} \times 6\,\mathrm{mm}$ and $5\,\mathrm{mm} \times 10\,\mathrm{mm}$. The first size analysed was infinitesimally small as this allowed the influence of implementing a single strain value, instead of an average of multiple strain values (as a strain gauge would generate), to be studied. The selection of the other three sizes was done such that a smaller size, the same size, and a larger size, than the strain gauge size from the experimental validation was studied. Figure 2 shows the effects of the strain gauge size on the amount of error compensation generated, for both test cases.

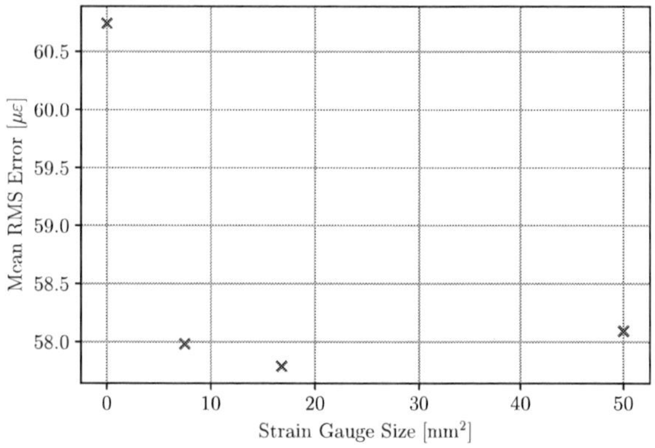

(a) Simple specimen

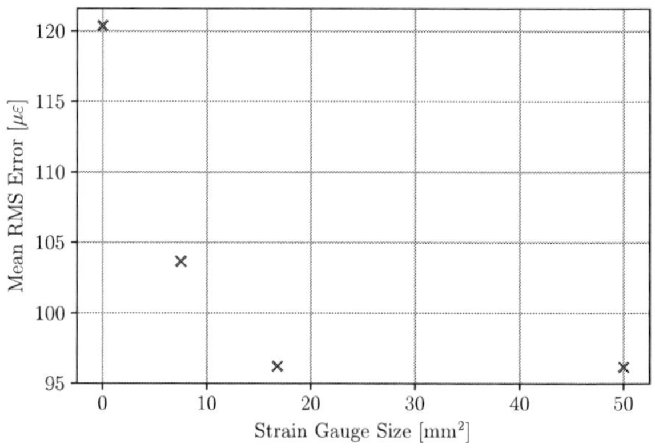

(b) Complex specimen

Figure 2: The influence which the strain gauge size used has on the error compensation method.

The results from this analysis show that varying the size of the strain gauges used significantly improved the error compensation, for both test cases. As shown in Figures 2a and 2b, increasing the strain gauge size led to reductions in the mean RMS error of almost $20\,\mu\varepsilon$ and almost $25\,\mu\varepsilon$ for the simple and complex specimen test cases, respectively. This improvement can be attributed to the averaged strain values, obtained from the large strain gauges, providing strain error values that are less susceptible to noise than the single-point strain values. The increased robustness to noise would in turn lead to a more accurate plane fit to the strain error values being achieved. From this analysis, it was evident that an appropriate strain gauge size of $2.8\,\mathrm{mm} \times 6\,\mathrm{mm}$ was used during the practical experiments, as this size produced the lowest mean RMS error for both test cases here.

2.5 Joint Effect of Number and Size of Strain Gauges Used

Lastly, the combined effect of both the number and the size of the strain gauges employed by the error compensation method was considered. The results obtained here are shown in Figure 3.

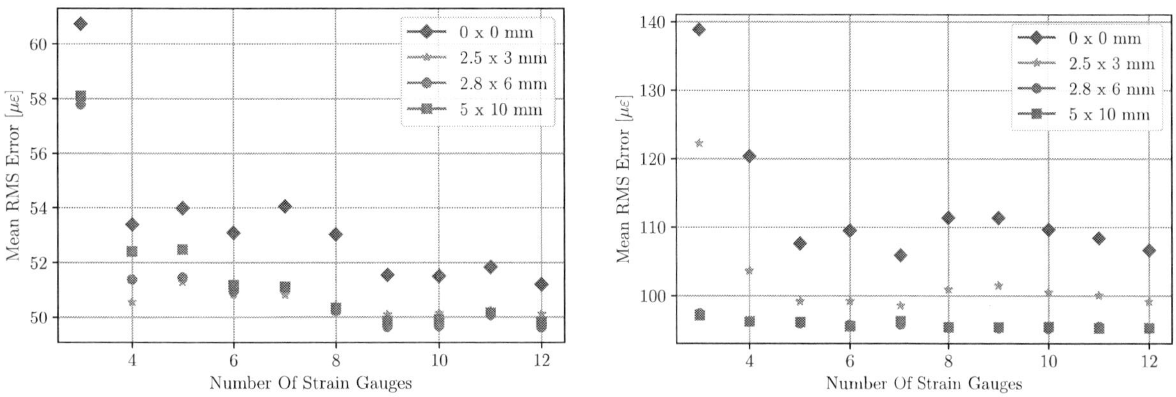

(a) Simple specimen

(b) Complex specimen

Figure 3: Influence of both the number and the size of strain gauges used by error compensation method.

Using Figure 3, similar conclusions can be made from the analysis of the combined influence of the strain gauge number and size as those made from the studies of their individual influences from before. In both test cases, when a strain gauge size greater than $0\,\mathrm{mm}^2$ was used, it led to a reduction in the mean RMS error. Furthermore, in all of the strain gauge numbers simulated, the $2.8\,\mathrm{mm} \times 6\,\mathrm{mm}$ size strain gauge led to the lowest mean RMS error, or close to the lowest mean RMS error. When considering the effect of the number of strain gauges, for the simple specimen test case, the mean RMS error decreased where the number of strain gauges used was more than three, for all of the strain gauge sizes. For the complex specimen test case, the mean RMS error decreased for strain gauge numbers greater than six, for the sizes of $0\,\mathrm{mm} \times 0\,\mathrm{mm}$ and $2.5\,\mathrm{mm} \times 3\,\mathrm{mm}$, and also when the strain gauge number simulated was above three, for the sizes of $2.8\,\mathrm{mm} \times 6\,\mathrm{mm}$ and $5\,\mathrm{mm} \times 10\,\mathrm{mm}$. For both test cases, when the strain gauge size of $2.8\,\mathrm{mm} \times 6\,\mathrm{mm}$ was simulated, an increase in the number of strain gauges beyond four strain gauges did not lead to a significant decrease in the mean RMS error. The results therefore show that a sufficient number of strain gauges were used in both of the test cases and that the size of the strain gauges used throughout the experimental validation was appropriate.

3 Further Analysis of Experimental Results

A variety of observations were made on the results of the practical experiments upon further examination of the data. Firstly, the realisation was made that the error compensation method's performance was not adequately demonstrated. This is due to the corrected 2D-DIC results being compared to the stereo-DIC results, instead of the strain gauge data. In this way, the error compensation was performed using the strain gauge data on the uncorrected 2D-DIC data and then the corrected 2D-DIC data was in turn compared to the stereo-DIC data. A more accurate representation of the compensation method's performance could be obtained by either comparing the corrected 2D-DIC data to full-field strain gauge data or by performing error compensation using stereo-DIC data. Since the full-field strain gauge data was not available and would have been impractical to obtain in reality, error compensation was instead performed using stereo-DIC data, on both test cases, which yielded the results shown in Figure 4.

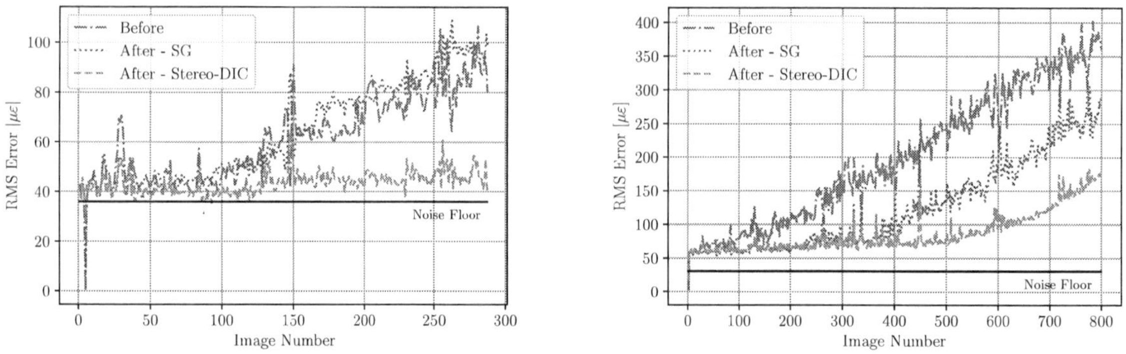

(a) Simple specimen test case

(b) Complex specimen test case

Figure 4: Example results of the RMS error difference between the 2D-DIC and stereo-DIC measurements. The plots show the RMS error before and after the application of the error compensation method, where correction was performed using both strain gauge and stereo-DIC data.

In Figure 4, the label "Before" refers to the RMS error before the application of the error compensation method, while "After - SG " and "After - Stereo-DIC" refers to the RMS error after the application of the error compensation method using strain gauge and stereo-DIC data, as the ground truth, respectively. For both test cases, when the error compensation was done using stereo-DIC virtual strain gauge data, it resulted in a noticeable improvement in the degree of error compensation that occurred, particularly in the simple specimen test case. This difference in results can be attributed to the stereo-DIC data now being used both as a reference and for correction.

In Figure 5, the strain data from a physical strain gauge, and from stereo-DIC, uncorrected 2D-DIC and corrected 2D-DIC virtual strain gauges, all measured at the same location, are compared to one another. These results show a clear improvement in the accuracy of the corrected 2D-DIC data, for both test cases. For the simple specimen test case, the difference is even clearer since before in Figure 4a the 2D-DIC data that was corrected using strain gauge data did not show an improvement. The reason the results here now show improved accuracy in the corrected 2D-DIC data is likely due to the corrected 2D-DIC data now being compared to the strain gauge data, which is also the data which it was corrected with.

When the data was further examined, it was also noted that in the complex specimen test case, the error compensation method kept the RMS error relatively constant for about the first half of the data, but the RMS error would then increase afterward, as shown in Figure 4b. This can be explained by both an increase in the specimen's out-of-plane motion throughout the test as well as a limitation inherent to the error compensation method. In the latter half of the measurement where high stresses would be present, particularly at stress concentrations, the test specimen would begin to plastically deform, in the form of necking for example. When this occurs, the surface of the specimen would no longer be planar, leading to out-of-plane motion and a non-planar strain error as a result. Since the error compensation method describes the strain error as a plane, the error compensation method provides an inaccurate estimate of the strain error, which in turn leads to less error being corrected. A depiction of how the out-of-plane motion during a test varies from more planar to less planar is shown in Figure 6, where the data was created by finding the difference between the measured out-of-plane motion data and a plane fit to that data.

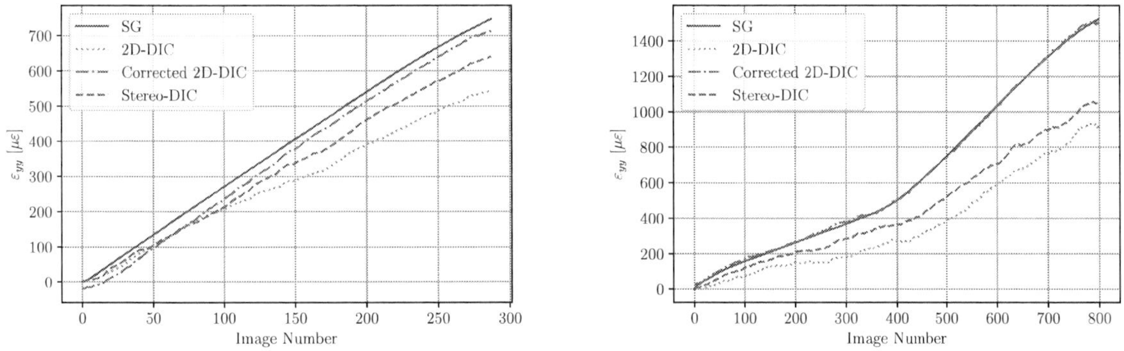

(a) Simple specimen test case

(b) Complex specimen test case

Figure 5: Example results of the strain measured by various sources compared to one another. For both plots, the data was obtained from the strain gauge and virtual strain gauge measurements at the same locations.

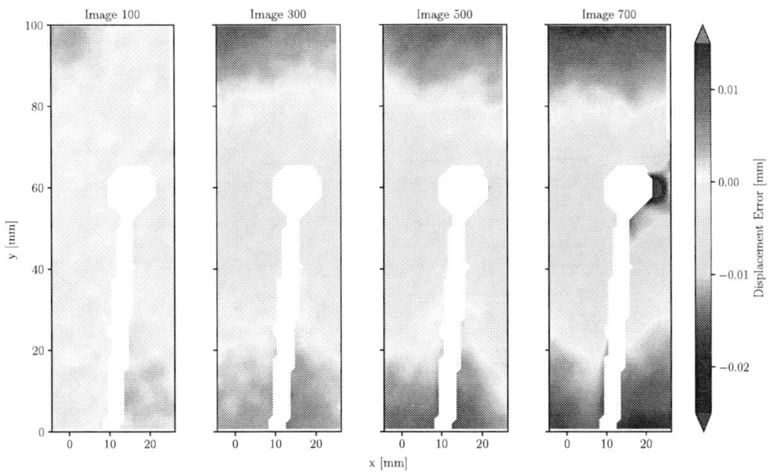

Figure 6: Example of how the planarity of the out-of-plane motion changes throughout a measurement, found by subtracting the out-of-plane motion data from a plane fit of the data.

4 Conclusion

This work is a further investigation into the functionality of the strain gauge-based error compensation method. A sensitivity analysis was done to determine the impact of varying the number and size of the strain gauges used on the error compensation method. The results showed that the number of strain gauges had an insignificant impact, while the size had a significant one, and it ultimately demonstrated that a sufficient number and size of strain gauges were used during the practical experiments. Further analysis of the data from the practical application of the error compensation method showed that while the error compensation method always improves measurement accuracy, it is unable to properly compensate for errors when non-planar out-of-plane motion occurs. The analysis also found that when the same data is used for both error compensation and for comparison of the results, a clear indication that the error compensation method improves a 2D-DIC measurement's accuracy is provided. The results from this work therefore serve as further support for the use of the strain gauge-based error compensation method as a method to improve the accuracy of 2D-DIC data.

Nomenclature

ε	Strain
$yy,\ total\ error$	Total strain error from out-of-plane motion in the normal y-direction
A, B, C	Unknown strain error equation constants
X & Y	Locations in the region-of-interest
2D	Two-Dimensional
DIC	Digital Image Correlation
FE	Finite Element
RMS	Root-Mean-Square
SG	Strain Gauge

References

[1] A. L. Hijazi. "A Novel Approach for the Determination of Surface Tilt Angles in Two-Dimensional Digital Image Correlation Experiments". In: *Experimental Mechanics* 60 (3 2020), pp. 267–282. ISSN: 17412765. DOI: 10.1007/s11340-019-00554-8.

[2] X. Xu et al. "High-Accuracy, High-Efficiency Compensation Method in Two-Dimensional Digital Image Correlation". In: *Experimental Mechanics* 57 (6 2017), pp. 831–846. ISSN: 17412765. DOI: 10.1007/s11340-017-0274-2.

[3] S. Yoneyama et al. "Lens distortion correction for digital image correlation by measuring rigid body displacement". In: *Optical Engineering* 45 (2 2006), p. 023602. ISSN: 0091-3286. DOI: 10.1117/1.2168411.

[4] M. A. Sutton et al. "The effect of out-of-plane motion on 2D and 3D digital image correlation measurements". In: *Optics and Lasers in Engineering* 46 (10 2008), pp. 746–757. ISSN: 01438166. DOI: 10.1016/j.optlaseng.2008.05.005.

[5] B. Pan et al. "Systematic errors in two-dimensional digital image correlation due to lens distortion". In: *Optics and Lasers in Engineering* 51 (2 2013), pp. 140–147. ISSN: 01438166. DOI: 10.1016/j.optlaseng.2012.08.012.

[6] L. Wittevrongel et al. "Evaluation of Methodologies for Compensation of Out of Plane Motions in a 2D Digital Image Correlation Setup". In: *Strain* 51 (5 2015), pp. 357–369. ISSN: 00392103. DOI: 10.1111/str.12146. URL: https://onlinelibrary.wiley.com/doi/10.1111/str.12146.

[7] B. Pan, L. Yu, and D. Wu. "High-accuracy 2D digital image correlation measurements using low-cost imaging lenses: Implementation of a generalized compensation method". In: *Measurement Science and Technology* 25 (2 2014). ISSN: 13616501. DOI: 10.1088/0957-0233/25/2/025001.

[8] P. Lava et al. "Error estimation in measuring strain fields with DIC on planar sheet metal specimens with a non-perpendicular camera alignment". In: *Optics and Lasers in Engineering* 49 (1 2011), pp. 57–65. ISSN: 01438166. DOI: 10.1016/j.optlaseng.2010.08.017. URL: http://dx.doi.org/10.1016/j.optlaseng.2010.08.017.

[9] C.-H. Visser, G. Venter, and M. Neaves. "A Strain-Gauge-Based Method for the Compensation of Out-of-Plane Motions in 2D Digital Image Correlation". In: *Mathematical and Computational Applications* 28.2 (2023), p. 40. ISSN: 2297-8747. DOI: 10.3390/mca28020040. URL: https://www.mdpi.com/2297-8747/28/2/40.

[10] A. Labuschagne. *Federov*. https://github.com/AWlappies/Federov. Accessed: 27 July 2023. 2023.

[11] P. de Aguiar et al. "Tutorial D-optimal designs". In: *Chemometrics and Intelligent Laboratory Systems* 30 (1995), pp. 199–210.

Multifidelity Surrogate Models: A New Data Fusion Perspective

Daniel N. Wilke[1]
nico.wilke@up.ac.za
[1] Emerging Engineering Technology (EET) Lab, Department of Mechanical Engineering, University of Pretoria, Pretoria, Hatfield, 0086

Abstract:

Multifidelity surrogate modelling combines data of varying accuracy and cost from different sources. It strategically uses low-fidelity models for rapid evaluations, saving computational resources, and high-fidelity models for detailed refinement. This method improves decision-making by addressing uncertainties and surpassing the limits of single-fidelity models, which either oversimplify or are computationally intensive. It blends high-fidelity data for detailed responses with frequent low-fidelity data for quick approximations, facilitating design optimisation in aerospace and automotive engineering while trading between precision and computational effort in materials science. In finance, it enhances risk management by integrating diverse data sources, and in environmental sciences, it aids in more accurate climate change and disaster simulations.

Despite progress in interpolation, regression, enhanced sampling, error estimation, variable fidelity, and data fusion techniques, challenges persist in selecting fidelity levels and developing efficient data fusion methods. Nonetheless, multifidelity surrogate modelling is a groundbreaking approach that increases accuracy, speeds up simulations, and enables complex analyses in various domains.

Keywords: Multifidelity, Surrogate, Modelling, Data Fusion

1. Introduction

Multifidelity surrogate modelling aims to combine data from various sources that may have varying accuracy and computational costs. The key idea is to use lower-fidelity models for many quick evaluations, conserving computational resources while incorporating high-fidelity models to refine predictions. This approach also helps to address uncertainties to enable more informed decision-making. It can overcome the limitations of single-fidelity models, which may oversimplify real-world physics or be too expensive to compute tractably, with Toal (2015) presenting some guidelines. Multifidelity surrogate modelling expedites design optimisation in engineering fields, including aerospace and automotive design, and trades precision for computational effort in materials science. Financial multifidelity models help manage risk more effectively by combining data from various sources of reliability. Moreover, it supports environmental modelling, enabling more accurate predictions for climate change and natural disaster simulations.

Zhang et al. (2018) and Fernández-Godino et al. (2019) employed linear regression to correct multifidelity surrogate models, underscoring the method's effectiveness in handling complex engineering problems. These studies, along with contributions from researchers like Choi et al. (2005), Robinson et al. (2008), and Yamazaki et al. (2010), showcase the application of these models in design optimisation and predictive analysis. The research demonstrates how multi-fidelity surrogate modelling can enhance the efficiency of reduced-order models, as shown by Lee et al. (2016). It provides strategies for mixed-variable optimisation using models, as Singh and Grandhi (2010) explored. Zimmermann et al. (2010) and Han et al. (2013) have explored gradient-enhanced surrogates within a multi-fidelity context, with Han et al. (2013) proposing an additional hybrid bridge function to enhance multi-fidelity surrogates.

MSM has seen notable advancements in aerodynamic modelling and optimisation strategies. For instance, Mackman et al. (2013) compare various adaptive sampling methods for generating surrogate aerodynamic models applied in aerospace engineering. Kuya et al. (2011) extend this application to the multi-fidelity surrogate modelling of experimental and computational aerodynamic datasets, illustrating the technique's versatility. In terms of optimisation, the works of Forrester et al. (2006, 2007) and Eldred et al. (2004, 2006) play a significant role in integrating surrogate models with partially converged computational fluid dynamics simulations and developing formulations for surrogate-based optimisation with data fit and multifidelity models. Choi et al. (2005) applied a multi-fidelity framework to supersonic applications. These advancements enhance the reliability and efficiency of surrogate models and broaden their applicability in various engineering domains.

In computational modelling and data analysis, multi-fidelity surrogate modelling (MFSM) and data fusion have emerged as valuable interdisciplinary areas, addressing the challenges of data's increasing complexity and scale. The fundamental concepts, methodologies, and applications provide a foundational understanding and insight into its growing significance, as Fernández-Godino's comprehensive review from 2023 outlines. The review provides detailed insight into the evolution and current state of multifidelity (MM) and multifidelity surrogate models (MSM), highlighting their significance in computational sciences.

This study proposes a new fusion approach to construct multi-fidelity surrogate models by merely constructing gradient-only surrogates that use only gradients and regression. Although omitted in this study, the proposed approach is easily extended to weighted regression to accommodate the heteroscedastic nature of spatial variance resulting from multi-fidelity data. Wilke (2016) proposed gradient-only surrogates to construct smooth surrogates for discontinuous functions with test problems presented in Snyman and Wilke (2018). Gradient-only surrogates for multi-fidelity data is limited to line search approximations investigated by Chae and Wilke (2019) and Chae et al. (2023).

2. Surrogate Strategies

Radial basis function surrogates with randomly sampled centres are considered in this study. Three surrogate types are considered, those using only function (f) values, functions, and gradients (f-g), and only gradients (g). Surrogates constructed using only gradients are not unique, as any constant offset of the surrogates fits equally well. Therefore, all gradient-only surrogates are translated to 0 as the lowest value. All studies utilise Gaussian kernels, as formulated by Snyman and Wilke (2018). The shape parameter is uniformly sampled on a log-scale between 10^{-4} and 10^{5} using 121 points. The shape parameter resulting in the lowest mean squared error on the training set is selected. Although prone to overfitting, we restrict the number of basis functions to be at least six times less than the number of observations, thereby mitigating the main risks of overfitting through regression, ensuring the ready reproduction of this study. For the interested reader, Snyman and Wilke (2018) and Correia and Wilke (2021) provide detailed discussions on improving the selection of the optimal shape parameter.

3. Experimental Setup

The foundational problem provides a simple yet effective illustration of the fundamental essence of the multi-fidelity problem. The study distils the core concepts and challenges associated with multi-fidelity modelling and optimisation by fitting a quadratic function. This example is a foundational reference point for understanding the broader implications of multi-fidelity approaches in more complex scenarios through the systematic variation of mini-batch sizes and their corresponding impact on function fitting and convergence. In addition to enhancing scientific reproducibility, it acts as a fundamental building block to help researchers and practitioners grasp the key principles and potential benefits of applying multi-fidelity techniques to various real-world problems.

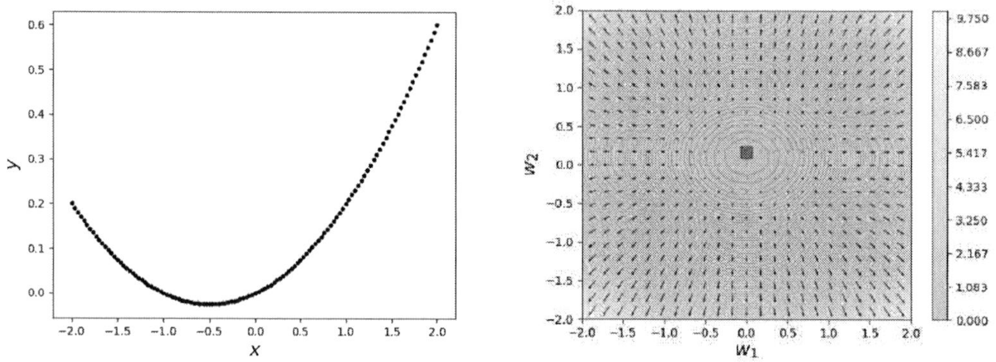

Figure 2 The full-batch training dataset (left), and the sampled Mean Squared Error loss surface for varying w_1 and w_2 between -2 and 2 (right). The full batch loss surface serves as reference for the mini-batch sampled loss surfaces. The red square indicates the lowest sampled function value.

The foundational problem addressed in this study centres around fitting a one-dimensional quadratic function of the form $w_1 x^2 + w_2 x$. The full-batch training data is generated by sampling a quadratic function of the form, $0.1 x^2 + 0.1 x$, uniformly across the interval [-2, 2] with 121 data points, as depicted in Figure 2.

The main aim is to construct a surrogate function of the Mean Squared Error (MSE) loss surface when sampling w_1 and w_2 over a 25x25 full-factorial grid. Each variable w_1 and w_2 varies between [-2, 2]. The surrogate functions are constructed by subsampling the full-batch training data using stochastic mini-batches to evaluate the MSE loss function and gradients. This resembles differences in model fidelity as biases in function values and sensitivity changes in gradients are captured in the sampled data as expected in any multi-fidelity system. More samples resemble higher fidelity data, while fewer samples represent lower fidelity data. The mini-batches considered have a maximum batch size of 3 and 30, where the number of samples per batch is randomly sampled from 1 to the maximum batch size. It is important to note that only three sampled function values are required to recover the quadratic function exactly. This is readily achieved using an optimality criterion solution for the least-squares problem in closed form (see Snyman and Wilke (2018) for a detailed derivation). However, a sampling approach introduces changes in biases and sensitivities between $[w_1, w_2]$ weights, introducing discontinuities (biases) and sensitivity changes (fidelity) between adjacent weights as expected from any multi-fidelity system. This renders the Mean Squared Error (MSE) loss surface discontinuous over w_1 and w_2, as shown in Figure 3.

Figure 3 depicts four figures as a 2x2 grid. Along the columns, this experimental computation uses mini-batches of varying maximum mini-batch sizes, specifically {3, 30} samples per batch. The maximum indicates the maximum sample per mini-batch uniformly sampled from 1 to the maximum mini-batch size. Each row represents a different set of samples, highlighting how the choice of mini-batch size impacts the behaviour of the MSE. In addition, a reference plot is provided at the bottom centre, representing the full batch loss surface, serving as a benchmark for comparison. This reference allows us to visually assess the performance of different mini-batch sizes relative to the full batch scenario. The red squares pinpoint the location with the lowest sampled function value.

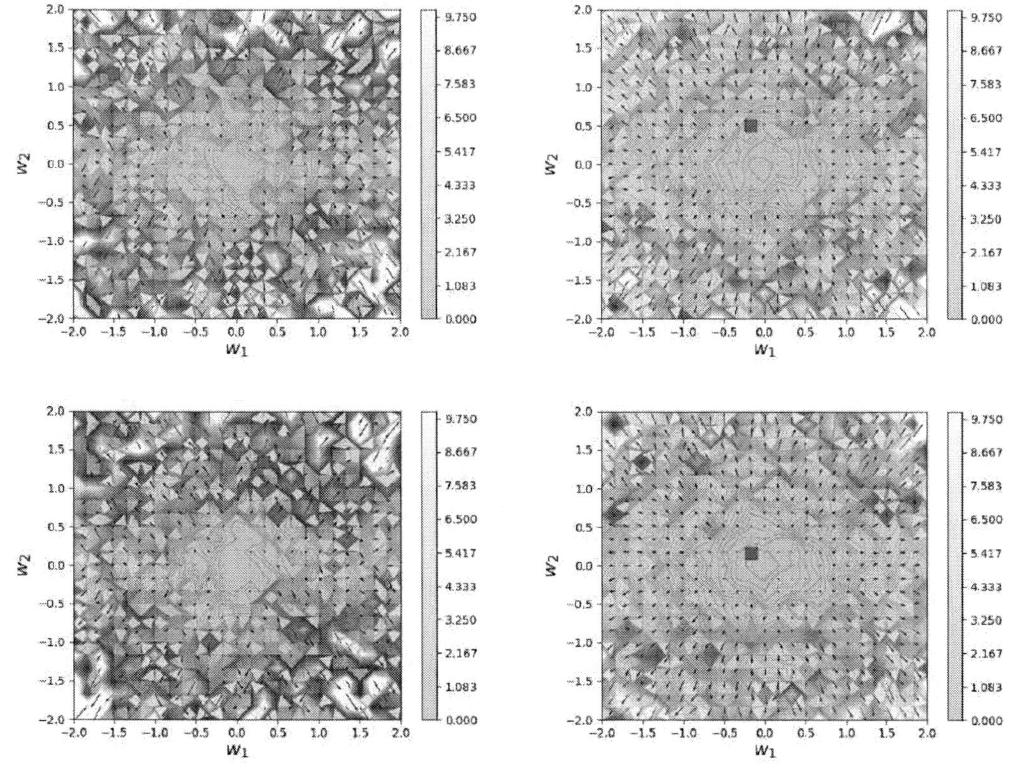

Figure 3 Mean Squared Error (MSE) training data sampled over a 25x25 full factorial grid for mini-batches of maximum sizes {3, 30} (over the columns). Two sampled surfaces (over the rows) indicate typical function and gradient value changes between samplings. The full batch loss surface is included as a reference (bottom centre). The red square in each subplot indicates the lowest sampled function value.

4. Results

The study investigates the Mean Squared Error (MSE) loss surfaces in Radial Basis Function (RBF) surrogate construction, examining mini-batch sizes and RBF centre counts for simulating multi-fidelity systems. The MSE loss surfaces are analysed for mini-batches up to size three and RBF centres numbering 1 and 100, exploring methods 'f', 'f-g', and 'g' in surrogate construction. Each method distinctly impacts the MSE loss surface, as shown in Figures 4 and 5. Surrogates built using gradients alone ('g') are positive definite in all cases. In contrast, using function values alone ('f') or combined with gradients ('f-g') may lead to negative definite functions for mini-batch sizes of 3 and 30. Highlighting how additional data, while relevant, can degrade fit quality, introducing local minima into the convex loss function. Thus, careful data curation is vital to maintain surrogate quality. The evidence demonstrates the benefits of using only gradients as a multi-fidelity fusion strategy, rendering gradients a simple and scalable data curation strategy when readily available and accurately computed.

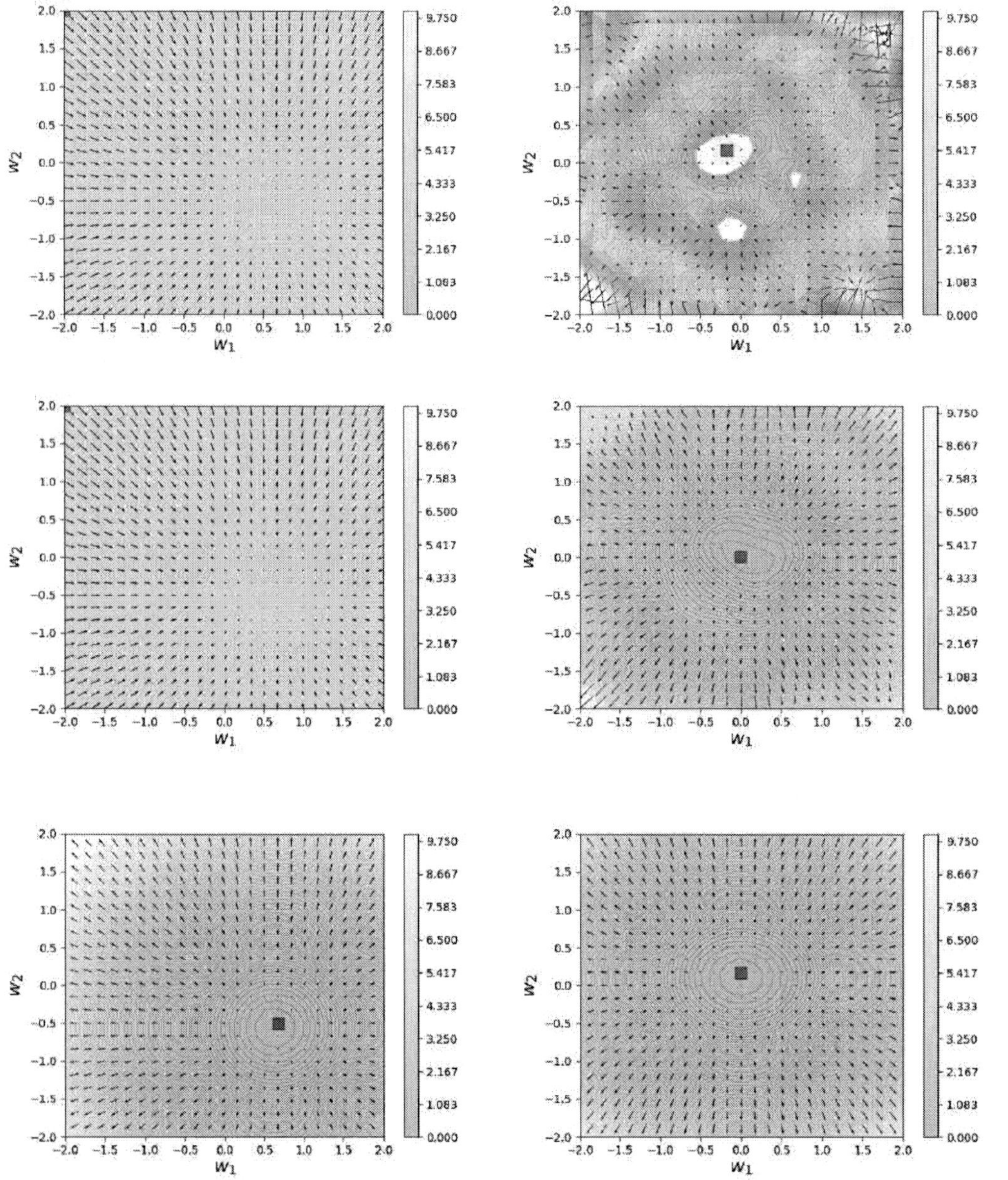

Figure 4 Mean Squared Error (MSE) loss surfaces for mini-batches of maximum size {3} (randomly sampled), and {1, 100} RBF centres (over the columns) when constructing RBF surrogates using {f, f-g, g} (over the rows).

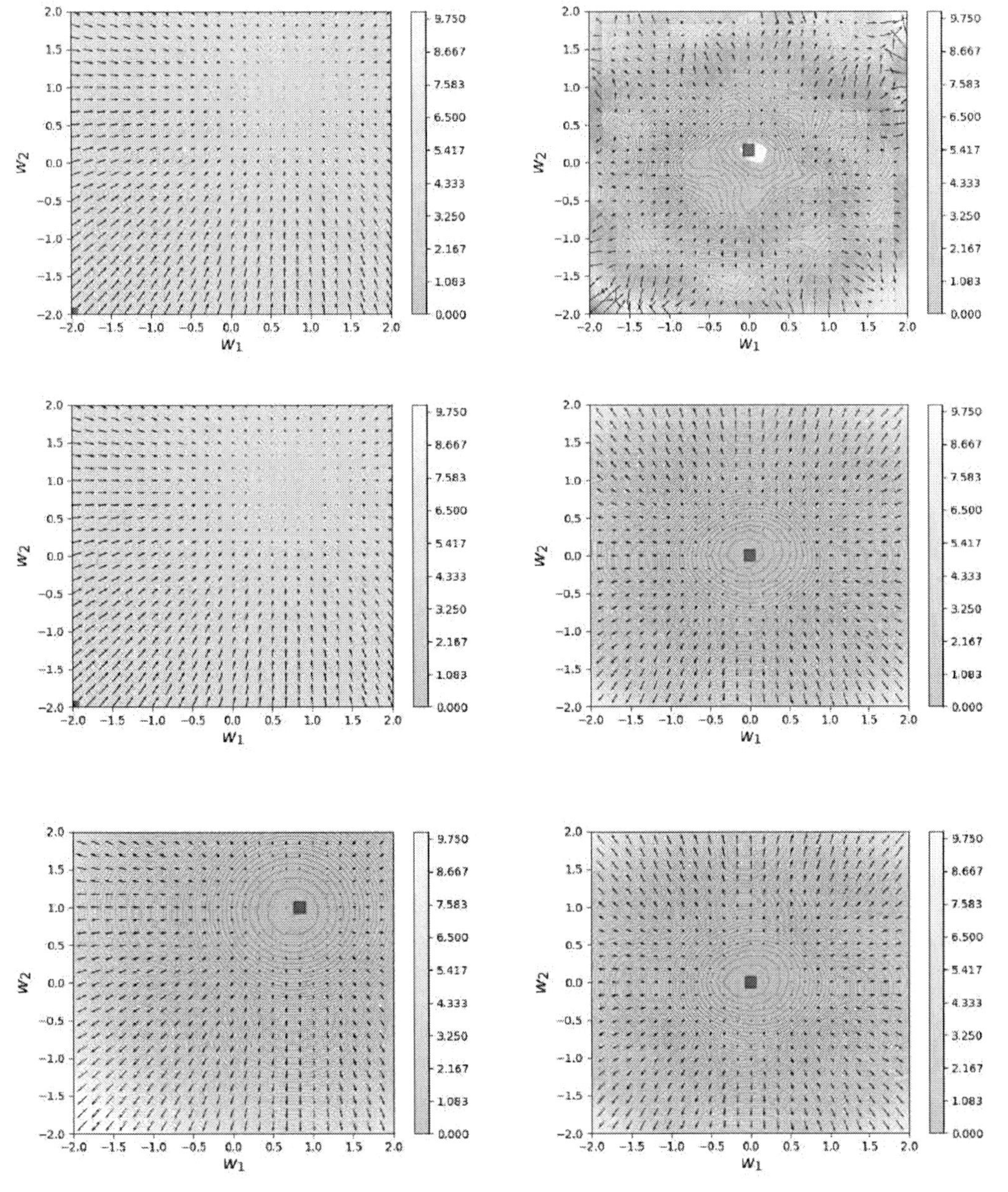

Figure 5 Mean Squared Error (MSE) loss surfaces for mini-batches of maximum size {30} (randomly sampled), and {1, 100} RBF centres (over the columns) when constructing RBF surrogates with {f, f-g, g} (over the rows).

5. Conclusion

In conclusion, this study represents a significant leap forward in understanding multi-fidelity surrogate models, offering a fresh and innovative perspective of prioritising gradients and omitting function values. The elegance of this approach is that multi-fidelity data fusion is achieved by constructing gradient-only surrogates. This key insight is demonstrated for a foundational problem involving the least squares fit of a basic quadratic function. The conventional notion that more data leads to better results in regression is challenged.

The generalisation of the perspective that gradient-only surrogates bring to the table is that this research highlights the importance of discerning seemingly applicable data between high-quality and low-quality data. By adopting this refined approach, this study has demonstrated that the fusion of carefully curated, high-quality data can often outperform the inclusion of vaster quantities of uncurated yet seemingly applicable data. This shift in paradigm challenges the often conventional notion and underscores the critical role of data quality assessment and curation in developing and optimising surrogate models.

The main takeaway from this study is that achieving better results is not merely a matter of accumulating more data but rather of making informed decisions about the quality and relevance of the data. This profound insight opens new avenues for research and practical applications, emphasising the need to prioritise data relevance and quality over quantity, ultimately leading to more efficient and effective multi-fidelity modelling and optimisation strategies.

References

Chae, Y., & Wilke, D. N. (2019). Empirical study towards understanding line search approximations for training neural networks. arXiv preprint arXiv:1909.06893.

Chae, Y., Wilke, D. N., & Kafka, D. (2023). Gradient-only surrogate to resolve learning rates for robust and consistent training of deep neural networks. Applied Intelligence, 53(11), 13741-13762.

Choi, S., Alonso, J., & Kroo, I. (2005). Multi-fidelity design optimization studies for supersonic jets using surrogate management frame method. In 23rd AIAA Applied Aerodynamics Conference, page 5077.

Correia, D., & Wilke, D. N. (2021). Purposeful cross-validation: A novel cross-validation strategy for improved surrogate optimizability. Engineering Optimization, 53(9), 1558-1573.

Eldred, M., Giunta, A., & Collis, S. (2004). Second-order corrections for surrogate-based optimization with model hierarchies. In 10th AIAA/ISSMO Multidisciplinary Analysis and Optimization Conference, page 4457.

Eldred, M., & Dunlavy, D. (2006). Formulations for surrogate-based optimization with data fit, multifidelity, and reduced-order models. In Proceedings of the 11th AIAA/ISSMO Multidisciplinary Analysis and Optimization Conference, number AIAA-2006-7117, Portsmouth, VA, volume 199.

Fernández-Godino, M. Giselle, Dubreuil, Sylvain, Bartoli, Nathalie, Gogu, Christian, Balachandar, Sivaramakrishnan, & Haftka, Raphael T (2019). Linear regression-based multifidelity surrogate for disturbance amplification in multiphase explosion. Structural and Multidisciplinary Optimization, 60, 2205–2220.

Fernández-Godino, M. G. (2023). Review of multi-fidelity models. Advances in Computational Science and Engineering, 1(4), 351-400.

Forrester, A. I., Bressloff, N. W., & Keane, A. J. (2006). Optimization using surrogate models and partially converged computational fluid dynamics simulations. Proceedings of the Royal Society of London A: Mathematical, Physical and Engineering Sciences, 462(2071), 2177–2204.

Forrester, A. I., Sóbester, A., & Keane, A. J. (2007). Multi-fidelity optimization via surrogate modelling. Proceedings of the Royal Society of London A: Mathematical, Physical and Engineering Sciences, 463(2088), 3251–3269.

Han, Z.-H., Görtz, S., & Zimmermann, R. (2013). Improving variable-fidelity surrogate modeling via gradient-enhanced kriging and a generalized hybrid bridge function. Aerospace Science and Technology, 25(1), 177–189.

Kuya, Y., Takeda, K., Zhang, X., & J. Forrester, A. I. (2011). Multifidelity surrogate modeling of experimental and computational aerodynamic data sets. AIAA journal, 49(2), 289–298.

Lee, S., Kim, T., Jun, S. O., & Yee, K. (2016). Efficiency enhancement of reduced order model using variable fidelity modeling. In 57th AIAA/ASCE/AHS/ASC Structures, Structural Dynamics, and Materials Conference, page 0465.

Mackman, T., Allen, C., Ghoreyshi, M., & Badcock, K. (2013). Comparison of adaptive sampling methods for generation of surrogate aerodynamic models. AIAA journal, 51(4), 797–808.

Robinson, T., Eldred, M., Willcox, K., & Haimes, R. (2008). Surrogate-based optimization using multifidelity models with variable parameterization and corrected space mapping. AIAA Journal, 46(11), 2814–2822.

Snyman, J. A., & Wilke, D. N. (2018). Practical Mathematical Optimization: Basic Optimization Theory and Gradient-Based Algorithms (Springer Optimization and Its Applications Series). Springer Cham.

Toal, D. J. (2015). Some considerations regarding the use of multi-fidelity Kriging in the construction of surrogate models. Structural and Multidisciplinary Optimization, 51(6), 1223–1245.

Wilke, D. N. How to get rid of discontinuities when constructing surrogates from piece-wise discontinuous functions. In J. Herskovits, editor, 5th International Conference on Engineering Optimization, Iguassu Falls, Brazil, 2016.

Yamazaki, W., Rumpfkeil, M., & Mavriplis, D. (2010). Design optimization utilizing gradient/Hessian enhanced surrogate model. In 28th AIAA Applied Aerodynamics Conference, page 4363.

Zimmermann, R., & Han, Z. (2010). Simplified cross-correlation estimation for multi-fidelity surrogate Cokriging models. Advances and Applications in Mathematical Sciences, 7(2), 181–202.

Zhang, Y., Kim, N.H., Park, C., & Haftka, R. T. (2018). Multifidelity surrogate based on single linear regression. AIAA Journal, 56(12), 4944–4952.

ABSTRACTS

Book of abstracts

Computational analysis of heat transfer augmentation through a rectangular dimple endwalls geometry

SA Aasa[1] and SA Shote[2]
[1]Department of Mechanical and Aeronautical Engineering, University of Pretoria.
[2]Department of Mechanical Engineering, Olabisi Onabanjo University Agoiwoye, Nigeria

Using a suitable turbulence generator to improve convective heat transfer is a simple and effective method. In this study, a baseline smooth surface is simulated in comparison to a geometrical protrusion surface. The protrusion is a pair of oval shapes inclined at a certain angle to each other. The study also referred to the performance and heat enhancement capabilities of the baseline smooth and protrusion surfaces at different channel heights of 5mm, 7mm and 10mm. The baseline and protrusion geometry surfaces are simulated at different velocities and Reynolds number (Re) of 600, 1000, 5000, 7000 and 15000. The results show that the Nusselt number of the protruded surface gives the corresponding higher value at different Reynolds number as compared to them baseline smooth surface. Also, the presence of protruded configurations on the plate surface can be seen to significantly improve heat transfer performance when compared to the smooth baseline surface. When the pressure drop and frictional drag force between the smooth baseline surface and the protruded surface are compared between the smooth baseline surface and the protruded surface, the protruded surface performs significantly better than the smooth baseline surface.

Temperature dependant mechanical property characterisation using digital image correlation and infrared thermography

M Blackwell[1], M Vos[2], S George[2], M Neaves[1], TH Becker[2]

Department of Mechanical and Mechatronic Engineering, Stellenbosch University.

Centre for Materials Engineering, University of Cape Town.

As applications for materials become increasingly demanding, high-performance materials and novel manufacturing techniques are being developed to meet those needs. Often, these materials and manufacturing methods are expensive, which makes conducting extensive material testing difficult. Therefore, it is necessary to develop specimen-efficient testing techniques allowing material properties to be extracted simultaneously in a range of testing conditions.

One such scenario is the high-temperature testing of additively manufactured (AM) Inconel 718. Inconel 718, a nickel-based superalloy, finds extensive applications in the aerospace and energy industries due to its exceptional high-temperature mechanical properties.

Accurate characterisation of the mechanical behaviour of AM Inconel 718, especially at extreme temperatures, is crucial to ensure the reliability and performance of components. A sudden drop-off in strength is known to occur in wrought Inconel 718 near 700° C. This is accompanied by embrittlement of the material between 650° C and 800° C. It is unclear if AM Inconel 718 will exhibit this same strength drop-off and embrittlement. This paper presents a test setup, which is used to determine the high-temperature mechanical properties over a range of operating temperatures by integrating full-field digital image correlation (DIC) and infrared thermography.

This is done by resistively heating a purpose-designed specimen using a Gleeble 3800 thermomechanical simulator to create a thermal gradient across the length of the specimen gauge region. DIC, which is a non-contact optical measurement method that enables accurate full-field displacement measurements of an object surface coated with a randomised speckle pattern, is used to capture displacement data of the specimen surface. The randomised speckle pattern is generated using heat-resistant paint. A high-powered low wavelength blue light source and narrow bandpass filters are used to eliminate radiation from the specimen, thereby improving the DIC accuracy. Simultaneously, a thermal camera is used to measure full-field thermal data. These two datasets are temporally synchronised using a common triggering system.

The acquired full-field displacement data from the DIC system is processed using the virtual fields method (VFM) to generate strain maps. This strain data is then spatially correlated with the thermal data which allows for the extraction of temperature-dependent properties such as Young's modulus, yield strength and ultimate tensile strength for a temperature range of more than 50 degrees and up to 100 degrees for certain parameters, all from a single tensile test and specimen.

This testing procedure has proven to be beneficial in testing materials for which a large range of temperature-dependent properties are required and where the high specimen cost limits the number of tests that can be feasibly conducted. Additionally, this test setup could potentially be used in its current state to measure other properties such as thermal expansion coefficients and observe material behaviour such as phase transformation which often result in thermal absorption/release and material expansion/contraction.

Including thermochemical equilibrium calculations into multiphysics models – a finite volume-based phase change model

AEJ Bogaers[1], JJ Swanepoel[1], WA Roos[1], JH Zietsman[1,2]
1 Ex Mente Technologies, Pretoria, South Africa
2 University of Pretoria, South Africa

Phase change is the process associated with matter changing from one state to another, such as a solid changing to liquid or gas, and forms the governing process of most high-temperature pyrometallurgical processes. Process material is introduced into a furnace as solids, allowed to melt, advect, diffuse and chemically react with one another to form the primary products of interest. Modelling phase change in these systems requires capturing the potentially large non-linearities in both the material properties and enthalpy changes, as a function of temperature.
The standard approach of solving the energy equation, when considering phase change, is to linearise the energy equation by splitting the system's enthalpy as the sum of its sensible enthalpy and latent energy. The latent energy associated with phase change can then be added as a source term to the energy equation. This approach has however been demonstrated to be inefficient when dealing with the severe non-linearities present in real pyrometallurgical systems.
In this paper, we present an alternative linearization of the energy equation. The temperature-enthalpy, and temperature-density relationships are locally linearized via Taylor series expansions, which are then introduced into a semi-discretised form of the energy equation. This new linearisation leads to approximately an order of magnitude improvement in convergence behaviour, when compared to the standard source-based approach. It further leads to a generalised form of the energy equation, which can be applied to any arbitrary phase change system, as long as gradients with respect to temperature can be computed.
The set of partial differential equations describing mass, momentum, and energy transport are solved using a fixed-grid, finite-volume solution approach, where each of the phases, or chemical component concentrations, are treated using a phase-fraction approach. The enthalpy and relevant material properties, as a function of temperature and chemical component concentrations, are obtained by directly coupling ChemApp into the finite-volume solver. ChemApp is a numerical thermochemical equilibrium calculator, based on the Gibbs free energy minimisation algorithm. Directly including a thermochemical equilibrium calculator does add a significant computational overhead to the overall solution process. The Gibbs free energy minimisation currently has to be performed several times per time step, for each of the computational cells. In its current form, depending on the complexity of the system, the cost of performing the equilibrium calculations can outweigh the cost of solving the set of partial differential equations. Despite this inefficiency, the solver was developed to serve as a testing platform for a newly developed equilibrium calculation accelerator, which will be implemented in the near future.

Calibration of Laser Absorptivity in CFD Modelling of LPBF In-Situ Alloying with Ray Tracing

MR Botete, DC Blaine, G Venter

Department of Mechanical and Mechatronic Engineering
Stellenbosch University, South Africa

Laser Powder Bed Fusion (LPBF) is a powerful additive manufacturing technique that allows the creation of complex geometries with superior mechanical properties. In-situ alloying during LPBF enables the fabrication of customised metal alloys with enhanced performance, by depositing different powders into the same build chamber [1]. However, it is challenging to produce high-quality, defect-free microstructures using LPBF in-situ alloying technique [2]. The most common defects, in addition to defects inherent to the LPBF process [3], are micro-segregation, unmolten powders, and high levels of porosity. These defects originate primarily from inhomogeneities within the feedstock mixture relative to the beam spot size. Therefore, it is essential to predict and control these defects to increase the stability and adaptability of the in-situ alloying and the broader LPBF process. This requires a deep understanding of the physical processes involved and computational simulations offer great assistance in this regard.

Review studies have outlined the methodologies and applications of LPBF process simulations [4–6]. Most work has focused on three areas: the behaviour of the melt pool which forms during the laser-powder interactions, the evolution of microstructures during solidification and cooling, and the overall part distortion and stress distribution. Currently, particular emphasis has been placed on the simulation of the LPBF in-situ alloying process, particularly in the first focus area concerning melt pool behaviour occurring at the mesoscale [7,8]. Understanding the melt pool behaviour in the context of LPBF in-situ alloying is key for defect prediction and process optimization.

In recent years, significant progress has been achieved regarding the mesoscale simulation of melt pool behaviour, effectively capturing the authentic physics associated with the LPBF process. This encompasses elements such as stochastic powder bed characteristics, melt pool dynamics, variations in surface tension, and vapour recoil pressure [9]. Nevertheless, achieving accurate simulation results remains a challenge due to the dozen parameters required as input. The most critical parameter is laser absorptivity which defines the fraction of incident laser energy absorbed by the material [10]. The laser absorptivity of metal powders can be measured in the laboratory using a variety of techniques. However, these laboratory measurements may not be accurate because they do not account for the dynamic conditions of the LPBF process [11]. A more accurate way to determine the laser absorptivity of metal powders for LPBF is to calibrate simulations against experimental data. This can be done by simulating a single-track LPBF in-situ alloying process and comparing the simulation results to the experimental results.

In this study, the laser absorptivity of the elemental mixture of Ti, Al, and V powders, prior to melting, is determined by simulating the energy input from the laser beam, using ray tracing modelling, within a CFD simulation that predicts the rise in powder particle temperature due to the energy input. The results are compared with absorptivity obtained by measuring the thermal response with pyrometry, within the LPBF environment. This will provide insight on the factors to consider when calibrating absorptivity for LPBF in-situ alloying simulations.

In-vitro and in-silico modelling of arteriovenous malformations for surgical planning

J Bougardt, W Ho, RT Paton and J Pillai
School of Mechanical, Industrial & Aeronautical Engineering, University of the Witwatersrand

Arteriovenous malformations (AVMs) are vascular abnormalities, which present as direct connections between high-pressure arterial and low-pressure venous flow regimes in the body, that is to say that there is no interconnecting capillary bed between the flows through which perfusion may take place. AVMs have a number of associated symptoms including various forms of ischaemia (strokes and seizures) and even high-output heart failure. The principal method of treating AVMs is known as transcatheter coil embolisation, which takes the form of the delivery of an occluding (flow blocking) coil to an affected area via a catheter. The evaluation of these coil embolisations is, at present, postoperative and highly subjective.

This project, which falls under the field of biomedical Engineering or more specifically biomechanics, aims to generate an objective metric surgeons may utilise to improve the outcome of patients undergoing coil embolisation procedures by computationally and experimentally modelling dominant vascular system characteristics such as patient-specific vasculature, non-Newtonian blood rheology and pulsatile blood flow while taking into consideration clot growth and applying these parameters to the study of the implantation of the embolising coil and the subsequent occlusion of the AVM vasculature pre-surgically. This can better inform clinicians of the methods which should be applied during the surgery such as what length of coil needs to be inserted and where the best point of occlusion is.

Some of the features developed include the synthesis of a fluid which mimics the non-Newtonian, specifically shear-thinning, behaviour of true blood, also known as a blood-mimicking fluid (BMF) as well as pulsatile UDF's which replicate the cyclical nature of heartbeat related pressure waveforms. The AVM geometry has been derived from affected patient MRI scans through the use of segmentation software and these have been 3D printed using stereolithographic resin (SLA) for the experimental model.

The effect of the embolisation coil has been modelled through the use of a porous zone. The coils have been subjected to rigorous experimental testing through the use of a constant pressure head system in order to derive the pressure drop associated with these coils as a function of velocity. This relationship was subsequently passed into the solver by deriving viscous and inertial resistance coefficients which have been used to tune the momentum sink term fed into the governing equations. Other developments include the observation of clotting within the experimental and computational models through the use of empirically derived functions for clot growth as a function of shear rate as well as the measurement of flow vectors within the experimental (3D SLA printed) AVM structures through the use of 4D MRI technology.

Multilayered floating absorber for interfacial evaporation with excellent salt rejection for solar thermal applications

AK Chandran and SCS Kumar
Department of Mechanical Engineering, National Institute of Technology Calicut, India

The traditional evaporation process, aided by bulk heating, hinders the solar-to-vapor conversion efficiency, leading to the advent of interfacial evaporation. Confining the heat into the evaporating surface, solar-driven interfacial evaporation, gained tremendous research interest as an efficient strategy to alleviate the freshwater scarcity by deployment in the solar thermal desalination units. The present work experimentally studies the effectiveness of floating absorbers with polyvinyl alcohol (PVA)-based wick material and the development of a triple-layered absorber capable of excellent salt rejection. The single-layer absorber cum wick material yields an evaporation rate of 1.7 $kg\, m^2 h^{-1}$ and 1.42 $kg\, m^2 h^{-1}$, respectively under 1 sun illumination in normal water and 3.5 % salt solution. While the double and triple layer absorbers in salt solution enhance the evaporation rate by 10% and 20%, respectively with a yield of 1.55 $kg\, m^2 h^{-1}$ and 1.7 $kg\, m^2 h^{-1}$ under similar illumination conditions. The effective diffusion of salt back to the water source through the wick material reveals the capability of the system to be a potential candidate for solar desalination applications.

Deformation Behaviour, Limitations in Design, Applications, and Additive Manufacturing of Hierarchical Honeycombs

MI Chibinyani[1], TC Dzogbewu[1], M Maringa[1] and AM Muiruri[2]
[1]Department of Mechanical and Mechatronics Engineering, Central University of Technology.
[2]Department of Mechanical and Manufacturing Engineering, Southern Eastern University of Kenya.

Hierarchical honeycomb structures are particularly advantageous in aerospace industries as a result of their ability to overcome challenges related to bend-dominated behaviour. These structures are ideal for aeronautical applications because of their capacity to integrate lightweight design, good mechanical properties, and efficient load distribution. However, the efficacy under applied loads is influenced by the type of polygon cell introduced at the vertices. The current literature has gaps in knowledge with reference to the highest order of hierarchy that can practically be achieved for hierarchical honeycomb structures with different cell shapes at the vertices. This paper presents an overview of hierarchical honeycomb structures with triangular, hexagonal, and circular cells substituted at the vertices, with the aim of examining their mechanical properties and how efficiently they can be designed. In the first section, existing, and relevant analytical models in literature, on the mechanical properties of hierarchical honeycomb structures are identified and discussed. These include Vogel's model for hierarchical honeycombs, the Gibson and Ashby models, and the Castigliano methods, which are typically employed for analytical modelling of the behaviour of hierarchical honeycomb structures. Though these analytical models are built using the relationships that exist between connecting vertices, they fail to account for the deformation behaviour at these points. This is followed by the second section, focusing on determination of the constraints in the order of hierarchy for these structures, particularly the geometric parameters. The highest order of hierarchy, presently feasible, is observed for a second-order hierarchical honeycomb structure with hexagonal cells at the vertices. Furthermore, the hierarchical structuring of honeycomb structures is limited by geometrical parameters, including the edge length ratio, cell angles, and the ratio of thickness to edge length. In the third section, a review of relevant applications for hierarchically built honeycomb parts in the aerospace industries is presented. Aerospace industries typically build lightweight structural parts, capable of absorbing high strain energies through the use of hierarchical honeycomb designs. It is observed in the paper that the hierarchical honeycomb structure comprised of triangular cell substructures at the vertices has the greatest capacity for absorption of energy of the three first-order hierarchical structures studied here. Finally, the efficacy, as well as challenges of using additive manufacturing for building hierarchical honeycomb structures is discussed, noting that additive manufacturing is challenged by the geometric intricacy and complexity of hierarchical honeycomb designs, as well as the minimum resolution that can be printed, a constraint that increases with increasing order of hierarchy.

The role of simulation in the development of a cost-effective biomimetic prosthetic hand

M de Wet, JH Kruger

School of Mechanical Engineering, North-West University,
Potchefstroom.

In developing countries where healthcare services are limited or prohibitively expensive, there is a grow- ing need for affordable, accessible and maintainable prosthetic limbs. When it comes to daily interaction with our surroundings, the arm and the hand is one of the most important limbs we have, with the loss of an arm or hand resulting in a large loss of functionality in daily life. The functionality of the human hand is especially difficult to replicate, consisting of 27 degrees of freedom and each hand representing 13% of the bones in the human body. Thus, this results in prosthetics that are usually complex and very ex- pensive, or prosthetics that are very simple approximations of the human hand and thus can't adequately provide the functionality that was initially lost.

This project aimed at creating a cost-effective upper limb prosthesis that restores major aspects of func- tionality. The final design followed principles of biomimetics, featuring a total of 14 degrees of freedom, controlled by a servo under-actuated tendon control system utilising high strength braided nylon cords. The prosthetic features a fully articulated thumb and wrist alongside finger abduction and adduction, fea- tures seldom found in commercial prosthetics. Controlled by an Arduino Mega Pro mini microcontroller, the prosthetic features 14 different servos with inbuilt clutches to eliminate the chance of servo burnout when gripping objects. Gripping is initiated when an ultrasonic distance sensor detects objects within the palm of the hand, with the grip being released by pressing a touch sensor on the arm. The hand features a prosthetics silicone glove to aid with the gripping of objects.

Pre-manufacture motion simulations using Siemens NX were conducted on the phalangeal digits of the prosthetic to ensure no collisions are present between the fingers of the prosthetic and any other part of the device. These mechanism simulations were also utilised to ensure the hand and fingers are capable of the predetermined ranges of motion. FEM simulations were performed on the main structural arm piece to ensure that the part is capable of enduring force loading that is necessary for performing daily tasks. During manufacturing, the GCODE produced by the additive slicing software was simulated before being run on the 3D printer to ensure there were no errors present or incorrect slicer settings. Functional testing was performed on the final prototype to validate parameters such as range-of-motion and grip strength with regards to the design specifications.

Thermal-hydraulic behaviour of a micro nuclear reactor during a DLOFC transient

CG du Toit[1], W Boyes[2] and JF Slabber[2]
[1] School of Mechanical Engineering, North-West University, Potchefstroom
[2] Department of Mechanical & Aeronautical Engineering, University of Pretoria, Pretoria

By using nuclear power integrated with renewable energy sources on a decentralized basis South Africa can achieve safe affordable base load energy. This can be achieved through the development of micro modular nuclear reactors, to supply energy to towns, small communities, mines, and processing plants. A High Temperature Advanced Micro Reactor (AMR) to address this need is under investigation and the philosophy is to design for inherent safety, using technology that has been developed and validated in previous HTR programs. The existing experience and expertise gained during the Pebble Bed Modular reactor (PBMR) project forms the basis of the concept. The AMR reactors are intended to be factory built to obtain good quality control and rolled out to various sites. Once the reactor has reached its end of life, it would be returned to a licensed organisation for refurbishment and refuelling. The AMR will be designed to produce 10MW of thermal power. The reactor configuration will employ prismatic graphite blocks for structural and moderator material, which are to be arranged to form a cylindrical core layout. The fuel assemblies will be silicon carbide tubes that house coated particle fuel, immersed in a lead-bismuth eutectic alloy (LBE). Each fuel assembly is to be contained in a boring within the graphite moderator that allows an annulus for cooling. There will be 420 fuel assemblies in the core using low enriched fuel. The helium coolant will enter the core at 320 °C and exit at 750 °C. Flownex SE has been used to set-up a systems network model of the reactor and study the thermal- hydraulic behaviour of the reactor under steady-state conditions and during a Depressurized Loss of Forced Cooling (DLOFC) event. This paper focuses primarily on the initial stages of the DLOFC transient. Specific attention is paid to the heat balance of the reactor, the mass flow rate through the reactor, the friction factors of the coolant channels, the pressure drop over the core and the fuel temperatures as functions of time. It is shown how the solids initially release accumulated heat, only to start absorbing heat again. The effect of the transitioning from turbulent to laminar of the flow through the bottom reflector on the pressure drop over the core, as well as the relationship between the pressure drop over the core and the mass flow rate through the core, are also discussed. It is also observed how the fuel temperatures after an initial sharp decrease are controlled by the graphite moderator temperatures.

Improved accelerometer placement for railway wheel flat detection using a single sensor

D Ellis and A Bekker
Department of Mechanical and Mechatronic Engineering, Stellenbosch University

Wheel flats are circumferential defects that form on railway wheels at discrete locations due to sliding incidents, brake malfunctions, or improper use. These flats can lead to increased noise pollution, rail infrastructure damage, and rolling stock maintenance. Often rolling stock operators rely on acoustic observations from their employees to raise these issues. This research focuses on locating the best wayside position to place a single accelerometer to measure the impulse caused by a wheelset with a flat. Digital signal processing techniques are used to analyse the vibration signal to identify the faulty wheelset on the train set. Placement on the sleepers and rail at different locations were considered. The location yielding the ideal position for a single sensor is found using beam and rail theory and dynamic simulations. These results are compared to field tests using a train with multiple wheel flats and the results shown.

Defining citrus pallet stability and failure modes: Developing tools for carton improvement

NJ Erasmus[1], CJ Coetzee1, TM Berry[2]
[1] Department of Mechanical and Mechatronic Engineering, Stellenbosch University.
[2] Citrus Research International, Department of Horticultural Sciences, Stellenbosch University.

The export of horticultural produce, especially citrus, is an ever-evolving and dynamic industry with constant regulation changes in terms of carton design and post-harvest management. This dynamic, however, offers opportunities for carton design innovations.

Extensive research has been conducted on the structural integrity of cartons using the finite element method. However, the true loads acting on a carton has not yet been measured and is largely unknown. Furthermore, the FEM models assume that the loads acting on the carton are evenly distributed and that the cartons are perfectly aligned. On the other hand, the stability of a stacked pallet under dynamic loading has been investigated for general goods, but not for horticultural produce.

This work aimed at addressing the shortcomings of existing studies and developed experimental tools to measure the load distribution acting on individual cartons inside a stacked pallet, and the whole pallet stability under dynamic load conditions. The tools include the A15C Load Measurement Device (A15C-LMD) and the Pallet Lateral Acceleration Tester (PLAT). The A15C-LMD was manufactured to have the same geometric dimensions as a standard A15C citrus carton. It was also equipped with four load cells to measure the total vertical force distribution on the carton. Replacing a carton in the pallet stack with this device, allowed for measuring the static and dynamic force acting on cartons at strategic positions in the stack. The PLAT was manufactured to enable repeatable dynamic testing, as described by the EUOMOS 40509 which is a standard for EU pallet stability and safe goods transportation. The dynamic tests were similar to the acceleration and deceleration of a truck.

These tools were used in unison to quantify the dynamic loading on an A15C carton within a stacked citrus pallet as well as the whole pallet stability. The distribution of the load on the four quadrants of the carton was measured for a number of positions of the carton in the stack. The stability of the stack was quantified in terms of relative stack displacement under dynamic loads. These measurements were repeated for different stacking configurations used in industry (column stack versus brick stack). Structural carton damage was observed and recorded for the different stack configurations and compared to standard box compression tests (BCT).

In summary, the relation between carton load, carton damage and pallet stability was quantified. These results and observation will in future be used to improve the FEM modelling of cartons to be a more accurate representation of real-world industry conditions.

Development of a spillway and sluice gate demonstration system

N Gertenbach and JH Kruger
School of Mechanical Engineering, North-West University, Potchefstroom.

Hydraulic control structures form vital components in modern-day infrastructure and engineering, from hydroelectric power generation to water storage and canal locks. With numerous applications, these hydraulic structures need to be designed with proper knowledge. Mechanical engineering students study fluid mechanics and will benefit from a practical experiment on hydraulic control structures. This paper covers the design, simulation, construction and testing of a test bench to conduct open-channel flow experiments for engineering education.
The flow phenomena incorporated in the designed apparatus were focused on the development of hydraulic jumps in the open channel and the transition from free to submerged flow at a sluice gate opening. CFD simulations using the Volume-Of-Fluids methodology in Ansys Fluent were validated with an analytical model as both indicated the formation of a hydraulic jump at specific gate openings and flow rates. The effects of three variables were then further characterised during subsequent simulations and experiments, namely the height of the gate opening, the flow rate and the stop log depth.
The testing and construction of the test bench was completed at North-West University and proved that the test bench was successful and could be used to investigate other hydraulic structures such as culverts, radial gates, weirs and stilling basins.

Numerical Modelling of Blood Flow During Syringing

J Goedhals[1,2], A de Villiers[1] and GJF Smit[1]
[1] Division of Applied Mathematics, Department of Mathematics, Stellenbosch University
[2] DSI-NRF Centre of Excellence in Mathematical and Statistical Sciences (CoE-MaSS), South Africa

Hemolysis, defined as the breaking open of red blood cells, can occur due to a range of factors which can be physical, chemical or biological in nature. Physical damage can manifest in scenarios such as emergency rapid blood transfusions conducted through syringing, a practice which may be employed in resuscitation procedures, particularly in rural hospitals. It has been demonstrated that such syringe-based transfusions result in noteworthy hemolysis, significantly surpassing the impact of pressure bag usage. This heightened hemolysis carries potential negative implications for the recipient of the transfusion.

The core objective of this project revolves around emulating the blood flow through a sudden contraction, simulating a syringe, to emulate and quantify the underlying mechanisms driving hemolysis. Employing numerical simulations alongside a variety of hemolysis models, the goal is to identify the specific flow conditions that are most conducive to hemolysis occurrence.

The modelling process commences with the depiction of blood flow as a Newtonian fluid traversing a sudden contraction, representing the syringe's action. This initial setup is subsequently expanded to account for non-Newtonian flow behaviours. To validate the modelling approach, a two-dimensional scenario is first simulated using the finite element software deal.II. This is then repeated and compared to results from numerical simulations using ANSYS Fluent. The scope extends further to encompass a three-dimensional setup, incorporating a representative syringe and hypodermic needle (illustrated in the model example below). The time history of shear stress along streamlines is then extracted to facilitate hemolysis analysis.

Cost Effective Ballistic Protection for Vehicles

JD Jones[1], JD Reinecke[2] and L Pambuka[1]
1 School of Mechanical, Industrial and Aeronautical Engineering, University of the Witwatersrand.
2 Landward Sciences Impact Area, Defence and Security Cluster, CSIR.

In the design of protected vehicles there is a constant trade-off between mobility, protection and cost. In order to protect against increasing threat levels, designers are usually required to use armour materials with increased mass and thickness. However, this has a negative effect on the vehicles mobility. Similarly, reducing the mass of armour materials for the same level of protection usually requires the use of more expensive materials, thus increasing the cost. The aim of this project was to investigate how the areal density of armour materials, used for vehicle protection against a NATO Level 3 ballistic threat, can be reduced whilst still maintaining the required level of threat protection, and optimising cost by exploring varying material layers in a composite armour laminate assembly.

This work used computational modelling to evaluate protection capabilities of various combinations of lower cost materials that were then manufactured and tested. The first test laminae panels were initially based on evaluating the published computational work of Rahman *et al.* This work proposed a low cost combination of Al-7075-T6 aluminium and Strenx 700 steel laminates that would provide ballistic protection against NATO Level 3 threat [1]. These proposed laminates computationally provided a reduction in aerial density of 12% compared to equivalent homogeneous amour steel plate. Additional composite laminates using other combinations of Strenx 700E Al-7075-T6 with Aramid and Dyneema panels were developed and computationally evaluated and assessed using the Richt-Ipson model. The laminated panels proposed by, and the additional laminate panels that were numerically evaluated, were then manufactured and subjected to ballistic tested against NATO level 3 (7.62x51 WC) armour piecing rounds.

None of the proposed and computationally verified laminate combinations provided the required ballistic protection. All test panels were easily perforated. Additional computational models were completed post test and provided limited correlation to the physical tests. The main reason for this is ascribed to the application and use of only published material parameters and the implementation of the failure model.

Influence of the length of a photovoltaic single-axis tracker mounting rail

JH Koekemoer[1], JR Bredell[1], G Venter[1] and J Reinecke[2]
1 Department of Mechanical & Mechatronic Engineering, Stellenbosch University.
2 Scatec Solar Africa Pty(Ltd), Cape Town.

Single-axis trackers are actuated structures often used in utility-scale photovoltaic (PV) electric power plants. These systems increase yield by reducing the angle of incidence relative to the incoming direct solar radiation. Wind loads on trackers can be significant and often govern their structural design. A critical component in the tracker is the mounting rail, which forms the structural support between the rotating horizontal torque tube and the PV modules. The rails are thin gauge cold-formed steel members with bolted connections. Commercial PV modules can accommodate a range of mounting rail lengths for one-in-portrait (1P) configurations. As the industry grows and develops tracker designs become more optimised to reduce cost. Recent design trends show a decrease in the length of mounting rails in trackers. The current research will investigate the influence of mounting rail length in a 1P tracker using linear static finite element (FE) analysis. The FE models include the rails as well as the PV modules. PV module material properties are calibrated using experimental force-displacement data and equivalent material properties are proposed. Both uniform and non-uniform distributed loads are considered to represent wind loads. The centre of pressure for the non-uniform load case is eccentric to the torque tube and is consistent with equivalent static code-based design loads. Stresses in the modelled components are reported and compared for various rail lengths.

Aerodynamic stability of the tilting PV wing of the NWU solar car

E Kruger, JH Kruger and J Bosman
School of Mechanical Engineering, North-West University, Potchefstroom.

The efficiency of a solar vehicle depends on the balance between power generation via its photovoltaic (PV) wing and the power usage by its electric motor for propulsion. Therefore, optimising the solar energy harnessing efficiency of the PV wing significantly improves the efficiency, and therefore competitive capability of said solar car.

A manner of optimising the solar radiation collection efficiency is having a PV wing that can tilt to dynamically align itself with the position of the sun. This was implemented on the Naledi2.0 solar car during the 2021 upgrade cycle performed on the car. However, this design modification comes with a subsequent uncertainty with regard to the lateral stability of the car, specifically during cross wind conditions. The inclination of the wing increases the projected vertical area facing into the crosswind, thereby increasing the lateral force exerted on the car by the wind. During cross-country racing conditions, bursts of natural or traffic-induced cross winds occur which induce a roll moment on the car. In cases where this moment is larger than the stabilising reaction moment imposed by the opposing wheels, the car would be considered unstable.

In order to quantify the influence of cross winds on the stability of the car, CFD simulations were performed using Ansys Fluent software. Xfoil software was first used to validate the Ansys simulation method using a simple airfoil geometry. An experimental procedure was then used to validate the CFD methodology for a non-rotated wing simulation, comparing the drag results obtained with acceptable uncertainty. The crosswind simulations followed, using the whole vehicle's geometry and by rotating the PV wing along a lateral axis. To account for the presence of both laminar and turbulent flow, a boundary layer transition model was used during simulation. As a baseline, the non-rotated wing crosswind case was simulated for the maximum wind speed found along the typical race path of the solar car. Additionally, for a range of wing angles, and both lateral wind directions, the crosswind simulations were performed and the stability with regards to roll-over moment was numerically investigated.

Experimental and numerical analysis of dilute gas-particle flow in a pipe section

JH Kruger and L le Grange
School of Mechanical Engineering, North-West University, Potchefstroom.

Many coal-fired power stations experience challenges related to the milling and conveying of Pulverised Coal (PC). Coal is milled down to a typical average particle size of 75um and then distributed through a network of pipes, using pneumatic transport, to individual burners feeding the boilers. These systems can be plagued by problems such as varying coal fineness, settling and choking of PC pipes and varying mass flow as the PC enters the burners. These varying flow conditions can cause costly operational problems such as high wear, fouling of boiler tubes leading to unplanned plant breakdowns as well as the emission of toxic pollutants due to poor combustion conditions.

A first step towards an improved PC milled quality and gas-particle flow stability is accurate and online measurement of particle mass flow and particle size distribution. A Particle Flow Test Loop (PFTL) was developed for the ultimate purpose of investigating settling velocities of PC in pipes. The PFTL includes a blower fan providing gas flow for the system, a solid-particle feeder for mixing particles with the gas, a number of long pipe sections and finally a cyclone separator and gas cleaning system. The facility was originally developed at the University of Cape Town (UCT) in 2016 and transferred to the North-West University (NWU) in 2020 for further research focussed on the evaluation of different gas-particle flow measurement techniques.

To enable these flow measurements, a new pipe test section was designed and analysed using CFD simulations. The multiphase software code Neptune CFD by EDF was used in conjunction with Ansys Fluent. The final design was manufactured and fitted to the PFTL and can accommodate measuring devices such as an iso-kinetic sampler as well as online laser-based devices. Two-phase CFD modelling was used to predict cross-sectional solid concentrations for comparison with experimental measurements of particle velocity and particle size distribution. In this paper, selected results from the gas-particle flow simulations are presented, together with comparisons to experimental data obtained in the PFTL test section.

Using experimental flow visualisation techniques to interpret computational fluid dynamics data

C Law[1] and J Nordström[2,3]
1 School of Mechanical, Industrial and Aeronautical Engineering, University of the Witwatersrand,
2 Department of Mathematics, Linköping University, Sweden.
3 Department of Mathematics and Applied Mathematics, University of Johannesburg.

The application of simulated shadowgram and Schlieren images to interpret Computational Fluid Dynamics (CFD) data is becoming a generally well-accepted method. Recent articles by Kumar et al. (2D) and Luthman et al. (3D) demonstrate the interpretation of CFD results using this. Tecplot has now added a Shadowgraph feature to its data analysis suite for post-processing of numerical data.
While the work of Kumar et al. and Luthman et al. are two of many recent examples, the technique has been around for some time and is used extensively within the engineering and shock wave research communities, as demonstrated by the work of Law and Skews.

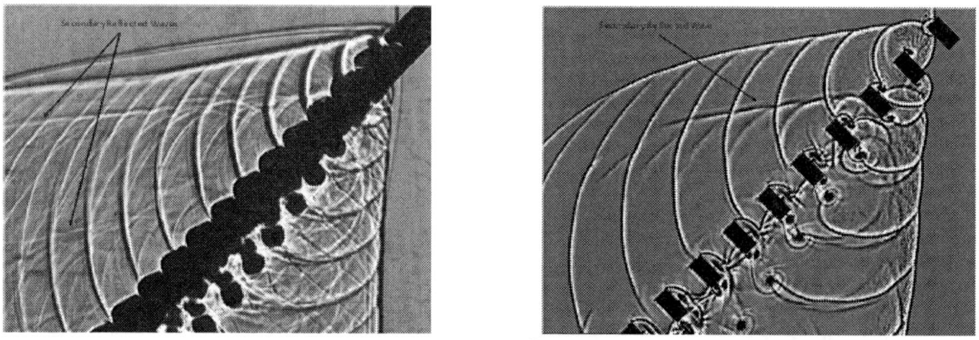

a. **b.**

Figure 1: Shadowgrams of a Mach 1.353 shock impacting a grid at 45.0° with a blockage ratio of 33%. **a.** experimental, **b.** numerical.

The Flow Research Unit (FRU) at the University of the Witwatersrand had a long history of employing flow visualisation techniques across many, varied flow regimes. As CFD became an increasingly important investigational tool for the compressible flow research being undertaken at the time, it was a logical step to begin applying experimental flow visualisation techniques
to CFD data. Initially this was done as a means to validate the numerical models by comparing them to experimental images as these were often the primary data. An example of this is shown in Fig.1 where a side by side comparison of experimental and numerical results are presented for a shock interacting with a rigid grid.
This paper will provide a brief historical overview of the use of schlieren, shadowgraph and interferometry techniques as applied at the FRU to numerical data. This builds on work presented by Law and aims to demonstrate the power of the technique in interpreting numerical compressible flow results.

A Framework for the Design of Soft Pneumatic Actuators Using Computational Tools.

PF Ligthart and MP Venter
Department of Mechanical and Mechatroinc Engineering, Stellenbosch University.

Soft robots possess significant potential in applications such as human-to- machine interfaces and delicate handling operations. Unfortunately, their development is hindered by the complex design space resulting from nonlinear material behaviour. This makes finite element simulations costly and difficult to interpret. This study utilizes a hierarchical design approach to create soft pneumatic actuating units by combining designer intuition with computational tools. Our approach integrates Physics Engines, Finite Element Analysis, Surrogate Modeling, Optimization, and parallel computing, orchestrating a process to efficiently design soft pneumatic actuators. The main goal is to create a replicating unit that can be assembled into a soft robot. Such a replicating unit has been previously developed in the Evogym Benchmark developed by researchers at Massachusetts Institute of Technology (MIT). The designed units follow idealised deformations and the unit was not realised. We demonstrate real-world deformations achieved through the shape, placement, and interplay of pressurized and unpressurised cavities, controlling the deformation process. This pneumatic setup enables pressurized cavities to create deformation while unpressurised cavities absorb deformation. A variable stiffness effect is achieved using this setup. Together these features allow for achieving complex deformation. Our process achieves realistic models that closely match the idealized models used in the Evogym Benchmark. This process creates a crucial link between cost-effective physics environments and the practical realization of the actuators. Our approach is versatile and holds great promise for advancing soft robotics through pneumatic actuation.

Measurement of Lamb wave phase velocity using low-cost equipment

PW Loveday[1] and P Fromme[2]
1 School of Mechanical, Industrial and Aeronautical Engineering, University of the Witwatersrand.
2 Department of Mechanical Engineering, University College London (UCL).

Guided wave ultrasound has the potential to inspect large volumes of structures from a single transducer location. The inspection of plate-like structures using Lamb waves is currently receiving significant research attention. Inspection based on guided waves is complicated by the nature of the wave propagation. This includes multiple modes of propagation, and the modes of propagation are dispersive, resulting in excitation signals becoming distorted as the wave propagates. This complication often necessitates the use of a combination of simulation and experimentation in this field of research.
The multi-mode and dispersive nature of guided waves requires the use of specialized excitation signals in experiments, and requires more complex equipment than that required in conventional ultrasonic testing. In addition, the measurement of properties such as phase velocity require the measurement of the response at precisely positioned measurement points and with high levels of repeatability. A measurement technique involving the excitation of the waves using a custom developed piezoelectric transducer bonded to the plate for exciting the waves and a scanning laser vibrometer for measuring the response at numerous positions, extending radially outward from the excitation location, has been shown to achieve very accurate measurement of phase velocity in metal and composite plates. Unfortunately, such measurement systems are expensive and are often unaffordable. This paper investigates what can be achieved with a low-cost measurement system costing less than USD 1000, which is believed to be accessible to most researchers.
The low-cost measurement system is based on a USB oscilloscope, which includes an arbitrary waveform generator and two channels of analogue input. Use is made of a custom developed piezoelectric transducer, bonded to the plate, for excitation of the waves. A low-cost power amplifier is used to drive the transducer. The response is measured by a receive transducer sequentially positioned at several measurement locations. Two options are investigated for the receive transducer. Firstly, a piezoelectric transducer is used. This type of transducer has a very good signal to noise ratio, but the less-than-ideal repeatability of the measurements has been found to limit the overall accuracy of the phase velocity measurement [1]. Secondly, the use of a custom developed electromagnetic acoustic transducer (EMAT) is investigated. These transducers are far less sensitive that the piezoelectric transducers but could offer improved repeatability. Measurements performed with the low-cost equipment will be compared to measurements performed with the scanning laser vibrometer-based system.

Enhancing Railway Condition Monitoring and Operational Efficiency through Real-Time Data Visualization

N Mabaso
Department of Industrial Engineering Student, Stellenbosch University

In the realm of modern railway engineering, the integration of predictive techniques and data analysis offers profound potential for advancing condition monitoring and maintenance efficiency—two pivotal factors that drive operational safety and effectiveness. This paper presents an approach centred on the utilization of Looker Studio, a data visualization platform, for real-time visualization within the railway domain. The approach utilizes data sourced from TrainTracer, a condition-monitoring digital solution installed on Gibela's X'Trapolis Mega trainsets. This methodology emphasizes data preprocessing, predictive modelling, and the creation of dynamic dashboards using Looker Studio, providing stakeholders with actionable insights for proactive condition monitoring.

The approach begins with highlighting specific challenges related to the collection, and ingestion of mechanical, and specifically railway, engineering data. The successive phase of data preprocessing addresses concerns of data quality, missing values, and outliers—factors that ensure the reliability of subsequent analyses conducted within Looker Studio.

Embedded within this approach is the strategic application of predictive modelling techniques, meticulously chosen to address the intricacies of condition monitoring within railway systems. Model training and validation ensure accuracy and reliability in projecting potential issues.

The cornerstone of this paper lies in the tailored utilization of Looker Studio, designed to bridge the gap between complex data analysis and informed decision-making. While the dashboards presented in this study are initially static, the paper delves into the investigation of ingestion strategies to render them live. These strategies involve the exploration of modern technologies for real-time data ingestion, such as event streaming platforms and data connectors. By leveraging these technologies, stakeholders gain access to up-to-the-minute insights, facilitating swift and informed decision-making.

Through real-world applications, the tangible impact of Looker Studio is displayed in enhancing condition monitoring within railway engineering. Leveraging TrainTracer data, the approach has the potential to pre-emptively identify potential equipment issues, streamline maintenance protocols, and minimize operational deviations—objectives that are integral to maintaining efficient railway operations.

The concept of continuous monitoring and maintenance of Looker Studio dashboards is foundational to the proposed approach, ensuring persistent reliability and accuracy. The integration of feedback loops from stakeholders fosters iterative refinements, aligning the Looker Studio dashboards with evolving operational prerequisites.

In summation, this paper introduces a suitable methodology that leverages Looker Studio to empower stakeholders with dynamic, data-driven dashboards, poised to enhance condition monitoring within the realm of railway engineering, utilizing data generously supplied by Gibela. By synthesizing predictive modelling and Looker Studio's visualization capabilities, this paper provides a toolkit for proactive decision-making and advancing operational efficiency within the railway domain, while also exploring cutting-edge technologies for live data ingestion.

Appropriate Mode of storage of Porcine heart myocardia without tempering with its passive mechanical properties.

I Mabuda[1], T Pandelani[1], H Ngwangwa[1], F Nemavhola[2]
1 Biomedical Engineering Research Group, Department of Mechanical Engineering, University of South Africa.
2 Faculty of Engineering and the Built Environment, Durban University of Technology.

One of the important aspects of mechanics of soft tissues is perseveration during and before mechanical testing. Different methods of storage specifically before testing of heart myocardium are utilised. It remains unknown as to what are the best parameters for preserving myocardium before testing after harvesting. The main purpose of the study is to evaluate the traditional methods used for preserving myocardium and to study the possible effect of preservation methods on the mechanical properties.
150 hearts harvested for a South African landrace pigs aged between 6 and 8 months and weighing from 200kg and upwards were collected from an abattoir based on their availability. In order to preserve tissue condition to a state very close to its physiological condition, whole hearts were packaged and transported in a cooler box at temperatures between -2$_{o}$C and -4$_{o}$C from the abattoir to the testing laboratory.
Each tissues specimen was tested in cyclic tension using the CellScale (UStretch, Wateroo, ON, Canada®) specifically designed for soft tissue mechanical testing. The saw-tooth function was used during the cyclic testing. A total testing length of 25 mm was consistently used throughout all the tests thereby leaving camping lengths of 7.5 mm at each end.
Before collecting data, a normal 0.1 N preload was applied to remove any tissue slackness. The tissues were subjected to a total of 3 cycles at a maximum strain of 40% for a duration of 6s per cycle. To maintain hydration and mimic the body temperature, saline 0.91% w/v of NaCl was placed in the bath and maintained for the duration of testing. The 40% strain was selected to be the physiological magnitude of mammal heart.
The results shows that Young's modulus, irrespective of the direction of stretching, the values of this index in all experimental groups after storage in the studied solutions do not significantly differ from the indices of native tissues. In conclusion, the mechanical properties of the porcine hearts do not change significantly after different storage compared to native materials.

A novel fluid-structure interaction model for sea ice dynamics

NNJ Mahomed[1], A Bogaers[2], CH Rycroft[3] and S Skatulla[1]
1 Polar Engineering Research Group, Department of Civil Engineering, University of Cape Town.
2 Multiphysics Modelling Group, Ex Mente Technologies
3 Department of Mathematics, University of Wisconsin-Madison, USA

Sea ice covers approximately 10% of the Earth's surface and modulates the sea-air heat and momentum exchange, playing a major role in the global climate. The seasonal sea ice advance and retreat of the Antarctic is one of the main contributors to the albedo and physical variations on the globe, yet adequate research on the region has been lacking due to its complex nature. The Antarctic Marginal Ice Zone (MIZ) is a highly dynamic area containing mobile round pancake ice floes surrounded by interstitial grease ice. The sea ice dynamics of this region influences the global climate due to the large surface area of ice moderating solar energy intake, thus research of this area is important to understand the global climate and predict long-term changes thereof. Accurate modelling of the interplay of temperature, waves, sea ice dynamics as well as wind and ocean currents is necessary, not only to predict sea ice dynamics and wave energy dissipation, but also to advance understanding of the governing mechanisms of sea ice growth and decay in the Antarctic MIZ. This work aims to provide an approach to better characterize the sea ice dynamics in the Antarctic MIZ. It studies ice floe motion and collisions due to wave action and ocean currents. The study uses a continuum approach that accounts for the heterogeneous ice cover composition in terms of geometrical layout of ice floes and interstitial grease ice. A novel monolithic Fluid-Structure Interaction (FSI) method for sea ice dynamics adopting the Reference Map Technique (RMT) is proposed and implemented in FEniCS. Traditional FSI models use a Lagrangian solid solver coupled with an Eulerian fluid solver. In contrast, this approach significantly reduces complexity by facilitating the combination of computational fluid dynamics and solid mechanics in the same mathematical framework, by describing the solid stress state in the Eulerian reference frame. This enables us to accurately describe the interaction between solid and fluid constituents, as well as solid-solid collisions. The model includes the respective material laws, Saint Venant-Kirchhoff hyperelasticity and viscoplasticity, describing the nonlinear solid-like deformation behaviour of ice floes and the fluid-like viscous behaviour of grease ice.

A High Order Stable Continuous Galerkin Scheme for Smooth and Discontinuous Fields

AG Malan[1] and J Nordstrom[2,3]
1 Industrial CFD Research Group, Department of Mechanical Engineering, University of Cape Town.
2 Dept. of Mathematics, Computational Mathematics, Linköping University, Sweden.
3 Dept. of Mathematics and Applied Mathematics, University of Johannesburg.

The continuous Galerkin finite element method (CGFEM) offers exceptional efficiency potential for the solution of advection diffusion systems.. What has been hindering this is a rigorous methodology which guarantees stability for both smooth and discontinuous fields. To deal with this we develop this first a provably stable high order CGFEM framework for the transient advection-diffusion equation in the presence of both smooth and discontinuous (shocks) fields. We achieve provable stability via employing the summation by parts (SBP) framework together with the weak imposition of boundary conditions. Non-physical oscillations over discontinuities is addressed by using artificial dissipation. Computational efficiency is achieved via Gauss-Labotto quadrature, yielding super-convergent accuracy of up to $O(p + 2)$ when using p^{th} order Lagrange polynomials for smooth fields, while WENO like behaviour is achieved over shocks.

Studying transvalvular aortic haemodynamics with simplified fluid-structure interaction and experiment

E Marais[1], MP Venter[1], R Laubscher[1] and P Herbst[2]
1 Department of Mechanical and Mechatronic Engineering, Stellenbosch University.
2 Division of Cardiology, Faculty of Medicine and Health Sciences, Stellenbosch University.

Globally cardiovascular diseases (CVDs) contribute to 32 % of deaths making it a leading cause of mortality placing a heavy burden on both the social and healthcare sectors. 75 % of these deaths occur in low- and middle-income areas such as Sub-Saharan Africa (SSA). The common CVD in SSA is rheumatic heart disease. It is often those young and middle-aged adults and in some cases children, suffer from rheumatic heart disease in SSA. Comparing this to high-income countries where it is most common for the elderly to suffer from CVDs, this disrupts the economies of SSA countries. Rheumatic heart disease stems from rheumatic fever, which is left untreated allowing bacteria, which form due to strep throat, to travel in the bloodstream and make its way to the aortic valve. Over time the bacteria causes stenosis stiffening the leaflets of the aortic valve leading to a lifelong disability or death.

Producing tools that can assist clinicians in studying the effects of these lesions and obstructions could allow clinicians to produce non-invasive personalised treatments. Currently, in computational precision medicine, fluid-structure interaction (FSI) simulations allow the consideration of fluid flow and leaflet deformations during operation. FSI simulations are accurate but have the disadvantage of lengthy simulation times and high computational requirements.

The present work investigates the max and mean pressure gradients as well as the flow rate through the aortic valve using simplified modelling techniques. These techniques consist of a lumped parameter model (LPM) which makes use of a 0-D dynamic valve model and a simplified 0-D 3-D FSI. The 0-D 3-D FSI makes use of the LPM but replaces the 0-D valve model with a 3-D finite element (FE) model of a generic aortic valve. Furthermore, the results from an in-vitro test making use of a silicone aortic valve model, based on the FEM model, will also be compared to the modelling techniques.

The findings suggest that the flexibility and dynamics of the 3-D valve influence both transvalvular pressure and flow rate producing pressures higher than that of the 0-D valve model. Additionally, these values align better with predetermined physiological ranges.

The effect of optimization model on the accuracy of constitutive model.

L Mathebela[1], R Sigwadi[2], TLekwana[1], H Ngwangwa[1], T Pandelani[1], F Nemavhola[3]
1 Biomechanics Research Lab, Department of Mechanical Engineering, University of South Africa.
2 Department of Chemical Engineering, University of South Africa.
3 Department of Mechanical Engineering, Durban University of Technology.

Characterizing the single curly fiber under tensile loading requires the large number of sampling and it is time consuming. This is due to material properties variations even from one doner. Selecting the optimizing method from the hyperfit that could yield the best results also is time consuming. The purpose of the study is to assess the effect of the optimizing methods on the accuracy of the model.
Ustretch tensile tester from the Cellscale with 44N Load *cell* was used for tensile testing of single curly hair Fibers. The paper frame technique has been adopted from the previous for clamping the single fibers on to the Ustretch machine. The N=10 of the 20 mm specimens were tested at the strain rate of 10^{-2}. s^{-1} under dry condition and room temperature (± 25 °C). The stress-strain curve was fitted on Hyperfit curve fitting software using the hyperelastic constitutive Mooney-Rivlin model. The effect of the accuracy of the material parameters C10, C01, R^2, R, NRMSE, NE, SS& SA was assessed using the constrained optimizing methods namely Gradient decent, Nelder Mead, Levenberg-maruardt, the: Nelder-Mead, Powell, CG & COBYLA .The coefficient of variation was also utilize to find the data variances per material parameters.
The eight sets of parameters namely C10, C01, R^2, R, NRMSE, NE, SS& SA from optimizing methods were presented for Mooney-Rivlin. The average and standard deviation amongst the optimizing methods for C10, C01, R^2, R, NRMSE, NE, SS& SA was found to be - 2370.347±218.660, 2937.106±242.770, 0.983±0.008, 0.993±0.004, 0.048±0.011, 0.038±0.005, 731.290±96.779 & 5576.293±2740.581 respectively. The material parameter C10 and C01 satisfy material stability criteria for the model on all the optimizing methods. However the Levenberg- Maruardt, Nelder-Mead Powell , CG & COBYLA methods yield the similar and better results than Gradient decent, Nelder Mead on all the material parameter. The coefficient of variation (CV) For C10, C01, R^2, R, NRMSE, NE, SS & SA was found to be 9.22,8.27,0.84,0.44,22.35,13.37,13.23 &49.15 within the range.

Optimizing driving behaviour for underground electric vehicles under varying loads

RF Meeser and DN Wilke
Department of Mechanical and Aeronautical Engineering, University of Pretoria

Reducing natural resource consumption is essential in moving towards a sustainable society. One such application field is in the mining sector where electric locomotives are used to haul tons of ore for many kilometres underground. Improving the efficiency of these vehicles will result in a higher machine utilization, increased profitability and reduced overall cost of ownership of these vehicles. To be able to optimize the energy consumption of such a vehicle it is of great importance to know the energy requirements for the vehicle.

The main contributing factors to energy consumption are the type of vehicle, the route being travelled, the gross combined mass and the velocity profile to complete the route in the logistical constraints imposed. Regenerative braking, various powertrain efficiencies and environmental conditions also play a significant role in the overall energy consumed on a route.

This project is broken down to three phases, namely: route identification, mass estimation and velocity profile optimization. A novel method of route identification is developed that can identify an underground route being travelled in real-time from a database without the use of external communication. The route is identified by searching for patterns in heading and altitude data by comparing these to the available database in a computationally simple method in real time. If the route cannot be identified, the data is stored as a new route in the database and used in future traversing of the route. From the recorded data the route topography is now available and used throughout the rest of the study. Extensive above-ground and limited underground tests confirm the usability of the strategy. A patented real-time mass estimation strategy is developed that makes use of a simple torsional load cell on the driving axle of the vehicle. This method makes mass estimation simpler as it doesn't require multiple load cells and doesn't need trailers to be instrumented either. The driving torque along with the route topography and vehicle dynamics equations are applied to Newton's second law of motion for the vehicle, with the equations rearranged to solve for the unknown mass of the vehicle rig. Test results show that the vehicle's gross combined mass can be estimated to be within 5% of the actual value. Now the only controllable variable in energy consumption for a specific route is the vehicle velocity profile. A strategy is developed that can robustly optimize the velocity profile for the vehicle on a known route with a known mass such that the overall energy consumed is reduced, whilst still adhering to the logistical requirements. The initial results show an energy reduction from 8.5 MJ to 5.8 MJ (32%) is obtained over a normal driver velocity profile for the 10 km test route. With a robust velocity optimizing strategy available, an optimizer is being developed that will be able to reduce the total cost of ownership for the electric locomotive based on battery and efficiency models of the various components in the electric locomotive driveline.

Prediction of Ship Motion Using Forecast Weather

M Melim, A Bekker and K Kruger
Department of Mechanical & Mechatronic Engineering, Stellenbosch University.

In the complex domain of maritime operations, the nuanced prediction of ship motion stands as a critical cornerstone, underpinning operational ef5iciency, strategic navigation, and the paramount safety of seafarers. Predicting ship motion provides insight into how vessels react to different sea conditions, highlighting the relationship between a ship's structural design and the variable patterns of the ocean. This insight facilitates the calculation of the motion sickness dose value (MSDV), a key metric determining seafarers' susceptibility to motion sickness, which can in5luence their health and overall performance. This research presents an in-depth approach for predicting ship motion by incorporating factors like weather predictions, wave spectra, and the ship's response amplitude operator (RAO). This approach, as depicted in the subsequent 5igure, starts with gathering weather forecast data, particularly focusing on wind speed and direction, and then using this data the wave spectrum is generated, allowing for the analysis of potential wave interactions. The RAO, in5luenced by factors such as vessel characteristics, load state, and navigational parameters, is integrated with the wave spectrum to form the ship's motion spectrum. The ships motion spectrum offers insights into potential ship behaviour under specific sea conditions, from which the power spectral density curve can subsequently be derived. The area under the PSD curve denotes the root mean square value, which, according to ISO 2631-1, determines the MSDV, and with this value, decision-making can be enhanced to improve seafarer well-being and performance.

Low-speed wind tunnel tests of two airfoils for standard class sailplane application

T Melman, J Bosman, and A Jonker
School of Mechanical Engineering, North-West University, Potchefstroom

A significant portion of aircraft drag is attributed to the aerodynamic performance of the wings, which is largely dependent on the geometric shape of the airfoils. For a new standard-class (no flaps) sailplane, a novel airfoil (OPT110) was developed through a genetic optimization algorithm that utilizes XFoil (vortex panel method) as the aerodynamic calculation engine. When compared to a baseline case, the sailplane with improved airfoils exhibited remarkable improvements in calculated average cross-country speeds.

Although 2D airfoil research is generally considered resolved by most aerodynamic academics, discrepancies between numerical analysis methods were observed. While using CFD to compare the differences in aerodynamic performance between the baseline and the new airfoil, the improvement is not as substantial as previously indicated by XFoil. Numerous CFD simulations were conducted, incorporating the best-known transition models and methodologies, which included far-field drag calculations in an attempt to achieve comparable results with XFoil, but without success.

When designing new aircraft, confidence in performance prediction is essential, and significant differences between the results of airfoil analysis methods create uncertainties that, in turn, impede aircraft development. It is for this reason that both the baseline and new airfoil sections were tested in a low wind-speed wind tunnel facility provided by the CSIR to determine their true performance characteristics. Wind tunnel models for both airfoil sections were manufactured using CNC-milled molds, incorporating pressure taps built into the skins, which are connected to a 64-channel miniature Scanivalve analogue pressure scanner. Drag was measured with a wake rake, and the test rig also integrates load cells to mechanically measure lift and drag.

These experimental results are expected to offer insights into the ability of XFoil and CFD to analyze aerodynamic characteristics of 2D airfoil sections designed for low-speed range sailplanes.

Biomechanical analysis: Advancing snake robot locomotion in complex environment

BL Moloi
Department of Mechanical and Mechatronic Engineering, Stellenbosch University

Inspired by biological snake, the snake robot can navigate complex environment due to their slenderness and flexible body. They can be used in complex environment such as boiler maintenance or search and rescue mission. However, for such task the robot must mimic the snake locomotion strategy and operate effectively given the payload limitation for these robots. The robot must be capable of navigating the complex environment and return to its starting position. Hence the need to optimize energy efficiency and adaptability. The common snake robot locomotion is due to the serpentine gait and is achieved by controlling the robot internal body shape producing a travelling wave. The locomotion is based on the anisotropic frictional property that mimic the skin of the snake. For a planar environment, the propulsive force produced is due to friction forces characterised by contact condition arising from sliding, collision of the body with the environment. However, in simulations the speed and net displacement shows that the serpentine gait causes a backward motion, thus reducing the speed of the robot. The problem worsens when the robot is placed in a complex/unknown environment where friction (with unknown friction coefficient) is not the only force acting on the body but also forces due to impact with obstacles. In some environments condition motion is not possible which is even true for biological snake. The current development and research in snake robots focused on understanding the snake locomotion rather than mimicking real snake to produce a robot that can complete a task.
Lagrangian mechanics is used to model the dynamics of a multibody system (snake skeleton) with multiple contacts where the propulsive force equation is computed and studied. Based on contact mechanics the resultant equations are used to proof by induction that there exist conditions for which the serpentine gait produce no net displacement. Computer simulations are performed without path following control to determine the propulsive force behaviour. In planar environment without impact from obstacles, the propulsive force oscillates, and the speed of the robot fluctuate as a result. However, the propulsive force behaviour shows that for a serpentine gait there are body shape motion that creates a backward motion. Furthermore, with the serpentine gait, there exist environmental setups where the snake robot is unable to move forward. Mapping the body shape with the propulsive forces enable the determination of the body shapes of the snake robot that produce backward motion. The backward motion is then eliminated through control design by adjusting the local stiffness behaviour of the contact zones. The robot speed for the new control strategy is higher and the snake robot can move forward for most complex environment. Knowing the resultant dynamics due to the contact forces allows for the design of a control strategy that produce optimal performance of snake robots in unknown environment.

Deep Reinforcement Learning in Frontier and Sampling-Based MAV Exploration of Unknown Environments

ME Moyo[1] and T Celik[2]
1 Mechanical, Industrial and Aeronautical Engineering, University of the Witwatersrand.
2 Electrical and Information Engineering, University of the Witwatersrand.

The challenge of effectively exploring unknown GPS and communication-denied environments using micro-aerial vehicles (MAVs) presents a significant hurdle for both civil and military applications. These applications include recreational use, inspection, reconnaissance, search and rescue, as well as information gathering in fields like mining, archaeology, civil engineering and defence. Traditional methods like frontier-based and sampling-based exploration provide partial solutions but come with their own limitations, particularly in dynamic and complex environments. This study introduces an integrated approach combining these two methods with Deep Reinforcement Learning (DRL) to facilitate rapid and effective exploration.

Frontier-based exploration methods perform well in identifying unexplored boundaries, known as frontiers, but often fail to adapt in rapidly changing environments. Conversely, sampling-based methods can tackle complex terrains. However, sampling-based techniques are computationally expensive and equally ineffective when exploring dynamic environments, and when using mobile platforms like aerial vehicles. To address these challenges, this paper proposes a novel hybrid framework that leverages both frontier-based and sampling-based exploration techniques, augmented with Deep Reinforcement Learning (DRL).

In our novel approach, frontier-based exploration is employed initially to identify and prioritise unexplored regions in the environment. The sampling-based approach is then used to explore these frontiers in finer detail. A semantic map is generated using a Simultaneous Localisation and Mapping (SLAM) module, which serves as the input to the Deep Reinforcement Learning (DRL) module. The MAV agent is trained in this DRL module to develop optimal policies for navigation, exploration and battery management. The perception inputs to the exploration techniques are observations from LiDAR and RGB camera sensors. The agent uses these optimal policies to select the most promising viewpoints for further exploration. This study also incorporates a Next-Best-View Planner (NBVP) to optimise the path for the MAV to reach these selected viewpoints.

A comprehensive exploration framework detailing this integrated strategy is presented in this paper. It begins by initialising a semantic map and the MAV's current state. The frontier and sampling-based methods update this map iteratively, which in turn influences the DRL agent's decision-making process.An optimal path is chosen based on multiple parameters like distance, information gain, and global coverage. The exploration policy is dynamically updated, preferring actions leading to significant new exploration and assigning lower rewards to actions that revisit already explored areas.

Preliminary results demonstrate that this integrated method outperforms traditional frontier-based and sampling-based techniques in terms of exploration speed, computational efficiency, and adaptability to dynamic changes. The study holds promise for applications beyond MAVs, including robotic and autonomous system navigation in inhospitable conditions. For example micro-aerial vehicles (MAVs) with autonomous exploration capabilities can replace traditional methods that involve human presence in hazardous environments, such as post-blast inspections of mine shafts.

This research contributes to the broader field of computational mechanics by introducing a robust, computationally efficient method for MAV navigation in challenging these environments. It provides a foundation for future work focused on the application of integrated exploration strategies in real-world conditions.

CFD analysis of flow inside the cyclone unit in the Particle Flow Test Loop

(Withdrawn)

H Muller and JH Kruger
School of Mechanical Engineering, North-West University, Potchefstroom.

To enhance the efficiency of electricity generation within a coal-fired power plant, it is important to have a comprehensive understanding of the particle flow dynamics of the pulverised coal-air mixture transported into the furnace for combustion to take place. Conditions in a coal power plant can be difficult and dangerous for measurements to take place. Therefore, Eskom under its power plant improvement research program (EPPEI) has developed the Particle Flow Test Loop (PFTL) to create more controllable conditions for measurements to take place and to understand the flow characteristics within the plant. The particle flow test loop was originally developed at the University of Cape Town and was moved to the North-West University in Potchefstroom in 2020 where research continues in gas-particulate flows.

The facility uses an open-loop cycle that contains a cyclone unit to separate solid particles from the carrier before exhausting clean air. The cyclone separator consists of a cylindrical shell with a tangential inlet where the air containing the solid particles enters. The tangential inlet forces the mixture to spiral around inside the cylindrical shell creating a vortex. The solid particles which are heavier than the air are forced to the outer walls and travel downward into the conical section forming an outer vortex. The increasing velocity in the outer vortex results in a centrifugal force on the particles that separates them from the air stream. When the air reaches the bottom of the cone, it begins to flow radially inwards and out the top as clean air while the solid particles fall into the dust collection chamber attached to the bottom of the cyclone.

To comprehensively grasp and characterise the flow patterns within the cyclone unit, computational fluid dynamics (CFD) simulations were conducted utilising Ansys fluent. Given the inherently turbulent nature of flow within the cyclone separator, an appropriate turbulent model was identified for accurate representation. The simulation process involved inputting fluid properties corresponding to test conditions and using pressure measurements obtained at the inlet, overflow outlet, and underflow outlet as boundary conditions. The calculated flow rates at the outlets of the cyclone separator were used for the validation of the simulation, comparing them with the measured flow rates conducted during the test procedure, enhancing the understanding of its operational behaviour. Simulations of particle flow behaviour and separation efficiency were then performed as part of the numerical investigation into the operation of the cyclone.

Angled expanding jet impingement in supersonic crossflow

S Myburgh and C Law
School of Mechanical, Industrial and Aeronautical Engineering University of the Witwatersrand,

Significant work has been done on the effect of both perpendicular and spanwise inclined uniform jet injection in supersonic cross flow. Recent articles by Sebastian et al. demonstrate the complex effects of asymmetry on the mixing region between the jet outflow and the freestream. This is further complicated if there is a velocity gradient in the free stream.

Implicit transitional SST turbulent simulations were performed to investigate the flow field around a streamwise inclined, asymmetrically expanding jet in supersonic cross flow at Mach number 3.5. Various expansion ratio configurations were explored, through which supersonic jets were injected into the same primary flow field conditions. The overexpanded nozzle geometry case is focused on here due to its unique flow features, with various shadowgraph and contour plots considered.

The interaction between the angled, nonuniform jet impinging on the free stream was observed. The region between the jet exit plane and free stream featured a strong mismatch in pressure and momentum. The pressure difference established a strong suction-pressure gradient within the nozzle, forming an overexpanded jet. Additionally, due to the angled injection, a separation bubble formed on the lower nozzle wall near the exit plane, acting as a wall to further narrow and deflect the overexpanded jet. At the upstream exit edge of the nozzle an expansion fan formed which accelerated the free stream, with a weak oblique shock wave intercepting the lower nozzle wall causing further separation. The mixing region of the free stream and jet was also of interest. The effect of the momentum difference between the jet and the free stream on the jet interaction due to the flow features will be explored further.

Three Dimensional Computational Fluid Dynamics Modelling of Pancake Sea Ice on Waves

A Nair[1,2], E Boje[1], and R Verrinder[1,2]
1 Department of Electrical Engineering, University of Cape Town.
2 Marine and Antarctic Research Centre for Innovation and Sustainability, University of Cape Town.

The Antarctic Marginal Ice Zone (MIZ) is a highly dynamic sea ice region at the interface between open ocean and consolidated pack ice. It contains various ice types, including pancake ice, which is characterised by a roughly circular shape and raised ridges. The seasonal evolution of the pancake ice size distribution is not well documented due to the high variability of the MIZ and difficulty in obtaining high resolution in situ data sets. Computational models are used to simulate the small scale interactions of pancake ice and waves. However, many of these models are limited to two dimensions, and approximate pancake ice as cylindrical. This paper presents a computational fluid dynamics solution of the six degree of freedom dynamic response of a single pancake ice floe in a wave field. This model accounts for drag effects above and below the water surface and removes the cylindrical shape assumption. The model was created with OpenFOAM, using the volume of fluid method and the overset mesh approach, to simulate pancake ice interacting with sinusoidal waves. Kinematic data for all six degrees of freedom were collected by running simulations of various pancake ice geometries and sizes on waves with different wavelengths chosen with respect to the relative diameter of the ice floe. A validation experiment using a wave flume is set to be conducted at the Council for Scientific and Industrial Research (CSIR), in Stellenbosch, to test the fidelity of the model. The simulation data will be compared to visual and inertial data collected with an artificial pancake ice floe in the wave flume. From the analysis of the simulation data, it was seen that the translational motions of different pancake ice sizes were largely similar and that most of the variation was visible in the rotations. It was noted that, while the surge and heave motions were sinusoidal as expected, the pitch motion displayed two modes. It is hypothesized that there are two main components in these responses: the waves frequency and the natural frequency of the geometry used. It is also hypothesized that the geometry of the pancake ice floe can be approximated from its dynamic behaviour. The study of the relationship between pancake ice size and shape and its response to waves can give us insight to the variability of the MIZ. In future work we hope to develop an algorithm that can approximate the size and geometry of pancake ice using the inertial measurements captured by buoys deployed on them. This will enable us to study the evolution of pancake ice size on the scale of individual floes.

Assessing regression capabilities of layer recurrent neural network with a scaled gradient conjugate training function

HM Ngwangwa
Department of Mechanical Engineering, University of South Africa

Approximation of forcing functions on mechanical systems remains a challenging problem involving the solution of inverse problems. However the system dynamics of most mechanical systems cannot be accurately determined, and even when they are available, they do not often render themselves amenable to calculation of their inverse forms. In such cases, the application of artificial neural networks to the approximation of the forcing functions may be expedient. The approximated functions may subsequently be applied to other numerical models for estimation of stresses and other related quantities of interest. In this paper, a layer recurrent artificial neural network using a scaled gradient conjugate algorithm was investigated on its capabilities to approximate the inputs to a system having different underlying transfer functions. The estimated inputs were intentionally constructed based on a Belgian waveform with steps that are spaced out along the time span in a random manner. The assessed transfer functions comprise the following: constant-valued functions, proportional (ramp) functions, integral functions, exponentially decaying functions, and harmonically varying transfer functions. All computations were conducted using Matlab. The results show that the artificial neural network was able to accurately approximate the inputs in the constant-valued functions, proportional functions, integral functions, and exponentially decaying functions. The inputs into the harmonically varying transfer functions were extremely hard to approximate. For this transfer function, it was found that the increase in the training data, reinitialization of the network weights and biases, and increase in the size of the artificial neural network did not improve the accuracy of the network. This is currently being further investigated by the researcher.

Investigation of inverse model for the estimation of impulsive loads

BM Nickerson
Department of Mechanical and Mechatronic Engineering, Stellenbosch University

Structures are often investigated in terms of their responses to given input loads. It may, however, sometimes be necessary to estimate the input loads for a structure based on the responses. This is referred to as an inverse problem, which can often be difficult to solve due said problems being ill-conditioned. This work investigates an inverse model originally developed to estimate propeller loading from measured shaft responses for an ice-going vessel. The model makes use of modal superposition to represent a structure as a combination of its mode shapes. The work focuses on whether the developed methods can be applied to other structures in a controlled laboratory environment, in an effort to assist with validation of the original model.

A Weakly-Compressible Two-Phase Formulation for Hydrogen Containment Modelling

MY Oomar[1], AG Malan[1], MD Wright[1], LC Malan[1], S Zaléski[2,3,4], and F Gambioli[5]

[1]Industrial CFD Research Group, Department Mechanical Engineering, University of Cape Town, South Africa
[2]Sorbonne Université, Institut Jean Le Rond d'Alembert UMR 7190, F-75005 Paris, France
[3]CNRS, Institut Jean Le Rond d'Alembert UMR 7190, F-75005 Paris, France
[4]Institut Universitaire de France, Paris, France
[5]Airbus Operations Ltd Pegasus House, Aerospace Avenue, Filton, Bristol, BS43 7PA, England, U.K

The design of cryogenic storage tanks rely on the accurate modelling of the self-pressurisation and sloshing process which occur during the storage and transportation of liquefied gases at extreme low temperatures. In this context, an accurate and efficient weakly-compressible Volume-of-Fluid (VoF) model based on the two-fluid formulation proposed by Baer & Nunziato [1] is developed and simplified via a non-dimensional analysis of a liquid-gas hydrogen system at saturation conditions. The resulting model treats the liquid as incompressible and the gas as compressible. The set of governing equations are solved via the implicit pressure based projection method in Elemental® [2, 3, 4] which is extended to include non-isothermal effects. Unlike traditional compressible flow solvers which employ a two-way coupled system for pressure and temperature, an implicit pressure solve is found sufficient to capture both sonic characteristics and thermal effects in the gaseous phase. The model is validated using academic and industrial relevant test-cases. These include a syringe problem to validate compressibility effects, a natural buoyancy case to validate non-isothermal effects and a classic flight-weight case [5, 6] to demonstrate the ability of the model to capture pressurisation for liquid-gas hydrogen. For all tested cases, an excellent correlation is seen between analytical and numerical solution. We also focus on a two-phase mass inflow pressurisation which is subjected to a violent acceleration. In particular, the results show flow features consistent with that obtained in literature where the developed solver is found to be both accurate and efficient for such industrial relevant problems.

Estimating Mechanical Properties of AM Lattices Using CT-Scans and FE Models

M Park[1], MP Venter[1] and A Du Plessis[2,3]
1 Department of Mechanical and Mechatronics Engineering, Stellenbosch University.
2 Department of Physics, Stellenbosch University.
3 Object Research Systems Inc, Montreal, Canada.

The prevalent reliance on expensive, time-consuming CT scans and traditional mechanical testing presents significant challenges in determining the mechanical properties of additively manufactured metallic lattices fabricated through laser powder bed fusion. This work integrates CT-scan data with finite element models to correlate process parameters (layer thickness, laser power) with mechanical performance. This correlation allows us to use interpolation methods to predict lattice properties without further CT scans. Two such methods, direct and indirect interpolation, are proposed. Direct interpolation correlates process parameters with mechanical performance, while indirect interpolation uses parameters and manufacturing defects to inform the finite element model. Numerical interpolation results obtained from both direct and indirect methods are compared with experimental results, with the targeted accuracy criterion set at 92% based on reproducibility error of experiment. The findings demonstrate promising accuracy rates for both methods, with the indirect interpolation method achieving a higher reliability with an accuracy rate of 94.68%, surpassing the target criterion. The direct interpolation method also achieves a respectable accuracy rate of 92.74%. Both interpolation methods offer cost-effective numerical solutions for material characterization and build quality monitoring in the context of additively manufactured metallic lattices. By reducing reliance on CT scans, this approach significantly saves time and costs associated with the analysis of mechanical properties, presenting a valuable alternative for industry applications.

Optical Laser Rail Wheel Wear Measuring and Detection Unit

AK Phillips, PDF Conradie
Department of Mechanical and Mechatronic Engineering, Stellenbosch University.

In the rail industry, the importance of the wheel-rail interface is widely acknowledged. Due to the high stresses in both the wheel and the rail material, surface wear, subsurface defects, and wheel profile wear is inevitable in normal operation. Therefore, standards and specifications are in place to control the technical requirements of railway wheels, which contributes to the safety of passengers.
As part of the condition-based maintenance tactic, regular inspections ensure that railway wheels are within specified wear limits and conform to the standards. Currently, inspections are done at set intervals and rely heavily on human involvement and manual equipment, which can lead to the late detection of defects with consequential effects. Therefore, more frequent contactless inspection on railway wheels is needed.
This paper uses a high-performance class 3R laser scanner to detect defects and wear on the railway wheel. The laser uses optical triangulation with a laser line projected on wheel surface via a linear optical system, and the diffusely reflected light is replicated on a sensor matrix based on the Scheimpflug principle. Using the requirements from the BS EN 15313:2016 wheel standard, the focus is on the wheel rim, where most defects and wear occur. Single and multiple two-dimensional images are generated to identify and quantify the wear and defects. The optical rail wheel wear measuring and detection unit is tested on a half-scale 3D printed wheel model. Defects from the BS EN 15313:2016 standard are replicated on the scale model, and the unit is used to identify the defects.
The research results could potentially be used in a trackside monitoring system that can inspect the profile of a railway wheel while the train is moving at low speeds.

Enhanced Data-Driven Hybrid Approach for Unavailable Damage Scenarios in Welded Rail Tracks with Multiple Reflections and Reflectors for Ultrasonic-Guided Wave Inspection

DA Ramatlo[1], DN Wilke[1] and PW Loveday[2]
[1]Department of Mechanical and Aeronautical Engineering, University of Pretoria.
[2]School of Mechanical, Industrial and Aeronautical Engineering, University of the Witwatersrand.

To develop reliable ultrasonic-guided wave monitoring systems, a significant amount of inspection data must be collected for each specific application scenario. While experimental investigations are fundamental, they can be time-consuming and costly, especially when it comes to real-life testing. Additionally, there is often a lack of experimental data that includes damage. In certain applications such as pipelines, it is possible to introduce artificial damage and perform laboratory experiments on the test structure. However, laboratory experiments in rail track applications are often neither possible nor meaningful, as reflections from short rail sections dominate the response. Therefore, generating synthetic data using modelling capabilities has become increasingly important.

This paper extends the application of a previously proposed deep learning strategy that utilizes a variational autoencoder (VAE) model to generate synthetic ultrasonic inspection data for welded railway tracks in the presence of multiple reflections and reflectors. The main objective is to create synthetic data that contains damage signatures at specified positions along the length of a rail track. The VAE model is trained to encode an input damage-free baseline signal and decode to reconstruct an inspection signal with damage by adding a damage signature on either side of the transducer by specifying the distance to the damage signature as an additional variable in the latent space.

The training data was produced from a physics-based model that computes virtual experimental response signals using the Semi-Analytical Finite Element (SAFE) and traditional finite element procedures. The VAE model successfully generated synthetic inspection data by fusing the response signal from a damage-free rail with the reflection from a crack model at specified distances from the transducer on either the right or left side. In addition to capturing the complex physics of the problem due to the interaction of multiple modes of propagation, dispersion and attenuation, the VAE was further able to capture the reverberations caused by the interaction of the damage reflection with multiple reflectors such as multiple welds in the rail.

Overall, this study highlights the practical application of this hybrid framework to generate synthetic data with damage signatures for unavailable scenarios in the presence of multiple reflections and multiple reflectors. The results show that it is possible to generate realistic inspection data that can be used to develop reliable ultrasonic-guided wave monitoring systems.

Computational modelling of alternating current conduction effects in packed beds

QG Reynolds[1,2] and AEJ Bogaers[3]
[1]Pyrometallurgy Division, Mintek, Randburg, South Africa.
[2]Department of Chemical Engineering, Stellenbosch University.
[3]Ex Mente Technologies, Pretoria, South Africa

Alternating current, or AC, is used as a source of electrical power in a wide variety of household and industrial applications. In particular it is used to provide resistive heating in certain types of pyrometallurgical smelting furnaces, where AC is passed between large graphite electrodes mounted into refractory-lined vessels in order to melt mined ores and perform thermochemical reactions with them. This process produces economically important commodities such as iron and steel, platinum-group metals, chromium, manganese, silicon, and many others. In order to heat the furnace, current must pass through the charge burden – this is a heterogeneous bed of particulate reductant and ore materials, with the ore and any reaction products in a solid or molten state depending on their location in the furnace. The reductant is typically a carbonaceous material such as metallurgical coke, and is electrically conductive at high temperatures.

Quantification of electrical conduction through packed particle beds is a non-trivial problem, and requires an understanding of particle settling and packing behaviour as well as surface contact phenomena. This is further complicated in the case of AC current, where the oscillatory time-varying nature of the electrical and magnetic fields can cause a redistribution of the current field – this is commonly observed as the "skin effect" in solid cylindrical conductors, where AC current is preferentially carried near to the conductor boundary rather than uniformly across the cross-section.

In the present work, the authors discuss numerical approaches for solving AC electromagnetic field problems in both the time and phasor domains. The methods are implemented in the OpenFOAM® computational mechanics framework and used to examine whether current redistribution effects are significant in packed beds of conductive particles. Random particle packings are produced using discrete element modelling, and used as a template for the generation of computational meshes and domains for the electromagnetic solvers. The effect of parameters such as particle size and electrical conductivity are studied in some simplified packed bed geometries. The results are also compared to those obtained from continuum models, a commonly-used approximation in which the packed bed is treated as a uniform material with averaged properties.

Geometric Acceleration of Thermochemical Equilibrium Calculations for Inclusion in Multiphysics and Process Models

WA Roos[1,2], AEJ Bogaers[1], JH Zietsman[1,2]
[1]Ex Mente Technologies, Pretoria, South Africa
[2]University of Pretoria, South Africa

Incorporating thermochemical equilibrium calculations into multiphysics and process models can provide significant insights into industrial processes that current modelling or measurements cannot. Equilibrium calculations are however, in general, omitted or incorporated in a simplified manner due to their computational expense. An accelerator algorithm was developed based on established thermochemical theory to include equilibrium calculations into models more efficiently.

A phase diagram is a geometrical representation of a thermochemical system and the proposed algorithm utilises this, together with the Gibbs phase rule, to construct a geometrical database of the system. Initially the database is empty. As a model requests thermochemical equilibrium properties at a system state (a coordinate consisting of temperature and system component concentrations), a computationally expensive equilibrium calculation is performed and the result stored in the database as a geometrical feature in the phase diagram known as a tie simplex. Tie simplices that are within specified temperature and composition tolerances of each other can be combined to locally discretise the phase diagram into small cells in the regions that the model accesses regularly. The model creates its own database as it progresses, known as in-situ discretisation, and no prior human knowledge or intervention is required.

As the database is populated more densely, the algorithm can start recalling discretisation cells and perform more computationally efficient interpolations instead of the more expensive equilibrium calculations. When a thermochemical equilibrium state is required at a given system state in a discretisation cell, interpolation can be performed, however, to remain thermodynamically consistent interpolation is performed in two steps. Firstly, by interpolating to constant non-compositional potentials such as temperature. This creates a section through the discretisation cell and all the possible tie simplices on this section has the same non-compositional potential, but different chemical potentials. The second step is to find the single tie simplex on this section that passes through the composition of the system state – it would therefore have the same chemical potential. To find this unique tie simplex, vector algebra, or more specifically Barycentric coordinates, were used to create a set of non-linear equations, that when solved, provide the tie simplex that perfectly passes through the system state. This tie simplex is then used to provide the model with the required interpolated thermochemical equilibrium properties.

Even though a thermochemical system's phase diagram increases in dimension as more system components are considered, the developed algorithm is generic and works in all dimensions due to the generality of phase diagram geometry, the Gibbs phase rule, and Barycentric coordinates. This work focuses on presenting the discretisation step and the derivation of the set of non-linear equations for the interpolation step with geometric examples. The algorithm's performance was tested on two industry-relevant systems and the levels of acceleration achieved and accuracy will also be presented.

Aeroacoustic simulation and improvement of a propeller for an electric self-launch sailplane

K Schutte, J Bosman, and JH Kruger
School of Mechanical Engineering, North-West University, Potchefstroom.

Propeller design for various aviation type applications has always been a challenging task where air-frame, engine and propeller properties must match in order to obtain adequate thrust at the required flight speeds. In some applications, such as electric sailplanes, propeller geometries are restricted and further complicate the design, because the system retracts into the fuselage and usually limits the radius of the propeller. Consequently, compromises must be made such as high rotational speeds when adequate thrust is needed. These compromises usually lead to secondary adverse effects such as inefficiency and excessive noise.

There are regulations for the noise levels associated with electric powered sailplanes and to certify these propulsion systems, they must adhere to specifications. This means that the propeller designer must also include noise levels as a parameter to consider and minimize. Aeroacoustic CFD models have the potential to predict noise levels of propellers, where the pressure fluctuations from the fluid flow simulation are used as sources and coupled with equations to describe the propagation of small perturbations in pressure and velocity that produces sound waves.

In this study, the accuracy of this design method is investigated and ultimately to be used to improve current propeller designs. An electric propulsion system of a sailplane was used as a baseline validation case, where the propeller produced significant noise levels when measured. Aeroacoustic simulations using Starccm+ including the Ffowcs Williams-Hawkings (FW-H) model, agreed well with the spectral analysis and noise levels of the radiated sound from the experimental measurements at various RPM settings.

As a second part of the study, various modifications to the geometry that have potential for reducing propeller noise have been applied to the baseline design and were subsequently analysed with CFD to evaluate the improvements. An adequate reduction in noise levels was indicated by the aeroacoustic analyses with the new propeller geometry which included tip airfoil modifications. The improved propeller was manufactured, and noise level measurements were taken at various rotational speeds. Experimental results correlated well with CFD calculations and showed adequate reductions in noise levels in comparison with the baseline propeller.

Comparing Source-Driven and Variance-Driven Statistical Learning Models to Generate and Analyse Guided Wave Ultrasound Time Series Signals

II Setshedi, DN Wilke and PW Loveday
Department of Mechanical & Aeronautical Engineering, University of Pretoria.

This study presents a comparison between source-driven and variance-driven statistical learning models for reconstructing and analysing rail-guided wave-based time series signals. Two linear statistical learning models are chosen, namely the variance-driven Principal Component Analysis (PCA) and the source-driven Independent Component Analysis (ICA), which are applied as linear autoencoders. We investigate their efficacy in reconstructing as well as capturing underlying features within simulated and experimental time series datasets.

The use of simulated datasets for training allows an investigation of the influence of using datasets that include: a single propagation mode and reflector; multiple modes of propagation and a single reflector; and multiple modes of propagation and multiple reflectors.

Initial results indicate comparable reconstruction capability between ICA and PCA. The best reconstruction occurs when using the multiple simulated reflectors dataset which is due to the higher feature and information density per data sample, as well as mode coupling between reflectors being present in the dataset.

For latent analysis using ICA the single-mode time series data yields the best organized latent space when compared to the other datasets and PCA on all datasets. This study shows the capability of linear autoencoders to not only train on simulated data to effectively learn and reconstruct both simulated and experimental time series signals but also the capability of using simulated data to improve latent interpretation.

Future studies will involve comparison with traditional methods used for interpreting guided wave-based measurements and non-linear machine learning methods with deeper architectures such as variational autoencoders (VAEs). We expect improved disentanglement of the latent spaces when applying source-driven non-linear machine learning methods.

CFD analysis of convection cooling of the NWU solar car PV wing

D Struwig, JH Kruger, and J Bosman
School of Mechanical Engineering, North-West University, Potchefstroom.

ABSTRACT

When considering the performance of solar vehicles, the efficiency of their solar PV cells is crucial. Several different factors have an impact on the efficiency of these cells, and require consideration when optimising the performance of the solar vehicle. One of the influential factors is the operating temperature of the PV cells. During operation, the surface temperature of the cells increases to between 60°C and 80°C, resulting in a decrease in solar panel efficiency for every degree above the standard operating temperature of typically 25°C.

Earlier designs of the solar car of the NWU relied on the air passing over the wing's top alone, to cool the PV cells. The latest model, Naledi 2.0, allows for airflow through the wing, intending to reduce PV cell temperature and increase PV panel and overall solar car efficiency. The influence of the drag of this internal flow and the PV panel temperature distribution is however not yet known.

Calculation of the drag and PV panel temperature required the setup of the correct mathematical models for flow and heat transfer, using Ansys Fluent software. For the simulations investigating the drag of the solar car, the momentum equations were employed, together with a boundary layer transition model to accurately predict the turbulence around the body. Drag simulations were performed for both internal and non-internal flow configurations to make sensible comparisons between the two cases. Heat transfer simulations required the integration of momentum, thermal and radiation equations to represent the energy and fluid flow domain simultaneously. To accurately represent the physical case, the correct energy sources and thermal boundary layers were selected, which entails the selection of the solar ray-tracing algorithm to accurately predict the irradiance of the sun at a given time and date, as well as the discrete ordinates radiation model to simulate the radiative heat transfer inside the wing geometry. Additional thermal boundary layers were also implemented to describe the geometry of the solar car accurately.

Simulations were validated against experimental measurements obtained during solar car operation. Experiments conducted included the determination of the drag coefficient of the solar car, as well as the temperature distribution across the PV wing, together with the effect of the temperature on the electric collection efficiency of the solar panels.

Start-up and Transient Response Characteristics of Natural Draft Direct Dry Cooling Systems at Various Scales

W Strydom, JP Pretorius, and R Laubscher
Department of Mechanical & Mechatronic Engineering, Stellenbosch University

Natural draft direct dry cooling systems (NDDDCSs) aim to combine the advantages of traditional cooling systems such as air-cooled condensers (ACCs) and indirect dry cooling systems. These systems offer benefits of reduced system complexity due to the absence of a separate shell-and- tube condenser, high thermal efficiency, as well as low parasitic power and operational costs. The NDDDCS aims to be a scalable cooling system, operating in large coal-fired power plants (900 MWt, large-scale), concentrated solar power (CSP) plants (100 MWt, medium-scale) and water desalination plants (1 MWt, small-scale). Whereas our previous work established the steady state performance levels at various scales, there is also a need to investigate the start-up characteristics and transient behaviour of the NDDDCS. Power plants operating as peaking or fast-response units may be subject to rapid starting times or load ramps as electricity demand acutely increases throughout a day, requiring an increase in cooling system heat rejection, and as such the response time of the NDDDCS must be investigated. This paper sets out to characterise the transient response of the NDDDCSs to start-up and load ramp conditions via a transient 1-D mathematical model. Results are presented for the large- and medium-scale NDDDCSs, and limiting factors are identified.

Nearfield dispersion modelling of discharged effluents under various hydrodynamic conditions

TL Surajpal, GPJ Diedericks, and GJF Smit
Division of Applied Mathematics, Department of Mathematical Science, Stellenbosch University.

As water scarcity continues to rise, many companies are responding by planning and constructing desalination plants along coastlines. These plants remove salt from seawater to produce freshwater, but during the process, brine, which is water with elevated salt content and higher temperatures than seawater, is returned to the sea. To ensure the environmental impact of these desalination plants is thoroughly assessed, engineers employ nearfield models for designing the intakes and outflows of these facilities. These models can vary in complexity, ranging from empirical to advanced Computational Fluid Dynamics (CFD) models.

To meet legislative requirements and safeguard marine life in the area, comprehensive environmental impact assessments are conducted to determine the dispersion of the effluents in the far-field, ensuring they do not have adverse effects. Precisely modelling the characteristics of the outfall plume at different mixing stages is crucial for accurately determining recirculation, spreading, and the resulting environmental impacts.

In this study, we investigated the use of ANSYS Fluent for the nearfield simulations, and additionally to incorporate far-field hydrodynamics. The CFD model proved to be effective in simulating a wide range of scenarios with considerable detail. This suggests that ANSYS Fluent could be a valuable tool for assessing the environmental impacts of desalination plant effluents on marine ecosystems. Further advancements in two-way coupling methods may enhance the accuracy of such assessments in the future.

Wake Redirection Techniques for Wind Farms

K Surujhlal[1], WH Ho[1], A Biobaku[1], and N Zang[2]
1 School of Mechanical, Industrial and Aeronautical Engineering, University of the Witwatersrand
2 School of Civil, Aerospace and Design Engineering, University of Bristol, United Kingdom

Since there is an increasing demand worldwide for renewable energy and specifically wind energy, the need for optimization of wind farms has increased. In a wind farm, the upstream turbine has several negative effects on the downstream turbine which leads to uneven loads exerted on the downstream turbine and reduced power outputs of the wind farm as a whole. Wake redirection techniques could reduce these negative effects and improve the overall power output of a wind farm. The wake redirection technique investigated in this study is yaw misalignment, which is a deflection of the wake of the upstream turbine to the right and left. The aim of this study is to understand the effect of an upstream turbine on a downstream turbine when both turbines are at non-zero yaw angles. The loads exerted on the downstream turbine and the wakes generated were investigated. A surrogate model was created using an existing dataset of a two-turbine array generated by the Simulator for Wind Turbine Applications (SOWFA) that uses Large Eddy Simulations (LES). The model was created by first creating a reduced order model (ROM) to allow for faster run times and use less computational resources without sacrificing significant accuracy. Dynamic Mode Decomposition (DMD) was used in order to obtain more detailed information about the wake flow such as the modal frequencies and growth or decay rates. An analysis was conducted on the modes generated from this reduced order model to determine the most suitable modes that should be used for reconstruction of the wake flow. These reconstructions of the wake flow using the DMD model were compared to the original SOWFA reconstructions and the deviations between the two were plotted. An example of a comparison at one time step is shown in Figure 1 below. Once the necessary flow field dynamics of the system were captured comprehensively, the DMD model was used to train a machine learning algorithm that is used to predict the dynamics of the system outside of the training and testing dataset. Wind tunnel experimental results were also generated to validate the model. Based on the reduced order model results and experimental observation, the effectiveness of the wake redirection technique will be evaluated.

Augmentation of measurement procedures for motion sickness assessment on ships

NC Taylor, A Bekker and K Kruger
Department of Mechanical and Mechatronic Engineering, Stellenbosch University.

Human exposure to vibration is known to occur in many environments, particularly in transportation where the whole human body is subjected to motion. Human responses to induced vibration have been a topic of study since early recognition of undesired related phenomena, including discomfort, injury and illness. On board ships, motion sickness is an unpleasant and, at times, debilitating response to low-frequency vessel motions. Symptoms of motion sickness, such as nausea, headaches and vomiting, can vary greatly in incidence and severity among individuals. The assessment of vibration understood to influence motion sickness is an important requirement in pursuit of minimising its unsought effects on passengers and crew.

ISO 2631-1 (1997) is a widely adopted and recommended standard for the evaluation of human exposure to whole-body vibration in transportation, devoting an annex to motion sickness assessment. The assessment procedure provides guidance for motion exposure measurement, signal processing and interpretation of resultant metrics. The measurement of motion exposure is advised to be at locations where individuals are in contact with the vibration-inducing surfaces, which provides practical challenges when multiple passengers and crew members are moving around many spaces in a ship. ABS (2021) suggest that selecting passenger spaces with the most extreme motion exposure could provide a fair representation of motion exposure. This would provide a measure of the upper bounds of motion an individual could be exposed to, but may not actually be, as they could prefer to frequent areas with lesser motion.

The proposed paper establishes a clear method of measurement location selection for motion sickness assessment on ships. The structure of the paper includes a detailed description of existing guidelines for measurement of vibration related to the evaluation of motion sickness. Directed by findings in literature and experience from a research campaign on board the SA Agulhas II, South Africa's polar supply and research vessel, four selection procedures are defined. The outcomes of the selection procedures on motion sickness assessment will be investigated using real-world ship motion measurements and human responses acquired on a voyage of the SA Agulhas II. Unique benefits and challenges of each procedure are to be discussed, considering factors such as correlation with human responses and viable measurement effort. The most fitting measurement location selection procedure is anticipated to offer practical guidance that augments existing strategies for motion exposure measurement critical to motion sickness assessment.

Making funding proposals more effective: Plain Language guidelines

A van Staden
English Department, University of Pretoria

To solve complex problems in the face of globalization and rapid technological advances and to find funding, scientists and engineers in various specialisations must often cross-collaborate in an interdisciplinary arena. They have to communicate across disciplines and across national, cultural and linguistic borders. Such communication is mainly in English, currently the global language of science, but a second language for many scientists. An important area of interdisciplinary peer-to- peer specialist communication is writing and reviewing funding proposals, formulated in terms of discipline-specific details, terminology and concepts. However, panels reviewing funding proposals seldom consist only of experts from the same field of expertise as the proposers.
This article offers a definition of Plain Language for funding proposals and eight Plain Language guidelines to assist scientists and engineers who hope to write more effective funding proposals, to ensure that a multi-disciplinary reviewing panel can discern the merit of a particular research topic. The aim is to increase both scientific precision and readability. These guidelines are derived from an exploratory, mixed-methods study on the use of Plain Language principles in interdisciplinary peer- to-peer science communication and how they can help to clarify funding proposals. The study included an extensive literature review, a survey and follow-up interviews with scientists from different specialisations.

Dynamic Simulation of a Nutating Grinding Mill

JJ van Tonder, JR Bredell, and C Coetzee
Department of Mechanical and Mechatronic Engineering, Stellenbosch University,

Mined minerals undergo comminution for the purposes of valuable ore extraction. One way this is achieved is by the nutation of a grinding chamber, where nutation refers to a motion similar to holding a conical flask by the rim and rotating one's wrist. The HICOM is an example of such a mill and achieves high breakage rates due to the intense centrifugal acceleration field generated within the grinding chamber. Although known for its efficient grinding action and compact design, the HICOM mill presents operational challenges affecting its mechanical reliability, with the primary causes being bearing life, dimensional tolerances and load path characterisation.

This study focuses on investigating the dynamics of the nutating grinding mill, with a specific focus on dynamic force response at the primary kinematic joints. The primary kinematic joints are the universal joint, located at the top of the mill chamber, and the joint connecting the chamber base to the driving torque arm that nutates the chamber. To perform the investigation, the study designed and built a laboratory experimental model called the NuMILL, which is kinematically equivalent to the HICOM. The NuMILL was used to joint force and torque across a set of operational conditions that served to validate subsequent simulation models. Two independent simulation methods were devised: a Multi-Body Dynamics (MBD) model and a Discrete Element Method (DEM) model. The underlying assumption with the two separate simulations was that the dynamic effects of the charge material (DEM simulation) could be superimposed on the dynamic effect of the NuMILL structure (MBD simulation), with the combination of the two yielding an accurate prediction of the total dynamic responses.

It was found that the isolated MBD model could not accurately predict the driving arm's torque. However, the combined DEM and MBD simulations showed great potential in being able to accurately simulate the NuMILL responses as compared to the DEM or MBD simulation in isolation. Furthermore, insight gained from the simulations and experiments led to the formulation of non-dimensional phase angle and centrifugal acceleration parameters that allow the estimation of dynamic force responses without the need for repeated MBD or DEM simulations. This reduced order parameter model, when applied to the HICOM, shows potential in serving as an on-site model that can guide operating conditions for load-critical scenarios to help increase the mechanical reliability of the HICOM.

A Strain Gauge-Based Method for Correcting Out-Of-Plane Motion Errors in 2D Digital Image Correlation

CH Visser, G Venter, and M Neaves
Department of Mechanical and Mechatronic Engineering, Stellenbosch University

Due to its superior accuracy, digital image correlation (DIC) measurements are typically conducted in the form of a multi-camera setup, referred to as stereo-DIC. However, scenarios still exist where single-camera 2D-DIC is preferred. It is both financially and computationally less expensive than stereo-DIC and if the accuracy of 2D-DIC can be improved, it could become a more suitable option for certain measurement scenarios, such as where an experiment's setup can only accommodate one camera to capture its field-of-view.

Errors caused by out-of-plane motion in 2D-DIC measurements are of particular concern since they are corrected in stereo-DIC, but result in significant inaccuracies in 2D-DIC (Sutton et al., 2008). A strain gauge-based method for compensation of strain errors due to out-of- plane motion in 2D-DIC data has therefore been proposed in previous work by the authors (Visser, Venter, and Neaves, 2023). The method was first developed theoretically and was then validated experimentally for two different test cases. The results illustrate that the error compensation method clearly improves 2D-DIC measurement accuracy. This improvement in accuracy makes the corrected 2D-DIC a more viable measurement option than before.

The current paper presents further work that has been conducted on this strain gauge- based error compensation method and its functionality. A sensitivity analysis was conducted to investigate the effect of strain gauge size and the number of strain gauges on the performance of the error compensation method. Although the influence of the number of strain gauges used was minimal for both test cases, a significant improvement in accuracy could be seen when using larger strain gauges as opposed to smaller ones. Furthermore, the error compensation method was studied to determine its limitations. It was found that the error compensation is reduced in situations where the amount of out-of-plane motion varies in a non-linear fashion over the region of interest, such as where a stress concentration is present. A comparison was also done to determine the effect of using virtual strain gauge values from stereo-DIC data (as opposed to measurements from physical strain gauges) to perform error compensation on 2D-DIC data. It was found that using stereo-DIC data provided better error compensation on 2D-DIC data than using physical strain gauge data, likely because the stereo-DIC data was the reference measurement for the 2D-DIC data.

Physics-Based, Hybrid, and Data-Driven Computational Mechanics Generative Modelling: Manifestation of Shared Concepts

DN Wilke
Department of Mechanical & Aeronautical Engineering, University of Pretoria

The convergence of physics-based, hybrid, and data-driven generative modelling in computational mechanics requires us to reflect on shared foundational concepts that allow us to understand the similarities and differences between these approaches, thereby, enabling us to enhance predictive accuracy and interpretability to expand our modelling capabilities. This study outlines some of the core concepts that unify these diverse methodologies.

Central to these approaches is the recognition that complex physical systems can be represented through mathematical abstractions. The abstractions determine the predictive capability of a generative model, as well as our ability to interpret the generative model itself in addition to the inputs to the model. Physics-based modelling starts with fundamental principles, deriving equations that encapsulate the underlying physical system. Hybrid modelling builds on these principles but acknowledges the shortcomings of theoretical principles supplementing them with empirical data to seamlessly integrate theoretical information and observed data. Data-driven generative models scatter an entire spectrum of modelling paradigms but ultimately aim to recover calibrated physics from data.

Uncertainty handling, quantification and implementation are distinctly different between approaches. Physics-based models inherently grapple with assumptions and simplifications, resulting in uncertainty regarding their accuracy in real-world scenarios, which is evident in the inadequate way basic parameter estimation is generally conducted. Although strides have been made in probabilistic modelling it remains isolated to a few fields. Hybrid modelling acknowledges uncertainty by incorporating data to calibrate and refine model parameters, acknowledging the limits of theoretical formulations. Data-driven techniques thrive on uncertainty, employing probabilistic methods to provide predictive insights even in data-scarce or noisy environments.

Validation and verification are essential cornerstones for predictive modelling. Physics-based models are verified by comparing predictions with established physical laws, while validated post calibration on experimental data. Hybrid models must be verified and validated against both theoretical and empirical benchmarks, while data-driven models, on the other hand, require rigorous validation against independent datasets as they often lack direct physical interpretability, necessitating careful consideration of their application domain.

Adaptability of models Physics-based models excel when fundamental equations are well-established, while hybrid models adapt to varying complexities by blending theory and data. Data-driven techniques, ever versatile, mould themselves to diverse datasets that can reveal intricate relationships. Together, they pave the way for a more comprehensive understanding of complex systems and provide a versatile toolkit for researchers and practitioners to tackle a wide array of challenges across engineering, materials science, and beyond.

Multifidelity Surrogate Modelling: A New Data Fusion Perspective

DN Wilke
Department of Mechanical & Aeronautical Engineering, University of Pretoria

Multifidelity surrogate modelling revolutionises predictive modelling by integrating data from various sources with distinct levels of accuracy and computational cost.

Conventional single-fidelity surrogate models often struggle to capture the complexities of real-world phenomena due to oversimplifications or limited accuracy. In addition, single-fidelity surrogate models can be expensive and computationally costly. Multifidelity surrogate modelling addresses these limitations by amalgamating data from multiple sources, each offering a different fidelity level. High-fidelity sources furnish detailed insights, while low-fidelity sources provide rapid approximations.

The hierarchy of fidelities forms the crux of multifidelity surrogates, allowing for a tiered integration of models based on their accuracy and computational demands that result in a unified and responsive surrogate model.

Multifidelity surrogate modelling enhances efficiency by leveraging low-fidelity models for quick evaluations, conserving computational resources. Simultaneously, the incorporation of high-fidelity data refines predictions, addressing the limitations of simplistic approximations. Furthermore, when a comprehensive perspective on prediction uncertainties is embraced, informed decision-making is fostered.

Multifidelity surrogate modelling spans diverse domains of application. In engineering, multifidelity surrogate models expedite design optimization processes without compromising accuracy, benefiting fields like aerospace and automotive design. In materials science, they enable the study of material properties with a balance between precision and computational effort.

While multifidelity surrogate modelling holds immense promise, challenges remain. Selecting appropriate fidelities for specific scenarios demands careful consideration, and developing efficient data fusion techniques across multifidelity sources remains an ongoing research endeavour. This study provides a new perspective on data fusion of multifidelity data, its rationale, and benefits in multifidelity surrogate modelling.

Multifidelity surrogate modelling stands at the forefront of predictive modelling innovation. By synergising disparate data sources, it creates a versatile framework that enhances accuracy, accelerates simulations, and facilitates insights across a multitude of disciplines. This approach not only overcomes the limitations of single-fidelity models but also opens the door to more sophisticated and informed analyses of complex systems that enable us to ask questions we would never have dared to ask.